AF587337

NEOCLASSICAL ANALYSIS: CALCULUS CLOSER TO THE REAL WORLD

NEOCLASSICAL ANALYSIS: CALCULUS CLOSER TO THE REAL WORLD

MARK BURGIN

Nova Science Publishers, Inc.
New York

For permission to use material from this book please contact us:
Telephone 631-231-7269; Fax 631-231-8175
Web Site: http://www.novapublishers.com

LIBRARY OF CONGRESS CATALOGING-IN-PUBLICATION DATA

Neoclassical analysis: calculus closer to the real world / Mark Burgin.
p. cm
Includes index.
ISBN-13: 978-1-60021-946-7 (hardcover)
ISBN-10: 1-60021-946-2 (hardcover)
1. Calculus. I. Burgin, M. S. (Mark Semenovich)
QA300.N35 2008
515--dc22

2007030967

Published by Nova Science Publishers, Inc. ✦ New York

CONTENTS

ANNOTATION

Neoclassical analysis extends methods of classical calculus to reflect uncertainties that arise in computations and measurements. In it, ordinary structures of analysis, that is, functions, sequences, series, and operators, are studied by means of fuzzy concepts: fuzzy limits, fuzzy continuity, and fuzzy derivatives. For example, continuous functions, which are studied in the classical analysis, become a part of the set of the fuzzy continuous functions studied in neoclassical analysis.

Aiming at representation of uncertainties and imprecision and extending the scope of the classical calculus and analysis, neoclassical analysis makes, at the same time, methods of the classical calculus more precise with respect to real life applications. Consequently, new results are obtained extending and even completing classical theorems. In addition, facilities of analytical methods for various applications also become more broad and efficient.

Neoclassical analysis is closely related to fuzzy set theory, set-valued analysis, and interval analysis.

The book presents the core of the neoclassical analysis on three levels in each chapter. At first, basic classical constructions of the conventional calculus, such as limits, continuous functions, derivatives, differentiable functions, and integrals, and their properties are considered. The next level gives an exposition of neoclassical, fuzzy extensions of the classical constructions for real functions and sequences, i.e., neoclassical analysis of real functions and sequences is constructed. The third level elevates elements of calculus from numerical functions to functions in metric and normed linear spaces and is presented at the end of the main chapters.

This facilitates reading of the book and makes it accessible to a wide range of readers. Even those who did not learn or forgot calculus can read this book as the beginning section of each chapter introduces the reader to the classical calculus. Even for those who studied calculus at a college or university, the expository section liberates such a reader from necessity to go from time to time to other sources. At the same time, mathematicians who know the classical calculus very well will be able to find many innovative information about a new, extended and more realistic calculus called neoclassical analysis.

It is possible to use this book for enhancing traditional courses of calculus for undergraduates, as well as for teaching separate courses for graduate and undergraduate students at colleges and universities.

PREFACE

How thoroughly it is ingrained in mathematical science that every real advance goes hand in hand with the invention of sharper tools and simpler methods which, at the same time, assist in understanding earlier theories and in casting aside some more complicated developments.
David Hilbert (1862 - 1943)

The intuitive mind is a sacred gift
and the rational mind is a faithful servant.
We have created a society that honors the servant
and has forgotten the gift.
Albert Einstein (1879-1955)

Calculus originated by Newton and Leibniz and developed by many generations of mathematicians has proved to be very useful and efficient in physics, engineering, biology and many other fields. However, the development of science and technology brought limitations of the classical calculus to attention of mathematicians, scientists and engineers.

Calculus that we teach in colleges and universities is based on infinite processes (limits, continuity, differentiation, and integration) and assumes infinite precision of objects and operations. At the same time, computation is always limited to a finite number of symbols and achieves full exactness in a very limited number of cases. For instance, because computers can operate only with finite entities of rational numbers, all computable functions are defined only for a finite set of points. Moreover, theoretical results related to computable or recursive real numbers show that computation has inherent limitations on precision. Consequently, computable functions cannot be continuous in the sense of the classical calculus. Computers, by their nature, imply imprecision and work with approximations of ideal mathematical structures.

Another cause of imprecision is measurement. Any real measurement does not provide absolutely precise results but gives only approximate values in the majority of cases. For instance, it is impossible to find out if any series of numbers obtained in experiments converges or a function determined by measurements is continuous at a given point. Moreover, modern physical theories imply (due, for example, to such a fundamental law of contemporary physics as the Principle of Uncertainty introduced by Werner von Heisenberg) that measurement has inherent limitations on precision. For example, if we take the rate of the particle position change, impossibility to measure it with absolute precision is one of the consequences of the Principle of Uncertainty. Similar relationship holds for all dynamic attributes (Herbert, 1985). Heisenberg relations guarantee that any experiment and measurement will contain a blind spot. As Koenderink (1990) writes, "the limit for whatever size to zero has never made any physical sense." Consequently, constructions and methods developed in the classical calculus and in analysis, as the extended form of calculus is called, are only approximations to what exists in reality. In many situations, such approximations have been giving a sufficiently adequate representation of studied phenomena. However, scientists and, especially, engineers have discovered many cases in which such methods did not work because classical approach is too coarse.

To eliminate this deficiency, mathematics develops new methods that allow one to treat inexactness, vagueness, and uncertainty with mathematical rigor and accuracy. One mathematical direction that deals with inexactness, vagueness, and uncertainty that exist in computation is *interval analysis* introduced by R. E. Moore (circa 1959). Interval analysis studies intervals instead of numbers and interval valued functions and operations instead of numerical functions and operations of classical mathematics. As a mathematical field, interval analysis is a part of *set-valued analysis,* which, in turn, is a part of named set theory.

Another mathematical direction that deals with inexactness, vagueness, and uncertainty, which are prevailing in our reality, is *fuzzy set theory.* Now fuzzy set theory and set-valued analysis are flourishing fields, which have produced numerous publications and conferences.

Neoclassical analysis is a new direction that suggests an original solution to these problems of imprecision and inexactness in real life. Neoclassical analysis is a synthesis of three mathematical fields: the classical calculus, set-valued analysis (in a broad sense that includes interval analysis), and fuzzy set theory. The aim of this synthesis is to extend the powerful technique of the classical calculus to a much broader scope, and to make this technique more relevant to situations in physics and computation. Neoclassical analysis has common features with all three fields that it synthesizes, as well as essential distinctions from them. Thus, in contrast to set-valued analysis and fuzzy set theory, neoclassical analysis studies ordinary structures of classical calculus, such as numbers, functions, sets, sequences, series, and operators. Methods of set-valued analysis are oriented at sets. Methods of fuzzy set theory are oriented at fuzzy sets. Neoclassical analysis, following the classical calculus, is interested in individual objects: numbers, points, functions, curves, etc.

At the same time, in neoclassical analysis, ordinary structures of the classical calculus are studied by means of fuzzy concepts: fuzzy limits, fuzzy continuity, fuzzy derivatives, etc. For example, continuous functions studied in the classical calculus become a part of the set of fuzzy continuous functions studied in neoclassical analysis.

One more goal of neoclassical analysis is to extend the powerful calculus technology to a broader scope of exact (crisp) objects, e.g., for functions and curves that are discontinuous. This new technology developed in neoclassical analysis allows one to build adequate mathematical models for discrete spaces and functions in such spaces, introducing flexible and scalable concepts of discrete continuity, differentiation, and integration. This is especially important for computation as it operates only with discrete sets.

In addition to better representation of computational structures, neoclassical analysis makes it possible to extend and, in some cases, even to complete many basic results of the classical calculus. These results produce a deeper insight into and better understanding of the classical theory. For instance, one of the core classical results is the Weierstrass theorem. It states that a continuous real function on a closed interval is bounded. It is known that the converse is not true, that is, boundedness does not imply continuity. Only introduction of fuzzy continuity made it possible to get a complete criterion of boundedness. Namely, it is proved that a real function on a closed interval is bounded if and only if it is fuzzy continuous. At the same time, there is a difference between bounded and fuzzy continuous functions in a general case.

In comparison with the three hundred years of the conventional calculus history, neoclassical analysis is a new mathematical field. That is why in this book, we develop it in a comprehensive way, with many detailed proofs and examples. Such attention to detail is important because similarity of neoclassical analysis to the classical calculus can be misleading. Some properties of classical and new structures are virtually the same, while others can be essentially different and even opposite. For instance, the sum and product of two continuous functions are always continuous. At the same time, the sum of two fuzzy continuous functions is always fuzzy continuous, while in some cases, the product of two fuzzy continuous functions is not fuzzy continuous.

This book represents neoclassical analysis in a form accessible to those who start studying the conventional calculus and, at the same time, provides a rigorous exposition of the new approach to calculus for experts.

We start each chapter with the basic classical constructions of the conventional calculus, such as limits, continuous functions, derivatives, differentiable functions, and integrals, and study their properties. The goal is to make exposition of this book self-contained, allowing the reader to learn calculus from the very beginning and demonstrating that neoclassical analysis is virtually not more complicated than the classical calculus. At the same time, the classical calculus is a part (actually, a subtheory) of neoclassical analysis, which has essentially broader scope than the classical calculus.

Exposition of material is aimed at three groups of readers. The form of the book is designed specifically for students who study calculus and instructors who teach calculus. Each chapter and many sections have an informal exposition of the main ideas. The majority of statements (theorems, lemmas, and propositions) have thorough proofs. Those that do not have proofs are relatively easy and may be used as exercises to achieve better understanding of the material. For better understanding given proofs and developing skills of problem solving, it is possible to recommend (Sollow, 1982; Polya, 1954; 1957; 1962; Wickelgren, 1974).

This book may be used for enhancing traditional courses of calculus for undergraduates, as well as for teaching a separate course for graduate or undergraduate students at colleges and universities. Some authors avoid in their textbooks such abstract ideas and constructions as metric spaces, norms, linear spaces, ordered systems, and topological spaces. In this book, all material related to such abstract issues is given only at the very end of some chapters. Thus, those readers who are not mathematicians and are not going to become professional mathematicians can easily skip these sections.

At the same time, it is possible to use this book for teaching such courses as "Foundations of Calculus," "Real Analysis," and "Calculus" aimed at mathematically oriented students. Detailed proofs, formalized definitions of the main constructions combined with an informal exposition of the main ideas provide convenient opportunities to learn calculus in both classical and innovative form. In the educational context, those statements that are given without proofs can be used as homework and classwork problems.

It is necessary to remark that exposition of the conventional calculus material is given in the book in a more comprehensive way than in the majority of textbooks on the calculus and analysis that are now used at colleges and universities. A student and a professor will be able to find a variety of theorems, propositions and lemmas that describe useful properties of classical structures from the calculus, such as limits, continuous functions, derivatives and integrals.

The second group of intended readers consists of research mathematicians who would be able to extend their research tools from the classical to neoclassical analysis. For them, there are many open problems and directions for further research at the end of the book.

The third group of tentative readers includes those who apply calculus in their work (physicists, engineers, economists, etc.) and are looking for more adequate means in their area. Consequently, a variety of readers will be able to find interesting and useful issues in this book if each reader chooses those topics that are of interest to her or to him.

It is necessary to remark that the research related to neoclassical analysis is so active that it is impossible to include all ideas, issues, and references, for which we ask the reader's forbearance.

Chapter 1

INTRODUCTION

It takes a very unusual mind to undertake analysis of the obvious.
Alfred North Whitehead (1861-1947)

To better understand neoclassical analysis, we start this Introduction (see Section 1.1) with the history of the classical calculus and analysis that has been developed from the calculus. Without too much detail, we virtually bring this history to the present day. Different sources were used for this historical overview. Internet sources, such as Wikipedia, MacTutor Mathematical biographies and historical articles of O'Connor and Robertson, were especially useful.

It is necessary to remark that the research in analysis and related fields of mathematics has been so active that it is impossible to include all ideas, issues, and references, as well as to write about all mathematicians contributed to this field, for which we ask the reader's forbearance.

Then (in Section 1.2), we show that in spite of successful application of the classical calculus to a huge diversity of theoretical and practical areas (such as physics and engineering, medicine and biology, astronomy and celestial mechanics, and so on), there is a gap between theoretical models that use calculus constructions, which are completely precise and real world phenomena, which are essentially vague and imprecise.

After this we proceed to a brief exposition of the main areas that are aimed at reflection of imprecision, vagueness, and uncertainty by rigorous mathematical means. In Section 1.3.1, we consider fuzzy set theory, its generalizations, and related structures, such as rough sets, which study imprecision, vagueness, ambiguity, and uncertainty by exact mathematical tools. To make exposition more uniform and to explicate the implicit structure of fuzzy sets and other generalizations of sets, exposition is based on the framework of named sets, which encompass all fuzzy set generalizations. In Section 1.3.2, we present elements of interval analysis. Section 1.3.3 gives a general overview of the research related to neoclassical analysis. Section 1.4 represents the general structure of the main body of this book

1.1. A Brief History of Calculus

Calculus required continuity,
and continuity was supposed to require the infinitely little;
but nobody could discover what the infinitely little might be.
Bertrand Russell (1872-1970)

The word *calculus* has two meanings in mathematics. The most popular in general mathematics understanding is that *Calculus* is a name that is now used to denote the field of mathematics that studies properties of functions, curves, and surfaces. As this is the most popular meaning in mathematics, we call it *the calculus*. Kline (1972) calls the calculus one of the two greatest creations in the whole history of mathematics.

Methods of the calculus are based on limit processes that define continuity and two main calculus operations: differentiation and integration. That is why the calculus usually is subdivided into two parts: *differential calculus* and *integral calculus*. These operations reduce search for many function properties to algorithmical calculations with these functions. This peculiarity explains the name *calculus* used for this field. This name originated from the Latin word meaning pebble because people many years ago used pebbles to count and do arithmetical calculations. The Romans used *calculos subducere* for "to calculate."

Thus, the calculus is called so because it provides analytic, algebra-like techniques, or means of computing, which apply algorithmically to various functions, curves, surfaces, etc. Many mathematical problems that had very hard solutions or even such problems that mathematicians had not been able to solve, after the calculus had been developed, became easily solvable by mathematics students. At the same time, the calculus is perceived only as the introductory level of a more general field of mathematics called *mathematical analysis* or simply, *analysis*. Analysis, as a rigorous extension of the calculus, is a major area in mathematics, with diverse applications in science, engineering, business, medicine and art.

Analysis has its beginnings in the rigorous formulation of calculus, is explicitly concerned with the notion of a limit, deals generally with infinite processes and includes such areas as real analysis and complex analysis. Some mathematicians also include differential equations in analysis.

Another mathematical meaning of the word *calculus* comes from mathematical logic where *calculus* is a formal system used for logical modeling of mathematical and scientific theories. A logical calculus consists of three parts: axioms, rules of deduction (inference), and theorems (cf. (Kleene, 2002; Mendelson, 1997)).

Here we are interested only in calculus as a field of mathematics, and in what follows, we call it simply the calculus.

It is necessary to remark that aspects of the calculus development related to problems of uncertainty, imprecision, and vagueness are not presented in this section because we discuss these problems in other sections of Introduction.

The way to the calculus was not easy. The main ideas that underpin the calculus had been developed over a very long period of time. It is generally assumed that the first steps in the direction of calculus were taken by Greek mathematicians, whose works display the beginnings of integration aimed at finding areas of curved regions and volumes of curved solids. Around 360 B.C.E., Eudoxus of Cnidus (circa 408 - 355 B.C.E.) used a ``method of exhaustion", close to the limiting concept of calculus, to find areas and volumes of curvilinear figures. It is easy to see that the method of exhaustion is an early, geometric form of integration. Some researchers think that even before Eudoxus, Antiphon the Sophist (born circa 460 B.C.E.) making an attempt to square the circle, developed a method of exhaustion around 430 B.C.E. Although it is not entirely clear how well he understood his own proposal. Antiphon proposed successively doubling the number of sides of a regular polygon inscribed in a circle so that the difference in areas would eventually become exhausted. However, there is evidence that like many mathematicians who invent a new technique, Antiphon did not formulate the method rigorously. Some sources indicate that Democritus of Abdera (circa 460 - 370 B.C.E.) and Hippocrates of Chios (470 - 400 B.C.E.) also elaborated some methods of integration.

Archimedes (circa 287-212 B.C.E.) used the method of exhaustion to find an approximation to the area of a circle. This is an early example of integration. It was used to approximate values of π. Results of Archimedes based on integration included calculations of the volume and surface area of a sphere, the area of a parabolic segment, the volume and area of a cone, the surface area of an ellipse, the volume of any segment of a paraboloid of revolution and a segment of an hyperboloid of revolution.

The traditional history of mathematics asserts that no further progress in this area was made until the 16th century. However, this is true only for Europe. After Greeks, some methods related to integration were further developed by mathematicians living in the Arabian Halifat. The most notable in this area were Thabit ibn Qurra (836–901) born at Harran (present Turkey) and his grandson Ibrahim ibn Sinan (908-946) whose works extended the approach of Archimedes and may be considered as steps taken towards the integral calculus. Abu Ali al-Hasan ibn al-Haytham (Latinized: Alhacen or (deprecated) Alhazen) (965 – 1039) developed a method for determining the general formula for the sum of integral powers, which was related to the development of the calculus. Mathematicians and scientists from the Halifat were influenced both by Greek and Indian mathematical achievements, while their works influenced medieval European science and mathematics. In particular, they brought different ideas, concepts, and structures (such as zero or negative numbers) to Europe from India.

India always has had great mathematical traditions. Indian mathematicians made many discoveries and inventions in different fields including those that are related to the calculus. In 1835, Charles Whish published an article in the *Transactions of the Royal Asiatic Society of Great Britain and Ireland*, in which he claimed that the work of the Kerala school from India laid the foundation for a developed system of the calculus. It was not until the 1940s however, that historians of mathematics verified Whish's claims and came to the conclusion that many elements of the calculus were developed in India. The 14th century Indian mathematician Madhava of Sangamagrama (1340-1425), along with other mathematicians of the Kerala School, studied infinite series, including power series, Taylor series, trigonometric series, and Maclaurin series, and utilized ideas close to convergence, differentiation, and integration. Later an astronomer of the Kerala School Jyestadeva (1500-1575) wrote his book *Yuktibhasa*. Some historians consider it as the first differential calculus text. This work was unique at the time for its exact proofs of the theorems it presented.

But this was not the beginning. Long before this, in 499 the mathematician-astronomer Aryabhata (circa 476–550) from India used some methods related to infinitesimals and differential equations (cf. Mathematical Analysis, Wikipedia). These results eventually led

Bhaskara II (1114-1185) to ideas that are essential to the development of differential calculus, including the earliest use of such concepts as derivative and differential coefficient. Using these concepts, he found the differential of the sine function, as well as the Earth's velocity in successive positions of its elliptical orbit around the Sun. One of the statements Bhaskara II made is what now mathematicians call the Rolle theorem, an important special case of the mean value theorem (cf. Chapter 4).

Some researchers go as far as to suggest that Keralese mathematics may have been transmitted to Europe and influenced the development of calculus. However, there is no evidence of direct transmission by way of relevant manuscripts, although there is evidence of methodological similarities, communication routes and a suitable chronology for transmission. Besides, knowledge of some derivatives and integrals, as it was in India, does not mean invention of the calculus, which includes a general technique for differentiation and integration, as well as diverse properties of these operations.

Nevertheless, claims of the total Indian priority in mathematics go much further. Two Indian authors, Lakshmikantham and Leela, claim in their book (2000) that the modern historiography for India was distorted by the mix-up of Chandragupta Maurya (reigned circa 1534–1500 B.C.E.) and Chandragupta (circa 327–320 B.C.E.) of the Gupta dynasty. As a result, Lakshmikantham and Leela come to the conclusion that the first known mathematician and astronomer from India, Aryabhata, was born in 2765 B.C.E., and not in 476 as it is assumed in the traditional history of mathematics. In addition, they claim that Aryabhata used infinitesimals and expressed an astronomical problem in the form of a basic differential equation long before than the ancient Greek mathematics even appeared.

Some research related to the calculus also did Chinese mathematicians. In particular, Liu Hui (fl. 3rd century) derived approximations of π and found the volume of a cylinder. Zu Chongzhi, or Tsu Ch'ung-chih, (429–500) also derived approximations of π and found the volume of a sphere.

In any case, the development of mathematics almost completely moved to Europe in the 17th century. According to Kline (1972), calculus was created to treat the major scientific problems of the 17th century, such as the study of motion, finding tangent lines, maximum-minimum problems, determination of the lengths of curves, areas of surfaces, volumes of solids, and centers of gravity of solid bodies. Here are the most important achievements of that period.

Johannes Kepler (1571-1630), in his work on planetary motion, found the area of sectors of an ellipse. His method consisted of representing sectors as sums of lines, another empirical form of integration. However, Kepler did not follow Greek rigor and was rather lucky to obtain the correct answer after making two canceling errors in this work. To calculate volumes of solids of revolution in his book *New Measurement of the Volume of Wine Casks* published in 1615, Kepler also approximated those solids with sums of numerous thin layers, each of which was a cylindrical disk. This is actually the way of how today triple integrals are evaluated (cf., for example, (Stewart, 2003)).

In 1635, Bonaventura Cavalieri (1598-1647) calculates volumes using infinitely small sections. In 1655, John Wallis (1616-1703) studies infinite series in *Arithmetic of Infinitesimals*. In 1658, Blaise Pascal (1623-1662), working on the sine function, "almost discovers" differential calculus.

Important contributions were also made by Isaac Barrow (1630-1677), René Descartes (1596-1650), Pierre de Fermat (1601-1665), Christiaan Huygens (1629-1695) and others.

Barrow, Fermat, Pascal, and Wallis are said to have discussed the idea of a derivative. René Descartes introduced analytic geometry in 1637, providing the foundation for calculus.

Fermat, among other things, is credited with an ingenious trick for evaluating the integral of any power function directly, thus providing a valuable clue to Newton and Leibniz in their development of the fundamental theorems of calculus. Fermat also investigated maxima and minima by considering when the tangent to the curve was parallel to the x-axis. He wrote to Descartes giving the method essentially as used today for finding maxima and minima by calculating when the derivative of the function was 0, although he did not introduce the concept of derivative. Just because of this, Lagrange wrote that Fermat was the inventor of the calculus.

James Gregory (1638-1675) was able to prove a restricted version of the second fundamental theorem of calculus. Barrow, with Newton assisting in preparation, published his book *Lectiones Geometricae* in 1670. It described methods for determining tangents to curves that closely resembled methods used in contemporary calculus textbooks, as well as techniques for finding lengths of curves. The book as whole was the culmination of the 17th century investigations leading toward calculus. Although some credit Barrow for inventing the calculus, his technique was essentially geometrical, as he argued that algebra is not even a part of true mathematics. This precluded Barrow from developing an algorithmic technique that constitutes the power of the calculus.

Around the same time, there was also some work in this area being done by Japanese mathematicians, particularly, by Seki Takakazu, also called Seki Kowa (1642-1708). He made a number of contributions in methods of determining areas of figures, extending the work of Archimedes (Smith and Mikami, 1914).

Now Isaac Newton (1642-1727) and Gottfried Wilhelm Leibniz (1646-1716) are usually credited with the invention, independently of one another in the late 1600s, of differential and integral calculus that led to the calculus as we know it today. One of their most important contributions was the fundamental theorem of calculus. In addition, Leibniz was very inventive with developing consistent and useful notation, techniques, and concepts. He introduced the d/dx notation for differentiation and $\int$ notation for integration, which are very popular in mathematics. Leibniz explained importance of notation for mathematics in his letter to Johann Bernoulli, writing, "As regards signs, I see clearly that it is to the interest of the Republic of Letters, and especially of students, that learned men should reach agreement on signs." (cf. (Burton, 1997)).

At the same time, Newton was the first to organize the field into one consistent subject, and also provided some of the first and most important applications, especially, of integral calculus to physics. Newton also influenced the calculus notation used today, introducing the $\dot{g}(x)$ and $\dot{y}$ notations, which are now used mostly for time derivatives, as opposed to slope or position derivatives. Later (in 1797) Joseph-Louis Lagrange transformed $\dot{g}(x)$ and $\dot{y}$ into the $g'(x)$ and y' notations for derivatives. Both notations are very common now in mathematics and physics.

Newton generalized the methods used at that time to draw tangents to curves and to calculate the area swept by curves. In the autumn of 1666, he developed the fluxional method as the first form of the calculus. However, Newton was often slow to publish his discoveries. Only in1669, Newton included his method for finding areas under curves in his work *On the Analysis of Equations Unlimited in the Number of Their Terms,* which circulated privately.

Although Newton wrote his book *The Method of Fluxions* around 1671, it did not appear in print until 1736, nine years after his death. John Wallis publishes method of fluxions in volume two of his *Mathematical Works* in 1693. It was the first publication of Newton's results on the calculus. Newton himself published a detailed exposition of his fluxional method only in 1704.

In 1673-75, Leibniz arrived independently at virtually the same method, which he called differential calculus, and wrote (1675) a manuscript where he introduced the modern notation $\int f(x)\,dx$ for integration for the first time. In addition, he also used the modern notation dx/dy for differentiation. However, as it was peculiar to that time, publications started only in 1684 when Leibniz published *Nova methodus pro maximis et minimis* ("A new method for maxima and minima as well as tangents, which is impeded neither by fractional nor by irrational quantities, and a remarkable type of calculus for this") in *Acta Eruditorum*, a journal established in Leipzig two years earlier. So, Leibniz started to publish his results on the calculus esentially earlier than Newton. The first paper contained the familiar dy/dx notation, the rules for computing the derivatives of powers, products and quotients. Everything was without proofs. Only six pages long, this text was very hard for understanding, and Jacob Bernoulli called it an enigma rather than an explanation. In 1686, Leibniz published the second paper, which contained his method of integral calculus, which he called '*calculus summatorius*', in an issue of *Acta Eruditorum*. The name the name integral calculus was suggested by Jacob Bernoulli in 1690.

Although Newton and Leibniz developed practically the same calculus, their approach was different due to distinctions in their mentality and goals. Newton's thinking was geometrical, rooted in problems and techniques from physics. Leibniz's mentality was more analytical, rooted in the linguistic context. As a result, Newton's calculus technique was geometrically oriented, aimed at problems in mechanics and optics, and based on physical intuition. For instance, Newton understood integration as finding fluents for a given fluxion. This directly implied that integration and differentiation were inverse operations. At the same time, Leibniz's calculus technique was algorithmically oriented, aimed at problems in mathematics, and based on operational intuition. For instance, Leibniz considered integration operationally as summation.

In the 18th century, there was a lot of debate over whether it was Newton or Leibniz who first "invented" calculus. This dispute, the "Newton versus Leibniz calculus controversy" between the German Leibniz and the English Newton, was at the heart of a rift in the English and continental mathematical communities. Some historians of mathematics even suppose that this controversy set back British mathematical analysis based on calculus and caused that mathematics in Britain was behind mathematics in Continental Europe for a long time. Due to this controversy, British approach was based on the geometric "calculus of fluxions" suggested by Newton, while in Continental Europe, algorithmic methods of Leibniz's differential and integral calculi were adopted. Algorithmic methods proved to be more efficient in this area.

Much of the controversy centered on certain early manuscripts of Newton's that Leibniz might have had access to. Newton began his work on calculus at least as early as 1666, giving plenty of time for this to occur, as Leibniz did not begin his work until 1676. Leibniz was in England in 1673 and 1676, and probably did see some of Newton's manuscripts. It is not known how much this may have influenced Leibniz. He declared that he was led to the invention of the calculus more by studying Pascal's writings than anything else. Both Newton and Leibniz claimed that the other plagiarized their respective works. In fact, they may very well have influenced one another, but it is now widely accepted that the two developed their ideas mostly independently.

After Newton and Leibniz, the development of the calculus was continued primarily in the continental Europe. At first, very few mathematicians worked in this area. The research of Newton and Leibniz was carried on by brothers from Switzerland Jacob (Jacques) Bernoulli (1654–1705) and Johann Bernoulli (1667–1748), the chief disciples of Leibniz. It must be understood that Leibniz 's publications on the calculus were very innovative and thus obscure to mathematicians of that time. As Burton writes (1997), marred by misprints and poor exposition, publications of Leibniz on the calculus encountered almost universal lack of understanding.

The Bernoulli brothers were the first who understood, understood, developed and applied Leibniz 's ideas in the calculus. They did an essential contribution. In 1696, Guillaume François Antoine l'Hospital (1661-1704), who had the title Marquis de St.-Mesme and was among the earliest pupils of John Bernoulli, published the first in Europe textbook in calculus *Analyse des infiniment petits*. Although he was a capable mathematician, important results included in his work in calculus were purchased from Johann Bernoulli and this fact was kept secret at the time.

The power of the calculus was greatly developed in the 18th century. Geometric and mechanical structures, ideas and problems predominated in the early calculus. Making emphasis on curves and movement represented by curves, the calculus provided coherence to what was otherwise a disparate collection of analytic techniques. The calculus development gradually established the subject on a purely analytic basis, making the geometry of curves and trajectories only one field of calculus applications.

The most prolific in the further calculus development was Leonhard Euler (1707–1783), a student of Johann Bernoulli. Euler was born at Basel, Switzerland, and worked both in Russia and Germany. He made decisive and formative contributions to calculus, integrating Leibniz's calculus and Newton's method of fluxions into a unified mathematical field and developing the calculus on a wide scale. Euler contributed to numerous areas of both pure and applied mathematics, including the calculus of variations, analysis, number theory, algebra, geometry, trigonometry, analytical mechanics, hydrodynamics, and the lunar theory (calculation of the motion of the moon). We owe to Euler the notation $f(x)$ for a function (1734), Σ for summation (1755), the notation for finite differences Δy and $\Delta^2 y$, differential operator D, which being applied to a function f, gives the first derivative Df, and many others. Some historians claim that his *Introductio in analysin infinitorum*, which appeared in 1748, is to the calculus what Euclid's *Elements* is to geometry and al-Khowarizmi's *Al-Jabr W'almuquabalah* is to algebra (cf. (Burton, 1997)). The *Analysis Infinitorum* was followed in 1755 by the *Institutiones Calculi Differentialis*, to which *Analysis Infinitorum* was intended as an introduction. This series of works was completed by the publication in three volumes in 1768 to 1770 of the *Institutiones Calculi Integralis*, in which the results of several of Euler's earlier memoirs on the same subject and on differential equations were included. This, like the similar treatise on the differential calculus, summed up what was then known on the subject, but many of the theorems were recast and the proofs improved.

Euler and his contemporaries made many important discoveries but had little concern with the logical foundations of the method that they were applying so successfully. Adrien-

Marie Legendre originally introduced the modern notation $\partial y/\partial x$ for partial differentiation in 1786, but immediately abandoned it. Only in 1841, Carl Gustav Jacob Jacobi adopts it as a conventional notation. Another milestone in the development of the calculus at that time was the appearance of *Théorie des fonctions analytiques* (*The Theory of Analytic Functions*), by Joseph Louis Lagrange, in 1797.

However, not everything went well with the calculus. In spite of the numerous achievements, the famous philosopher George Berkeley (1685-1753) severely attacked it in 1734 for its lack of a rigorous foundation and even disputed the logic on which it was based. He taunted the proponents of calculus with "submitting to authority, taking things on trust, and believing points inconceivable." As notes Burton (1997), these were precisely the charges against religious people and Berkeley tried to restore regard for religion by showing that even in mathematics there are theories based on faith.

This polemic caused essential effort made to tighten the reasoning in the calculus. Colin Maclaurin (1698-1776) attempted to put the calculus on a rigorous geometrical basis but his attempt was not successful. Only Augustin Louis Cauchy (1789-1857) developed a rigorous foundation for the calculus. Cauchy grounded derivatives and integrals of functions, as well as operations with them via the concept of limits defined for real and complex numbers. For instance, Cauchy demonstrated that the definite integral could be defined as the limit of a characteristic sum and was independent of the definition of the derivative. In such a way, Cauchy build for continuous functions what we call now the Riemann integral. It is interesting to know that, according to Burton (1997), the first prominent mathematician to suggest that limits were fundamental in calculus was Jean le Rond d'Alembert (1717–1783). He wrote in his article in *Encyclopėdie* published in 2nd half of the 18th century that differentiation of equations consists simply in finding the limits.

Bernard Placidus Johann Nepomuk Bolzano (1781–1848) and Karl Theodor Wilhelm Weierstrass (1815-1897) also contributed to the foundations of the calculus, helping to clarify it and making more precise. As the calculus studied numerical functions, a firm foundation for the calculus required a firm foundation for the system of real numbers. This foundation was developed mostly by Julius Wihelm Richard Dedekind (1831–1916) and Georg Ferdinand Ludwig Philipp Cantor (1845-1918). Cantor's research in the calculus brought him to one of the greatest discoveries in mathematics – he built *set theory*, which gave the first rigorous foundation and tools for working with infinite sets. In the 20th century, set theory originated by Cantor became the foundation of the major part of mathematics.

Although foundations were important for the calculus and their development gave an impetus to the area, the main concern of mathematicians was advancement of calculus itself and its application. Many mathematicians from different countries have contributed to the continuing development of the calculus and applied it to a huge diversity of problems. In this brief historical overview, we can only mention some of the most outstanding contributors.

In addition to those outstanding mathematicians whose work in the calculus was already indicated here, in the 19th century, we see such giants as Carl Friedrich Gauss (1777-1855), Niels Henrik Abel (1802-1829), Joseph-Louis Lagrange (1736-1813), Carl Gustav Jacob Jacobi (1804-1851), Johann Peter Gustav Lejeune Dirichlet (1805-1859), Georg Friedrich Bernhard Riemann (1826-1866), Adrien-Marie Legendre (1752-1833), and Charles Hermite (1822-1901). Such mathematicians as Pierre Simon Laplace (1749-1827), Jean Baptiste Joseph Fourier (1768-1830), Siméon-Denis Poisson (1781-1840), and Mikhail Vasilievich Ostrogradski (1801-1862) were noted for their applications of the calculus.

Here are some of the main achievements in the area of the calculus and related fields during this period. Lagrange created the calculus of variations and established the theory of differential equations. Riemann extended the concept of the Cauchy integral and defined (in 1854) the integral in a way that does not require continuity. Now his construction is called the Riemann integral. Legendre, Abel and Jacobi developed the theory of elliptic functions and integrals. Dirichlet (in 1829) gave a definition of a numerical function that is still used today. Namely, a variable y is a function of a variable x when each value of x in a given interval is corresponded to a unique value of y. Somewhat later, Nikolai Ivanovich Lobachevsky (1792–1856) gave a similar definition of a function as a correspondence between two sets of real numbers. Fourier suggested that any function can be expanded in a series of sines and proved this for some functions. Though in general this result is not correct, his observation that a discontinuous function could be represented as a sum of infinite series was a breakthrough and the question of determining when a function can be represented as a sum of its Fourier series has been fundamental till our time. Fourier also explicitly used a differential operator, when he wrote D for the Laplacian and D^2 for its square. Weiestrass originated the movement known as the arithmetization of analysis, is considered the world's greatest analyst during the last third of the 19th century, and due to this, is sometimes called "the father of modern analysis" (Burton, 1997). Hermite proved that the number e is *transcendental* and solved the general quintic equation and, utilizing methods from the calculus, namely, elliptic functions. Eduard Heine (1821-1881), a student of Weierstrass, gave the modern ``epsilon-delta" definition of a limit in 1872.

Like Newton, Laplace used the calculus to develop a concise system of celestial mechanics. He gave an exposition of this system in the *Traité de Mécanique Céleste*, published in five large volumes over 26 years (1799-1825). Some reviewers praised this treatise as "the highest point to which man has yet ascended in the scale of intellectual attainment." (cf. (Burton, 1997)).

As it is often done in physics, Laplace was satisfied coming to a result that seemed correct and inserting the optimistic note, "It is easy to prove" or "Thus it plainly appears." According to Burton (1997), the American astronomer Nathaniel Bowditch (1773-1838), who translated four of five volumes of the treatise into English, recollected that when he encountered such a note, he had to intensively work for hours to fill up the chasm and find out and show how it plainly appears.

However, the original technique of Newton and Leibniz based on fluxions, differentials and other infinitesimals, although being extensively used, still remained ungrounded. In his critic of calculus, Berkeley wrote:

> *"And what are those fluxions? The velocities of evanescent increments. And what are these same evanescent increments? They are neither finite quantities, nor quantities infinitely small, nor yet nothing. May we not call them ghosts of departed quantities?"*

Berkeley's questions remained unanswered until the middle of the 20th century.

It is also necessary to remark that the development of the calculus and algebra made possible solution of three famous problems formulated by ancient Greek mathematicians. Greek asked to find methods for the *quadrature of* (or *squaring*) *a circle* (i.e., for an arbitrary given circle to build a square with the same area), the *duplication of a cube* (i.e., for an arbitrary given cube to build a cube with the doubled volume), and the *trisection of an angle*.

Tradition has it that Plato (429-348 B.C.E.) demanded that these tasks be performed with straightedge and compass only. For more than 2000 years, mathematicians were unable to solve these problems. As Burton writes (1997), these problems have remained landmarks in the history of mathematics, a source of stimulation and fascination for amateurs and scholars alike through ages. The first of these problems was solved by Ferdinand von Lindeman (1852-1939), who proved in 1882 that π is a transcendental number and thus, it is impossible to square a circle with straightedge and compass only. In 1837, Pierre Wantzel (1814-1848) gave the first rigorous proof of the impossibility of trisecting any given angle and duplicating any given cube with straightedge and compass only. However, these results did not influence further development of mathematics and were similar to sport records.

Thus, in spite of such a critic and existing problems with its foundations, the calculus was widely used and became a very powerful tool in mathematics and physics. More and more mathematicians contributed to the calculus. Further advances were mostly considered not as the development of the calculus itself but as the progress of fields that emerged from the calculus, such as real analysis, complex analysis, differential equations, integral equations, and later functional analysis, operator theory, tensor and vector analysis, measure theory, analysis on manifolds, analysis on supermanifolds, numerical analysis, calculus of variations, harmonic analysis, and so on. It is interesting to note that integration has grew into a specific theory – integration theory, while differentiation remained, as before, only a basic operation in analysis. Different kinds of analysis have their specific areas. *Real analysis* studies real numbers, their sequences, series, and real functions, whith an emphasis on functions. *Complex analysis* studies complex numbers, their sequences, series, and complex functions, whith an emphasis on functions. *Tensor* and *vector analysis* studies tensors and vectors, their sequences, series, and functions. The main concern of *harmonic analysis* is periodic functions (harmonics, waves), such as sin x or cos x, and how they synthesize other functions. *Functional analysis* studies spaces of functions and functions on these spaces (functionals and operators). *Analysis on manifolds and supermanifolds* is concerned with functions on manifolds and supermanifolds. *Numerical analysis* studies of algorithms and methods for solving problems of continuous mathematics, such summation of series, finding zeroes and derivatives of functions, and evaluating integrals.

As a result, it is possible to consider calculus in a strict sense and in a broad sense. In a strict sense, the calculus is the discipline that is studied under this name in colleges and universities. In a broad sense, calculus is the discipline that includes not only the calculus, but also mathematical fields that emerged from the calculus (real analysis, complex analysis, functional analysis, measure theory, calculus of variations, etc.).

The development of the calculus both in a broad sense and in a strict sense has continued to our time. At the end of the 19th century and in the 20th century, we see many mathematicians who made important contributions to the mathematics in general and to the calculus, in particular. It is possible to name such outstanding mathematicians as Jules Henri Poincaré (1854-1912), David Hilbert (1862-1943), John von Neumann (originally, János Lajos Margittai von Neumann) (1903-1957), Andrey Nikolaevich Kolmogorov (1903-1987), Thomas Joannes Stieltjes (1856–1894), Félix Edouard Justin Émile Borel (1871-1956), Henri Léon Lebesgue (1875-1941), Frigyes Riesz (1880-1956), Hermann Klaus Hugo Weyl (1885-1955), Alfréd Haar (1885-1933), Stephan Banach (1892-1945), Norbert Wiener (1894–1964), Henri Cartan (1904-), Jean Dieudonné (1906-1992), Mark Grigorjevich Krein (1907-1989), Sergei L'vovich Sobolev (1908-1989), Israil Moiseevic Gelfand (1913-), Laurent Schwartz

(1915–2002), Kiyosi Ito (1915-), Abraham Robinson (1918-1974), and Yisrael Robert John Aumann (1930-).

Here are some of contributions to the calculus during this period. One of the most innovative mathematicians was Poincaré. He was a mathematician and a scientist preoccupied by many aspects of mathematics, physics and philosophy. Due to this, he is often described as the last universalist in mathematics. In the areas related to the calculus, he developed the concept of automorphic functions, worked in the fields of differential equations and multiple integrals, and is considered the originator of the theory of analytic functions of several complex variables. In applied mathematics and natural sciences, Poincaré had important results in optics, electricity, telegraphy, capillarity, elasticity, thermodynamics, celestial mechanics, potential theory, quantum theory, electromagnetic theory, and cosmology. In particular, his research of celestial mechanics included the three-body-problem, which originated the theory of dynamical systems. Due to his papers on the dynamics of the electron, Poincaré is acknowledged as a codiscoverer, with Albert Einstein and Hendrik Lorentz, of the special theory of relativity.

Borel created the first effective theory of the measure of sets of points. This marked the beginning of the modern theory of functions of a real variable, as well as the theory of measure and measure integration. The Lebesgue measure and integral introduced by Lebesgue occupy a place between the most important achievements of the calculus and analysis. Borel also developed a systematic theory for a divergent series. Hadamard introduced the word "functional" in 1903. Results published by Riesz in 1910 indicate the beginning of operator theory. He is also considered one of the founders of functional analysis. Banach developed fully axiomatic approach to the calculus in infinite dimensional spaces in his 1920 doctoral dissertation. He also took a further step in abstraction in 1932 moving from inner product spaces to normed spaces. Banach obtained many fundamental results in functional analysis. For instance, three cornerstones of functional analysis are Hahn-Banach theorem on extensions of linear functionals, Banach-Steihaus theorem on convergence of bounded linear operators and Banach theorem on existence of an inverse operator.

Functional analysis is a mathematical field whose basic subject is the study of infinite-dimensional vector spaces and their mappings. It is interesting that at the beginning of its existence functional analysis evoked a certain amount of skepticism. It looked like a repetition familiar facts of classical analysis in a new language. Moreover, the research in this area was interesting but did not give anything essentially new (Kantorovich, 1987). Later the apparatus of functional analysis was enriched, its results grew more profound and new objects and facts were discovered. So, it became clear that this is a new and fundamental part of research in mathematical analysis. Now many mathematicians assume that functional analysis is a field that has, in some sense, the same importance as analysis (Kantorovich, 1987).

On higher levels of analysis, the main calculus operations – differentiation and integration – have been developed into extremely abstract structures - differential and integral operators, functionals, and differential forms. Differentiation became so formal that it came to general algebra, giving birth to a new field in algebra called differential algebra. It studies differential rings and algebras, in which differentiation is a specific mapping (namely, an endomorphism of the additive group with additional restrictions) of a ring/algebra into itself.

Necessity to consider derivatives of arbitrary continuous and even of some discontinuous functions resulted in the creation of distribution theory. It was one of the main achievements in mathematics of the 20th century. It originated from some mathematically ungrounded but physically efficient techniques when at first Oliver Heavyside (1850-1925) and then Paul Adrien Maurice Dirac (1902-1984) used such a distribution as the delta-function $\delta(x)$. Independently, Sobolev considered generalized functions (later called distributions by Schwartz) in a form of generalized solutions for differential equations. Schwartz developed mathematical theory of distributions. This theory made delta-function a legal mathematical object – a functional of a special type. It is interesting that at the beginning of its existence distribution theory evoked a certain amount of skepticism. For instance, Sobolev did not develop a theory of generalized functions, although as a talented mathematician, he was able to do this. Nevertheless, the attitude changed and many mathematicians contributed to this important area. For instance, such a talented mathematician as Gelfand is notable for his results in distribution theory and functional analysis, making also various contributions to analysis applications and many other areas of pure and applied mathematics. Mikusinski and Sikorski developed another version of distribution theory – the sequential approach (Antosik *et al*, 1973). Now distributions (generalized functions) form one of the main tools of contemporary theoretical physics. Several theories that extend distribution theory have been developed: new generalized functions (Fisher, 1969; 1971; Rosinger, 1980; 1987; Colombeau, 1982; 1984; 1985; Li Bang-He and Li Ya-Qing, 1985; Oberguggenberger, 1986; 1992; Burgin, 1987; Egorov, 1989; 1990), extrafunctions and hyperdistributions (Burgin, 1990, 1991, 2002, 2004b).

Many results in the calculus development came from applications. In such a way, integration of set-valued functions and correspondences came from calculus applications in statistics (Kudo and Richter) and economics (Aumann and Debreu). A calculus over supermanifolds and the theory of suparmanifolds itself came from physics (Salam and Strathdee, Wess and Zumino, Leites and Berezin).

A new development of the calculus foundations came around 1960. Abraham Robinson introduced nonstandard analysis (Robinson, 1961; 1966). It answered all questions of Berkeley and made *infinitesimals* and *differentials* grounded mathematical concepts. While in the calculus foundations developed by Cauchy infinitesimals or infinitely small values are represented by sequences converging to 0, nonstandard analysis introduces infinitesimals as actual infinitely small numbers. Generalizations of nonstandard analysis have been used in various areas such as functional analysis, number theory, probability, dynamical systems, and mathematical economics, demonstrating its usefulness for constructing mathematical models for diverse phenomena (Cutland, 1988). Some preliminary form of an infinitesimal calculus with infinitely small and infinitely big numbers was suggested in (Schmieden and Laugwitz, 1958; Laugwitz, 1961).

It is necessary to remark that, in some sense, nonstandard analysis is special case of nonarchimedean analysis (cf. (van Rooij, 1978; Schneider, 2001)) as spaces of hyperreal and hypercomplex numbers from nonstandard analysis are nonarchimedean fields. Nonarchimedean analysis started in the works of Monna (1943; 1946) on nonarchimedean linear spaces. The basic difference between nonarchimedean and classical analysis is that classical analysis studies functions and related structures based on fields of real and complex numbers, while nonarchimedean analysis studies functions and related structures based on fields with a valuation (valued fields). The main interest is in the case when these fields are nonarchimedean and have an ultrametric valuation.

Important achievements in the area of integration were introduction of path integration by Richard Feynman (1961) and gauge integration by Ralph Henstock (1955) and Jaroslav

Kurzweil (1957). Both structures gave birth to new theories with a lot of researchers participating in their development. Path integration has become a principal tool of theoretical physics although it still lacks comprehensive mathematical foundation. Gauge integration provides a transparent unifying schema for a variety of integral constructions in mathematics (for Cauchy, Darboux, Riemann, Lebesgue, Denjoy, Perron and some other integrals). On the one hand, the gauge integral is more general than all these integrals. On the other hand, its complexity is virtually the same as complexity of the most popular Riemann integral.

Lipman Bers (1914 -1993) developed a theory of pseudo-analytic functions. Mark Krein made extensive contributions to the development of analysis in infinite dimensional spaces and general functional analysis. Elias Menachem Stein (1931-) and Charles Louis Fefferman (1949-) are known for their work in mathematical analysis.

The development of the calculus in the broad sense has been connected to introduction of different new mathematical structures, such as topological spaces, manifolds, fibre spaces, supermanifolds or hypermanifolds, in which calculus operations have been defined. Such leading mathematicians as Henri Poincaré, Felix Hausdorff (1868–1942), Stefan Banach, Oswald Veblen (1880-1960), John Henry Constantine Whitehead (1904–1960), Jean Leray (1906-1998), Karl Johannes Herbert Seifert (1907–1996), Saunders Mac Lane (1909–2005), Norman Steenrod (1910-1971), Samuel Eilenberg (1913-1998), and Abraham Robinson (1918–1974) made essential contribution to this development of new structures for analysis.

As a result, contemporary branches of the calculus and analysis have naturally expanded into very abstract structures.

For instance, integration of differential forms takes functions as differential forms of degree 0 and defines integration in the following way (cf. Wikipedia).

A modern understanding of a differential form of degree k is a smooth section of the k-th exterior power of the cotangent bundle of a manifold. An important property of differential forms is that at any point p on the manifold, a k-form gives a multilinear map from the k-th exterior power of the tangent space at p to R.

Differential forms of degree k are integrated over k dimensional chains. If $k = 0$, this is just evaluation of functions at points. Other values of $k = 1, 2, 3, \ldots$ correspond to line integrals, surface integrals, volume integrals etc.

Let

$$\omega = \Sigma\, a_{i1, \ldots, ik}(x) \mathrm{d}x_{i1} \wedge \ldots \wedge \mathrm{d}x_{ik}$$

be a differential form and S be a set for which we wish to integrate over, where S has the parameterization

$$S(\mathrm{u}) = (x_1(\mathrm{u}), \ldots , x_k(\mathrm{u}))$$

for u in the parameter domain D. Then the integral of the differential form over S is defined as

$$\int_S \omega = \int_D \Sigma\, a_{i1, \ldots, ik}(S(u))\ \partial(x_{i1}, \ldots , x_{ik}) / \partial(u_1, \ldots , u_k)\ \mathrm{du}$$

where $\partial(x_{i1}, \ldots , x_{ik}) / \partial(u_1, \ldots , u_k)$ is the determinant of the Jacobian. *

The calculus as a powerful, fruitful and popular mathematical theory gave birth to a diversity of mathematical fields, such as mathematical analysis, real analysis, measure theory, complex analysis, differential equations, ergodic theory, integral equations, functional analysis, operator theory, integral transforms, operational calculus, harmonic analysis, representation theory, tensor and vector analysis, numerical analysis, calculus of variations, global analysis, analysis on manifolds, differential geometry and topology, p-adic analysis, nonstandard analysis, and theory of distributions. The success of the calculus has been extended over time to these fields. In addition, methods of the calculus are successfully used in other mathematical fields: probability theory, number theory, optimization theory, algebra, geometry, and topology.

The development and use of the calculus has had wide reaching effects on nearly all areas of modern living. It, in essence, underlies nearly all of the sciences, especially, mechanics, physics in general, and economics. Virtually the majority of modern developments such as construction of buildings and roads, shipbuilding, aviation, communication, radio, TV, computers, and other technologies make fundamental use of the calculus and fields originated from it. Many mathematical formulas used for ballistics, heating and cooling, elasticity, optimization, engineering, optimal control, mathematical programming, and other practical areas were worked out through the use of the calculus and its apparatus entered into foundations of these disciplines.

To present a more structured picture of the development of the calculus in a broad sense, we consider three mathematical structures that play the role of principal parameters of the calculus as a mathematical discipline.

1. *Spaces* in which functions studied by the calculus are defined.
2. *Functions* studied by the calculus.
3. *Operations* applied to these functions.

The last of these parameters is naturally divided into two parts (subparameters):

3.1. Differentiation.
3.2. Integration.

Considering the calculus as a kind of mathematical technology, we can see that this parametrization corresponds to the main components of technology (Burgin, 2003a).

The process of the development is represented in the following figures and tables. In figures, the lowest level represents the beginning of the process and each next level reflects a step in the development. A brief history (or more exactly, historical pointers) for this development is given in the tables that go after the figures. Tables go from earlier achievements to later contributions.

It is necessary to note that the development of the calculus in a broad sense has been so fast and extensive that the figures and tables given below provide only a simplified and approximate picture of this development. However, they represent principal achievements and pivotal advances of the history of the calculus in a broad sense. In addition, so many mathematicians contributed to the expansion of calculus in broad sense, that it is not always easy to find who was the first to introduce this or that structure, idea or concept. For all this, we ask the reader's and mathematicians' forbearance.

Spaces:

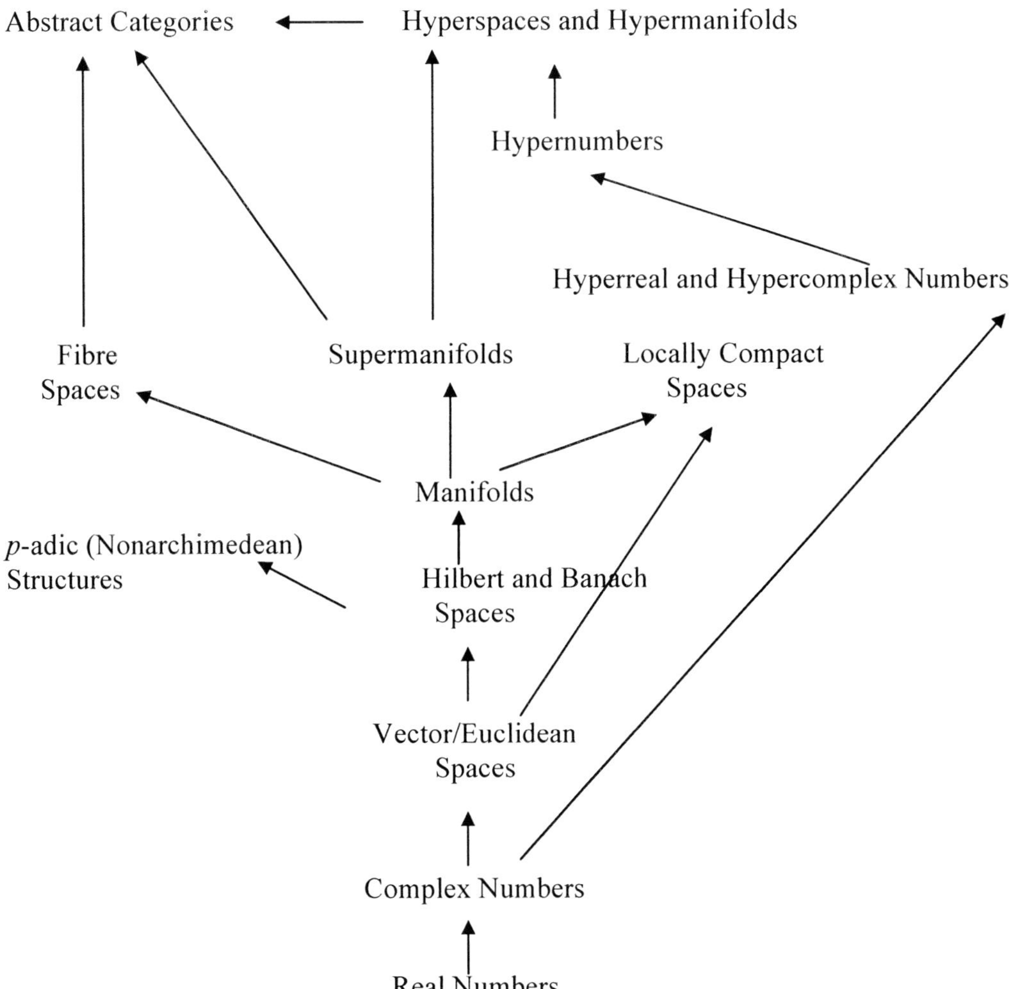

Figure 1.1. The development of spaces in which functions studied by the calculus are defined (from numbers to manifolds and hypernumbers).

Table 1.1. The development of spaces in which functions studied by the calculus are defined (from numbers to manifolds and hypernumbers)

Structures (Spaces)	Who introduced it in general and to the calculus/analysis, in particular
Real Numbers Theory of real numbers	Mathematicians from Kerala School (15th and 16th centuries) René Descartes (introduced the terms *real number* and *imaginary number*) Johannes Kepler, Bonaventura Cavalieri, John Wallis, Pierre de Fermat, Blaise Pascal, Isaac Barrow, Isaac Newton, Wilhelm Leibniz, and Seki Kowa (17th century) Karl Weierstrass (1863), Richard Dedekind (1876)
Complex Numbers	Leonhard Euler (1740s)
Vector/ Linear/ Euclidean Spaces	Arthur Kelly, Hermann Grassman (1840s) Giuseppe Peano (axiomatic form; 1888)
Tensor Spaces	William Rowan Hamilton (introduced the term *tensor*, 1846) Gregorio Ricci-Curbastro (general construction in the context of *absolute differential calculus*, around 1890)
Functional Spaces	Vitto Volterra (1887)
Metric Spaces	Maurice Frechet (1906)
Hilbert Spaces Normed Linear Spaces and Banach Spaces	David Hilbert* (different particular cases, 1906), Maurice Frechet (1907-1908), Erhard Schmidt (1908) John von Neumann (axioms for *abstract* Hilbert spaces, 1929) Frigyes Riesz (implicitly, 1818) Stephan Banach, H. Hahn, and E. Helly (1920s)
Manifolds	Carl Friedrich Gauss (considered surface as a two-dimensional manifold, however, without giving an explicit concept of a manifold) Bernhard Riemann (considered manifolds of different dimensions and introduced the German term, *Mannigfaltigkeit*, in *Grundlagen für eine Allgemeine Theorie der Functionen,* published (posthumously) in 1867; the term was translated by Clifford as "manifoldness") Henri Poincaré (studied and defined three-dimensional manifolds, 1995) Hermann Weyl (gave an intrinsic definition for differentiable manifolds, 1911-1913) H. Kneser (combinatorial manifolds, 1924) Oswald Veblen and John Whitehead (a general formal definition, 1932) Whitney (a definition of a smooth manifold, 1936)
Topological Spaces Topological Vector Spaces	Frigyes Riesz (sequential topology, 1909), Felix Hausdorff (general topology, 1914) Andrey Kolmogorov (general topological vector spaces, 1934), John von Neumann (general topological vector spaces, 1935)
Fibre Spaces Bundles, Sheaves, Fibrations	Herbert Seifert (decomposable surfaces, which are some kinds of fibre spaces, 1932) Whitney (sphere spaces as a kind of fibre spaces and a general idea of fibre spaces, 1935) Norman Steenrod (made a distinction between fibre spaces and fibre bundles, 1944) Jean Leray (sheaves, 1945)

Table 1.1. Continued

Structures (Spaces)	Who introduced it in general and to the calculus/analysis, in particular
p-adic numbers	A. F. Monna (1943)
p-adic (nonarchimedean) linear spaces	A. F. Monna (1943)
Categories	Samuel Eilenberg and Saunders Mac Lane (1945)
p-adic (Nonarchimedean) Banach Spaces	Edward Beckenstein (1968), Lawrence Narici (1968)
Hyperreal and Hypercomplex Numbers	Abraham Robinson (1961)
Superspaces Supermanifolds	Abdus Salam and J.A. Strathdee (in the context of super symmetry, 1974), Julius Wess and Bruno Zumino (in the context of super symmetry, 1974) Dimitry Leites (algebraic supervarieties, 1974) Felix A. Berezin and Dimitry Leites (smooth supermanifolds, 1975)
Operator Spaces	Edward G. Effros (1986)
Hypernumbers	Mark Burgin (1987)
Hyperspaces and Hypermanifolds	Mark Burgin (1990)

* There is a story (cf. (Smithies, 1997)) that in 1909 Hermann Weyl, in a lecture in Hilbert's Gottingen seminar, referred to "Hilbert space", by which he meant the space of sequences (x_1 , x_2 , x_3 , ... , x_n , ...) such that $\Sigma_{i=1}^{\infty} x_i < \infty$. Hilbert interrupted him to ask 'Tell me, please, what *is* a Hilbert space?'.

Functions:

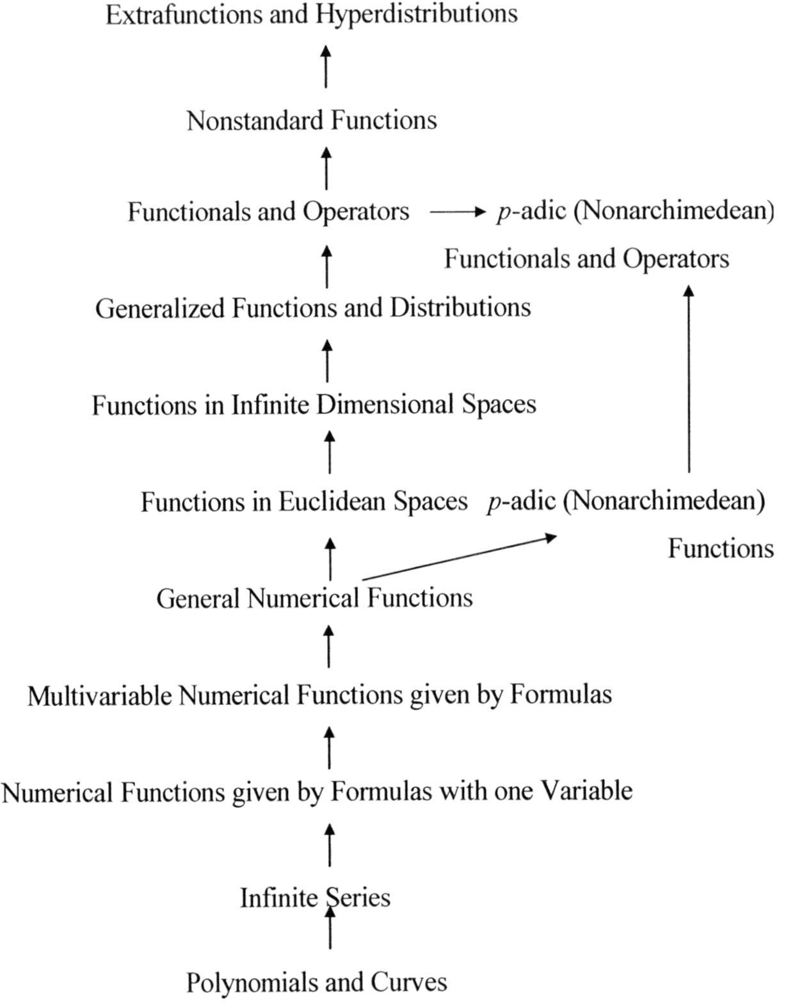

Figure 1.2. The development of functions studied by the calculus (from numerical functions to distributions and extrafunctions).

Table 1.2. The evolution of functions studied by the calculus

Structures (Functions)	Who started to use it in the calculus/analysis
Polynomials and Curves	Mathematicians from Kerala School (15th and 16th centuries) Johannes Kepler, Bonaventura Cavalieri, John Wallis, Pierre de Fermat, Blaise Pascal, Isaac Barrow, Isaac Newton, Wilhelm Leibniz, and Seki Kowa (17th century)
Infinite Series	Mathematicians from Kerala School (15th and 16th centuries) such as Madhava of Sangamagrama (1340-1425), Jyestadeva (1500-1575) John Wallis, Isaac Newton, Wilhelm Leibniz, and James Gregory (17th century)
Real Functions given by Formulas with one Variable	Gottfried Wilhelm Leibniz (first used the mathematical term *function*, 1673) Leonhard Euler (1748)
Functions of Complex Variables	Leonhard Euler (1740s)
Multivariable Numerical Functions given by Formulas	Leonhard Euler (1748)
General Numerical Functions	The notion of a function in a general form as a relation first occurred in the 14th century in the schools of natural philosophy at Oxford and Paris. More explicit understanding was achieved by Galilei Galileo (1564-1642) in his studies of motion. Gustav Lejeune Dirichlet (1829), Nikolai Ivanovich Lobachevsky (1838)
Functions in Vector/Euclidean Spaces	Arthur Kelly, Hermann Grassman (1840s)
Functions in Infinite Dimensional Spaces	Vito Volterra (in function spaces, 1920 – 1922) David Hilbert (1906), Maurice Frechet (1907-1908), Erhard Schmidt (1908), Frigyes Riesz (1818) Eduard Helly (1912) Hans Hahn (1927) Stephan Banach (1920; published 1932)

Table 1.2. Continued

Structures (Functions)	**Who started to use it in the calculus/analysis**
Functionals and Operators	The usage of the word *functional* goes back to the calculus of variations, implying a function whose argument is a function. Variational calculus was largely the creation of Leonhard Euler, the Bernoulli family, and Joseph-Louis Lagrange. The transforms of Pierre-Simon Laplace and Joseph Fourier are now some of the most studied and used kinds of operators on spaces of functions. Fourier also was the first to explicitly use a differential operator. Oliver Heaviside (created a systematic operational calculus, 1880-1887) Salvatore Pincherle (functional calculus, 1897) Jacques-Salomon Hadamard (introduced the term *functional*, 1903) Maurice Frechet (1906) (used the term *functional* in his published work) The concept of an algebra of operators made its appearance in series of articles culminating in a book by Frigyes Riesz (1913) Vito Volterra (used functionals in an explicit form, 1920–1922) John von Neumann (used operators in an explicit form, 1929) Stefan Banach (used operators in an explicit form, 1932)
Generalized Functions and Distributions	Sergey Sobolev (as generalized solutions to differential equations, 1935-1936) Laurent Schwartz (as a theory of functionals, 1945) Jan Mikusinski and Roman Sikorski (sequential approach, 1955-6)
p-adic functions	A. F. Monna (1943)
Nonstandard Functions	Abraham Robinson (1961)
Extrafunctions and Hyperdistributions	Mark Burgin (1991; 2004)

Derivatives:

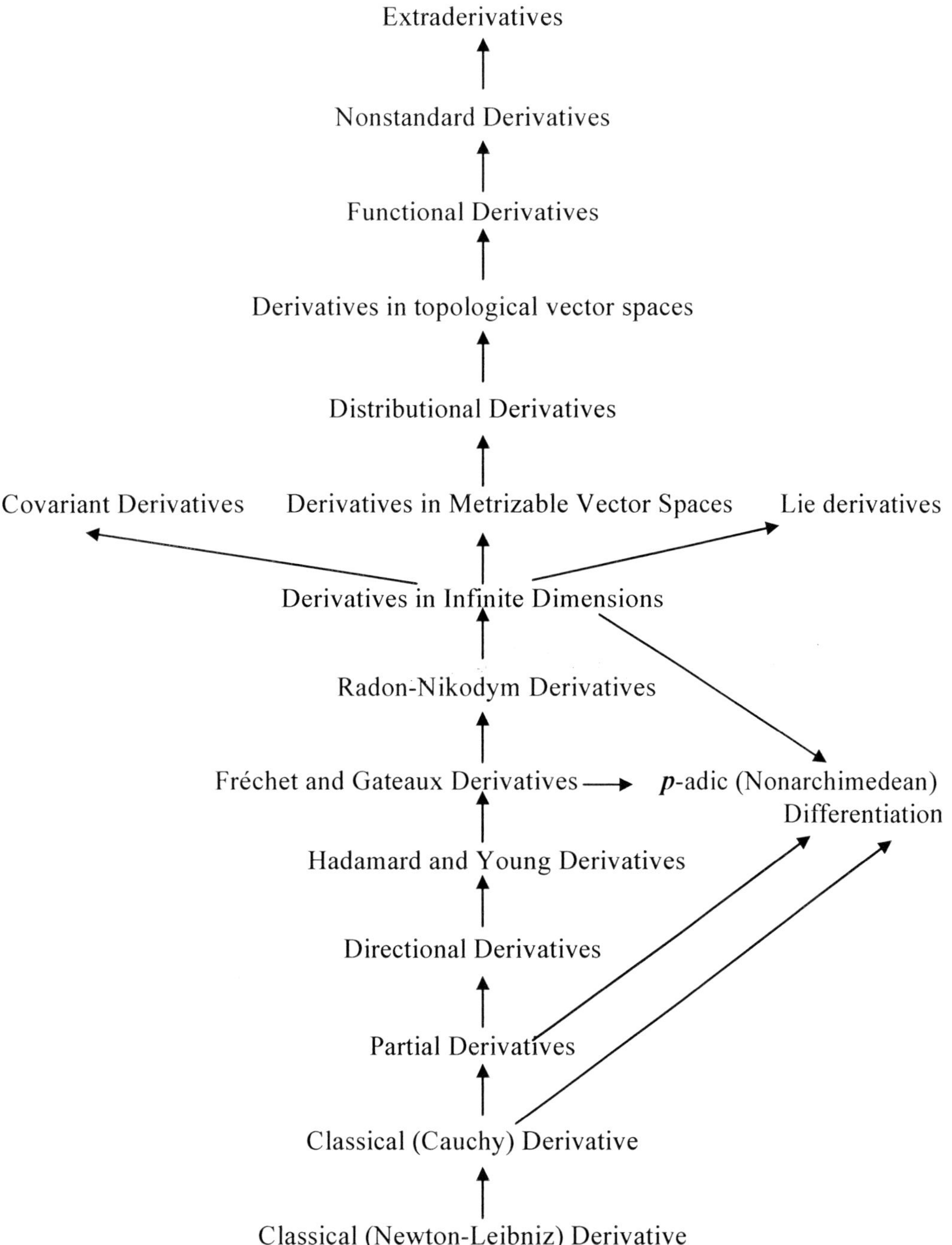

Figure 1.3. The development of differentiation (from classical derivatives to extraderivatives, functional and nonstandard derivatives).

Table 1.3. The evolution of differentiation

Structures in Differentiation	Who introduced it to the calculus/analysis
Classical (Newton-Leibniz) Derivatives	Bhaskara II (1114-1185), Jyestadeva (1500-1575) Isaac Newton and Wilhelm Leibniz (end of the 17th century)
Partial Derivatives	Isaac Newton and Wilhelm Leibniz (end of the 17th century) Adrien-Marie Legendre (introduced notation $\partial y/\partial x$ (1786), but immediately abandoned it) Carl Gustav Jacob Jacobi (adopted this notation, 1841) William Rowan Hamilton (used the term *partial derivative*, 1834)
Classical (Cauchy) Derivatives	Jean le Rond d'Alembert (used informally; 2nd half of the 18th century) Augustin Cauchy (1823)
Variational Derivative	Vitto Volterra (1887)
Directional Derivatives	Edwin Bidwell Wilson (1901)
Young Derivative and Hadamard Derivative	William Henry Young (1910) Jacques Hadamard (1923)
Fréchet Derivative and Gateaux Derivative	Maurice Fréchet (1925) René Gâteaux (1913)
Radon-Nikodym Derivatives	Johann Radon (1913 in R^n) Otton Marcin Nikodym (1930 for the general case)
Derivatives in Infinite Dimensional Spaces	Aristotle Michal (differentiability in infinite dimensional spaces, 1938) Alfred Fröhlicher (1981) and Andreas Kriegl (1982) the more adequate differential calculus in infinite dimensional spaces
Covariant Derivatives	Gregorio Ricci-Curbastro and Tullio Levi-Civita (Riemannian and pseudo-Riemannian geometry; 1900) Jean-Louis Koszul (in a general case of vector bundles; 1950)
Lie derivatives	Władysław Ślebodziński (1931)
p-adic (nonarchimedean) differentiation	Bernard M. Dwork (1973), Carl S. Weisman (1977), Wim H. Schikhov (1978)
Derivatives in Metrizable Vector Spaces	M.M. Balanzat (1949)
Distributional Derivatives	Laurent Schwartz (1945)
Derivatives in Topological Vector Spaces	M.M. Balanzat (1960)
Nonstandard Derivatives	Abraham Robinson (1961)
Derivatives in Grassman Algebras, Superspaces, and Supermanifolds	M.M. Vivier (1956), Felix A. Berezin (1961)
Extraderivatives	Mark Burgin (1993)
Differentiation in abstract categories	Valery A. Lunts and Alexander L. Rosenberg (1996-97) Vladimir Molotkov (2005)

Integrals (I):

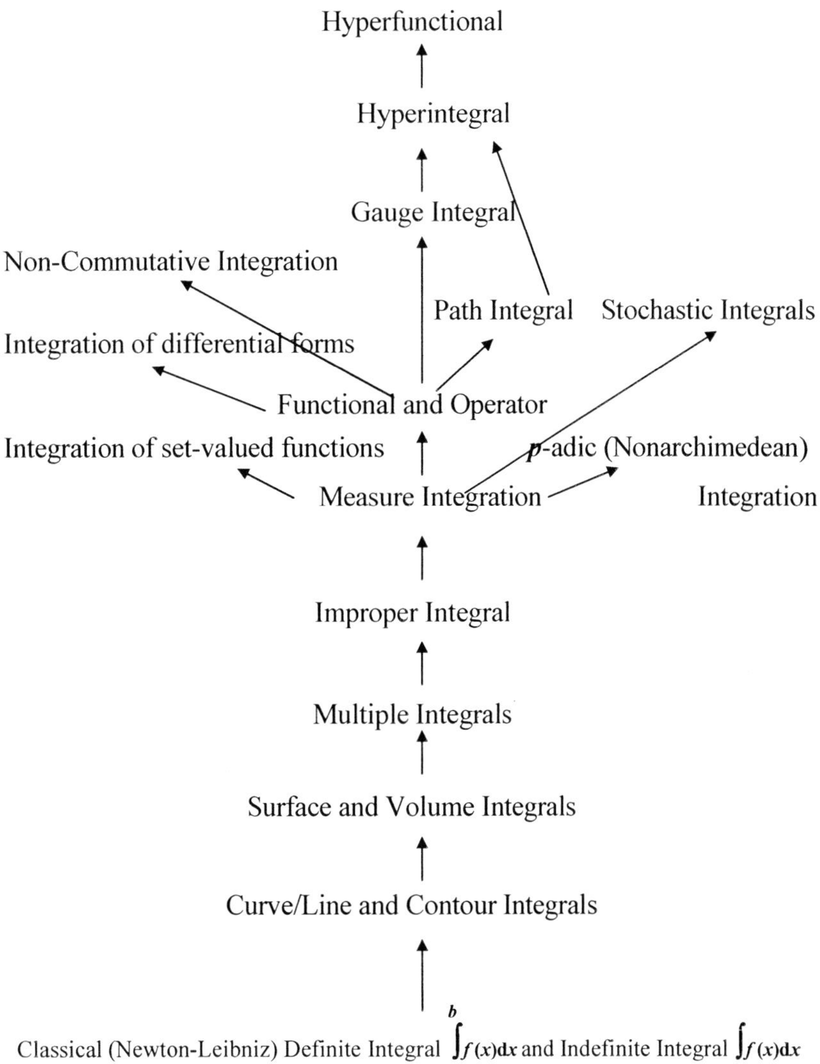

Figure 1.4. The first dimension of the development of integration (from classical integrals to path and gauge integral, to functionals and hyperfunctionals).

Table 1.4. The evolution of integration: structural aspect

Structures in Integration	**Who introduced it to the calculus/analysis**
Classical (Newton-Leibniz) Indefinite Integral $\int f(x)\mathrm{d}x$	Isaac Newton and Wilhelm Leibniz (end of the 17th century)
Definite Integral	Pierre-Simon Laplace (1779).
Curve/Line and Contour Integrals	Alexis Claude Clairaut (1743) Augustin Louis Cauchy (1829) James Clerk Maxwell (used the term *line integral*, 1873)
Surface and Volume Integrals	George Green (1828)
Multiple Integrals	Joseph-Louis Lagrange (wrote about double integrals in a letter to Euler; 1756) Leonhard Euler (published a theory of double integrals; 1769-1770) Joseph-Louis Lagrange (introduced triple integrals; 1772)
Improper Integrals	Leonhard Euler (introduced Gamma-function, which is an improper integral; 1729 - 1730)
Measure Integration	Henri Leon Lebesgue (with the Lebesgue measure, 1902) Johann Radon (general approach, 1913)
Integral of a Vector-Valued Function	Salomon Bochner (1933)
Integration over Topological Groups	Alfréd Haar (1933)
Functionals	Vito Volterra (1920 – 1922)
Functional Integration	Norbert Wiener (in the study of Brownian motion; 1923)
Stochastic Integrals	Norbert Wiener (1923) Kiyosi Ito (1944) D. L. Fisk (1963) Ruslan L. Stratonovich (1966)
p-adic (nonarchimedean) Integration	François Bruhat (1962), F. Thomas (1962)
Integration of Set-Valued Functions Integration of Correspondences	Considerations of integration of set-valued functions goes back to Hermann Minkowski (1864-1909) (cf., (Artstein and Burns, 1975) H. Kudo (1954), H. Richter (1963) integration of set-valued functions in connection with statistical problems Yisrael Robert John Aumann (integration of set-valued functions in connection with economical problems, 1965) M. Hukuhara (integration of set-valued functions, 1967) Gerard Debreu (integration of correspondences in connection with economical problems, 1967)

Table 1.4. Continued

Structures in Integration	Who introduced it to the calculus/analysis
Path Integrals	Richard Feynman (1961; based on an idea of Paul Adrien Maurice Dirac, 1933)
Integration over Grassman Algebras, Superspaces, and Supermanifolds	Isaak Markovich Halatnikov (in implicit form, 1954), Abdus Salam and Paul Taunton Matthews (in implicit form, 1954) Felix A. Berezin (in Grassman Algebras, 1961)
Gauge Integrals	Ralph Henstock (1955) and Jaroslav Kurzweil (1957)
Non-Commutative Integration	Alain Connes (1979; 1985)
Hyperintegrals	Mark Burgin (1990)
Hyperfunctionals	Mark Burgin (1991)
Motivic Integration	Maxim Kontsevich (1995)

Integrals (II):

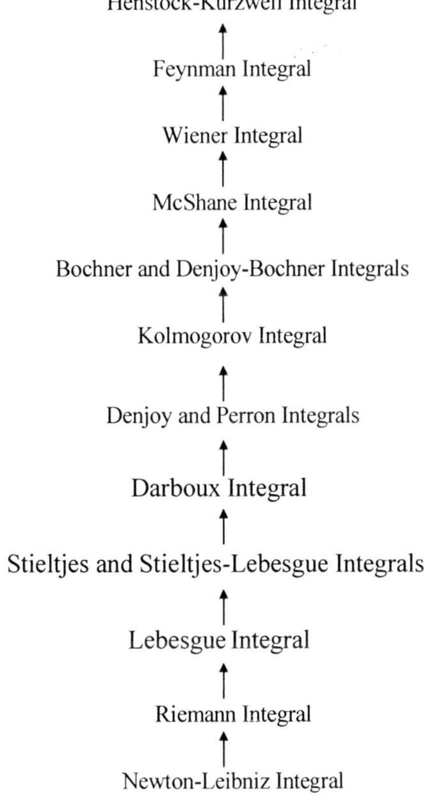

Figure 1.5. The second dimension of the development of integration (from Newton-Leibniz to Lebesgue, to Denjoy-Perron, to Henstock-Kurzweil integrals).

Table 1.5. The evolution of integration: personal aspect

Structures in Integration	**Who introduced it to the calculus/analysis**
Newton-Leibniz Integral	Newton and Wilhelm Leibniz (end of the 17th century)
Cauchy Integral Riemann Integral	Augustin Cauchy (for continuous functions, 1823) Bernhard Riemann (1854)
Darboux Integral	Jean Gaston Darboux (1879)
Stieltjes Integral	Thomas Stieltjes (1894)
Lebesgue Integral	Henri Lebesgue (1902)
Denjoy Integral Perron Integral	Arnaud Denjoy (1912) Oskar Perron (1914)
Daniell Integral	Percy Daniell (1918)
Kolmogorov Integral	Andrei Kolmogorov (1925)
Wiener Integral	Norbert Wiener (1923)
Bochner Integral	Salomon Bochner (1933)
Ito Integral	Kiyosi Ito (1944)
Henstock-Kurzweil Integral	Ralph Henstock (1955) and Jaroslav Kurzweil (1957)
Feynman Integral	Richard Feynman (1961; based on an idea of P. A. M. Dirac, 1933)
Aumann-Hukuhara Integral	Robert Aumann (integration of set-valued functions in connection with economical problems, 1965) M. Hukuhara (integration of set-valued functions, 1967)
McShane Integral	Edward James McShane (1973)

1.2. A Gap between Calculus and Real Life

As far as the laws of mathematics refer to reality,
they are not certain;
and as far as they are certain,
they do not refer to reality.
Albert Einstein (1879-1955)

Although this may seem a paradox,
all exact science is dominated by the idea of approximation.
Bertrand Russell (1872-1970)

Mathematics is an efficient tool for modeling real world phenomena. However, in its essence, mathematics is opposite to real world because mathematics is exact, rigorous and abstract, while real things and systems are imprecise, vague, and concrete.

This is especially true for the calculus. An important peculiarity of the calculus is that in it everything is considered with an ultimate precision. While this is realistic for whole and integer numbers, even rational numbers, not speaking about real and complex numbers, can sometimes cause a lot of troubles when we try to get an absolutely precise representation. For instance, computers are working in the binary number system. That is why, even such simple fractions as 1/3 or 1/7 cannot be represented with complete precision.

The situation becomes even more complicated when operations that involve infinite processes are applied. At the same time, this is an inherent feature of the calculus because its main operations, differentiation and integration, are based on such infinite processes. The cause of this is that such basic for the calculus concepts as limit and continuity also involve infinite processes. Now there are new forms of calculus, such as nonstandard analysis (Robinson, 1961; 1966; 1967; Davies, 1977) or theory of hypernumbers and extrafunctions (Burgin, 1993a; 1995a; 2002; 2004), in which differentiation and integration are determined without infinite processes related to limits. However, in these approaches, infinity still exists but becomes implicit in a similar way as it exists implicitly in real numbers. For instance, when we use the number π, we have only its name, general definition, some approximations, and algorithms that allow to compute π with arbitrary finite precision. As a result, some mathematicians have expressed an opinion that almost all real numbers do not really exist. When Lindemann proved that the number π was transcendental, Leopold Kronecker (1823-1891) was said to have express his attitude by saying to Lindemann: "Of what use is your beautiful proof, since π does not exist!"

There are even more problems with real numbers. It is possible to show that almost all real numbers are transcendental, uncomputable, and cannot even be named. Indeed, all real numbers are either algebraic or transcendental. All algebraic real numbers, that is real numbers that are roots of polynomials with whole coefficients, are denumerable. As there is a continuum of real numbers, almost all real numbers are transcendental. There are only a countable number of algorithms. So, all computable numbers are denumerable. Thus, almost all real numbers are uncomputable.

Emile Borel in 1927 pointed out that when we consider a real number as an infinite sequence of digits, then we could put an infinite amount of information into a single number, building the "know-it-all" number r. At first, we build a real number q such that q contains every English sentence in a coded form. Indeed, each sentence can be coded by a block of decimal digits. All English sentences are ordered and these blocks go after the decimal point in q one after another in the same order. In particular, q contains every possible true/false question that can be asked in English. Then we determine a real number r with the following structure. If the n-th block of q translates into a true/false question, then the n-th digit of q after the decimal point is equal to 1 if the answer to the question is "true", and equal to 0 if the answer is "false". If the n-th block of q does not translates into a true/false question, then we set the n-th digit of r after the decimal point equal to 3. As any sequence of decimal digits defines some real number, the described number r exists Thus, using r, it is possible to answer every possible true/false question that has ever been asked, or ever will be asked, in English. Borel calls this number r an unnatural real number, or an "unreal" real. This argument was further developed in (Burgin, 2005). It was demonstrated that a possibility to operate with arbitrary real numbers makes it possible to compute any function on words in a finite alphabet. In some sense, everything becomes computable. This contradicts our intuition, as well as the practice of computing.

Borel (1952) devotes a whole book to discuss another problem with real number existence, namely that of "inaccessible numbers". As Borel defines, an accessible number as a number that can be described as a mathematical object in some language. The problem is that it is permissible only to use a finite process or finite description of a process, e.g., algorithm, to describe a real number. Thus, it is possible to describe integer numbers easily enough by an algorithm that generates the decimal representation of an arbitrary integer number. Algorithmic description of integer numbers allows us to do the same with rational numbers. For instance, it is possible to represent a rational number either as a pair of integer numbers or by specifying (in an algorithmic form) the repeating decimal expansion. Hence, integer and rational numbers are accessible. We can get a similar representation for some other real numbers but not for all. For example, if we take the, so called, Liouville's transcendental number, then it can be described (built) by an algorithm that puts 1 in the place $n!$ and 0 elsewhere. A finite way of specifying the n-th term in a Cauchy sequence (cf. Section 2.1) of rational numbers gives us a finite description of the resulting real number. However, as Borel pointed out, there are only a countable number of such descriptions or algorithms. At the same time, there are uncountably many (continuum of) real numbers and continuum is much bigger than any countable number. So, the natural conclusion is that the majority of real numbers are inaccessible and thus, it is impossible to operate with them. This argument was further developed by Chaitin (2006). Thus, we see that there are problems with the classical calculus because it assumes a possibility to exactly operate with all real numbers.

In general, it is possible to operate with transcendental real numbers only using their symbolic names, like π or e, or their reasonable approximations, but not their exact values. Thus, imprecision related to infinity does not disappear whatever means we use to operate with real numbers. In new forms of calculus, such as nonstandard analysis (cf., (Robinson, 1966)) and the theory of hypernubers and extrafunctions (cf., (Burgin, 2002)), this imprecision only comes to a more basic level of number definitions and we still need new mathematical means to deal with this imprecision.

In addition, the necessity to launch investigation and implementation of fuzzy principles in the classical analysis, while studying ordinary functions, is caused by several practical reasons. One of the most important of them is connected with properties of measurements. Applications of mathematics in general and the calculus, in particular, include measurement and computation. Computational practice and physical theory show that both measurement and computation are inherently imprecise and vague. Any real measurement is not absolutely precise. It provides only approximate results. For instance, it is impossible to find out if any series of numbers obtained in experiments converges or a function determined by measurements is continuous at a given point. Actually, all such functions are discrete.

As a result, we encounter approximation processes in physical experiments because any measurement has a finite precision. Consequently, traditional continuous functions become approximations of what is actually measured and observed.

The problem is even deeper. It has been apparent since early times that there is something different between the mathematical properties of the real numbers and the quantities of measurement in physics at small scales. Riemann himself remarked on this disparity even as he constructed the formalism, which would be used to describe the space-time continuum for the next century of physics (cf., for example, (Gibbs, 1998)). When even the best intruments measure a distance or time interval, it is impossible to declare the result to be rational or irrational no matter how accurate the measurement manages to be. Furthermore, by the Bekenstein theorem (Bekenstein, 1981), there is a limit to the amount of information in a finite volume of space, while, as we have seen, a real number can contain an infinite amount of information.

We know that in the classical calculus and analysis, neighborhoods of points (intervals, balls, lengths, volumes, etc.) and their parameters can be arbitrary small. In physics this assumption never holds. As Landau and Lifschitz write in their Course of Theoretical Physics (1987):

> "... any small volume element (dV) in the fluid is always supposed so large that it still contains a very great number of molecules. Accordingly, when we speak of infinitely small elements of volume, we shall always mean those which are "physically" infinitely small, i.e., very small compared with the volume of the body under consideration, but large compared with the distance between molecules."

It is interesting to know that it is possible to mathematically formalize concepts of "physically" infinitely big and "physically" infinitely small numbers and other entities (lengths, areas, volumes, etc.) by means of non-Diophantine arithmetics (Burgin, 1997; 2007).

However, not only limit brings forth a disparity between the classical calculus and scintific reality. Let us consider another example of a concept from analysis such as convexity.

Definition 1.2.1. A set X in a linear space L is called *convex* if for any points a and b from X and any number λ from the interval [0, 1], the point $(1 - \lambda)a + \lambda b$ also belongs to X.

However, as Phu writes (2005) it is hard to find a material object that is really convex because most of physical things consist of separate tiny particles – atoms and molecules. For this point of view, such objects cannot be connected either. Therefore, what we assume as convex or connected in real life does not possess mathematical properties of convexity or connectivity.

Physicists are always limited by the precision of their experiments, the sophistication of the mathematics they use to model the studied systems, and by the very complexity of those systems. Most of the universe is simply too complicated to exactly treat it with the conventional mathematics. However, physicists ignore most of the complexities of reality, and instead work with relatively simple models.

Moreover, modern physical theories imply (cf., for example, such a fundamental law of contemporary physics as the Principle of Uncertainty, which was introduced by Werner von Heisenberg) that measurement has inherent limitations on precision for all dynamic attributes (Herbert, 1985). Heisenberg relations guarantee that any experiment and measurement will contain a blind spot. For example, if we take the rate of position change, impossibility to measure it with absolute precision is one of the consequences of the Principle of Uncertainty, which was introduced by Heisenberg. Besides, there are cases when exact rate exists, it is feasible to measure it, but it is impossible to calculate the value of this exact rate. As Koenderink (1990) writes, "the limit for whatever size to zero has never made any physical sense."

It is necessary to remark that it is also possible to explaine measurement uncertainty by the General Measurement Uncertainty Principle. Informally, this principle states that measurement of a dynamic characteristic has intrinsic uncertainty that depends on the rate of the characteristic change.

To derive an exact form of the General Measurement Uncertainty Principle, we make a natural conclusion that a dynamic characteristic (say, P) changes with time. Measurement of P is not performed at a single moment of time. It takes some interval of time $[t_0, t_1]$ to perform the measurement and get the result. Let us assume that the least length of the interaction time in measurement of P is equal to l and V is the rate of the change of P. We are not speaking about the instant rate of change, as it is possible that such rate does not exist. As a rule, the instant rate as the value of the derivative exists only in mathematics. In practice, we always have some average rate and can the value of the derivative only as some approximation to this average rate. Let the rate V of the change of P be equal to the ratio p/k where the number p represents units in which P is measured and the number k represents units of time. For instance, the rate of the mass change can be 3/5 grams per second.

If the rate is p/k and time is l, then a possible change is $(p/k)l$. This allows us to formulate the following conjecture.

The General Measurement Uncertainty Principle. The value of a dynamic characteristic P cannot be, in general, measured with the precision better than $(p/k)l$.

In addition, as Krylov (1979) and Hayes (2003) explain, outside the realm of pure mathematics the cost of maintaining very high exactness is seldom justified. Nothing in the physical world can be measured with such precision anyway. The situation is aggravated because, as it is written in (Petrucci, et al, 2001), all measurements are subject of error. There are systematic and random errors.

It is useful to remark that when dealing with uncertainty, it is often necessary to know boudaries of this uncertainty. For instance, according to contemporary standards, a measurement result is complete only when accompanied by a quantitative estimate of its uncertainty. This estimate is required in order to decide if the result is adequate for its intended purpose and to ascertain if it is consistent with other similar results. Over the years, many different approaches to evaluating and expressing the uncertainty of measurement results have been used.

All these issues show that constructions and methods developed in the classical analysis are only approximations to what exists in reality. In many situations such approximations have been giving a sufficiently adequate representation of studied phenomena. However, scientists and, especially, engineers have discovered many cases in which such methods did not work because classical approach is too coarse.

Even more, complexity of contemporary physics resulted in a situation when physicists encounter very fundamental physical models which are not amenable to exact treatment (Ydri, 2001). To overcome these limitations, fuzzy physics has been developed (Balachandran, et al, 2000; Balachandran and Vaidya, 2001; Ydri, 2001).

In addition, functions that are used in engineering and science are often not differentiable and even not continuous. At the same time, mathematical technique, for example, calculus or optimization theory, is based on operation of differentiation. Consequently, a problem arises what to do in a case when classical calculus is not applicable. Besides, even when a function is continuous or differentiable, we are able to get only approximations to its values in practical work, both in measurements and computations. Consequently, the exact technique of classical and new directions in analysis is not relevant to this situation (Kosko, 1993).

Some approaches to these problems are suggested by non-smooth analysis, distribution theory, and theory of extrafunctions. However, each of them has its limitations. Non-smooth analysis takes into account only Lipshutz functions and considers only extreme analogies of classical derivatives (Clark, 1983). Distribution theory (Schwartz, 1950-1951), in general, does not provide values of distributions at separate points, while these values are important for engineering problems. Theory of extrafunctions gives such values, but it is even more general than distribution theory and uses numbers that are more abstract than real numbers (Burgin, 2002; 2004).

Other causes for imprecision come from computation. Although, it is supposed that numerical computation is a precise methodology in contrast to qualitative methods, this is true only in a very few cases because computers work only with finite subsets of rational numbers. Computation is always limited to a finite number of symbols and achieves full exactness in a very limited number of cases. For example, all computable by computers functions are defined only for a finite set of points because computers operate only with rational numbers absolute values of which are bounded above and bounded below. Consequently, computable functions cannot be continuous in the sense of the classical calculus. Computers, by their nature, imply imprecision and work with approximations of ideal mathematical structures.

As a result, we have only approximations to theoretically defined functions, limits and derivatives. For example, according to the mathematical definition, if a point is the limit of some sequence, then elements of this sequence come to this limit infinitely close. In contrast to this, if we have a number in computer, its distance to all other computer numbers cannot be less than some small interval. Thus, a computed sequence may converge only approximately.

For limit processes, computation adds its uncertainty to the vagueness of initial data. As Chaitin (1999) writes, the fact is that in mathematics, for example, real numbers have infinite precision, but in the computer precision is finite. In some cases, this discrepancy between theoretical schemes and practical actions changes drastically outcomes of a research resulting in uncertainty of knowledge. For instance, as remarked the great mathematician Henri Poincaré, series convergence is different for mathematicians, who use abstract mathematical procedures, and for astronomers, who utilize numerical computations (cf. (Blehman, et al, 1983)).

These problems become principal concern for modern physical theories in which physical systems are described by chaotic processes and modeled by computer simulation. Taking into account the fact that chaotic solutions are obtained by computations, physicists ask (Cartwrite and Piro, 1992; Gontar, 1997) whether chaotic solutions of the differential equations that model physical systems reflect the dynamic laws of nature represented by these equations or whether they are solely the result of an extreme sensitivity of these solutions to numerical procedures and computational errors. For example, the definition of chaos, as Gontar emphasizes, makes it problematic to use the apparatus of the conventional differential calculus to describe chaotic motion mathematically (Gontar, 1997; Gontar and Ilin, 1991).

Moreover, not only contemporary computers and measuring devices are imprecise, but such devices cannot be absolutely precise in principle. For measurement, as we have seen, it follows from the Heisenberg's uncertainty principle. For digital computation, it follows from the fact that the set of all real numbers is uncountable, while the set of all computable real numbers is countable (cf. (Rice, 1951; Freund, 1983)). The situation with real numbers becomes even more aggravated due to the fact that not only algorithms but also linguistic means of number descriptions have their limitations and cannot represent all real numbers as there are inaccessible real numbers in the sense of (Borel, 1952).

In addition, scientists frequently come to a situation when the used continuous model does not work and it is necessary to build a discrete model. The development of discrete models started with the idea that everything (in nature) is built of atoms introduced by outstanding philosophers Democritus and Levkipus from ancient Greece. For a long time this idea was considered false due to the fact that scientists were not able to go sufficiently deep into the matter. Nevertheless, the development of scientific instruments and experimental methods made possible to discover such micro-particles that were and are called atoms although they possess very few of those properties that were ascribed to them by ancient philosophers. Later the discrete corpuscular theory of light competed with the continuous wave theory of light. One of the main proponents of the corpuscular theory of light was Newton.

Discrete structures also emerge in other sciences. For instance, contemporary biology assumes that inheritance and changes have a discrete character as they are controlled by discrete elements, genes.

One more example from physics is the, so-called, Barkhausen effect discovered in 1919 (Vonsovskii, 1974; Burke, 1986). This effect has been thoroughly studied, analyzed, explained, and utilized as a tool for investigation of many properties of ferro- and ferrimagnetic materials. There is a reasonable theoretical model of that effect. Its essence is as follows. If a ferro- or ferrimagnetic sample is being magnetized in an external magnetic field, the magnetization of the sample is increasing, along with the increase of the external magnetizing field. However, even if the magnetizing field is increasing in a continual way, the magnetization of the sample is increasing via thousands of small discontinuities ("Barkhausen jumps"). In other words, an impact of the field the change of which is described by a continuous function produces such changes that may be adequately represented only by a fuzzy continuous function.

Many features of the Barkhausen effect has been studied, including the distributions of Barkhausen discontinuities over their duration, and over their amplitude, and over their shape etc. Much is known about the mechanism of those "jumps" and about their relationship to many other properties of the sample, such as demagnetizing factor, saturation magnetization, remnant magnetization, etc. However, there is no such a continuous function that provides for a sufficiently correct description of the phenomenon. The difficulty is not mathematical but is caused by the physical nature of the process.

There are numerous examples of similar situations. In many cases development of measurement methodology and achieving in such a way higher than it was in the past precision of measurement and better than it was in the past understanding of physical phenomena helped to discover natural discontinuous processes that seemed continuous for a long period of time.

Physicists show that if they try to use an interferometer, or simple time of flight measurements to determine locality, they get the answer that the minimal distance measureable is the Planck length. So, there really is a sense in which distance shorter than the Planck length has no meaning (Calmet, Graesser, and Hsu, 2004).

These problems brought science to Planck scale physics (cf., for example, (Ring, *et al*, 1995; Requardt, 1998; Albrecht and Skordis, 2000; Brandenberger and Martin, 2001) or as some call it, string scale physics. It studies physical phenomena in space-time domains determined by the fundamental length and time intervals. In the domains at this scale, named after Max Planck, one of the founders of quantum theory, many properties of physical space and time are essentially different. As Requardt writes (1998), "starting from the working hypothesis that both physics and the corresponding mathematics have to be described by means of discrete concepts on the Planck scale, one of the many problems one has to face in this enterprise is to find the discrete protoforms."

Now when physics becomes more and more sophisticated, there is an active discussion whether our space-time is continuous or discrete. As Motl writes, "the idea that space-time could be discrete has been a recurring one in the scientific discourse of the twentieth century. A survey of just a few examples reveals that discrete space-time can actually mean many things and is motivated by a variety of philosophical or theoretical influences." As a result, a popular hypothesis in contemporary physics claims that there is a minimal length interval and minimal time interval connected to the Planck constant so that space, time, and matter all have to be discrete ('t Hooft, 1996; 1999).

Discrete spaces are also basic structures for pattern recognition, image analysis, and related areas. The mathematical challenge lies in the fact that a digital image is built in a lattice-point approximation of an Euclidean space $\boldsymbol{E}^2$. The image on the screen consists of a finite number of separate pixels. The deficiencies of the Euclidean space $\boldsymbol{E}^n$ as a framework for describing and reasoning about topological and geometric computation have, particularly in the past ten years, caused attention of many researchers. A major and fundamental source of problems is the mismatch between the discrete nature of computational objects and their theoretical description in terms of the continuous space $\boldsymbol{E}^n$ with the natural topology in it. Divergence between computational behavior and theoretical description shows up in rounding errors in floating point computation, which propagate into any area that uses geometric algorithms. The infinite precision afforded by $\boldsymbol{E}^n$ as the mathematical model is very often not appropriate for describing computation over data obtained using finite precision real-world measurement. Another fundamental issue is that $\boldsymbol{E}^n$ very often does not accord with the ways in which humans normally perceive, describe and reason about space. For instance, people do not perceive points in mathematical sense, nor in everyday reasoning do they think of spatial objects as sets of such mathematical points.

This results in blurring the border between discrete and continuous in people's perception. When a movie is demonstrated on the screen, the changes are not continuous – the film is simply a series of still photographs (frames or shots), each eventually slightly different than the next one. The frames are projected on the screen in rapid sequence, with dark intervals in between. When the rate is 24 frames per second, as it is today in movies, we see continuous movement. Moreover, to achieve perception, each frame is flashed on and off three times (cf., for example, (Atkinson, et al, 1990)). Thus, psychological continuity is, in effect, discrete in a finer scale and essentially different from mathematical continuity.

As a result, some types of human-computer interaction are considered continuous (cf., for example, (Sutcliffe, 2003; Connolly, et al, 2006)) although computers work in discrete time – whatever high frequency of processors is achieved and whatever primitive operations are performed, the whole process consists of discrete units.

There are related problems of continuity in software engineering, particularly in testing and certifying programs. As Hamlet writes (2002), most engineering artifacts behave in a continuous fashion, and this property is generally believed to underlie their dependability. In contrast, software systems do not have continuous behavior, which is taken to be an underlying cause of absence of dependability. The theory of software reliability has been questioned because technically the sampling on which it is based applies only to continuous functions. Consequently, it is necessary to have a technique that combines properties of continuous and discontinuous functions.

There are other areas where neither discrete models nor classical continuous models with absolute precision properly work. For instance, Steimann (2001) demonstrated that medicine needs new structures and theories to bridge the gap between discrete world of reasoning and continuity of reality.

Thus, exactness of mathematical objects projected onto real systems and processes becomes an illusion, and we begin to understand that constructions and methods developed in the classical analysis are only approximations to what exists in reality. In many situations, such approximations have been giving a sufficiently adequate representation of studied phenomena. However, scientists and, especially, engineers have discovered a lot of cases in which such methods did not work because classical approach is too rough (Zimmermann, 1991; Klir and Bo Yuan, 1995).

For example, as Petzold (2001) writes, there are three common problems in modeling: (1) identifying the "correct" model, (2) estimating parameters from experimental data, and (3) determining which of many parameters contribute most to the uncertainties in the prediction of the model. The issue is not if there are uncertainties, they are always present, the real challenge is to identify those components of the model that have the most influence on the prediction.

When scientists develop models of different phenomena, they encounter these problems. One of the most popular models in science and engineering is a system of differential equations. Differential equations are also used in economics and sociology. These and many other mathematical models utilize limit processes. For example, derivatives in differential equations are constructed as special limits of functions or points. Continuous functions and the calculus, differential equations and topology, all are based on limits and continuity. However, when we perform computations and measurements, we can do only finite number of operations and consequently, achieve only approximate results. At the same time, mathematical technique, e.g., calculus and optimization theory, are based on operation of differentiation.

All made considerations bring us to an important conclusion.

Classical models work well only up to some precision. Then they fail. In spite of this, they create an illusion of infinite precision, which does not exist in reality. There is a gap between rigorous and infinitely precise mathematics and imprecise, vague, and uncertain real world. As a result mathematicians encountered a necessity to represent this imprecision and vagueness by precise mathematical means to work in fuzzy environment of the real world. New methods and constructions are necessary to take into account such more sophisticated effects in different systems. It becomes clear that natural science, as well as computer science, need new methods to deal with existing imprecision, vagueness, ambiguity, and uncertainty. To achieve this goal, new mathematical theories have been introduced and studied.

1.3. Filling the Gap

Curiosity begins as an act of tearing to pieces or analysis.
Samuel Alexander (1859-1938)

To deal with uncertainty, mathematicians started to develop probability theory in the 16th century and later mathematical statistics. Now both theories are flourishing, growing, and considered as principal mathematical fields with a big variety of applications. However, tools of these theories treat only uncertainty that emerges from multiple events and expresses itself in the behavior of a big multiplicity of objects. In addition, correct application of probabilistic and statistical methods is subjected, as a rule, to rather restricted conditions.

Another trend of research investigated approximations and developed a huge variety of techniques for building them in different situations. For instance, for many differential equations, it is rather difficult or even impossible to find exact solutions of Cauchy problems. As a result, more often than not approximate methods of solution are employed. Construction, analysis and implementation of approximate solutions for differential and functional equations have been the major focus of an enormous amount of activity since the very beginning of the calculus. However, all this development has been going in the setting of conventional numbers and functions, while the essence of numerical approximations demanded new structures.

That is why new areas of mathematics, such as the most developed fuzzy set theory, interval analysis, rough set theory, neoclassical analysis and some others, were created in the second half of the 20th century.

1.3.1. Fuzzy Set Theory

There is nothing worse than a sharp image of a fuzzy concept.
Ansel Adams (1902-1984)

One of the most popular mathematical approaches to problems of uncertainty and imprecision is fuzzy set theory. Fuzzy sets were introduced by Lofti Asker Zadeh in 1965. The aim was to get better mathematical models for real-life systems and processes, as well as better techniques for human reasoning and decision-making, than the conventional set theory allowed by constructing a more realistic set theory. To achieve this goal, Zadeh considered generalizations of sets that allow graded membership of their elements. Namely, he assumes that elements can have different grades of membership in a set. His main argument was that "classes of objects encountered in real physical world do not have precisely defined criteria of membership" (Zadeh, 1965). This approach also reflects situations in which our knowledge about membership is incomplete.

Definition 1.3.1 (Zadeh, 1965). A *fuzzy set A in a set U* is defined by its a membership function μ_A: $U \rightarrow [0,1]$ of A where $\mu_A(x)$ is interpreted as the degree of membership in A of an element $x \in U$. Thus, a fuzzy set A is the triad $(U, \mu_A, [0,1])$.

Traditionally such a fuzzy set A is defined by the pair (A, μ_A) because it is assumed that U is always a universal set, that is, all theory is developed and all fuzzy sets are built for one chosen set U. However, as we know from set theory (cf., for example, (Fraenkel, and Bar-Hillel, 1958)), there is no one set that contains all other sets. That is why it is more rigorous to include the set U in definitions of a fuzzy set and related constructions. Nevertheless, for applications, it is usually possible to have a universal set U in which all fuzzy sets are defined.

Let us consider some basic constructions from fuzzy set theory. Similar to the conventional set theory, various operations are introduced for fuzzy sets (compare with Appendix A). Three basic operations with fuzzy sets, which transform given fuzzy sets into a new fuzzy set, are defined by constructing the membership function for the new fuzzy set, which is the result of operation.

Definition 1.3.2. The membership function of the *union* of two fuzzy sets (A, μ_A) and (B, μ_B) is defined as the maximum of the two individual membership functions. Namely, we have

$$(A, \mu_A) \cup (B, \mu_B) = (C, \mu_C) \text{ where } \mu_C(x) = \max \{\mu_B(x), \mu_A(x)\}$$

Definition 1.3.3. The membership function of the *intersection* of two fuzzy sets (A, μ_A) and (B, μ_B) is defined as the minimum of the two individual membership functions. Namely, we have

$$(A, \mu_A) \cap (B, \mu_B) = (D, \mu_D) \text{ where } \mu_D(x) = \min \{\mu_B(x), \mu_A(x)\}$$

Definition 1.3.4. The membership function of the *complement* of a fuzzy set (A, μ_A) is defined as the negation of the membership function of A. Namely, we have

$$(A, \mu_A) = (F, \mu_F) \text{ where } \mu_F(x) = 1 - \mu_A(x)$$

Fuzzy sets are characterized by their level sets. Let $\alpha \in [0, 1]$.

Definition 1.3.5 (Zimmermann, 2001). The set $A = \{ x \in U ; \mu_A(x) \geq \alpha \}$ is called the α–*level set* of the fuzzy set $A = (U, \mu_A, [0,1])$.

An important concept, especially, for the fuzzy set calculus, is the concept of a fuzzy number.

Definition 1.3.6 (Averkin *et al.*, 1986). A fuzzy set $A = (\boldsymbol{R}, \mu_A, [0,1])$ is called a *fuzzy number*.

However, there are different definitions of a fuzzy number. Some authors give more restricted definitions, adding such features as normality and convexity to Definition 1.3.6.

Definition 1.3.7 (Zimmermann, 2001). A fuzzy set $A = (U, \mu_A, [0,1])$ is called *normal* if $\sup\{ \mu_A(x); x \in U \} = 1$.

Definition 1.3.8 (Furukawa, 1996) A fuzzy set $A = (\boldsymbol{R}, \mu_A, [0,1])$ is called a *fuzzy number* if it is normal, there is exactly one $a \in \boldsymbol{R}$ for which $\mu_A(a) = 1$, and the membership function $\mu_A(x)$ is nondecreasing on $(-\infty, a]$ and nonincreasing on $[a, +\infty)$.

Definition 1.3.9 (Zimmermann, 2001). A fuzzy set $A = (U, \mu_A, [0,1])$ is called *convex* if its membership function $\mu_A(x)$ satisfies the following condition: $\mu_A(ax + (1 - a)y) \geq \min \{\mu_A(x) , \mu_A(y) \}$ for any number $a \in [0,1]$ and all $x, y \in U$.

Proposition 1.3.1 (Zimmermann, 2001). A fuzzy set $A = (U, \mu_A, [0,1])$ is convex if and only if all its α–level sets are convex.

Definition 1.3.10 (Zimmermann, 2001). A fuzzy set $A = (\boldsymbol{R}, \mu_A, [0,1])$ is called a *fuzzy number* if it is convex, normal, there is exactly one $a \in \boldsymbol{R}$ for which $\mu_A(a) = 1$, and the membership function $\mu_A(x)$ is piecewise continuous.

These constructions show essential differences between sets and fuzzy sets. First, while sets have only one property – their cardinality (for finite sets, cardinality is the number of elements in a set), fuzzy sets have many other properties. Here we see such properties as convexity and normality. In addition, fuzzy sets have cardinality, relative cardinality, and different indices of fuzziness, such as entropy (Zimmermann, 2001). Second, taking one crisp structure, it is possible to build several fuzzy generalizations. Third, while set theory is built as a collection of axiomatic theories and deals only with sets and their properties, fuzzy set theory is mostly developed in a constructive way and contains fuzzified forms of almost all major mathematical fields: algebra and topology, logic and the theory of algorithms, calculus and differential equations and so on.

Taking the concept of a fuzzy set, it is possible to develop a new kind of mathematics based on fuzzy sets in the same way as traditional mathematics is based on conventional sets. Some call this field by the name "fuzzy mathematics," while others (trying to avoid ambiguity in the meaning of the term "fuzzy mathematics") simply include everything that is based on fuzzy sets into fuzzy set theory. As the second approach is more popular, here we call this area of mathematics by the name fuzzy set theory. In fuzzy set theory, traditional objects of mathematics called crisp (such as sets, functions, numbers, groups, and relations) are changed to their fuzzy versions. For instance, conventional or crisp, sets are changed to fuzzy sets, while crisp numbers are changed to fuzzy numbers. This has divided mathematics into two parts: in one of them, which is much larger than the other, traditional (or crisp, in the terminology of the fuzzy set community) structures are studied, while the second part encompasses fuzzification of these structures.

Traditionally, fuzzy set theory has been developing in three directions (Kosko, 1993; Zimmermann, 2001; Klir and Bo Yuan, 1995). The first direction has been aimed at fuzzification of different classical mathematical structures and studying properties of these fuzzy objects. This development in many aspects has been parallel to the classical mathematics and fuzzy set theory expanded in almost all areas of mathematics. In such a way, fuzzy sets, fuzzy logic, fuzzy numbers, fuzzy topologies, fuzzy groups, fuzzy integrals, fuzzy relations, fuzzy measures, fuzzy languages and grammars, fuzzy algorithms and so on were introduced and studied. The main calculus operations, differentiation and integration, were also extended to fuzzy structures. Usually, one conventional, crisp structure has several fuzzy counterparts. For instance, one type of classical integration can be converted into integration of a fuzzy function over a crisp interval of conventional numbers, integration of a crisp function over a fuzzy interval of conventional numbers, integration of a fuzzy function over a fuzzy interval of conventional numbers, and integration of a crisp function over an interval of fuzzy numbers.

The second direction takes or elaborates fuzzy structures and applies them to different practical problems. In such a way, the concept of a fuzzy set was introduced and step by step fuzzy set theory has found many useful applications to problems of artificial intelligence (AI), pattern recognition, decision-making, operation research and many others. For instance, fuzzy sets became very popular in medical applications (cf., for example, (Adlassnig, 1986; Teodorescu, et al, 2001; Axer, et al, 2001; Mahfouf, et al, 2001; Steimann, 2001)). Fuzzy logic became a powerful tool in solving many practical problems.

When fuzzy set theory matured, the third direction appeared aimed at inner problems of this theory. An example of such a problem is the problem of axiomatization of fuzzy set theory.

Nevertheless, in spite of numerous applications and very fast growth of research and researchers, fuzzy set theory has aroused various objections in the professional community. While there have been generic complaints about the "fuzziness" of the process of assigning values to the membership function in fuzzy sets and multiplicity of introduced fuzzy structures, claims that fuzzy set theory is reducible to probability theory, perhaps the most cogent criticisms was expressed by Haack (1979). She argues that there is no need in fuzzy logic and consequently, in fuzzy set theory. However, all these objections have not hindered the development of fuzzy set theory.

Fuzzy set theory and, especially, fuzzy logic are rooted in multi-valued logics. In multi-valued logics (also called, many-valued or multiple-valued logics), there are more than two truth values (Rescher, 1969; Epstein, 1993; Malinowski, 1993; Gottwald, 2001).

Our conventional logic, which was originated by Aristotle, has two truth values: True and False. At the same time, Aristotle was the first known logician who did not fully accept the law of the excluded middle, suggesting that this law did not all apply to future events. This idea leads to multi-valued logics, but Aristotle never developed them. The opinion that logic ought to use more than two truth-values naturally emerged in medieval discussions of determinism and was re-examined by Charles Santiago Sanders Peirce (1839-1914), Hugh MacColl (1837-1909), and Nikolai Alexandrovich Vasiliev (1880–1940) in the first decade of the 20th century.

The development of the formalized multi-valued logic is usually related to 1920 when the Polish logician and philosopher Lukasiewicz created a ternary logic, using a third value "possible" to deal with Aristotle's paradox of the sea battle. Almost at the same time (in 1921), the American mathematician Post also introduced the formulation of additional truth degrees. In 1922, Lukasiewicz extended his ternary logic to a logic with $n + 1$ truth values. In 1930, Lukasiewicz and Tarski built a logic with infinitely many truth values. In 1932, Gödel showed that intuitionistic logic is not a finitely valued logic. In 1960's, logicians already studied logics in which truth values were taken from a compact topological space (cf. (Chen and Keisler, 1966)). One of such logics is the fuzzy logic introduced by Zadeh (1975c). There are also other fuzzy logics.

Today a big quantity of mathematicians is working in fuzzy set theory. Many books and thousands of papers have been published since introduction of fuzzy sets. Researchers that work in this area have organized many conferences and congresses.

It is interesting that before Zadeh determined fuzzy sets, Kubinski introduced and studied the notion of a vague term and the notion of an unsharp set (Kubinski, 1958; 1960). He applied the notion of a vague term to sets in which transition from full membership to non-membership was gradual. Such sets have no sharp boundaries.

Besides, at the same time as Zadeh introduced his fuzzy sets, several similar constructions were suggested by different mathematicians. In 1965, Salii (1965) defined a more general kind of structures called L-relations in a general context of abstract algebra. At the same time, Klaua (1965; 1966; 1967; 1967a) built two versions of a universe of many-valued sets. In both cases, the membership degrees were restricted to a finite set of equidistant points of the unit interval [0, 1]. It is a more realistic model of fuzzy sets because, as we have seen in Section 1.2, computation and measurement do not allow one to work with arbitrary real numbers. Then Jahn (1969) extended Klaua's model allowing all possible values from the interval [0, 1]. In a series of papers, Klaua and other authors developed further the model of many-valued sets (Klaua, 1970; 1970a; 1972; 1973; Schwartz, 1972; Maydole, 1975).

However, fuzzy sets in the sense of Zadeh have prevailed over other models because what Zadeh proposed is very much a paradigm shift. There was an essential opposition to this shift, but in spite of this, this approach at first gained acceptance in the Far East and its successful application has ensured its adoption around the world.

Some authors defined fuzzy sets in more general way than Zadeh did. Here we give some examples of such extensions.

Definition 1.3.11 (McVicar-Whelan, 1977). A *fuzzy set A in a set U* is the triad $(U, \mu_A, [-1,1])$, where $\mu_A: U \to [-1,1]$ is a membership function of A and $\mu_A(x)$ *is the degree of membership in A* of $x \in U$.

Definition 1.3.12 (Cai Wen, 1984). *A fuzzy set A in a set U* is the triad $(U, \mu_A, (-\infty, +\infty))$, where $\mu_A: U \to (-\infty, +\infty)$ is a membership function of A and $\mu_A(x)$ is the degree of membership in A of $x \in U$.

Ten years later after he introduced fuzzy sets, Zadeh came to the conclusion that fuzzy sets, as they were originally defined, do not allow a researcher to fully model imprecision, inexactness, and uncertainty. To improve this situation, Zadeh introduced type-2 fuzzy sets and higher order (type-n) fuzzy sets (1975).

Let us denote by FS the set of all fuzzy sets in a universal set U.

Definition 1.3.13. A *type-2 fuzzy set A in a set U* is a triad (U, μ_A, FS), where FS is the membership space, $\mu_A: U \to FS$ is a membership function of A and $\mu_A(x)$ is the degree of membership in A of $x \in U$.

Assuming that ordinary fuzzy sets have type 1, we have the following general definition for $n > 1$.

Definition 1.3.14. A *type-n fuzzy set A in a set U* is a triad $(U, \mu_A, FS(n-1))$, where the set $FS(n-1)$ of all type-$(n-1)$ fuzzy sets is the membership space, $\mu_A: U \to FS(n-1)$ is a membership function of A and $\mu_A(x)$ is the degree of membership in A of $x \in U$.

The most general of all fuzzy set definitions is given by Hans Jürgen Zimmermann (2001). It includes many generalizations of fuzzy sets, such as type-n fuzzy sets, L-fuzzy sets, vector-valued fuzzy sets and several others.

Definition 1.3.15. *A fuzzy set A in a set U* is a triad (U, μ_A, M), where M is the membership space, μ_A: $U \to M$ is a membership function of A and $\mu_A(x)$ is the degree of membership in A of $x \in U$.

Remark 1.3.1. The definition 1.3.15, in which M may be an arbitrary set, does not preserve the main idea of fuzzy set theory that the values of the membership function $\mu_A(x)$ reflect to what extent elements from the universal set U, which is the domain of $\mu_A(x)$, belong to the fuzzy set A. To reflect the extent of membership, the codomain I of $\mu_A(x)$ has to be, at least, partially ordered. At the same time, structures that have the form A = (U, μ_A, I), and in which I is an arbitrary set are considered in the theory of named sets and are called set-theoretical named sets (Burgin, 2004).

There are many mathematical constructions similar to or more general than fuzzy sets. Since Zadeh introduced fuzzy sets in 1965, a lot of new theories treating imprecision, inexactness, ambiguity, and uncertainty have been introduced. Some of these theories are extensions of fuzzy set theory, while others try to mathematically model imprecision and uncertainty in a different way (cf., for example, (Burgin and Chunihin, 1997; Kerre, 2001)). The variety of these constructions includes rough sets (Pawlak, 1982), vague sets (Wen-Lung Gau and Buehrer, 1993), flou sets (Gentilhomme), L-fuzzy sets (Goguen, 1967), intuitionistic fuzzy sets (Atanassov, 1983), underdetermined sets (Narinyani, 1980), real-valued fuzzy sets (Blizard, 1989), interval-valued fuzzy sets (Sambuc, 1975), interval-valued intuitionistic fuzzy sets (Atanassov, 1994; 1999), valued sets, blurry sets (Smith, 2004), and neutrosophic sets (Smarandache, 1999).

To achieve better understanding of mathematical tools for dealing with imprecision, inexactness, ambiguity, and uncertainty, here we give a general picture and explain some of these generalizations.

Let L be a complete lattice, i.e., a partially ordered set with operations sup and inf.

Definition 1.3.16. (Gogen, 1967). An *L-fuzzy set A in a set U is a triad* (U, μ_A, L), where μ_A: $U \to L$ is a membership function of A and $\mu_A(x)$ is the degree of membership in A of $x \in U$.

When $U \subseteq X \times Y$, we have L-fuzzy relation in the sense of Salii (1965).

Let B be a Boolean lattice.

Definition 1.3.17. (Brown, 1971). *A B-fuzzy set A in a set U* is a triad (U, μ_A, B), where μ_A: $U \to B$ is a membership function of A and $\mu_A(x)$ is the degree of membership in A of $x \in U$.

Intuitionistic fuzzy sets, which were introduced by Atanasov (1986; 1999) are represented either by two named sets or by a named set in which the naming relation is not a function.

Definition 1.3.18. (Atanasov, 1986; 1999). An *intuitionistic fuzzy set A in a set U* is a tetrad (U, μ_A , ν_A , [0,1]), where μ_A: $U \to$ [0,1] is a membership function of A , ν_A: $U \to$ [0,1] is a non-membership function of A, $\mu_A(x)$ is the degree of membership of $x \in U$, $\nu_A(x)$ is the degree of non-membership of $x \in U$, and $0 \le \mu_A(x) + \nu_A(x) \le 1$ for any $x \in U$.

Remark 1.3.2. An intuitionistic fuzzy set A may be represented as the pair (A_m , A_n) of fuzzy sets: the membership fuzzy set $A_m = (U, \mu_A, [0,1])$ and the non-membership fuzzy set $A_n = (U, \nu_A, [0,1])$, for which $A_m + A_n = (U, \mu_A + \nu_A, [0,1])$ is also a fuzzy set.

Valued sets are also an extension of the traditional set concept. In addition to the membership relation in a set, a member of a valued set also has a distinct value.

Essentially different approach, which treats imprecision, vagueness, ambiguity, and uncertainty by exact mathematical tools and is related to fuzzy set theory, is the theory of rough sets introduced by Pawlak (1982; 1991). Just like fuzzy set theory, rough set theory addresses the topic of dealing with imperfect knowledge. The introduction of rough sets was motivated by the practical needs in classification and concept formation with incomplete information. It is different from and, in some sense, complementary to fuzzy sets and their direct generalizations. Recently different authors made an attempt to combine both theories into a more flexible and expressive framework for modeling and processing incomplete information (Radzikowska and Kerre, 2002; Cornelis, et al, 2003).

An interesting attempt to build a general setting for different approaches to fuzziness is done by Lin (1995; 1996). As he writes (Lin, 1996), various theories have been proposed to mathematically represent idea of fuzziness and each theory is driven or motivated by specific applications and thus, is adjusted to a specific context. To overcome these limitations and "to capture and defuse the conflicts among existing fuzzy theories", Lin introduced the concept of a soft set, which allowed him to represent different kinds of fuzzy sets in a unified system.

A soft set is defined, similar to in the classical case of fuzzy sets, by its membership function FX: $U \to M$. The difference is that a specific structure is introduced into the total union of membership functions on the universe U. Utilizing different structures, Lin gets many kinds of soft sets, or as he continues to call them, fuzzy sets. The main classes are: W-softsets, F-softsets, P-softsets, R-softsets, FP-softsets, and FF-softsets.

There are other approaches in mathematics related to fuzzy set theory. One of them is the theory of Boolean valued models (Bell, 2005). Another is multiset theory, which models such properties of real objects as indistinguishability (Aigner, 1979; Knuth, 1997; 1998).

All these structures, including valued sets, rough sets, soft sets, and intuitionistic fuzzy sets, have the form A = (U, f_A, I), which is called a named set or fundamental triad in the theory of named sets (cf. Appendix A). The main idea formalized in the concept of a named set is that elements of a set (actually, any objects with which people are dealing) have names. Thus, in the first approximation, a named set is a set with elements that have names. Consequently, all fuzzy sets and their generalizations are special cases of named sets in which names are taken from some relevant mathematical structures (Burgin, 1990; 2004c).

For instance, L-fuzzy sets in the sense of Gogen are named sets in which the set of names is a complete lattice L. Fuzzy sets in the sense of McVicar-Whelan are named sets in which the set of names I is the interval [-1,1]. Fuzzy sets in the sense of Cai Wen are named sets in which the set of names is the interval $(-\infty, +\infty)$ and all relations on it. Intuitionistic fuzzy sets are named sets in which the naming relation is not a function but a relation that corresponds two numbers to each element from the support.

Even more (cf., for example, (Burgin and Chunihin, 1997; Chunihin, 1997)), named sets provide more extended and flexible means than fuzzy sets for representing and investigating uncertainty.

1.3.2. Interval Analysis

The interval between a cold expectation and a warm desire
may be filled by expectations of varying degrees of warmth
or by desires of varying degrees of coldness.
Samuel Alexander (1859-1938)

The main reason for the development of interval analysis and its part, *interval arithmetic*, is that numerical mathematics does not always give us the solution to the problem of accuracy estimation in computation (cf., for example, (Moore, 1966; Alefeld, and Herberger, 1983; Kearfott and Kreinovich, 1996; Neumaier, 2001)). Error estimation methods of numerical mathematics are often very complicated, and require difficult mathematical techniques. Because of that, algorithms for which such methods are known form a small subset in the set of all algorithms that are used for data processing. New data processing problems appear every day, and new data processing algorithms are being designed to solve these problems. For a reasonable period of time, no error estimation method is known for many data processing algorithms. Interval analysis is aimed at taking care of these problems with numerical methods.

Thus, the main area of the interval analysis applications is numerical computations and the main idea is to work with intervals instead of numbers. Since it is suggested to perform computations with intervals, the entire area is called *interval computations*. Interval computations are a useful tool for *validated computations*, i.e. computations with guaranteed accuracy taking into account various sources of error, from imprecise data to rounding errors during computer calculations.

As Hayes (2003) writes, interval arithmetic is not a new idea. Even before interval arithmetic came to computation, intervals instead of numbers were used in the theory of measurement. As it is explained in Section 1.2, measurement always has a finite precision. So, intervals better, in some sense, represent results of measurements than separate numbers. The first who described intervals as a result of measurement was Norbert Wiener: in 1914, he applied intervals to measuring distances, and in 1921, to measuring time.

Ten years later, Rosalind Cicely Young (1931) developed an algebra of many-valued quantities where operations with intervals and other sets of real numbers were defined. Twenty years later, Paul Dwyer (1951) considered similar operations with intervals called "range numbers" in his textbook on linear algebra.

A few years later, interval arithmetic was set forth as a tool for numerical computations by Mieczyslaw Warmus (1956) from Poland, Teruo Sunaga (1958) from Japan and Ramon E. Moore (1959) from the United States. The most developed form of interval computations appeared in the Ph.D. dissertation of R. E. Moore that was defended at Stanford in 1962. Later he wrote the first monograph in this area (Moore, 1966).

After this different researchers contributed to the interval analysis. In 1990s, it was an outburst of activity related to interval computations. A new international journal *Interval Computation* has been launched in 1991. Later its title was changed to *Reliable Computing*. Several international conferences on interval computations were organized.

It is necessary to remark that Moore's version of the interval analysis has been the most popular, in part because he emphasized solutions to problems of numerical computation, but also because he has continued for more than four decades to develop this area and to promote interval methods.

Now implementations of interval arithmetic are available both as specialized programming languages and as libraries that can be linked to a program written in a standard language. There are even interval spreadsheet programs and interval calculators.

However, one thing the interval community has been ardently seeking - support for interval algorithms in standard computer hardware – was not achieved. Most modern processor chips come equipped with circuitry for floating-point arithmetic. Analogous built-in facilities for interval computations are technologically feasible, but manufacturers have not chosen to provide them because people are acustomed to doing computations with numbers and not with intervals.

The main idea of interval computation is to use intervals instead of numbers in numerical computation and arithmetic calculations. Intervals used in such computations are called interval numbers.

Definition 1.3.19. An *interval number* is a real, closed interval of the form $[x, z]$ or $[a, b]$ where x, z , a, and b are real numbers and the following inequalities $x \leq z$ and $a \leq b$ are true.

There are following operations with interval numbers:

Addition: $[a, b] + [c, d] = [a + c, b + d]$.
Subtraction: $[a, b] - [c, d] = [a - d, b - c]$.

Multiplication is defined in a more sophisticated way:

$$[a, b] \times [c, d] = [\min \{ac, bc , ad, bd\}, \max \{ac, bc , ad, bd\}].$$

Examples of addition:

$[1, 2] + [5, 7] = [6, 9]$
$[-2, 2] + [3, 5] = [1, 7]$

Examples of multiplication:

$[1, 2] \cdot [5, 7] = [5, 14]$
$[-2, 2] \cdot [3, 5] = [-10, 10]$
$[-1, 2] \cdot [-3, -2] = [-6, 3]$

Division of interval numbers is even more difficult to rigorously determine than multiplication. This is a consequence of the possibility of a zero divisor. With conventional numbers, if a machine tries to perform an operation such as $7 \div 0$, the error is obvious, and the system software will prevent the user from doing this. An interval division in such cases when the second interval contains zero, e.g., $[3, 7] \div [-5, 5]$, has the same problem, but in disguise.

Some properties of interval numbers are the same as properties of real numbers. For instance, addition of interval numbers is commutative and associative. At the same time, many properties of operations with numbers do not hold for the interval arithmetic. For instance, for numbers, we have $a - a = 0$. For intervals, this is not always true. For instance, $[1, 2] - [1, 2] = [-1, 1]$. In general, an interval has no additive inverse. The distributive law $a(b + c) = ab + ac$ also fails for intervals and so on.

Definition 1.3.20. An *interval function* is an interval-valued function of one or more interval arguments.

In other words, if $\boldsymbol{I}$ is the set of all intervals from $\boldsymbol{R}$, then an interval-valued function of one interval argument is a mapping $F: \boldsymbol{I} \to \boldsymbol{I}$. An interval-valued function of n interval arguments is a mapping $F: \boldsymbol{I}^n \to \boldsymbol{I}$ where $\boldsymbol{I}^n$ is the n-th direct power of $\boldsymbol{I}$.

Let $f: \boldsymbol{R}^n \to \boldsymbol{R}$ and $F: \boldsymbol{I}^n \to \boldsymbol{I}$.

Definition 1.3.21. The function F is called an *interval extension* of f if for any a from Dom f, we have

$$F([a, a]) = [f(a), f(a)]$$

Definition 1.3.22. An interval function f is called *inclusion monotonic* if for any $I_1, I_2, \dots, I_n, J_1, J_2, \dots, J_n \in \boldsymbol{I}$, the inclusions $I_i \subseteq J_i$ that are true for all $i = 1, 2, 3, \dots n$ imply

$$f(I_1, I_2, \dots, I_n) \subseteq f(J_1, J_2, \dots, J_n),$$

or in other words, a function F is *inclusion monotonic* if for any A, B from $\boldsymbol{I}^n$, we have

$$A \subseteq B \text{ implies } F(A) \subseteq F(B)$$

Theorem 1.3.1. (Moore, 1966). If a function F is an inclusion monotonic interval extension of a function f, then for any A from $\boldsymbol{I}^n$ and any a from Dom f, we have

$$a \in A \to f(a) \in F(A)$$

An important class of interval functions consists of inclusion monotonic functions.

Definitions of interval operations imply the following results (Moore, 1966).

Lemma 1.3.2. Addition of intervals is inclusion monotonic.

Lemma 1.3.2. Subtraction of intervals is inclusion monotonic.

Lemma 1.3.2. Multiplication of intervals is inclusion monotonic.

Applying mathematical induction, we have the following result.

Theorem 1.3.2. Any interval polynomial determines an inclusion monotonic function.

In interval analysis and arithmetic, intervals are represented by their names as formal expressions $[a, b]$ and rules for operating with such expressions are given. Thus, as it is often done in mathematics, it is possible to give different interpretations for these expressions. In the context of dealing with errors and uncertainties of computation, the most natural thing is to look at $[a, b]$ as standing for some definite but unknown value x such that $a \leq x \leq b$. Another possibility is to interpret $[a, b]$ as the set of all real numbers between a and b, namely, as a closed interval (segment) of the real line. Another possibility is to treat the expression $[a, b]$ as denoting a new kind of number, in much the same way that two real numbers a and b combine to specify the complex number $a + ib$ (where i represents the square root of -1). The latter interpretation is actually used in interval computations, introducing just another data type, compatible with other kinds of data.

It is useful to remark that all three interpretations are formed by specific named sets. In the first case, $[a, b]$ is a name for an unknown real number x. In the second case, $[a, b]$ is a name for a closed interval of real numbers. In the third case, $[a, b]$ is a name for a number of a new kind.

It is necessary to mention that not all expressions that are used in computations can be evaluated based on interval arithmetics. Here are some constraints that prohibit the interval evaluation: denominator of a fraction should not include 0, arguments of the square root or logarithm should not include negative values, arguments of the tangent should not include $\pi/2$, or arguments of the functions arcsin x or arccos x should be inside $[-1,1]$. If one of these restrictions is violated, a fatal error will be generated in numerical computations with intervals. However, estimation of data accuracy naturally produces intervals that violate these restrictions in a hidden form. For instance, if it necessary to compute x/y where $x = 5$, $y = 0.1$, and accuracy is 0.2. This gives us the interval $[4.8, 5.2]$ for x and the interval $[-0.1, 0.3]$ for y. The second interval includes 0 and division by it is forbidden. These restrictions do not occur in neoclassical analysis as it deals with numbers and not intervals.

It is necessary to mention that interval analysis is closely related to *set-valued analysis* where sets are used as values of functions instead of numbers (cf. (Aubin and Frankowska, 1990)). The most developed part of set-valued analysis studies *differential inclusions*, which have various applications, for example, in control problems (cf. (Aubin and Cellina, 1984)).

To conclude, it is possible to state that invented and reinvented several times, interval analysis has never quite made it into the mainstream of numerical computing, and yet it has never been abandoned or forgotten either (Hayes, 2003).

1.3.3. Neoclassical Analysis

When a truth is necessary, the reason for it can be found by analysis,
that is, by resolving it into simpler ideas and truths
until the primary ones are reached.
Gottfried Leibniz (1646-1716)

Neoclassical analysis, being similar to the classical calculus and analysis, studies properties of functions, curves, surfaces, bodies, and akin objects. The main difference is in those concepts and constructions that are used for these studies. In neoclassical analysis, ordinary structures of analysis, such as numbers, functions, curves, and operators, are studied by means of fuzzy, or approximate, concepts, such as fuzzy limits, fuzzy continuity, fuzzy convergence, fuzzy supremum, fuzzy infimum, fuzzy maximum and minimum, and fuzzy operations, such as fuzzy differentiation and fuzzy integration.

In general, when we have some concept C, it is usually defined by a system *cond*(C) of some conditions. For instance, a function is binary relation between two sets that satisfies the condition that any element from the first set is corresponded to at most one element from the second set (cf. Appendix A).

If we allow approximate satisfaction of conditions from the set *cond*(C), we obtain a fuzzy counterpart of the concept C. In such a way, fuzzy limits are built from limits or fuzzy derivatives from derivatives as basic structures of neoclassical analysis.

The goal is to reflect, measure, model, and work with imprecision, vagueness, uncertainty, ambiguity and incompleteness that emerge in constructions (such as limit or continuity) and operations (such as differentiation and integration) of the classical calculus.

To understand what is a fuzzy, or approximate, concept, we consider the following example.

One of the basic concepts in the classical calculus and analysis is the concept of a limit (cf. Sections 1.1 and 2.1). It is used as the base for grounding operations of differentiation and integration.

Let $r \in \boldsymbol{R}^+$ and $l = \{a_i ; i = 1, 2, 3, \dots\}$ be a sequence of real numbers.

Definition 1.3.23. A number a is called a *limit* of a sequence l (it is usually denoted by $a = \lim_{i\to\infty} a_i$) if for any $\varepsilon \in \boldsymbol{R}^{++}$ the inequality $| a - a_i | < \varepsilon$ is valid for almost all a_i , i.e., there is such n that for any $i > n$, we have $| a - a_i | < \varepsilon$. A sequence that has a limit converges.

This definition includes the following condition:

for any $\varepsilon \in \boldsymbol{R}^{++}$, there is such n that for any $i > n$, we have $| a - a_i | < \varepsilon$ (1.1)

Often conditions in mathematical constructions and theories are represented by axioms written in some specific logical language. For instance, it is possible to write condition (1.1) as a formal expression:

$$\forall \varepsilon \in \boldsymbol{R}^{++} \exists n \in \boldsymbol{N} \forall i > n (| a - a_i | < \varepsilon) \quad (1.2)$$

Informally, a is a limit of a sequence l if for any arbitrarily small ε, the distance between a and all but a finite number of elements from l is smaller than ε.

To obtain a fuzzy concept of convergence and limit, we change arbitrarily small ε to a finitely small value $r + \varepsilon$. Such a change gives us concepts of a fuzzy limit and fuzzy convergence.

Definition 1.3.24. A number a is called an *r-limit of a sequence* l (it is denoted by $a = r\text{-}\lim_{i\to\infty} a_i$ or $a = r\text{-}\lim l$) if for any $\varepsilon \in \boldsymbol{R}^{++}$ the inequality $| a - a_i | < r + \varepsilon$ is valid for almost all a_i , i.e., there is such n that for any $i > n$, we have $| a - a_i | < r + \varepsilon$. A sequence that has an r-limit r-converges.

Here is a formal representation of the condition that defines an r–limit a:

$$\forall \varepsilon \in \boldsymbol{R}^{++} \exists n \in \boldsymbol{N} \forall i > n (| a - a_i | < r + \varepsilon) \quad (1.3)$$

Fuzzy concepts include exact concepts as their special cases. For example, limits are special cases of fuzzy limits and continuous functions studied in the classical analysis become a part of the set of fuzzy continuous functions studied in neoclassical analysis.

It is important that neoclassical analysis does not only allow deviations from standard concepts, but also provides a measure to what extent such deviations are acceptable.

Neoclassical analysis extends the scope of the classical analysis making, at the same time, methods of analysis more precise because fuzzy concepts better reflect reality.

Consequently, new results are obtained extending and even completing classical theorems. In addition, facilities of analytical methods for various applications also become more broad and efficient.

For instance, neoclassical analysis (and scalable topology, which is rooted in neoclassical analysis) provide mathematical models not only for conventional discrete spaces, but for more sophisticated discrete spaces. Conventional discrete spaces are (usually, orthogonal) grids with two main features: a) distances between neighboring points are equal, and b) there is nothing between neighboring points. Quantum discrete spaces have different nature. To show this, Wheeler suggested that the Planck length marked the quantum boundary where space and time become so violently scrambled that ordinary notions of measurement stopped making sense. He calls the result "*quantum foam*" and describes it in the following way (Wheeler, 1998):

> *"So great would be the fluctuations that there would literally be no left and right, no before and no after. Ordinary ideas of length would disappear. Ordinary ideas of time would evaporate."*

Neoclassical analysis gives means for working with spaces and functions defined in such spaces where gaps between points or small neighborhoods of points have unusual or even unknown structure.

At the same time, classical results for limits, functions, their integrals and derivatives, which can be found in various textbooks (cf., for example, (Stewart, 2003; Goldstein, et al, 1987; Shenk, 1979; Ribenboim, 1964; Fihtengoltz, 1955)), become direct corollaries of the corresponding results for fuzzy limits and derivatives. These corollaries, as a rule, do not need additional proofs. In a broader context of fuzzy continuity, fuzzy limits, and fuzzy derivatives, it is possible to extend many results of the classical mathematical analysis. In some cases, such a transition to a fuzzy context provides for completion of some basic results of the classical mathematical analysis. For example, one of the basic results of the classical calculus, the Weierstrass Theorem, states that any continuous function on a closed interval is bounded. The converse is not true. So, continuity is only a sufficient but not a necessary condition for boundedness of a function. However, fuzzy continuity makes attainable to obtain a complete criterion of boundedness of a function. Namely, as it is proved in (Burgin and Šostak, 1994), *a function on a closed interval is bounded* ***if and only if*** *it is fuzzy continuous*.

A similar situation existed for bounded sequences of real numbers. One of the basic results of the classical calculus states that any converging sequence is bounded. The converse is not true. So, convergence is only a sufficient but not a necessary condition for boundedness of a sequence. However, fuzzy convergence makes attainable to obtain a complete criterion of boundedness of a sequence. Namely, as it is demonstrated in (Burgin, 2001), *a sequence is bounded* ***if and only if*** *it fuzzy converges*. This result may look as a paradox, but it is necessary to keep in mind that fuzzy convergence may be very fuzzy or fuzzy to a little extent. Neoclassical analysis provides means for both measuring this fuzziness and choice of a satisfactory convergence.

Neoclassical analysis does not only bring new results, which extend and eventually complete their classical analogues, but it also produces deeper insights and a better understanding of the classical theory. In addition, neoclassical analysis introduces new useful concepts for the classical mathematical analysis and computational mathematics. For example, fuzzy derivatives studied in Chapter 4, correspond to main algorithmic schemes for computation of derivatives of real functions. Really (cf., for example, (Burden and Faires, 2001)), there are different kinds of formulas for such computations. If we compare these formulas to fuzzy derivatives, we can see that the forward-difference formula for computation corresponds to right fuzzy derivatives, the backward-difference formula for computation corresponds to left fuzzy derivatives, and the three-point formula for computation corresponds to two-sided fuzzy derivatives. Other computational formulas for derivatives induce additional constructions of fuzzy derivatives.

The main stimulus for the development of neoclassical analysis was an intention to lessen the gap between rigorous and infinitely precise calculus and imprecise, vague, and uncertain real world, to provide efficient methods and constructions for dealing with imprecision, vagueness and uncertainty that emerge from unlimited processes, involving numbers. There is a prevalent illusion that if we take crisp (exact) structures, everything has to be always exact. It is not true for unlimited processes related to sequences and series. One of sources of fuzziness is limit constructions and infinite processes, which constitute the core of calculus and topology. However, conditions that define these constructions can be true only in abstract mathematics. As researchers state (cf., for example, (Blehman, et al, 1983)), in science and engineering, the notion of convergence for sequences and series is essentially different from convergence that we find in mathematical theories. A more realistic, than classical, and closer to real life and science approach is developed in neoclassical analysis.

It is necessary to remark that in many cases, properties of classical concepts and corresponding fuzzy concepts are in essence different. In some cases, this difference is crucial. For instance, we know such a property of continuous functions that if $g(x)$ and $f(x)$ are continuous functions defined on $\boldsymbol{R}$, then their product $g(x)\cdot f(x)$ is also a continuous function on $\boldsymbol{R}$. At the same time, the product $h(x)\cdot k(x)$ of fuzzy continuous functions $h(x)$ and $k(x)$ can be not a fuzzy continuous function on $\boldsymbol{R}$.

In other cases, there are similarities between properties, but the properties of fuzzy concepts are usually more sophisticated than corresponding properties of classical concepts. For instance, we know such a property of continuous functions that if $g(x)$ and $f(x)$ are continuous functions, then $g(x) + f(x)$ is also a continuous function. However, if $g(x)$ and $f(x)$ are r-continuous functions, then $g(x) + f(x)$ is not, in general, an r-continuous function. It is only a $2r$-continuous function. Thus, research in neoclassical analysis cannot be reduced to a simple transition of classical results to a fuzzy context.

The classical calculus, fuzzy set theory, and interval analysis (or even more general, set-valued analysis) form three roots of neoclassical analysis. The technique of neoclassical analysis is an advanced form of the calculus technique. The approach of neoclassical analysis is based on the idea of fuzzy structures central in fuzzy set theory. The intention of neoclassical analysis is to measure imprecision, vagueness, and uncertainty in way that has similarities with the interval analysis approach.

Neoclassical analysis as a specific mathematical area was singled out in 1995 (Burgin, 1995). However, the situation with neoclassical analysis was similar to the situation with the calculus. As we have seen in Section 1.1, different results related to the calculus (such as results of Archimedes, Aryabhata, and later Bhaskara II, Madhava, Jyestadeva, Kepler, Cavalieri, Wallis, Fermat, Pascal, etc.) appeared much earlier than the creation of the calculus as a separate mathematical field took place. In a similar way, research related to neoclassical analysis had emerged much earlier than neoclassical analysis was formed.

One of the first problems related to neoclassical analysis was considered by Ulam in 1940. He introduced approximate homomorphisms. Namely, an *approximate homomorphism* in a category of groups is a mapping f from a group A to a group B with a metric $\mathbf{d}$ such that $f(xy)$ is not necessarily equal to $f(x)f(y)$, but must be "close" to $f(x)f(y)$, i.e., the distance $\mathbf{d}(f(x)f(y), f(xy))$ has to be less that some small number δ for all x and y from A. It is possible to consider approximate homomorphisms in different categories, e.g., category of rings, of topological spaces, etc.). Examples of approximate homomorphisms in the category of topological spaces are fuzzy continuous mappings, while an example of approximate homomorphisms in the category of series is fuzzy summation studied in neoclassical analysis.

Ulam also formulated the problem whether there was a strict homomorphism $g: A \to B$ which approximates any approximate homomorphism f in the sense that $f(x)$ is always "close" to $g(x)$. In 1941, Hyers gave a partial solution to this problem. This result and Ulam stability were further extended in various directions (cf., Rassias, 1978; 1991; 1993; Skof, 1983; Forti, 1987; Gajda, 1991; Isac and Rassias, 1993; Găvruta, 1994; Borelli and Forti, 1995; Jung, 1996; 1997). Recent results on approximate homomorphisms are presented in (Kanovei and Reeken, 2000).

One of the basic directions in neoclassical analysis started with the important work of Klee (1961) who introduced and studied r-continuous (also known as ε-continuous or nearly continuous) mappings of a topological space into a metric space where r and ε denoted some positive real numbers. Later different authors studied nearly (or approximately) continuous mappings introduced by Klee. In this work, Klee also defined a stability condition for r-continuous (or nearly continuous) mappings. Namely, a compact space X satisfies the stability condition if for any nearly continuous mapping f of X into itself, there is a point in X nearly invariant (or fuzzy invariant in the context of neoclassical analysis) under f. Such a point is called an approximate or fuzzy fixed point. Approximate or fuzzy fixed points were studied in (Klee, 1961), as well as in the work of Yandl (1965).

Existence of solutions of equations is a central problem in many areas of mathematics. One of the powerful tools to prove existence of solutions is fixed-point theorems. Due to this, the concept of a fixed point is one of the unifying concepts for the whole mathematics. For instance, fixed point theory has become an efficient tool for studying nonlinear differential and integral equations. Fixed point theory also has many applications beyond mathematics: in control theory, convex optimization and economics. As Warnock (1968) writes, fixed-point theorems proved very useful for solving physical problems. However, in the majority of situations, it is impossible to find an exact solution and only approximate or fuzzy solutions are accessible. For instance, physicists use approximate fixed points to characterize regions of relative stability (Di Bari and Foot, 2000). As a result, the theory of approximate or fuzzy fixed points has become one of the most popular areas in neoclassical analysis.

Let us consider a mapping $f: X \to X$. Here X can be a set with any structures, e.g., a topological space, or without any structure.

Definition 1.3.25. An element x from X is called a *fixed point* of the mapping f if $f(x) = x$.

For a fuzzy fixed point this condition is true only to some extent.

Nadler (1968; 1969) studied fixed points for ε–nonexpansive r-continuous and uniformly r-continuous multifunctions and functions, introducing a local concept of an ε–nonexpansive mapping.

Definition 1.3.26. For a given $\varepsilon > 0$, a (multivalued) mapping $f: X \to X$ is called ε–*nonexpansive* if for any x and y from a metric space X, $\mathbf{d}(x, y) < \varepsilon$ implies $\mathbf{d}(f(x), f(y)) \leq \mathbf{d}(x, y)$.

This concept is similar to the concept introduced in (Edelstein, 1964) and is dual to a fuzzy concept of an ε*–nonexpansive mapping.

Definition 1.3.27. For a given $\varepsilon > 0$, a (multivalued) mapping $f: X \to X$ is called ε*–*nonexpansive* if for any x and y with $\mathbf{d}(x, y) > \varepsilon$ from a metric space X, we have $\mathbf{d}(f(x), f(y)) \leq \mathbf{d}(x, y)$.

It is also possible to introduce local fuzzy concept of an (ε, γ)–nonexpansive mapping.

Definition 1.3.28. For a given $\varepsilon > 0$, a (multivalued) mapping $f: X \to X$ is called (ε, γ)–*nonexpansive* if for any x and y with $\gamma > \mathbf{d}(x, y) > \varepsilon$ from a metric space X, we have $\mathbf{d}(f(x), f(y)) \leq \mathbf{d}(x, y)$.

Both ε- and ε*-nonexpansive mappings are special cases of (ε, γ)-nonexpansive mappings.

Utilizing the same technique as Nadler (1968; 1969) used in his works, it is possible to prove similar results about existence of fuzzy fixed points for ε*–nonexpansive and (ε, γ)–nonexpansive r-continuous and uniformly r-continuous multifunctions and functions. Fedeli and Pelant (1991) applied the concept of r-continuous function to problems of multifunctions.

Interest in approximate or fuzzy fixed points has been always very active. Due to this interest, an extensive research in what was called later neoclassical analysis has been aimed at obtaining different fuzzy or approximate fixed-point theorems. Many researchers contributed to this area: Muenzenberger (1968), Muenzenberger and Smithson (1968), Wong (1976), Dugungji and Granas (1982), Reich (1983), Kirk (1986; 1997; 1998; 2003), Carbone and Marino (1987), Reich and Shafrir (1987) Idzik (1988), Sanjurjo (1989a), Sine (1989), Kirk and Martinez-Yanez (1990), Cromme and Diener (1991) Hankin and Hunt (1992), Marino and Pietramala (1992), Gajek, Jachymski and Zagrodny (1995), Burgin (1997a), Cromme (1997), Janiš (1997), Phu and Truong (2000), Branzei, Morgan, Scalzo, and Tijs (2003), Chidume and Zegeye, (2003; 2004), Kuczumow (2003), Tijs, Torre and Branzei (2003a) Li, Rychlik, Szidarovszky and Chiarella (2003), Phu (2003; 2003d; 2004), Wisnicki (2003), Chen (2004) Chidume and Zegeye, 2004), Khamsi (2004), Kohlenbach and Leustean (2005), and Reich and Zaslavski (2006).

ε-continuous mappings find interesting applications in topology. Research in this area was started by Klee and Yandl (1974). In particular, they introduced an important concept of an ε–homotopy of a topological space into a metric space. Namely, an ε–*homotopy* of a topological space T into a metric space L is an ε-continuous mapping $h: T \times [0, 1] \to L$.

Klee and Yandl (1974) developed some of the basic parts of the theory of absolute retracts in a proximate form where continuous functions are replaced by nearly continuous functions, fixed points by nearly fixed points, etc. They argued that notions of nearly continuous functions and nearly fixed points are natural in connection with the mathematical modeling of physical problems, where there may be strong empirical evidence for the near continuity of a map even though its full continuity cannot be tested experimentally. The same is true for nearly fixed points. Moreover, the conclusions as well as the hypotheses are naturally of a proximate nature in many applied situations.

Interesting results in this direction were obtained by Felt (1974) and Sanjurjo (1988; 1992). They used ε-continuous mappings and ε–homotopy to study the shape category of compact spaces. Borsuk (1968) built the theory of shapes for classification of compact metric spaces, taking into account their global properties and neglecting the local ones.

In (Giraldo, et, al, 2001), the concept of ε-continuity is used to prove that the Čech homology groups are in fact inverse limits of certain homology groups. It gives an intrinsic characterization of Čech homology, eliminating all the external elements, like embeddings, inverse sequences of simplicial complexes, etc.

An interesting concept of a ε-small multivalued mapping (binary relation) is introduced and studied in (Sanjurjo, 1991, 1992, 1994). This concept is a fuzzy extension of the concept of a function. Indeed, a function f from X to Y is a binary relation between sets X and Y in which any element from X is corresponded to at most one element from Y, i.e., Diam $(F(x)) = 0$.

Definition 1.3.29. A multivalued function (binary relation) $F: X \to Y$ between metric spaces is called ε-*small* if $\text{Diam}(F(x)) < \varepsilon$ for any x from X.

Informally, it means that such a multivalued function is close to a function, or it is approximately a function because the equality $\text{Diam}(f(x)) = 0$ is characteristic for any function f, i.e., a binary relation f between metric spaces X and Y is a function (or more exactly, a partial function) if and only if $\text{Diam}(f(x)) = 0$ for any x from X.

Grace (1977, 1986) and Grace and Vought (1989, 2003) studied refinable maps and their relations to proximate fixed points. The authors used many concepts from neoclassical analysis, such as ε-continuous functions, proximate fixed points, ε-functions and some others.

In neoclassical analysis, some deviations from an exact mathematical concept are allowed because in practice it is often impossible to check exact conditions from the concept definition. For instance, a function $f: X \to Y$ is one-to-one if for any y from Y, the inverse image $f^{-1}(y)$ consists of one point. A possible fuzzy counterpart of this concept is the concept of an ε-function.

Definition 1.3.30 (Grace, 1977). A function $f: X \to Y$ is called an ε-*function* if for any y from Y, the diameter $\text{Diam}(f^{-1}(y))$ is less than ε.

It is possible to attribute explicit beginning of the research in neoclassical analysis to the papers (Burgin and Šostak, 1992; 1994), where measures of discontinuity and fuzzy continuous functions were introduced and studied, and papers (Hu, Klee, and Larman, 1989; Söllner, 1991; Hartwig, 1992; Phu, 1993; 1995; 1995a), where fuzzy (or roughly, in the terminology of the authors) convex functions were introduced and studied.

Approximately at the same time, Cromme and Diener (1991) defined two global measures of discontinuity. Namely, if M is a subset of a normed linear space, $B_r(x)$ is a ball with radius r and center x and $f: M \to M$ is a mapping, then:

$$\delta(f) = \sup \lim_{x \in M} \sup \lim_{r \to 0} \sup_{y \in Br(x)} \| f(x) - f(y) \|$$

and

$$\delta'(f) = \sup \lim_{x\in M} \sup \lim_{r\to 0} \sup_{y\in B_r(x)\setminus\{x\}} ||f(x) - f(y)||$$

They demonstrated that if M is compact and convex, then there is a point $a \in M$ such that $||f(a) - a|| \leq \delta(f)$ and for any $q > \delta'(f)$, there is a point $b \in M$ such that $||f(b) - b|| \leq q$.

Thus, different concepts and structures from the classical analysis were extended to their fuzzy counterparts and studied in a diversity of papers. Common approach and ideas united this research, and it was natural that neoclassical analysis was singled out as a separate field of mathematics in (Burgin, 1995).

This gave a new impetus to the development of the field and many concepts, structures, and theories from the classical calculus and analysis have acquired a fuzzy form in the context of neoclassical analysis. The main results of this development are: the theory of limits of sequences and functions (Burgin, 2000; Burgin and Kalina, 2005); theory of statistical limits of sequences (Burgin and Duman, 2006); theory of continuity and continuous functions (Burgin and Šostak, 1992; 1994; Burgin, 1995, 1999; Burgin and Glushchenko, 1997a; 1998a; 1998b), systems of criteria for extrema of functions (Burgin, 2003; 2004), neoclassical differential calculus (Burgin, 2001; 2001b; 2004) and neoclassical integral calculus (Burgin, 2007). Transition to fuzzy concepts in analysis brought forth new structures and concepts. Examples are measures of convergence (Burgin and Duman, 2006), continuity defects and discontinuity measures (Burgin and Šostak, 1992; 1994; Burgin and Glushchenko, 1998b; Burgin, 1999), and measures of differentiability (Kalina and Šostak, 2006).

A variety of problems of neoclassical analysis were studied by Martin Kalina and his school in the context of fuzzy nearness relation and fuzzy continuity. This relation was introduced by Kalina (1997) and further developed by Janiš (1998) and Dobrakovova (1998; 1999). They elaborated an original and interesting direction in neoclassical analysis. For instance, Kalina (2001) introduced the concept of fuzzy limit and fuzzy convergence based on the fuzzy nearness relation and built a calculus of nearness-based limits.

An interesting approach to fuzzy differentiation of real functions was suggested by Kalina (1998; 1999; 1999a) and Janiš (1999). Their construction for differentiation is based on the concepts of fuzzy continuity from (Burgin and Šostak, 1992; 1994; Burgin, 1995) and nearness on the set $\boldsymbol{R}$ of real numbers, which was introduced by Kalina (1997). Many interesting results were obtained in this direction. In particular, in the context of neoclassical analysis, Janiš (1999) extended such important results as Rolle and Lagrange mean value theorems. He introduced and studied a nearness derivative $int(D_\alpha)$ for functions in spaces of real numbers with a nearness. It is a set-valued function in contrast to fuzzy derivatives, which are considered in this book and are point-valued functions.

In (Janiš, 1997), ε-fixed points (fuzzy fixed points) are studied for fuzzy functions that are defined on a metric spaces and take values in fuzzy numbers and for fuzzy continuous functions in the sense of neoclassical analysis.

Topological properties of nearness relations are studied by Dobrakovová (2001). In (Kalina and Dobrakovová, 2002; Kalina, 2004), the concept of nearness relations is generalized to linear spaces and topological properties of nearness-based convergence are studied.

A variety of problems of neoclassical analysis were studied by Phy and his school. The main attention has been paid to fuzzy, or rough, in their terminology, fixed points and convexity. They call their area of research by the name *rough analysis*, a concise overview of which is given by Phy (2005). Phu (2001; 2003; 2004) studied rough convergence or r-convergence, r-limits, convergence degrees, and Cauchy degrees of sequences. Namely, if a sequence r-converges, then r is its convergence degree, while the Cauchy degree gives an inner characterization of r-convergent sequences.

Rough continuity (or r_X-r_Y-continuity) is studied for general mappings of normed linear spaces in (Phu, et al, 2000) and for linear operators in (Phu, 2001). One more useful and important concept of neoclassical analysis - fuzzy or rough convexity of functions - was studied in (Hartwig, 1992; 1996; Hu, et al, 1989; Phu, 1993; 1995; 1995a; 1997; Phu and Hai, 1996; 2005; Hai, 2001; Hai and Phu, 1999; 2001; Phu, Hai, and An, 2003; Söllner, 1991). Obtained properties of roughly or fuzzy convex functions have various applications in economics. As it is known, many functions suitable for solving economical problems are convex or concave. At the same time, functions that come from real life often have jumps and thus, cannot be convex or concave in the classical sense. However, such functions are fuzzy or roughly convex and thus, properties of fuzzy or roughly convex functions are essential for economical problems. Rough convexity is a kind (or actually, produces several kinds) of generalized convexity (cf. (Hartwig, 1983; Komlosi, et al, 1994)).

Aiming at representation of uncertainties and imprecision and extending the scope of the classical calculus and analysis, neoclassical analysis makes, at the same time, methods of the classical calculus more precise with respect to real life applications. Consequently, new results are obtained extending and even completing classical theorems. In addition, facilities of analytical methods for various applications also become more broad and efficient. For example, recently, psychologists and computer scientists built a new model for motion perception of human beings and animals (Johnston, and Clifford, 1995; McOwan, *et al*, 1999). This model is based on temporal and spatial filters, which utilize fuzzy derivatives of Gaussian functions (Young, 1985). The authors of the model present biological evidence that the form of the cell receptive fields in early visual cortex may be accurately modeled by fuzzy derivatives. Fuzziness is the consequence of noise in accepted signals. To deal with this noise, biologists use an alternative construction of fuzzy derivative. However, constructions from neoclassical analysis allow one to build more adequate models for psychological phenomena and it looks beneficial to apply neoclassical analysis to the problems of visual information processing in the brain.

In addition to this, the technique developed in neoclassical analysis makes it possible to eliminate discrepancies existing in numerical analysis. The problem is that computations are realized on finite machines, while many processes of mathematics, such as differentiation and integration, demand the use of a limit, which is an infinite process. As a consequence, correct algorithms based on classical methods of calculus, when implemented, turn into unreliable programs.

Neoclassical analysis treats such processes more adequately than the classical analysis, suggesting an alternative to interval analysis, which is also aimed at computational mathematics. This is demonstrated by applications of neoclassical analysis to problems of numerical computations and control (Burgin and Westman, 2000). In particular, new methods allow one to solve the problem that was formulated by Poincaré, who wrote (cf. (Blehman, et al, 1983)) that series convergence is different for mathematicians, who use abstract mathematical procedures, and for physicists and astronomers, who utilize numerical computations. Neoclassical analysis provides mathematical structures to formalize and study by rigorous methods those approaches to convergence that exist in physics, economics and other sciences.

Consequently, the new technique provides for a better than does classical analysis utilization of numerical computations for artificial intelligence, especially, in the case when uncertainty of computation is multiplied by the uncertainty of input information. For example, in (Burgin and Glushchenko, 1997; 1998) neoclassical analysis is applied to problems of decision-making. Similar constructions are utilized for solving such practical problems as controlled imprecision in data transfer and processing, controlled transactions deadline overruns (Saad-Bouzefrane and Sadeg, 2000) and management of uncertainty in network disconnections (Amanton, Sadeg and Saad-Bouzefrane, 2001). Fuzzy continuous functions are used for object classification in (Burgin, 1993).

Structures from neoclassical analysis, such as fuzzy limits, fuzzy sums of series, and fuzzy continuous functions, are also used for the development of algorithmic tools in general (Burgin, 2001h) and for computer simulation, in particular (Burgin and Westman, 2000; Burgin, 2001a). These structures allow one to define new types of algorithms oriented to better and more adequate computer modeling continuous processes and systems.

Another approach in theoretical computer science related to neoclassical analysis is based on models of real number computation that include possible imprecision. Boldi and Vigna (1998) introduced a version of the finite-dimensional machine over the field of real numbers $\boldsymbol{R}$, called a δ-uniform finite-dimensional machine over $\boldsymbol{R}$. In this machine, exact tests are not allowed, and the test for equality with 0 is replaced with a test for membership in an arbitrary ball around 0. The condition of δ-uniformity reduces the full power of such machines nearly to the level of Turing machines. Approximate computations studied by Boldi and Vigna involve three new types of real functions: pointwise δ-approximable, uniform δ-approximable, and computable δ-approximable.

Chadzelek and Hotz (1999) introduced and studied *δ–**Q**-machines*, which take real numbers as inputs and use infinite converging computations on more and more precise rational roundings of their inputs. It is demonstrated that infinite converging computations with rational numbers can simulate finite computations with real numbers. In addition, the authors study such problems as computability of solutions of ordinary differential equations (ODE) and decidability of stability problems for dynamical systems defined by ODEs.

The approach to continuity of discrete functions suggested in digital topology (Rosenfeld, 1979) also belongs to neoclassical analysis as continuity of functions in discrete topology is a kind of (q, r)-continuity studied in Chapter 3. Digital topology has many applications. It is used in pattern recognition, image analysis, and related areas (cf., for example, (Rosenfeld, 1979; 1986; Boxer, 1994; Latecki and Prokop, 1995)), and applied to software design and utilization (Hamlet, 2002).

Results from neoclassical analysis also find their applications in economics, game theory and social sciences (Bula, 1996; 2003; Bula and Weber, 2002; Branzei, et al, 2003; Scalzo, 2005; Bula and Rika, 2006).

1.4. STRUCTURE OF THE BOOK

There's only one solution: look at the map.
Umberto Eco (1932-)

There are four main chapters, in which the basic concepts of neoclassical analysis are introduced and studied: fuzzy limits in Chapter 2, fuzzy continuous functions in Chapter 3, fuzzy differentiation in Chapter 4, and fuzzy integration in Chapter 7. Chapters 5, 6, and 8 consider applications of neoclassical analysis. In Chapter 5, monotone and fuzzy monotone functions are studied by means developed in previous chapters. In Chapter 6, criteria for fuzzy minimums and maximums are obtained based on properties of fuzzy derivatives. In Chapter 8, dynamical systems are studied in the context of neoclassical analysis. This setting provides for an essentially wider scope of applications, as well as for more adequate and flexible analysis of real systems and processes. Almost all results are obtained for real functions of one variable and then exposition goes straight to functions in metric and normed linear spaces. This encompasses multivariable and complex functions, as well as functions in infinite dimensional spaces. However, it would be interesting to develop neoclassical analysis specifically for these situations.

Exposition in the book goes from foundations (fuzzy limits and fuzzy continuous functions) to basic operations (fuzzy differentiation and fuzzy integration) to applications (properties of monotone functions, criteria for fuzzy minimums and maximums, and fuzzy dynamical systems). Each chapter and many sections have an informal exposition of the main ideas.

All basic concepts are thoroughly explored by using different constructions for their definition. It is demonstrated how different constructions can bring mathematicians to the same mathematical object. For instance, it is possible to define continuity by utilizing limits, neighborhoods and by the most popular in calculus (ε, δ)-definition. For real functions, all these definitions result in the same concept of continuity. However, in a more general situation of topological spaces, these three definitions bring us to different structures. Neighborhood continuity is directly extended to continuous mappings of topological spaces. Limits define continuity only for sequential topology, while the (ε, δ)-definition can be extended only to metric spaces but not to general topological spaces.

The book presents the core of neoclassical analysis on three levels. At first, on the first level, the basic classical constructions of the conventional calculus, such as limits, continuous functions, derivatives, differentiable functions, and integrals, are considered and their properties are studied. It is necessary to remark that exposition of the conventional calculus material is given in this book in a more comprehensive way than in the majority of textbooks on the calculus and analysis that are now used at colleges and universities. A student will be able to find different theorems, propositions and lemmas that describe useful properties of classical structures from the calculus, such as limits, continuous functions, derivatives and integrals. Those statements (theorems, propositions, corollaries, and lemmas) that are given without proofs or have only some directions how to build a proof are, as a rule, relatively easy to prove and are left to the reader as exercises.

Presenting classical results, we preserve the names of these results used in mathematical literature, e.g., the First Weierstrass Theorem, Fermat theorem, Rolle theorem, Mean Value theorem, Comparison Test or Cauchy criterion. Their fuzzy counterparts (when they exist) mostly preserve these names, e.g., the First Fuzzy Weierstrass Theorem, Fuzzy Fermat Theorem, Fuzzy Rolle theorem, Fuzzy Mean Value theorem, fuzzy Comparison Test or fuzzy Cauchy criterion.

The second level gives an exposition of neoclassical, or fuzzy, extensions of the classical constructions for real functions and sequences, i.e., neoclassical analysis of real functions and sequences is constructed. The third level elevates calculus from numerical functions to functions in metric spaces and is presented at the end of two main chapters. At this level, exposition is more formal. Unlike the previous levels, the given exposition of this level does not have many examples and explanations.

This three-tier approach to the calculus exposition is aimed at concept understanding facilitation. Understanding is very important. As Stewart states (2003), nearly everybody agrees that the emphasis of calculus teaching should be on understanding concepts.

In addition to the three-tier exposition, to facilitate reader comprehension, we go from a simpler case of a fuzzy concept to the general case of the same concept. For instance, at first, we study r-limits of functions, which are closer to conventional limits of functions, and only then (r, q)-limits. At first, we study r-continuous functions, which are closer to continuous functions, and only then (r, q)-continuous functions.

This simplifies reading of the book and makes its accessible to a wide range of readers. Even those who did not learn or learned and forgot calculus can read this book as the beginning section of each chapter introduces the reader to the corresponding structures and results of the classical calculus. To make things simpler, each chapter and many sections have an informal exposition of the main ideas, while the Appendix contains a compendium of basic concepts from calculus foundations.

At the same time, for those who studied calculus at a college or university, the expository section of each of the main chapters liberates such a reader from necessity to consult other sources (textbooks, monographs, etc.) from time to time. At the same time, mathematicians who know the classical calculus very well will be able to find many new things in the book about a new, extended and more realistic calculus called neoclassical analysis. It is necessary to remark that some topics are only outlined for further research. Consequently, the book opens many directions for a more extensive study. Different problems and directions for further research are formulated in Conclusion.

Some of classical results are given with detailed proofs to demonstrate that proofs of their fuzzy counterparts have the same complexity. In other cases, classical results are obtained as direct corollaries of the corresponding fuzzy results and given without proofs. This shows that form the beginning, it is possible to study the calculus and especially, analysis in the form of neoclassical analysis, for which classical analysis is a special case and a subtheory.

At the end of the book, there is an Appendix, in which basic concepts and denotations from set theory, logic, algebra, and topology, as well as important constructions based on numbers and numerical functions are given. Usually analysis textbooks start with a rigorous exposition the theory of natural, rational and real numbers and sometimes even with set theory. However, the main goal of this book is a new field in calculus and analysis - neoclassical analysis. That is why, here (in Appendix) we give only main definitions and constructions related to numbers, sets, and functions and do not go into detail as this material has been repeated in dozens of different textbooks. It is possible to use Appendix in two ways. The reader either can, at first, read this material and then to go to the main text of the book or can read the main text, going from time to time to Appendix in case she or he encounters some unknown denotation or concept.

ACKNOWLEDGEMENTS

Many wonderful people have made contributions with my efforts with this work. I am especially grateful to Frank Columbus and Maya Columbus from Nova Science Publishers, for their encouragement and help in bringing about this publication. In developing ideas in neoclassical analysis, I have benefited from conversations with many friends and colleagues who have communicated with me on problems of calculus and functional analysis. Thus, I am grateful for the interest and helpful discussions with those who have discussed with me these problems. Credit for my desire to write this book must go to my academic colleagues. Their questions and queries made significant contribution to my understanding and further development of new approaches in analysis. I would particularly like to thank many fine participants of the Applied Mathematics Colloquium, Analysis Seminar, Topology Seminar at UCLA, Mathematics and Statistics Department Colloquium at California State Polytechnic University, Pomona, and Meetings of the American Mathematical Society for extensive and helpful discussions on neoclassical analysis that gave me much encouragement for further work in this direction. My collaboration with Oktay Duman, Vitalii Glushchenko and Martin Kalina gave much to the development of ideas and concepts of neoclassical analysis. I would also like to thank the Departments of Mathematics and Computer Science in the School of Engineering at UCLA for providing space, equipment, and helpful discussions.

Chapter 2

FUZZY LIMITS

It appears evident, then that there is a distinct limit,
as regards length, to all works of literary art ...
Edgar Allan Poe (1809-1849)

It is not possible to wait for inspiration,
and even inspiration alone is not sufficient.
Work and more work is necessary.
P.I. Tchaikovsky (1840-1893)

Concepts of a limit and convergence are basic for the whole calculus. In this chapter, we consider sequences and their limits of different kinds. We start (Section 2.1) with classical limits, which serve as a base built by Cauchy for calculus foundations. Various properties of classical limits are studied as it is done in the majority of calculus and analysis textbooks. The goal is to make exposition of this book self-contained, allowing the reader to learn calculus from the very beginning. To achieve this goal, proofs of the main results for limits are given. One more goal of this detailed exposition is to demonstrate that neoclassical analysis is not essentially more complicated than the classical calculus. At the same time, obtained in neoclassical analysis results show that neoclassical analysis contains the classical calculus as a part and encompasses a substentially broader scope than the classical calculus. The main results of Section 2.1 are Theorems 2.1.1, 2.1.4, 2.1.6, 2.1.7, 2.1.8, and 2.1.10.

However, as we have discussed in Introduction, analysis of computation and measurement procedures shows that the classical mathematical idealization can be irrelevant or insufficiently adequate to various applications. Classical models work well up to some precision. Then they fail. In spite of this, they create an illusion of infinite precision, which does not exist in reality. Consequently, natural science, as well as computer science, needs new methods to deal with such imprecision. To achieve this goal, the concept of a limit is extended to the concept of a fuzzy limit or r-limit in Section 2.2.

In conventional convergence of sequences, elements of the sequence are coming closer and closer to the limit point, concentrating in an arbitrarily small neighborhood of this point. However, measurement and numerical computations do not allow one to test this. In fuzzy convergence, it is sufficient that elements of the sequence are concentrating in some small, but not shrinking to zero neighborhood of the fuzzy limit. Taking sufficiently big neighborhood, we can test fuzzy convergence. Taking an appropriate size of such a neighborhood, we can get a good approximation of a conventional limit or to realistically estimate that the sequence fuzzy converges. In Section 2.2, we formalize these ideas.

It is necessary to remark that what is small and what is big depends on the situation or area of consideration. For instance, for microphysics, one millimeter is an extremely big distance, while in astronomy and astrophysics, one thousand miles is an extremely small distance. For a butterfly, one month is an extremely big time period, while in geological time, one thousand years is an extremely small time period.

This situation is well understood in numerical analysis, where (cf. Wikepedia) "the algorithm is said to be *backward stable* if the backward error is small for all inputs x. Of course, "small" is a relative term and its definition will depend on the context. Often, we want the error to be of the same order as, or perhaps only a few orders of magnitude bigger than, the unit round-off."

In a broader context of fuzzy limits and fuzzy convergence, it is possible to extend many results of the classical mathematical analysis. In some cases, such a transition to a fuzzy context makes it possible to complete some basic results of the classical calculus. For instance, one of the basic results of the classical calculus states that any converging sequence is bounded. The converse is not true. So, convergence is only a sufficient but not a necessary condition for boundedness of a sequence. However, fuzzy convergence makes attainable to obtain a complete criterion of boundedness of a sequence. Namely, as it is demonstrated in this section, *a sequence is bounded if and only if it fuzzy converges*. This result may look as a paradox, but it is necessary to keep in mind that fuzzy convergence may be very fuzzy or fuzzy to a little extent. Neoclassical analysis provides means for measuring this fuzziness and for a choice of satisfactory level of convergence.

The main results of Section 2.2 are concepts of an r-limit and r-convergence of sequences and Theorems 2.2.2, 2.2.5, 2.2.6, 2.2.7, and 2.2.8.

After a thorough study of fuzzy limits of sequences, we go to fuzzy limits of functions in Section 2.3. This prepares a base for the study of continuous and fuzzy continuous functions in Chapter 3. The main results of Section 2.3 are concepts of an r-limit of a function at a point and Theorems 2.3.1, 2.3.5, 2.3.7, 2.3.8, 2.3.12, and 2.3.13.

In Section 2.4, we consider fuzzy structures that emerge in computations and experiments from such fundamental mathematical objects as sequences and series of numbers and vectors. Namely, we study conditional fuzzy limits, conditional fuzzy convergence, and fuzzy nearness convergence. The main results of Section 2.4 are Theorems 2.4.2, 2.4.3, 2.4.5 - 2.4.7.

Conditions on sequences allow one to determine, for example, the rate of convergence. Thus, these concepts can reflect not only fuzziness of computational and observational processes, but also such an important property as the rate of convergence. If in a theory, it is necessary to know whether a sequence is convergent or not, for practical matter, it is important with what rate the sequence under consideration converges. Only methods with high convergence rate are useful for numerical computations. It means that conditional convergence represents computationally oriented versions of the classical calculus.

Relations between fuzzy conditional limits, fuzzy conditional convergence, fuzzy nearness limits, and fuzzy nearness convergence are considered in this section based on results of Burgin and Kalina (2005).

In Section 2.5, sets and fuzzy sets of fuzzy limits are constructed and studied. It is demonstrated, for example, that the set of all r-limits of a sequence is either empty or a closed interval. The main results of Section 2.5 are Theorems 2.5.1 - 2.5.4 and Lemma 2.5.3.

In Section 2.6, the concept of a fuzzy limit of a set of sequences is introduced and studied. This concept brings us to its crisp counterpart - a limit of a set of sequences, which is new for the conventional theory of limits although it allows mathematicians to reconsider some classical results of the calculus. Some of such results are obtained in this section. The main results of Section 2.6 are Theorems 2.6.1. - 2.6.3.

In Section 2.7, series, their summability and fuzzy summability are considered. We develop a more rigorous exposition of series than standard textbooks on calculus. In calculus, as a rule (cf., for example, (Ross, 1996; Stewart, 2003; or Fihtengoltz, 1955)), a distinction between a series and its sum is not made. For instance, in all textbooks, it is usually written (cf., for example, (Stewart, 2003)):

$$\Sigma_{i=1}^{\infty} a_i = S \text{ where } S = \lim_{n\to\infty} S_n \text{ where } S_n = a_1 + a_2 + \dots + a_n.$$

It means that a series and its sum are treated as the same object. However, there is a necessity to make a distinction between a series and its sum because there are different definitions of a sum of a series. Here we assume that a series has the form $\Sigma_{i=1}^{\infty} a_i$, its classical sum is denoted by $sum\Sigma_{i=1}^{\infty} a_i$ and its fuzzy sums are denoted by $r\text{-}sum\Sigma_{i=1}^{\infty} a_i$. Summation is considered as one of operations (namely, a topological operation) in series. Examples of other operations are addition and scalar multiplication of series. This approach to series is aimed at making calculus teaching in colleges and universities more rigorous and transparent.

Fuzzy summability extends the power of summability techniques and, for example, allows one to make mathematically meaningful some operations that originators of the calculus performed with classically divergent series, such as $1 - 1 + 1 - 1 + \dots$.

The main objects of Section 2.8 are statistical fuzzy limits and the corresponding convergence. It is demonstrated how these concepts are connected to problems of mathematical statistics. This section is based on results obtained by Burgin and Duman (2006).

Section 2.9 extends the theory of fuzzy limits and convergence to metric spaces and normed linear spaces. The majority of results obtained in previous sections are valid for metric spaces or for normed linear spaces when operations with sequences are considered. However, we demonstrate this only for some of the most important properties of sequences.

To conclude this introduction, it is necessary to remark that as an informal notion "fuzzy limit" is used in different fields. It is very popular in scientific literature. It is applied when astronomers and astrophysicists discuss galaxies (Stoughton, et al, 2002) and in spectral investigation of the universe (McCandliss, 2003), as well as when physicists discuss states of matter (Newton BBS). In (Buhmann, 2001), fuzzy limits are considered in problems of pattern recognition, clustering and empirical risk estimation. Moulton (2003) discusses problems of acoustics and treats the threshold of hearing as a fuzzy limit below which we cannot detect acoustic energy. Fuzzy limits are utilized in such areas as image processing (Ceccone, 2001), communication network technology (Hwang, et al, 2004), and environmental engineering (Partlow, et al).

In social sciences, fuzzy limits, as an informal notion, are very frequent. For instance, Barán, et al, (1999) write that "poverty extends continuously towards a fuzzy limit."

In a natural way, fuzzy limits emerge in numerical computations. As Moddemeijer (2006) writes, an important problem is to decide in how many iterations an adequate estimate is reached. There are different kinds of criteria for this.

1. *Numerical convergence criterion*: the iterations proceed until the relative change in the estimated parameter (parameter vector) is negligible. This negligibility is a fuzzy limit that depends on the numerical precision of the computer, the numerical stability of the algorithm, and the preferred accuracy of the parameter estimates.
2. *Statistical convergence criterion*: the iterations proceed until the change in the estimated parameter (parameter vector) is negligible with respect to the accuracy of the parameter estimates. This negligibility is also a fuzzy limit.

Similar situation for convergence diagnostics is described by Fonnesbeck (2004), who studied convergence diagnostics for Markov Chain Monte Carlo (MCMC) sampling technique. The theory of fuzzy limits developed in neoclassical analysis and presented in this chapter allows researchers to formalize such situations.

We also see many converging processes in science. For example, the numerically calibrated geologic time scale has been continuously refined since approximately the 1930s, although the amount of change with each revision has become smaller over the decades (Harland *et al*, 1990). With new information about subdivision or correlation of relative time, as well as with new measurements of absolute (physical) time, the dates applied to the geological time scale have been continuously changing. Experts think that because of continual refinement, none of the values depicted in the geological time scale should be considered definitive, even though some have not changed significantly in a long time and are very well constrained. The overall duration and relative length of these large geologic intervals is unlikely to change much, but the precise numbers may "wiggle" a bit as a result of new data (MacRae, 1996-1997). This process is perfectly modeled by fuzzy convergence of number sequences.

2.1. Classical Limits of Sequences

When knowledge is limited - it leads to folly...
When knowledge exceeds a certain limit, it leads to exploitation.
Abu Bakr (573-634)

2.1.1. General Concepts

An infinite sequence (or simply, a sequence, as we consider only infinite sequences) of real numbers is usually treated as a function $l: \boldsymbol{N} \to \boldsymbol{R}$. For instance, the sequence $l = \{ 1, ½ , 1/3 , \dots , 1/n , \dots \}$ is represented by the function $l(n) = 1/n$. In general, a sequence of real numbers has the form $l = \{a_i \in \boldsymbol{R}; i \in \omega\}$ or $l = \{a_i ; i = 1, 2, 3, \dots\}$. This form shows that sequence is a set of elements enumerated by natural numbers. Here we adopt the latter more naive but rigorous definition.

A number a is said to be the limit of a sequence $l = \{a_i ; i = 1, 2, 3, ...\}$ if the numbers a_i approach a indefinitely as n increases. This intuition is formalized by the following precise definition.

Definition 2.1.1. A number a is called a *limit* of a sequence l (it is denoted by $a = \lim_{i\to\infty} a_i$, $\lim_{i\to\infty} a_i = a$, $\lim l = a$ or $a = \lim l$) if for any $\varepsilon \in \boldsymbol{R}^{++}$ the inequality $| a - a_i | < \varepsilon$ is valid for almost all a_i , i.e., there is such n that for any $i > n$, we have $| a - a_i | < \varepsilon$.

Informally, this definition tells us that a is a limit of a sequence l if for an arbitrarily small ε, the distance between a and all but a finite number of elements from l is smaller than ε.

Definition 2.1.2. When a sequence l has a limit, l is called *convergent* and it is said that l *converges* to its limit. Otherwise, l is called *divergent* and it is said that l *diverges*.

Example 2.1.1. Let $l = \{1/i; i = 1, 2, 3, ...\}$.Then $\lim l = 0$. A rigorous proof goes through the following steps. Taking some positive real number ε, we find a natural number n such that $1/n < \varepsilon$. Then for any $i > n$, we have $| 0 - a_i | = | a_i | = 1/i < 1/n < \varepsilon$. So, conditions from Definition 2.1.1 are satisfied, implying that $\lim l = 0$.

Example 2.1.2. Let $l = \{ (-1)^i(1/i); i = 1, 2, 3, ...\}$.Then $\lim l = 0$. A rigorous proof goes through the following steps. Taking some positive real number ε, we find a natural number n such that $1/n < \varepsilon$. Then for any $i > n$, we have $| 0 - a_i | = | a_i | = 1/i < 1/n < \varepsilon$. So, conditions from Definition 2.1.1 are satisfied, implying that $\lim l = 0$.

Definition 2.1.3. A number b is called a *partial limit* or *accumulation point* of a sequence $l = \{a_i \in \boldsymbol{R}; i = 1, 2, 3, ...\}$ (it is denoted by $b = \text{plim } l$) if there is a subsequence h the sequence l such that $a = \lim h$.

Example 2.1.3. Let $l = \{1/i + (-1)^i; i = 1, 2, 3, ...\}$. This sequence does not have a limit, but 1 and –1 are its partial limits. A rigorous proof for the number 1 goes through the following steps. Taking the subsequence $h = \{1/i + 1; i = 2, 4, 6, ...\}$ of the sequence l and some positive real number ε, we find a natural number n such that $1/n < \varepsilon$. Then for any $i > n$, we have $| 1 - a_{2i} | = 1/i < 1/n < \varepsilon$. So, all conditions from Definition 2.1.3 are satisfied, implying that 1 is a limit of h and partial limit of l.

In the classical mathematical analysis, some cases of partial limits (such as $\underline{\lim}\, l = \lim\inf l$ or $\overline{\lim}\, l = \lim\sup l$) are considered. A general case of partial limits is treated in (Randolph, 1968) where partial limits, i.e., partial 0-limits by Lemma 2.2.4, are called subsequential limits.

Accumulation points play important role in topology where they are considered not only for sequences but also for arbitrary sets in topological spaces (cf., for example, (Bers, 1957) or (Kelly, 1957)).

Partial limits of sequences are used in the theory of automata and computation. For instance, there are such computing devices that are called timed automata and work with infinite words. These automata reflect temporal behavior of finite-state reactive programs such as communication and synchronization protocols. Examples of timed automata are Büchi (1960) and Muller (1963) automata.

Definition 2.1.4. A *Büchi automaton A* is a finite automaton input of which is an infinite sequence $l = \{a_i \in \Sigma; i = 1, 2, 3, ... \}$ of symbols. Each input sequence l generates a sequence $\sigma(l)$ of states of A that is called a run of A. As it is usual for finite automata, a subset F of the set Q_A of states of A is chosen, elements of which are called finite or accepting states. An input sequence l is accepted if some states from F occur infinitely often in $\sigma(l)$.

Definition 2.1.5. A *Muller automaton A* is a finite automaton input of which is an infinite sequence $l = \{a_i \in \Sigma;\ i = 1, 2, 3, \ldots\}$ of symbols. Each input sequence l generates a sequence $\sigma(l)$ of states of A that is called a run of A. In contrast to Büchi automata, a set F of subsets of the set Q_A of states of A is chosen. Elements of F are called finite or accepting sets. An input sequence l is accepted by A if the set of all states repeating infinitely often in $\sigma(l)$ belongs to F.

Informally, the accepting states of these automata designate states through which a run must pass infinitely often. More formally, on an infinite word, every run of an infinite automaton will visit some set of states infinitely often. An infinite automaton accepts a word if and only if this set intersects the set F of accepting states. In such a way, Büchi and Muller automata define ω–regular languages. While accepting a word, a run need not visit every accepting state infinitely often; it only needs visit some accepting state infinitely often. It means that accepting states have to be partial limits of all states in an automaton run. It is made precise and formalized by the following proposition.

Let the set Q_A is considered as a topological space with the discrete topology.

Proposition 2.1.1. a) An input sequence l is accepted by a Büchi automaton A with a set F of final states if and only if $\sigma(l)$ has a partial limit in F.

b) An input sequence l is accepted by a Muller automaton A with a set F of accepting sets if and only if the set of all partial limits of $\sigma(l)$ belongs to F.

Timed automata provide the foundation for theories and techniques for system analysis techniques such as computer-aided verification. Using timed automata, we can model desired properties of systems and protocols. An example of such a property is "whenever the system receives a request, it eventually produces an acknowledgment." Then, given a model M of a system, we can see whether every run of the system satisfies the property by checking whether the language produced by M is accepted by the automaton for the given property.

The untiming construction for timed automata forms the basis for verification algorithms for the branching temporal logics, as well as for realizability checking and synthesis of reactive software modules (Vardi and Wolper, 1986; 1994).

Important special cases of partial limits are one-sided limits.

Definition 2.1.6. A number a is called a *left* (*right*) *limit* or *limit from the left* (*right*) of a sequence l if there are infinitely many a_i from l such that $a_i \le a$ ($a_i \ge a$), and for any $\varepsilon \in \boldsymbol{R}^{++}$ the inequality $| a - a_i | < \varepsilon$ is valid for almost all a_i such that $a_i \le a$ ($a_i \ge a$), i.e., there is such n that for any $i > n$, we have $| a - a_i | < \varepsilon$ for all such a_i.

A right limit a of l is denoted by $a = +\lim_{i\to\infty} a_i$, $+\lim_{i\to\infty} a_i = a$, $+\lim l = a$ or $a = +\lim l$ and a left limit b of l is denoted by $b = -\lim_{i\to\infty} a_i$, $-\lim_{i\to\infty} a_i = b$, $-\lim l = b$ or $b = -\lim l$.

There are sequences that do not converge but have both right and left limits. For instance, the sequence $l = \{ (-1)^i ; i = 1, 2, 3, \ldots \}$ does not have a limit but its right limit is 1 and its left limit is -1.

At the same time, there are bounded sequences that do not have right and left limits.

For instance, the sequence $l = \{ 1 + (1/i);\ i = 1, 2, 3, \ldots\}$ does not have left limits, while the sequence $l = \{ a_i = 1 - (1/i);\ i = 1, 2, 3, \ldots\}$ does not have right limits.

It is also possible to introduce infinite limits of sequences.

Definition 2.1.7. The element ∞ (- ∞) is called a limit of a sequence l (it is denoted by $\infty = \lim_{i\to\infty} a_i$, $\lim_{i\to\infty} a_i = \infty$, $\lim l = \infty$ or $\infty = \lim l$ and $-\infty = \lim_{i\to\infty} a_i$, $\lim_{i\to\infty} a_i = -\infty$, $\lim l = -\infty$ or $-\infty = \lim l$, respectively) if for any $c \in \boldsymbol{R}^{++}$ the inequality $a_i < c$ ($a_i < -c$) is valid for almost all a_i , i.e., there is such n that for any $i > n$, we have $a_i < c$ ($a_i < -c$).

For instance, $\lim_{i\to\infty} a_i = \infty$ when $a_i = 10i$ and $\lim_{i\to\infty} a_i = -\infty$ when $a_i = -i^2$.

Traditionally, a sequence is called divergent when its limit is equal to ∞ or $-\infty$.

2.1.2. General Properties of Limits

Theorem 2.1.1. A limit of a sequence is unique (if this limit exists).

Proof. Let us consider a sequence $l = \{a_i \in \boldsymbol{R};\ i = 1, 2, 3, \dots \}$ and assume that there are two numbers a and b that satisfy conditions from Definition 2.1.1. Then either $a - b = k > 0$ or $b - a = r > 0$. We consider only the first situation because the second situation is symmetric. By Definition 2.1.1, we have that for any $\varepsilon \in \boldsymbol{R}^{++}$, the inequality $| a - a_i | < \varepsilon$ is valid for almost all $i \in \omega$ and the inequality $| b - a_i | < \varepsilon$ is valid for almost all $i \in \omega$. Consequently, $a - b = a - a_i + a_i - b \leq | a - a_i | + | a_i - b | < 2\varepsilon$. Then taking ε smaller than one half of k, we come to a contradiction represented by the following sequence of inequalities:

$$k < a - b < 2\varepsilon < k$$

Theorem is proved.

By a similar argument, we prove the following result.

Theorem 2.1.2. A right (left) limit of a sequence is unique (if this limit exists).

Definitions directly imply three following results.

Proposition 2.1.2. If a is a right limit of a sequence l and b is a left limit of l, then $a > b$.

Proposition 2.1.3. A sequence converges if and only if it has both right and left limits and they coincide.

Lemma 2.1.1. A sequence all elements of which are equal to some number q converges to q.

Proposition 2.1.4. If $a = \lim l$ and $a > b$ ($a < c$), then $a_i > b$ ($a_i > c$) for almost all a_i from l.

Indeed, let $a - b = k$ and $\varepsilon = k/2$. Then by the definition of the limit, for almost all a_i from l, we have $|a - a_i | < \varepsilon$. This implies that almost all a_i from l are larger than $a - k/2$. Consequently, $a_i > b$ for almost all a_i from l. as $b = a - k$. The second part when $a < c$ is proved in similar way.

As a colorary, we get the following result.

Proposition 2.1.5. If $a = \lim l$ and $a > 0$, then $a_i > 0$ for almost all a_i from l.

Let us consider two sequences $l = \{a_i \in \boldsymbol{R};\ i = 1, 2, 3, \dots \}$ and $h = \{b_i \in \boldsymbol{R};\ i = 1, 2, 3, \dots \}$.

Theorem 2.1.3. If $a = \lim l$, $b = \lim h$ and $a_i \leq b_i$ for almost all $i = 1, 2, 3, \dots$, then:

a) if $0 \leq a_i$ for almost all $i = 1, 2, 3, \dots$, the sequence l is monotone, and the sequence h converges, then the sequence l also converges;

b) if $0 \leq a_i$ for almost all $i = 1, 2, 3, \ldots$ and the sequence l diverges for some r, then the sequence h also diverges;

c) $a = \lim l$ and $b = \lim h$ imply $a \leq b$.

Proof. a) If the sequence h converges, then the sequence l is bounded. By Theorem 2.1.8, any monotone bounded sequence converges.

b) If the sequence l diverges, then its elements go to infinity as they are all positive. As elements of the sequence h are larger than corresponding elements of the sequence l, then elements of the sequence h also go to infinity, i.e., the sequence h diverges.

c) If $a > b$, then $a > b + \varepsilon$ for some $\varepsilon > 0$ and by Proposition 2.1.4, $a_i > b + \varepsilon$ for almost all a_i from l. At the same time, almost all b_i from h are less than $b + \varepsilon$ because $b = \lim h$. This contradicts to the condition $a_i \leq b_i$ for almost all $i = 1, 2, 3, \ldots$ and shows that $a \leq b$.

Theorem is proved.

By a similar argument, we can prove the following result.

Proposition 2.1.6. If $a_i \leq b_i$ for almost all $i = 1, 2, 3, \ldots$, then $+\lim l \leq +\lim h$ and $-\lim l \leq -\lim h$.

Taking all elements in one of the sequences l or h equal to the same number, we derive from the Theorem 2.1.3 the following result.

Proposition 2.1.7. If $a = \lim l$ and $a_i > b$ ($a_i < c$) for almost all a_i from l, then $a \geq b$ ($a \leq c$).

Virtually, any course of the calculus (cf., for example, (Ribenboim, 1964; Fihtengoltz, 1955; Stewart, 2003)) contains the following classical result (Theorem 2.1.4).

Let $l = \{a_i \in \boldsymbol{R};\ i = 1, 2, 3, \ldots\}$ and $h = \{b_i \in \boldsymbol{R};\ i = 1, 2, 3, \ldots\}$. Then their sum $l + h$ is equal to the sequence $\{ a_i + b_i;\ i = 1, 2, 3, \ldots\}$, their difference $l - h$ is equal to the sequence $\{ a_i - b_i;\ i = 1, 2, 3, \ldots\}$ and their scalar product $l \cdot h$ is equal to the sequence $\{ a_i \cdot b_i;\ i = 1, 2, 3, \ldots\}$,. If $q \in \boldsymbol{R}$, then $ql = \{q\cdot a_i \in \boldsymbol{R};\ i = 1, 2, 3, \ldots\}$ and $q + l = \{q + a_i \in \boldsymbol{R};\ i = 1, 2, 3, \ldots\}$.

Theorem 2.1.4. If $a = \lim l$ and $b = \lim h$, then:

a) $a + b = \lim (l + h)$;
b) $a - b = \lim (l - h)$;
c) $qa = \lim (ql)$ for any $q \in \boldsymbol{R}$;
d) $q + a = \lim (q + l)$ for any $q \in \boldsymbol{R}$;
e) $a \cdot b = \lim (l \cdot h)$.

Proof. a) Let $a = \lim l$ and $b = \lim h$. Then by Definition 2.1.1, for any $\varepsilon \in \boldsymbol{R}^{++}$, there is such n that for any $i > n$, we have $| a - a_i | < \varepsilon/2$ and there is such m that for any $i > m$, we have $| b - b_i | < \varepsilon/2$. Taking $p = \max \{m, n\}$, we have $| a - a_i | < \varepsilon/2$ and $| b - b_i | < q + \varepsilon/2$ for any $i > p$. Consequently,

$$|(a + b) - (a_i + b_i)| = |(a - a_i) + (b - b_i)| \leq |(a - a_i)| + |(b - b_i)| < \varepsilon$$

for any $i > p$. By Definition 2.1.1, it means that $a + b = \lim(l + h)$ as $l + h = \{ a_i + b_i;\ i = 1, 2, 3, \ldots\}$.

b) Let $a = \lim l$ and $b = \lim h$. Then by Definition 2.1.1, for any $\varepsilon \in \boldsymbol{R}^{++}$, there is such n that for any $i > n$, we have $| a - a_i | < \varepsilon/2$ and there is such m that for any $i > m$, we have $| b - b_i | < \varepsilon/2$. Taking $p = \max \{m, n\}$, we have $| a - a_i | < \varepsilon$ and $| b - b_i | < \varepsilon/2$ for any $i > p$.

Consequently, $|(a - b) - (a_i - b_i)| = |(a - a_i) + (-(b - b_i))| \leq |(a - a_i)| + |(b - b_i)| < \varepsilon$ for any $i > p$. By Definition 2.1.1, it means that $a - b = \lim(l - h)$ as $l - h = \{ a_i - b_i; i = 1, 2, 3, \ldots\}$.

c) Let $a = \lim l$ and $k \in \boldsymbol{R}$. Then by Definition 2.1.1, for any $\varepsilon \in \boldsymbol{R}^{++}$, there is such n that for any $i > n$, we have $| a - a_i | < \varepsilon/k$. Consequently, $| ka - ka_i | = |k|\cdot| a - a_i| < |k|\cdot r + \varepsilon$ for any $i > n$. By Definition 2.1.1, it means that $ka = |k|\cdot r\text{-}\lim (kl)$ as $kl = \{k\cdot a_i ; i = 1, 2, 3, \ldots\}$.

Part (d) is a consequence of the part a) when all elements of the sequence h are equal to q.

Proof of (e) is similar to the proof of (a) because for sufficiently small $\varepsilon>0$, we have $\varepsilon^2<\varepsilon$.

Theorem is proved.

By a similar argument, we can prove the following result.

Theorem 2.1.5. If $a = +\lim l$ and $b = +\lim h$ ($a = -\lim l$ and $b = -\lim h$), then:

a) $a + b = +\lim (l + h)$; ($a + b = -\lim (l + h)$);
b) $qa = +\lim (ql)$ ($qa = -\lim (ql)$) for any $q \in \boldsymbol{R}$.
c) $q + a = +\lim (q + l)$ ($q + a = -\lim (q + l)$) for any $q \in \boldsymbol{R}$.

Remark 2.1.1. These properties of right and left limits are similar to the properties of limits. However, the statement

$a - b = +\lim (l - h)$; ($a - b = -\lim (l - h)$),

which is similar to part (b) of Theorem 2.1.4, is not true in general for right and left limits as the following example demonstrates.

Example 2.1.4. Let $h = \{-1 - 1/i ; i = 1, 2, 3, \ldots\}$ and $l = \{ 2 - 1/2i ; i = 1, 2, 3, \ldots\}$. Then $2 = -\lim l$ and $-1 = -\lim h$. At the same time, $l - h = \{ 3 + 1/2i ; i = 1, 2, 3, \ldots\}$ and the number $2 - (-1) = 3$ is not the left limit of the sequence $l - h$ because this sequence does not have elements less than 3.

Proposition 2.1.8. The following conditions are equivalent:

1. a sequence l does not converge;
2. some subsequence of l diverges;
3. the diameter $\mathrm{Diam}(\{a_i; a_i \in l\})$ is infinite.

We do not give here a proof, as this result is a direct consequence of its fuzzy counterpart (Proposition 2.2.1).

Definition 2.1.5. A sequence $l = \{a_i ; i = 1, 2, 3, \ldots\}$ is called *bounded above* if there is a number M such that

$$a_i < M \text{ for all } i = 1, 2, 3, \ldots$$

Definition 2.1.6. A sequence $l = \{a_i ; i = 1, 2, 3, \ldots\}$ is called *bounded below* if there is a number m such that

$$a_i > m \text{ for all } i = 1, 2, 3, \ldots$$

Definition 2.1.7. A sequence $l = \{a_i ; i = 1, 2, 3, \ldots\}$ is called *bounded* if it is a bounded both above and below.

Theorem 2.1.6. Any convergent sequence l is bounded.

Proof. Let us take a convergent sequence $l = \{a_i ; i = 1, 2, 3, \dots\}$ and assume that it is not bounded. A sequence can unbounded either from above or from below or from both sides. For convenience, we consider the first case. Two others are treated in a similar way. It means that for any number n, there are infinitely many elements a_i larger than n. We take an element b_1 from l that is larger than 1. Then we take an element b_2 from l that is larger than 2 and so on. Continuing this process, we obtain a subsequence of the sequence l that tends to infinity. By Proposition 2.1.1, the sequence l also diverges. This contradiction completes the proof.

However, there are bounded sequences that do not converge. An example of such a sequence is $l = \{ (-1)^i ; i = 1, 2, 3, \dots\}$.

Theorem 2.1.7 (the Bolzano-Weierstrass Theorem). Any bounded sequence (bounded infinite set of real numbers) has a convergent subsequence.

Proof. Let us take a bounded sequence $l = \{a_i ; i = 1, 2, 3, \dots\}$. Boundedness means that there is a number C such that $- C < a_i < C$ for all $i = 1, 2, 3, \dots$, i.e., all a_i belong to the interval $[- C, C]$. Then either $[0, C]$ contains infinitely many elements from l or $[- C, 0]$ contains infinitely many elements from l. We denote the interval with by I_1 and take an element b_1 from l that belongs to I_1. Then we divide I_1 into two equal parts. At least, one of them contains infinitely many elements from l. We denote the interval with by I_2 and take an element b_2 from l such that $b_2 \neq b_1$ and b_2 belongs to I_2. It is possible because I_2 contains infinitely many elements from l. Continuing this process, we obtain a sequence of nested intervals $I_1 , I_2 , \dots , I_n , \dots$ the length of which tends to zero and a sequence of numbers $h = \{b_n \in \boldsymbol{R}; n = 1, 2, 3, \dots\}$ such that $b_n \in I_n$ for all $n = 1, 2, 3, \dots$ By properties of real numbers (cf. Appendix B)), there is a point a that belongs to all intervals I_n. This point is the limit of the sequence h.

Indeed, given a number $\varepsilon \in \boldsymbol{R}^{++}$, there is $n \in \omega$ such that the length of I_n is less than ε. By construction of the sequence h, we have $b_i \in I_n$ for all $i > n$ As the point a also belongs to the interval I_n, we have $|a - b_i| < \varepsilon$ for all $i > n$.

Theorem is proved because the number ε is arbitrary and any infinite bounded set of real numbers contains a bounded sequence.

Theorem 2.1.8. A bounded monotone sequence l always converges.

Proof. By Theorem 2.1.7, a bounded sequence $l = \{a_i ; i = 1, 2, 3, \dots\}$ has a subsequense h that converges to some number a. Then for any $\varepsilon > 0$, if the inequality $| a - a_i | < \varepsilon$ is valid for almost all elements a_i from h, then the inequality $| a - a_i | < \varepsilon$ is valid for almost all elements a_i from l as l is a monotone sequence. Really, if the sequence l is increasing, then almost all elements of l are larger than or equal to any given element from l. If the sequence l is decreasing, then almost all elements of l are less than or equal to any given element from l. Thus, the sequence l converges to a.

Definition 2.1.1 implies the following result.

Proposition 2.1.9. If $a = \lim l$, then $a = \lim k$ for any subsequence k of l.

Theorem 2.1.9. Any sequence l has a monotone subsequence.

Proof. If l contains infinitely many members equal to the same number, then these members form a monotone subsequence of l. If l does not contain such elements, then it is possible to take a subsequence h of l such that all elements from h are distinct.

There are two options: either h is unbounded or bounded. In the first case, we assume, for convenience, that h is unbounded above. Then for each element there is a larger element and we can find a strictly monotone subsequence tending to ∞. When h is unbounded below, we can find a strictly monotone subsequence tending to $- \infty$.

In the second case, when h is bounded, there is a converging subsequence k of h. If $a = \lim k$, then either there are infinitely many elements of k larger than a or there are infinitely many elements of k less than a. Both situations are also possible for the subsequence k at the same time, i.e., k can have infinitely many elements less than a and infinitely many elements larger than a.

In the first case, for any element c from k that is larger than a, almost all elements from k are less than c because k converges to a. This allows us to find an infinite decreasing subsequence of k and thus, subsequence of l. Indeed, we take one element c_1, then there is an element c_2 smaller than c_1. Then there is an element c_3 smaller than c_2 and so on.

In the second case, for any element c from k that is smaller than a, almost all elements from k are larger than c. This allows us to find an infinite increasing subsequence of k and thus, subsequence of l.

Theorem is proved.

Note that not every sequence has a strictly monotone subsequence.

2.1.3. Convergence Tests and Criteria

The main criterion of sequence convergence is the Cauchy criterion (cf., for example, (Ribenboim, 1964; Fihtengoltz, 1955)).

Definition 2.1.8. A sequence $l = \{a_i ; i = 1, 2, 3, \ldots\}$ of real numbers is called *fundamental* or a *Cauchy sequence* if for any $\varepsilon \in \boldsymbol{R}^{++}$, there is $n \in \omega$ such that for any $i, j \geq n$ we have $| a_j - a_i | < \varepsilon$.

Theorem 2.1.10 (the Cauchy Criterion for sequences). A sequence l converges if and only if it is fundamental.

Proof. Necessity. Let $a = \lim l$ and $\varepsilon \in \boldsymbol{R}^{++}$. Then by definition, there is a number $n \in \omega$ such that for any $i > n$, we have $| a - a_i | \leq \varepsilon/2$. Consequently, for any $i, j > n$, we have

$$| a_i - a_j | \leq | a - a_i | + | a - a_j | \leq 2(\varepsilon/2) < \varepsilon.$$

Thus, l is a fundamental sequence.

Sufficiency. Let l be a Cauchy sequence and $k_m = 1/m$. If we fix some number m, then there is a number n, which is dependent on m and thus, denoted by $n(m)$, such that for all $i, j > n(m)$, we have $| a_i - a_j | < 1/m$. That is, all points a_i with $i > n(m)$ belong to a closed interval I_m. Really, let us put $\mathrm{T}_m = \{a_i ; i > n(m) \}$, $b = \sup \mathrm{T}_m$, and $c = \inf \mathrm{T}_m$. Then all a_i belong to the interval $I_m = [c, b]$ for all $i > n(m)$.

Let us estimate the length of this interval I_m. Suppose that $| b - c | > 1/m$. It means that for some $h \in \boldsymbol{R}^+$, we have $| b - c | > 1/m + h$. At the same time, there are such elements a_p and a_q, for which $p, q > n(m)$, $| b - a_p | \leq (1/3)\cdot h$, and $| c - a_q | \leq (1/3)\cdot h$ because b is the supremum and c is the infimum of all elements from the set T_m. Consequently, for these a_p and a_q, we have

$$| b - c | \leq | b - a_p | + | a_p - a_q | + |a_q - c | <$$

$$(1/3)\cdot h + 1/m + (1/3)\cdot h = 1/m + (2/3)h < 1/m + h$$

This contradicts our supposition that $| b - c | > 1/m + h$. Thus, the length of $I_m = [c, b]$ is not larger than $1/m$.

We can choose these intervals I_m so that the inclusion $I_{m+1} \subseteq I_m$ will be valid for all $m \in \omega$. In such a way, we obtain a sequence of nested closed intervals $\{ I_m; m \in \omega\}$. By the properties of the space $\boldsymbol{R}$ the intersection $I = \bigcap_{m=1}^{\infty} I_m$ is non-empty. As the length of intervals I_m tends to zero, the intersection I consists of one point d and thus, the sequence l converges to d as all a_i belong to the interval $I_m = [c, b]$ for all $i > n(m)$.

Theorem is proved.

Remark 2.1.2. Bolzano formulated this criterion of convergence in 1817 and Cauchy independently formulated and rigorously proved it in 1823.

Let us consider three sequences $l = \{a_i \in \boldsymbol{R}; i = 1, 2, 3, \dots \}$, $k = \{c_i \in \boldsymbol{R}; i = 1, 2, 3, \dots \}$, and $h = \{b_i \in \boldsymbol{R}; i = 1, 2, 3, \dots \}$.

Theorem 2.1.11 (the **Squeeze Theorem**). If $a_i \leq b_i \leq c_i$ for almost all $i = 1, 2, 3, \dots$ and the sequences l and k converge to the same number a, then the sequence h also converges to a.

We do not give here a proof, as this result is a direct consequence of its fuzzy counterpart.

Theorem 2.1.11 allows us to prove the following result.

Proposition 2.1.10. If $l = \{a_i \in \boldsymbol{R}; i = 1, 2, 3, \dots \}$, $h = \{b_i \in \boldsymbol{R}; i = 1, 2, 3, \dots \}$, and $0 \leq | a_i | \leq | b_i |$ for almost all $i = 1, 2, 3, \dots$, then:

a. if the sequence l diverges and tends to infinity, then the sequence h diverges and tends to infinity;
b. if the sequences h converges to 0, then the sequence l also converges to 0.

Indeed, if the sequence l diverges and tends to infinity, then absolute values $| a_i |$ tend to ∞. Consequently, absolute values $| b_i |$ tend to ∞. It means that the sequence h diverges and tends to infinity. Note that any of these seuences can tend both to ∞ and $-\infty$.

If the sequences h converges to 0, then absolute values $| b_i |$ converge to 0. By Theorem 2.1.11, absolute values $| a_i |$ converge to 0. Consequently, the sequence l also converges to 0.

Remark 2.1.3. However, it is possible that $0 \leq a_i \leq b_i$ for almost all $i = 1, 2, 3, \dots$, the sequences h converges, but the sequences l diverges. Really, the sequence $l = \{ ½ + (-1)^i ¼; i = 1, 2, 3, \dots \}$ diverges, while the sequence $h = \{ 1 + (1/i); i = 1, 2, 3, \dots \}$ converges and all elements of h are larger than any element of l.

Proposition 2.1.11. If $0 \leq a_i$, b_i and $a_i/b_i \leq c$ for almost all $i = 1, 2, 3, \dots$ and some positive number c, then:

a. if the sequences l diverges and tends to infinity, then sequences h diverges and tends to infinity;
b. if the sequences h converges to 0, then sequences l also converges to 0.

Indeed, if $a_i/b_i \leq c$ for almost all $i = 1, 2, 3, \dots$, then $0 \leq a_i \leq cb_i$. Thus Proposition 2.1.11 follows from Proposition 2.1.10 as the sequence $l = \{a_i \in \boldsymbol{R}; i = 1, 2, 3, \dots \}$ converges if and only if the sequence $cl = \{ ca_i \in \boldsymbol{R}; i = 1, 2, 3, \dots \}$ converges.

2.2. ABSOLUTE FUZZY LIMITS OF SEQUENCES

Don't limit yourself.
Many people limit themselves to what they think they can do.
You can go as far as your mind lets you.
What you believe, remember, you can achieve.
Mary Kay Ash (1918-2001)

2.2.1. General Concepts

Let $r \in \boldsymbol{R}^+$ and $l = \{a_i ; i = 1, 2, 3, \dots\}$ be a sequence of real numbers.

Definition 2.2.1. A number a is called an *r-limit of a sequence l* (it is denoted by $a = r\text{-}\lim_{i\to\infty} a_i$ or $a = r\text{-}\lim l$) if for any $\varepsilon \in \boldsymbol{R}^{++}$, the inequality $| a - a_i | < r + \varepsilon$ is valid for almost all a_i , i.e., there is such n that for any $i > n$, we have $| a - a_i | < r + \varepsilon$.

Informally, a is an r-limit of a sequence l if for an arbitrarily small ε, the distance between a and all but a finite number of elements from l is smaller than $r + \varepsilon$. In other words, a number a is an r-limit of a sequence l if for any $\varepsilon \in \boldsymbol{R}^{++}$ almost all a_i belong to the interval $(a - r - \varepsilon, a + r + \varepsilon)$.

Definition 2.2.2. A sequence l that has an r-limit is called *r-convergent* and it is said that *l r-converges* to its r-limit a.

It is denoted by $l \to_r a$.

Thus, r becomes a measure of convergence for l.

Example 2.2.1. Let $l = \{1/i; i = 1, 2, 3, \dots\}$. Then 1 and -1 are 1-limits of l; $0 = 0\text{-}\lim l$; and ½ is a (½)-limit of l, but 1 is not a (½)-limit of l. Thus, the sequence l 1-converges and 0-converges.

Example 2.2.2. Let $l = \{(-1)^i; i = 1, 2, 3, \dots\}$.Then this sequence does not have a limit. At the same time, 1 and -1 are 2-limits of l because all elements from this sequence starting with the second one are less than $-1 - \varepsilon$ and smaller than $1 + \varepsilon$ for any positive number ε. For the same reason, 0 is a 1-limit of l, but 1 is not a 1-limit of l as this sequence has infinitely many elements that are smaller than $1 - 1 - ¼ = - ¼$. Thus, the sequence l 2-converges and 1-converges, but does not 0-converge.

Example 2.2.3. Let us consider sequences $l = \{1 + 1/i; i = 1, 2, 3, \dots\}$, $h = \{1 + (-1)^i; i = 1, 2, 3, \dots\}$, and $k = \{1 + [(1 - i)/i]^i ; i = 1, 2, 3, \dots\}$. Sequence l has the conventional limit equal to 1 and many fuzzy limits (e.g., 0, 0.5, 2 are 1-limits of l). Sequence h does not have the conventional limit but has different fuzzy limits (e.g., 0 is a 1-limit of h, while 1, -1, and ½ are 2-limits of h). Sequence k does not have the conventional limit but has a variety of fuzzy limits (e.g., 1 is a 1-limit of k, while 2, 0, 1.5, 1.7, and 0.5 are 2-limits of k).

Thus, we see that many sequences that do not have the conventional limit have lots of fuzzy limits.

Remark 2.2.1. In the denotation $a = r\text{-}\lim_{i\to\infty} a_i$ or $a = r\text{-}\lim l$, the symbol "=" means "is", that is, the number a is an r-limit of a sequence l. However, it is incorrect to write $r\text{-}\lim_{i\to\infty} a_i = a$ because, in a general case, an r-limit of a sequence is not unique. In a similar way, it is possible to write "*Gone with the Wind*" is a novel or the novel is "*Gone with the Wind*", but it is incorrect to write a novel is "*Gone with the Wind*".

When $r = 0$, r-limit coincides with the conventional limit of a sequence as the following result demonstrates.

Lemma 2.2.1. For any sequence l, we have $a = \lim l$ if and only if $a = 0\text{-}\lim l$.

In other words, the sequence l converges if and only if it 0-converges.

Indeed, if we take $r = 0$ in Definition 2.2.1 of an r-limit, we get Definition 2.1.1 of the conventional limit.

This result shows that the concept of an r-limit is a natural extension of the concept of a (conventional) limit and r-convergence is a natural extension of the concept of a (conventional) convergence. We can see that the concepts of an r-limit and r-convergence essentually extend the conventional constructions of a limit and convergence (cf. Examples 2.2.2 and 2.2.3).

Definition 2.2.3. a) A number a is called a *fuzzy limit* of a sequence l if it is an r-limit of l for some $r \in \boldsymbol{R}^+$.

b) a sequence l *fuzzy converges* and is called *fuzzy convergent* if it has a fuzzy limit.

Some readers who are not very familiar with the initial ideas of fuzzy set approach in mathematics may wonder why r-limits are called fuzzy limits. There are, at least, three reasons for this. First, because this concept (of a fuzzy limit) introduces gradations (gradual values) in the concept of an ordinary (crisp) limit. Second, the number r in an r-limit gives only some (very often it is fuzzy) estimation to what extent a point may be called a limit of sequence. It is possible that an r-limit of a sequence is also a q-limit of the same sequence for $q < r$. Third, the concept of r-limit generates a fuzzy limit set of a sequence (cf. Section 2.5).

Remark 2.2.2. There are informal definitions of a limit and it is interesting that in some cases such informal definitions are better formalized by the construction of a fuzzy limit than by the construction of the classical limit. For instance, in the book (Ross, 1996), we have the following description:

The "limit" of a sequence $l = \{a_i ; i = 1, 2, 3, \dots\}$ *is a real number a that the values* a_i *are close to the number a for large values of i.*

Mathematically interpreted, "close" means that the values a_i belong to some small neighborhood of a for large values of i. Consequently, formalizing this notion, we come to the definition of r-limit for some small number r.

Example 2.2.4. Let us consider sequences $l = \{1 + 1/i; i = 1, 2, 3, \dots\}$, $h = \{1 + (-1)^i; i = 1, 2, 3, \dots\}$, $k = \{1 + [(1 - i)/i]^i ; i = 1, 2, 3, \dots\}$, and $u = \{1 + i; i = 1, 2, 3, \dots\}$. The sequence l has the conventional limit equal to 1 and many fuzzy limits (e.g., 0, 0.5, 2 are 1-limits of l). Thus, the sequence l converges in the classical sense and r-converges for any number r. The sequence h does not have the conventional limit but has different fuzzy limits (e.g., 0 is a 1-limit of h, while 1, -1, and ½ are 2-limits of h). The sequence k does not have the conventional limit but has a variety of fuzzy limits (e.g., 1 is a 1-limit of k, while 2, 0, 1.5, 1.7, and 0.5 are 2-limits of k). The equence u has neither conventional nor fuzzy limits

If we look at divergent sequences, we see that they diverge with a different pattern. For instance, divergence of the sequence $l = \{ (-1)^n ; n = 1, 2, 3, \dots \}$ is dissimilar from divergence of the sequence $h = \{ n ; n = 1, 2, 3, \dots \}$. An important peculiarity of the parameter r introduced in Definition 2.2.1 for characterization of fuzzy limits is that it measures not only convergence but also divergence.

Definition 2.2.4. A sequence l *r-diverges* and is called *r-divergent* if it does not have any r-limit.

Example 2.2.5. Let us consider the sequence $l = \{(-1)^i; i = 1, 2, 3, \ldots\}$.This sequence 1-converges but (½)-diverges and (1/3)-diverges.

In addition to fuzzy limits, one more construction also plays an important role in neoclassical analysis and scalable topology (Burgin, 2004a; 2005a; 2006). This is the construction of partial fuzzy limits or fuzzy accumulation points.

Definition 2.2.5. A number a is called a *partial r-limit*, *subsequential r-limit* or *r-accumulation point* of a sequence l (it is denoted by $a = r\text{-plim } l$) if for any $\varepsilon \in \boldsymbol{R}^{++}$ the inequality $| a - a_i | < r + \varepsilon$ is valid for infinitely many elements a_i from l.

When $r = 0$, partial r-limits coincides with accumulation points of a sequence as the following result demonstrates.

Lemma 2.2.2. For any sequence l, a number a is an accumulation point of l if and only if $a = 0\text{-plim } l$.

Examples of partial r-limits are r-supremums and r-infimums of sequences.

Let $r \in \boldsymbol{R}^+$ and X is a set of real numbers.

Definition 2.2.6. The *r-supremum* of X (denoted by r-sup X) is any number a (when such a number exists) that $a + r$ is larger than or equal to any number in X and for any $\varepsilon \in \boldsymbol{R}^{++}$, there is an element x from X such that $| a - x| < r + \varepsilon$. When there is no such a number, the r-supremum $r\text{-sup } X = \infty$.

Example 2.2.6. $¾ = ¼\text{-sup } \{ 1 - 1/i ; i = 1, 2, 3, \ldots \}$ and $¾ = ½\text{-sup } \{ 1 + 1/i ; i = 1, 2, 3, \ldots n\}$.

Definition 2.2.7. The *r-infimum* of X (denoted by r-inf X) is the number a (when such a number exists) that $a - r$ is smaller than or equal to any number in X and for any $\varepsilon \in \boldsymbol{R}^{++}$, there is an element x from X such that $| a - x| < r + \varepsilon$. When there is no such a number, the r-infimum $r\text{-inf } X = -\infty$.

Example 2.2.7. $¾ = ¼\text{-inf } \{ 1 + 1/i ; i = 1, 2, 3, \ldots \}$ and $¾ = ½\text{-inf } \{ 1 - 1/i ; i = 1, 2, 3, \ldots n\}$.

When $r = 0$, r-supremum coincides with the supremum and r-infimum coincides with the infimum of the same set as the following result demonstrates.

Lemma 2.2.3. For any sequence l, we have $a = \sup X$ ($b = \inf X$) if and only if $a = 0\text{-sup } X$ ($b = 0\text{-inf } X$).

Important special cases of partial limits are one-sided limits.

Definition 2.2.8. A number a is called a *left* (*right*) *r-limit* or *r-limit from the left* (*right*) of a sequence l if there are infinitely many a_i from l such that $a_i \leq a + r$ ($a_i \geq a - r$), for any $\varepsilon \in \boldsymbol{R}^{++}$ the inequality $| a - a_i | < r + \varepsilon$ is valid for almost all a_i such that $a_i \leq a + r$ ($a_i \geq a - r$), i.e., there is such n that for any $i > n$, we have $| a - a_i | < r + \varepsilon$ for all such a_i.

A right limit a of l is denoted by $a = +r\text{-lim}_{i\to\infty} a_i$ or $a = +r\text{-lim } l$ and a left limit b of l is denoted by $b = -r\text{-lim}_{i\to\infty} a_i$, $-r\text{-lim}_{i\to\infty} a_i = b$, $-r\text{-lim } l = b$ or $b = -r\text{-lim } l$.

There are sequences that do not r-converge but have both right and left r-limits. For instance, the sequence $l = \{ (-1)^i \cdot 3; i = 1, 2, 3, \ldots\}$ does not have a 1-limit but 3 is its right 1-limit and -3 is its left 1-limit. Moreover, number 3 is its right 0-limit and -3 is its left 0-limit.

Moreover, here are sequences that do not r-converge, do not have right and left limits, but have right and left r-limits.

Example 2.2.8. Let us consider the sequence $l = \{ a_{2i} = 1 + (1/2i), a_{2i-1} = 2i - 1; i = 1, 2, 3, \ldots\}$. Then this sequence does not r-converge, does not have right and left limits, but has left r-limits for any $r > 0$. For instance, 0.999 is a –0.01-limit of l.

Example 2.2.9. Let us consider the sequence $l = \{ a_{2i} = 1 - (1/2i), a_{2i-1} = 1 - 2i ; i = 1, 2, 3, \ldots\}$. Then this sequence does not r-converge, does not have right and left limits, but has right r-limits for any $r > 0$.

It is also possible to define infinite fuzzy limits. Let $r \in \boldsymbol{R}^{++}$.

Definition 2.2.9. ∞ $(-\infty)$ is an *infinite r-limit* of l if almost all elements a_i are bigger (less) than $1/r$ $(- 1/r)$.

Example 2.2.10. ∞ is an infinite 0.1-limit of the sequence $l = \{10 + 1/i; i = 1, 2, 3, \ldots\}$, infinite k-limit of the sequence $h = \{i ; i = 1, 2, 3, \ldots\}$ for any positive k, and infinite ½-limit of the sequence $k = \{2 + (i - 1)/i ; i = 1, 2, 3, \ldots\}$.

Remark 2.2.3. In the domain of real numbers, there are only two infinite limits. At the same time, in the realm of real hypernumbers (Burgin, 2002; 2004b), there are many different infinite hypernumbers, which can be treated as limits of sequences of real numbers.

Remark 2.2.4. Specific kinds of infinite limits and their finite approximations are utilized in computer science for program verification methods that are based on a general notion of approximate satisfaction of properties (Nitsche and Ochsenschläger, 1996; Ochsenschläger, Repp and Rieke, 1998; 2000). In the context of neoclassical analysis, approximate satisfaction of properties means satisfaction of fuzzy properties and infinite limits of system behaviors can be represented with some measure of fuzziness by finite system behaviors. These finite approximations are fuzzy limits that substitute unachievable ideal limits.

2.2.2. General Properties of Fuzzy Limits

Let $r, q \in \boldsymbol{R}^{+}$ and $l = \{a_i ; i = 1, 2, 3, \ldots\}$ be a sequence of real numbers.

Lemma 2.2.4. If $a = r\text{-lim } l$, then $a = q\text{-lim } l$ for any $q > r$.

Indeed, if the inequality $| a - a_i | < r + \varepsilon$ is valid for almost all a_i from l, then the inequality $| a - a_i | < q + \varepsilon$ is also valid for almost all a_i from l.

Theorem 2.2.1. If $a = r$-lim and $a > b + r$, then $a_i > b$ for almost all a_i from l.

Proof. Let $a = r\text{-lim } l$, and $a > b + r$, then for some positive number p, we have $a - b > r + p$. Let us take $\varepsilon = ½p$. Then the inequality $| a - a_i | < r + \varepsilon$ is valid for almost all a_i from l. Consequently, we have $a_i - b = a_i - a + a - b \geq - | a_i - a | + a - b > - r - \varepsilon + (a - b) > - r - \varepsilon + r + p = p - \varepsilon > 0$ for almost all a_i from l. Thus, $a_i > b$ for almost all a_i from l.

Theorem is proved.

Taking $r = 0$ in Theorem 2.2.1, we have the following classical result.

Corollary 2.2.1 (cf., for example, (Ribenboim, 1964; Fihtengoltz, 1955)). If $a = \lim l$ and $a > b$, then $a_i > b$ for almost all a_i from l.

Taking $b = 0$ in Corollary 2.2.1, we have the following classical result.

Corollary 2.2.2 (cf., for example, (Ribenboim, 1964; Fihtengoltz, 1955)). If $a = \lim l$ and $a > 0$, then $a_i > 0$ for almost all a_i from l.

Corollary 2.2.3. If $a_i \leq q$ for almost all a_i from l and $a = r\text{-lim } l$, then $a \leq q+r$.

Proof. When $a = r\text{-lim } l$ and $a > q+ r$, by Theorem 2.2.1, we have $a_i > q$ for almost all a_i from l. As by the initial assumption of the corollary, $a_i \leq q$ for almost all a_i from l, we cannot have $a > q+ r$. The by the principle of excluded middle, $a \leq q+ r$.

Corollary 2.2.3 is proved.

Taking $r = 0$ in Corollary 2.2.3, we have the following classical result.

Corollary 2.2.4 (cf., for example, (Ribenboim, 1964; Fihtengoltz, 1955)). If $a_i \leq q$ for almost all a_i from l and $a = \lim l$, then $a \leq q$.

Definition 2.2.1 implies the following result.

Lemma 2.2.5. If $a = q\text{-}\lim l$, then $a = q\text{-}\lim k$ for any subsequence k of l.

Let $r \in \mathbf{R}^+$, $l = \{a_i \in \mathbf{R}; i = 1, 2, 3, \ldots\}$, $h = \{b_i \in \mathbf{R}; i = 1, 2, 3, \ldots\}$, $k = \{c_i \in \mathbf{R}; i = 1, 2, 3, \ldots\}$, and the sequence l is the disjoint union of the sequences h and k, that is, elements of l are taken from h and k without repetition. For example, if $a_{2i} = b_i$ and $a_{2i+1} = c_i$ for all $i = 1, 2, 3, \ldots\}$, then l is the disjoint union of h and k.

Lemma 2.2.6. $a = r\text{-}\lim l$ if and only if $a = r\text{-}\lim h$ and $a = r\text{-}\lim k$.

Proof. Necessity follows from Lemma 2.2.3. Sufficiency follows from the fact that if almost all elements from h satisfy some condition P and almost all elements from k satisfy the same condition P, then almost all elements from l satisfy the condition P.

Lemma is proved.

Let $r \in \mathbf{R}^+$, $l = \{a_i \in \mathbf{R}; i = 1, 2, 3, \ldots\}$, and $h = \{a_i \in \mathbf{R}; i = m, m + 1, m + 2, m + 3, \ldots\}$. Then Definition 2.2.1 implies the following result.

Lemma 2.2.7. $a = r\text{-}\lim l$ if and only if $a = r\text{-}\lim h$.

Informally, it means that changing a finite number of elements of a sequence does not change a fuzzy limit of this sequence (if such fuzzy limit exists) because with respect to a fuzzy limit, we can ignore any finite part of this sequence. As limits are particular cases of fuzzy limits, this is also true for limits, i.e., changing a finite number of elements of a sequence does not change its limit (if such a limit exists).

Proposition 2.2.1. The following conditions are equivalent:

1) a sequence l has no finite fuzzy limits;
2) some subsequence of l diverges;
3) some subsequence of l has no finite fuzzy limits;
4) the diameter $\mathrm{Diam}(\{a_i; a_i \in l\})$ of the sequence l is infinite.

Proof. 1) $\to$ 2). By Definition 2.2.2, if a sequence l has no finite fuzzy limits, then for any number $n = 1, 2, 3, \ldots$, there is an element a_{in} from l such that $a_{in} > n$. Then the subsequence $h = \{ a_{in} \in \mathbf{R}; i_n = 1, 2, 3, \ldots\}$ diverges.

2) $\to$ 3). If a sequence diverges, then by Definitions 2.2.1 and 2.2.2, it has no finite fuzzy limits.

3) $\to$ 1) by Lemma 2.2.5.

4) $\to$ 2). When the diameter $\mathrm{Diam}(\{a_i; a_i \in l\})$ of the sequence l is infinite, then it is possible to build a fuzzy divergent subsequence in l.

2) $\to$ 4). As the diameter of the sequence l is larger than or equal to the diameter of any its subsequence and the diameter of a fuzzy divergent sequence is infinite, the diameter $\mathrm{Diam}(\{a_i ; a_i \in l\})$ of the sequence l is also infinite.

Proposition is proved.

Theorem 2.2.2. A sequence l fuzzy converges if and only if it is bounded.

Proof. Necessity. Let us take a fuzzy convergent sequence $l = \{a_i ; i = 1, 2, 3, \ldots\}$ and assume that it is not bounded. A sequence can be unbounded either from above or from below or from both sides. For convenience, we consider the first case. Two others are treated in a similar way. It means that for any number n, there are infinitely many elements a_i larger than

n. We take an element b_1 from l that is larger than 1. Then we take an element b_2 from l that is larger than 2 and so on. Continuing this process, we obtain a subsequence of the sequence l that tends to infinity. By Proposition 2.2.1, the sequence l has no fuzzy limits. This contradiction completes the proof.

Sufficiency. Let us consider a bounded sequence $l = \{a_i ; i = 1, 2, 3, \dots\}$. Then there are such numbers c and d such that $c < a_i < d$ for all $i = 1, 2, 3, \dots$. In this case, the number $b = c + ½(d - c)$ is an r-limit of the sequence l where $r = ½(d - c)$, that is, the sequence l fuzzy converges.

Theorem is proved.

Theorem 2.2.2 gives a criterion for boundedness of a sequence, while classical results (cf. Theorem 2.1.6) give only sufficient conditions (Ribenboim, 1964; Fihtengoltz, 1955). This result shows close ties between fuzzy convergent sequences and bounded sequences. However, the theory of fuzzy limits cannot be reduced to the theory of bounded sequences for three reasons. First, there is no theory of bounded sequences. Second, fuzzy limits give characterization of fuzzy converging sequences (an area of their density), while bounded sequences do not have such a characteristic. Third, the theory of fuzzy limits differentiates the scale of fuzzy convergence by the parameter r as a measure of convergence.

It is interesting to remark that Theorem 2.2.2 incorporates a fuzzy version of the Bolzano-Weierstrass theorem (cf. Section 2.1).

Theorems 2.1.7 and 2.2.2 imply the following result.

Corollary 2.2.5. Any fuzzy convergent sequence has a convergent subsequence.

Proof. If l is a fuzzy convergent sequence, then it is bounded by Theorem 2.2.2. By Theorem 2.1.7, any bounded sequence has a convergent subsequence. Thus, l has a convergent subsequence. Corollary is proved.

Let $l = \{a_i \in \mathbf{R}; i = 1, 2, 3, \dots\}$ be a bounded sequence.

Theorem 2.2.3. If for any convergent subsequence k of l, we have $a = r\text{-lim } k$, then $a = r\text{-lim } l$.

Proof. Let us assume that the condition of the theorem is satisfied, but $a \neq r\text{-lim } l$. Then there is $\varepsilon \in \mathbf{R}^{++}$ such that for infinitely many elements a_i, we have $| a - a_i | > r + \varepsilon$. Let us take these elements $a_{i1}, a_{i2}, \dots, a_{in}, \dots$. According to our assumption, the sequence $k = \{a_{in} \in \mathbf{R}; i = 1, 2, 3, \dots\}$ is bounded, as a subsequence of a bounded sequence. Consequently, by Theorem 2.1.7, h has a convergent subsequence $h = \{a_{in} \in \mathbf{R}; i = 1, 2, 3, \dots\}$. If $d = \lim k$, then $| a - d | \geq r + \varepsilon$ as for all a_i, we have $| a - a_i | > r + \varepsilon$. According to Definition 2.2.1, the point a is not an r-limit of the sequence l. This contradicts our assumption and by the principle of excluded middle, concludes the proof.

Remark 2.2.3. Not all properties of ordinary limits are the same as properties of fuzzy limits. For example, an r-limit may be not unique (cf., Examples 2.2.1-2.2.3). In the same way, for ordinary (exact) limits, we have the following result: if $a_i < b_i$ for almost all $i \in \omega$ and $a = \lim a_i$, $b = \lim b_i$, then $a \leq b$. For fuzzy limits, the resulting inequality is not always true as the following example demonstrates.

Example 2.2.11. Let us consider two sequences $l = \{ a_n = 1/n ; n = 1, 2, 3, \dots \}$ and $h = \{ b_n = -1/n ; n = 1, 2, 3, \dots \}$. In these sequences, $a_n > b_n$ for all $n = 1, 2, 3, \dots$. At the same time, $½ = 1\text{-lim } h$ and $-½ = 1\text{-lim } l$, i.e., a fuzzy limit of l is smaller than a fuzzy limit of h.

However, it is possible to prove that fuzzy limits have many properties similar to corresponding features of conventional limits.

Let $r, q \in \boldsymbol{R}^{+}$, $l = \{a_i \in \boldsymbol{R}; i = 1, 2, 3, \ldots\}$, and $h = \{b_i \in \boldsymbol{R}; i = 1, 2, 3, \ldots\}$.

Theorem 2.2.4. If $b_i - a_i > r + q$ for almost all $i \in \omega$, $a = r\text{-lim } l$, and $b = q\text{-lim } h$, then $b \geq a$.

Proof. Let $a = r\text{-lim } l$ and $b = q\text{-lim } h$. By Definition 2.2.1, for any $\varepsilon \in \boldsymbol{R}^{++}$, the inequality $|\, a - a_i \,| < r + \varepsilon$ is valid for almost all $i \in \omega$ and the inequality $|\, b - b_i \,| < q + \varepsilon$ is valid for almost all $i \in \omega$. Then $a_i - r - \varepsilon < a < a_i + r + \varepsilon$ and $b_i - q - \varepsilon < b < b_i + q + \varepsilon$ for almost all $i \in \omega$. Consequently,

$$a - b < (a_i + r + \varepsilon) - (b_i - q - \varepsilon) = (a_i - b_i) + r + q + 2\varepsilon.$$

As $b_i - a_i > r + q$, we have $a_i - b_i < -r - q$. Thus, $a - b < -r - q + r + q + 2\varepsilon = 2\varepsilon$ where ε is an arbitrary small number.

Let us assume that $b < a$. Then $a - b = k > 0$. Then taking ε smaller than one half of k, we come to a contradiction.

Theorem is proved.

As a direct corollary (when $r = q = 0$), Theorem 2.2.4 gives such a classical result as Theorem 2.1.3.c.

Proposition 2.2.2. If $a_i \leq b_i$ for almost all $i \in \omega$, $a = r\text{-lim } l$, and $b = q\text{-lim } h$, then $b + q + r \geq a$.

Proof. Let us consider the difference $a - b$. As $a = r\text{-lim } l$ and $b = q\text{-lim } h$, then by Definition 2.2.1, for any $\varepsilon \in \boldsymbol{R}^{++}$, the inequality $|\, a - a_i \,| < r + \varepsilon$ is valid for almost all $i \in \omega$ and the inequality $|\, b - b_i \,| < q + \varepsilon$ is valid for almost all $i \in \omega$. Consequently, for almost all $i \in \omega$, properties of operations with real numbers give us

$$\begin{gathered} a - b = a - a_i + a_i - b_i + b_i - b = \\ (a - a_i) + (a_i - b_i) + (b_i - b) \leq \\ r + \varepsilon + q + \varepsilon = r + q + 2\varepsilon \end{gathered}$$

because $a_i - b_i \leq 0$ for almost all $i \in \omega$. As this is true for any $\varepsilon > 0$, this gives us $a - b \leq r + q$ and $b + q + r \geq a$.

Proposition is proved.

Proposition 2.2.3. A monotone sequence l r-converges for some $r > 0$ if and only if it converges in the classical sense.

Proof. The condition to be convergent is sufficient for r-convergence as, by Lemma 2.2.2, any convergent sequence r-converges for all $r > 0$. The condition is necessary as, by Theorem 2.2.2, any r-convergent sequence is bounded and any bounded monotone sequence l converges by Theorem 2.1.8.

Proposition is proved.

However, this result does not mean that for monotone sequences, we do not need fuzzy limits as the classical limit always exists. We often are able to find exactly only some approximation to the classical limit and this approximation is a fuzzy limit. For instance, π is the limit of the monotone sequence $\{ a_n ; n = 1, 2, 3, \ldots \}$ where a_n is half of the perimeter of a regular polygon with n sides inscribed in a circle with the unit radius. As we know, it is possible to compute number π with the precision 10^{-6} or 10^{-9}, but it is impossible to compute number π with absolute precision.

Definitions imply the following result.

Proposition 2.2.4. If for some $r \in \boldsymbol{R}^{++}$, we have $a - r \leq a_i \leq a + r$ for almost all a_i, then $a = r\text{-lim } l$.

Lemma 2.2.8. If $a = r\text{-lim } l$ and $|b - a| = p$, then $b = q\text{-lim } l$ where $q = p + r$.

Indeed, if for any $\varepsilon \in \boldsymbol{R}^{++}$, the inequality $| a - a_i | < r + \varepsilon$ is valid for almost all a_i, then the inequality $| b - a_i | < q + \varepsilon$ is valid for almost all a_i as $| b - a_i | \leq | b - a | + | a - a_i |$. This means that $b = q\text{-lim } l$.

Let $r \in \boldsymbol{R}^{+}$, $l = \{a_i \in \boldsymbol{R}; i = 1, 2, 3, \ldots \}$, $h = \{b_i \in \boldsymbol{R}; i = 1, 2, 3, \ldots \}$, $k = \{c_i \in \boldsymbol{R}; i = 1, 2, 3, \ldots \}$, and the sequence l is the disjoint union of the sequences h and k, that is, elements of l are taken from h and k without repetition. For example, if $a_{2i} = b_i$ and $a_{2i+1} = c_i$ for all $i = 1, 2, 3, \ldots$ then l is the disjoint union of h and k.

Lemma 2.2.9. A number $a = r\text{-lim } l$ if and only if $a = r\text{-lim } h$ and $a = r\text{-lim } k$.

Lemma 2.2.10. ∞ $(-\infty)$ is the limit (in the classical sense, cf. Section 2.1 and (Ribenboim, 1964; Fihtengoltz, 1955)) of l if and only if it is an r-limit of l for any $r > 0$.

Lemma 2.2.11. ∞ $(-\infty)$ is an r-limit of $l = \{a_i \in \boldsymbol{R}; i = 1, 2, 3, \ldots\}$ if and only if 0 is a r-limit of the inverse sequence $h = \{a_i^{-1}; i = 1, 2, 3, \ldots\}$.

Lemma 2.2.12. If ∞ is an r-limit of l, then any q-limit of l is bigger than $(1/r) - q$.

Indeed, if $b = q\text{-lim } l$ and $b < (1/r) - q$, then there are infinitely many elements from l that are less than $1/r$. This contradicts the condition that $\infty = r\text{-lim } l$ (cf. Definition 2.2.9), and concludes the proof of Lemma 2.2.12.

Let $0 < r < 1$.

Corollary 2.2.6. If ∞ is an r-limit of l, then any r-limit of l is positive.

Proposition 2.2.5. If ∞ is an r-limit of l and b is a q-limit of l, then:

a) $1/r \leq b - q$ implies that ∞ is a $1/(b-q)$-limit of l;
b) $1/r > b - q$ implies that l has a $0.5\cdot(q + b - 1/r)$-limit.

Proof. a. By the initial condition, $b = q\text{-lim } l$. It means that for any $\varepsilon > 0$ and almost all a_i from l, we have $| a_i - b | < q + \varepsilon$. Thus, $a_i > b - q + \varepsilon$ for almost all a_i from l. By Definition 2.2.9, it means that $\infty = (1/(b - q - \varepsilon))\text{-lim } l$.

b. Indeed, if $\varepsilon > 0$, then almost all a_i from l belong to the interval $[1/r, b + q + \varepsilon)$. Taking the midpoint $(1/2)(1/r + b + q)$ of the interval $[1/r , b + q]$, we have

$$|(1/2)(1/r + b + q) - a_i | \leq (1/2)(b + q - 1/r) + \varepsilon$$

As ε is an arbitrary positive number, it means that $(1/2)(1/r + b + q)$ is an $(b + q - 1/r)$-limit of l.

Proposition is proved.

Proposition 2.2.6. If $a = r\text{-lim } l$ and $| a- b| = p$, then $b = (r+p)\text{-lim } l$.

Indeed, for any $\varepsilon > 0$, we have

$$| a_i - b | \leq | a_i - a | + | a - b | \leq r + \varepsilon + p = (r + p) + \varepsilon$$

for almost all a_i from l. Thus, $b = (r + p)$-lim l.

It is possible to perform different operations with sequences of numbers. Consequently, it is interesting to know how these operations change fuzzy limits.

Let $l = \{a_i \in \boldsymbol{R}; i = 1, 2, 3, \ldots\}$ and $h = \{b_i \in \boldsymbol{R}; i = 1, 2, 3, \ldots\}$. Then their sum $l + h$ is equal to the sequence $\{ a_i + b_i; i = 1, 2, 3, \ldots\}$ and their difference $l - h$ is equal to the sequence $\{ a_i - b_i; i = 1, 2, 3, \ldots\}$.

Theorem 2.2.5. If $a = r$-lim l and $b = q$-lim h, then:

a) $a + b = (r + q)$-lim $(l + h)$;
b) $a - b = (r + q)$-lim $(l - h)$;
c) $qa = |q|\cdot r$-lim (ql) for any $q \in \boldsymbol{R}$ where $ql = \{q \cdot a_i ; i = 1, 2, 3, \ldots\}$;
d) $a^2 = r(2|a| + r)$-lim l^2 where $l^2 = \{a_i^2 \in \boldsymbol{R}; i = 1, 2, 3, \ldots\}$;
e) $q + a = r$-lim $(q + l)$ for any $q \in \boldsymbol{R}$ where $q + l = \{ q + a_i ; i = 1, 2, 3, \ldots\}$.

Proof. a) Let $a = r$-lim l and $b = r$-lim h. Then by Definition 2.2.1, for any $\varepsilon \in \boldsymbol{R}^{++}$, there is such n that for any $i > n$, we have $| a - a_i | < r + \varepsilon/2$ and there is such m that for any $i > m$, we have $| b - b_i | < q + \varepsilon/2$. Taking $p = \max\{m, n\}$, we have $| a - a_i | < r + \varepsilon/2$ and $| b - b_i | < q + \varepsilon/2$ for any $i > p$. Consequently,

$$|(a + b) - (a_i + b_i)| = |(a - a_i) + (b - b_i)| \leq |(a - a_i)| + |(b - b_i)| < r + q + \varepsilon$$

for any $i > p$. By Definition 2.2.1, it means that $a + b = (r + q)$-lim $(l + h)$ as $l + h = \{ a_i + b_i; i = 1, 2, 3, \ldots\}$.

b) Let $a = r$-lim l and $b = r$-lim h. Then by Definition 2.2.1, for any $\varepsilon \in \boldsymbol{R}^{++}$, there is such n that for any $i > n$, we have $| a - a_i | < r + \varepsilon/2$ and there is such m that for any $i > m$, we have $| b - b_i | < q + \varepsilon/2$. Taking $p = \max\{m, n\}$, we have $| a - a_i | < r + \varepsilon$ and $| b - b_i | < q + \varepsilon/2$ for any $i > p$. Consequently,

$$|(a - b) - (a_i - b_i)| = |(a - a_i) + (- (b - b_i))| \leq |(a - a_i)| + |(b - b_i)| < r + q + \varepsilon$$

for any $i > p$. By Definition 2.2.1, it means that $a - b = (r + q)$-lim $(l - h)$ as $l - h = \{ a_i - b_i; i = 1, 2, 3, \ldots\}$.

c) Let $a = r$-lim l and $k \in \boldsymbol{R}$. Then by Definition 2.2.1, for any $\varepsilon \in \boldsymbol{R}^{++}$, there is such n that for any $i > n$, we have $| a - a_i | < r + \varepsilon/k$. Consequently, $| ka - ka_i | = |k|\cdot| a - a_i| < |k|\cdot r + \varepsilon$ for any $i > n$. By Definition 2.2.1, it means that $ka = |k|\cdot r$-lim (kl) as $kl = \{k \cdot a_i ; i = 1, 2, 3, \ldots\}$.

d) Let $a = r$-lim l. Then by Definition 2.2.1, for any $\varepsilon \in \boldsymbol{R}^{++}$, there is such n that for any $i > n$, we have $| a - a_i | < r + \varepsilon/k$ where $k = 2|a| + 2r$. Let us consider $| a^2 - a_i^2 |$. We have

$$\begin{gathered} | a^2 - a_i^2 | = |(a - a_i)(a + a_i) | \leq \\ | a - a_i | \, | a + a_i | \leq (r + \varepsilon/k)(2|a| + r + \varepsilon/k) = \\ r(2|a| + r) + (\varepsilon/k)(2|a| + r) + r(\varepsilon/k) = \\ r(2|a| + r) + (\varepsilon/k)(2|a| + 2r) = r(2|a| + r) + \varepsilon \end{gathered}$$

Thus, $a^2 = r(2|a| + r)$-lim l^2.

Part e) is a consequence of the part a) when all elements of the sequence h are equal to q. Theorem is proved.

Corollary 2.2.7. If $a = +r$-lim l and $b = +q$-lim h ($a = -r$-lim l and $b = -q$-lim h), then:

a) $a + b = +(r + q)$-lim$(l + h)$ ($a + b = -(r + q)$-lim$(l + h)$);

b) $a - b = +(r + q)$-lim$(l - h)$ ($a + b = -(r + q)$-lim$(l + h)$);

c) $qa = +|q|\cdot r$-lim (ql) ($qa = -|q|\cdot r$-lim (ql)) for any $q \in \boldsymbol{R}$ where $ql = \{q \cdot a_i ; i = 1, 2, 3, \dots\}$;

d) $q + a = +r$-lim $+(q + l)$ ($q + a = -r$-lim $+(q + l)$) for any $q \in \boldsymbol{R}$ where $q + l = \{ q + a_i ; i = 1, 2, 3, \dots\}$.

With other direct corollaries, Theorem 2.2.5 gives such classical results as Theorems 2.1.4, 2.1.5, and the following property of limits.

Corollary 2.2.8. If a = lim l, then a^2 = lim l^2.

Remark 2.2.7. The statement c) from Theorem 2.2.5 shows that in contrast to the conventional convergence, r-convergence is not invariant with respect to changes of the scale. Namely, when the scale changes (e.g., becomes ten times larger, that is, when the initial scale is 1 in to 100 mi., then after the change it becomes 1 in to 1000 mi.) the measure r of convergence changes correspondingly (e.g., r decreases and becomes 10 times smaller). Is such absence of invariance good or bad? It is good because it opens new opportunities allowing one to reflect changes of the scale and to use these changes to adjust the measure of convergence.

At the same time, the property of fuzzy convergence is invariant with respect to changes of the scale, that is, if a sequence fuzzy converges in one scale, then it fuzzy converges in any other scale.

Corollary 2.2.9. If a is a fuzzy limit of l and b is a fuzzy limit of h then $(a + b)$ ($(a - b)$ and ka) is a fuzzy limit of the sequence $(l + h)$ (of the sequences $(l - h)$ and kl, respectively).

Corollary 2.2.10. The sum and difference of fuzzy convergent sequences is a fuzzy convergent sequence.

However, the sum or difference of r-convergent sequences is not necessarily an r-convergent sequence.

Remark 2.2.8. Theorem 2.2.5 shows that fuzzy limit is a linear multivalued operator on the linear space of all sequences.

Let us consider the *scalar product of sequences*. If $l = \{a_i \in \boldsymbol{R}; i = 1, 2, 3, \dots\}$ and $h = \{b_i \in \boldsymbol{R}; i = 1, 2, 3, \dots\}$, then their scalar product $l \cdot h$ is equal to the sequence $\{ a_i \cdot b_i; i = 1, 2, 3, \dots\}$.

Theorem 2.2.6. If $a = r$-lim l and $b = q$-lim h, then $a \cdot b = (r \cdot q + |b| \cdot r + |a| \cdot q)$-lim $(l \cdot h)$.

Proof. Let us consider some $\varepsilon \in \boldsymbol{R}^{++}$. Then there is $\delta \in \boldsymbol{R}^{++}$ such that $(r + q + |b| + |a| + \delta)\delta < \varepsilon$. By the definition of fuzzy limits, we have the following property:

$$\exists\, n \in \omega\ \forall i > n\ (\,|\, a - a_i \,| < r + \delta\,) \text{ and } \exists\, m \in \omega\ \forall i > m\ (\,|\, b - b_i \,| < q + \delta\,).$$

Let us take $p = \max\{m, n\}$. Then $\forall i > p\ (|\, a - a_i \,| < r + \delta\,)$ and $\forall i > p\ (|\, b - b_i \,| < q + \delta\,)$. This and properties of the absolute value for any $i > p$ imply the following sequence of equalities and inequalities

$$| a_ib_i - ab | = | a_ib_i - a_ib + a_ib - ab | \leq$$
$$| a_ib_i - a_ib| + | a_ib - ab | \leq | a_i |\cdot| b_i - b | + | a_i - a|\cdot| b | <$$
$$| a_i |\cdot(q + \delta) + | b |\cdot(r + \delta) = | a_i - a + a|\cdot(q + \delta) + | b |\cdot(r + \delta) \leq$$
$$(| a_i - a| + | a|)\cdot(q + \delta) + | b |\cdot(r + \delta) = |$$
$$a_i - a|\cdot(q + \delta) + | a|\cdot(q + \delta) + | b |\cdot(r + \delta) <$$
$$(r + \delta)\cdot(q + \delta) + | a|\cdot(q + \delta) + | b |\cdot(r + \delta) =$$
$$r\cdot q + \delta\cdot q + \delta\cdot r + \delta^2 + | a|\cdot(q + \delta) + | b |\cdot(r + \delta) =$$
$$r\cdot q + \delta\cdot q + \delta\cdot r + \delta^2 + | a|\cdot q + | a|\cdot\delta + | b |\cdot r + | b |\cdot\delta =$$
$$r\cdot q + | a|\cdot q + |b |\cdot r + \delta\cdot q + \delta r + \delta^2 + | a|\cdot\delta + | b |\cdot\delta =$$
$$(r\cdot q + | a|\cdot q + |b |\cdot r) + (q + r + \delta + | a| + | b |)\cdot\delta <$$
$$(r\cdot q + | a|\cdot q + |b |\cdot r) + \varepsilon.$$

This means that $a\cdot b = (r\cdot q + |b|\cdot r + |a|\cdot q)\text{-lim}(l\cdot h)$.

Theorem is proved.

Corollary 2.2.11. If $l^2 = \{a_i^2 \in \boldsymbol{R};\ i = 1, 2, 3, \ldots\}$ and $a = r\text{-lim } l$, then $a\cdot a = (r\cdot r + |a|\,r + |a|\,r)\text{-lim}(l\cdot l)$, or $a^2 = r(2|a| + r)\text{-lim } l^2$.

When $b = \text{lim } h$, we have $b = 0\text{-lim } h$. Thus, Theorem 2.2.6 gives us the following result.

Corollary 2.2.12. If $a = r\text{-lim } l$ and $b = \text{lim } h$, then $a\cdot b = |b|\cdot r\text{-lim}(l\cdot h)$.

When $b = \text{lim } h$ and $a = \text{lim } l$, Thus, Corollary 2.2.11 implies the following result.

Corollary 2.2.13. If $a = \text{lim } l$ and $b = \text{lim } h$, then $a\cdot b = \text{lim } (l\cdot h)$.

2.2.3. Fuzzy Convergence Tests and Criteria

There are different criteria of sequence convergence developed in the calculus. Let us consider criteria of sequence fuzzy convergence.

Definition 2.2.10. a) A sequence l is called *r-fundamental* if for any $\varepsilon \in \boldsymbol{R}^{++}$ there is $n \in \boldsymbol{N}$ such that for any $i, j \geq n$, we have $| a_j - a_i | < 2r + \varepsilon$.

b) A sequence l is called *fuzzy fundamental* or a *fuzzy Cauchy sequence* if it is r-fundamental for some $r \in \boldsymbol{R}^{+}$.

Lemma 2.2.13. If $r \leq p$, then any r-fundamental sequence is p-fundamental.

Lemma 2.2.14. A sequence l is fundamental (in the ordinary sense, i.e., it is a Cauchy sequence) if and only if it is 0-fundamental.

Lemma 2.2.15. A subsequence of an r-fundamental sequence is r-fundamental.

Theorem 2.2.7 (the Extended Cauchy Criterion for sequences). A sequence l has an r-limit if and only if it is r-fundamental.

Proof. Necessity. Let $a = r\text{-lim } l$ and $\varepsilon \in \boldsymbol{R}^{++}$. Then by the definition there is a number $n \in \omega$ such that for any $i > n$, we have $| a - a_i | \leq r + \varepsilon/2$. Consequently, for any $i, j > n$, we have $| a_i - a_j | \leq | a - a_i | + | a - a_j | \leq 2r + 2(\varepsilon/2) < 2r + \varepsilon$. Thus, l is an r-fundamental sequence.

Sufficiency. Let l be an r-fundamental sequence and $k_m = 1/m$. If we fix some number m, then there is a number n, which is dependent on m and thus, denoted by $n(m)$, such that for all $i, j > n(m)$, we have $| a_i - a_j | < 2r + (1/m)$. That is, all points a_i with $i > n(m)$ belong to a closed interval I_m. Really, let us put $\mathrm{T}_m = \{a_i ;\ i > n(m) \}$, $b = \sup \mathrm{T}_m$, and $c = \inf \mathrm{T}_m$. Then all a_i belong to the interval $I_m = [c, b]$ for $i > n(m)$.

Let us estimate the length of this interval I_m. Suppose that $| b - c | > 2r + (1/m)$. It means that for some $h \in \boldsymbol{R}^+$, we have $| b - c | > 2r + (1/m) + h$. At the same time, there are such elements a_p and a_q, for which $p, q > n(m)$, $| b - a_p | \leq (1/3)\cdot h$, and $| c - a_q | \leq (1/3)\cdot h$ because b is the supremum and c is the infimum of all elements from the set T_m. Consequently, for these a_p and a_q, we have

$$| b - c | \leq | b - a_p | + | a_{p'} - a_q | + |a_q - c | <$$
$$(1/3)\cdot h + (2r + (1/\mathrm{m})) +(1/3)\cdot h =$$
$$2r + (1/m) +(2/3)h < 2r + (1/m) + h.$$

This contradicts our supposition that $| b - c | > 2r + (1/m) + h$. Thus, the length of $I_m = [c, b]$ is not larger than $2r + (1/m)$.

We can choose these intervals I_m so that the inclusion $I_{m+1} \subseteq I_m$ is valid for all $m = 1, 2, 3, \dots$. In such a way, we obtain a sequence of nested closed intervals $\{ I_m; m = 1, 2, 3, \dots \}$. The space $\boldsymbol{R}$ is a complete metric space (cf. Appendices B and C). Consequently, the intersection $I = \bigcap_{m=1}^{\infty} I_m$ is non-empty. That is, the interval I consists of one point d or I is a closed interval having the length not larger than $2r$ as the length of I_m is not larger than $2r + (1/m)$. When $I = \{d\}$, the sequence l converges, and thus, by Lemma 2.2.11, l has an r-limit.

When I is a non-trivial interval, we consider its midpoint e. Then for any number $k \in \boldsymbol{R}^{++}$, there is some $I_m \supseteq I$ such that $k > (1/m)$, $| e - e_m | < (1/3)\, k$ for the center e_m of I_m and almost all a_i belong to I_m. Besides, $| e_m - a_i | < r + (1/2m)$ when $i > n(m)$. Consequently, as $k > (1/m)$, we have

$$| e - a_i | \leq | e_m - a_i | + | e - e_m | < (r + (1/2m)) + (1/3)k < r + (1/2)k + (1/3)k < r + k$$

when $i > n(m)$, i.e., the point e is an r-limit of l.

Thus, in both cases, the sequence l has an r-limit.

Theorem is proved.

From Theorem 2.2.7, we obtain the following result.

Theorem 2.2.8 (the General Fuzzy Convergence Criterion for sequences). A sequence l fuzzy converges if and only if it is fuzzy fundamental.

Theorem 2.2.7 and Lemmas 2.2.1 and 2.2.14 imply such classical result as the Cauchy Criterion for sequences, which states that the sequence l converges if and only if it is fundamental.

This result and Lemma 2.2.14 demonstrate that the concept of fuzzy convergence is a natural extension of the concept of conventional convergence.

From Theorem 2.2.7 and Proposition 2.2.1, we obtain the following result.

Proposition 2.2.7. The following conditions are equivalent:

1) a sequence l is not fuzzy fundamental;
2) the sequence l is not bounded;
3) some subsequence of l has no fuzzy limits;
4) ∞ or $-\infty$ is the partial limit of l.

There are other tests that allow one to find whether a given sequence r-converges. Here we consider some of them.

Let us consider three sequences $l = \{a_i \in \mathbf{R}; i = 1, 2, 3, \dots \}$, $k = \{c_i \in \mathbf{R}; i = 1, 2, 3, \dots \}$, and $h = \{b_i \in \mathbf{R}; i = 1, 2, 3, \dots \}$.

Theorem 2.2.9 (the Fuzzy Squeeze Theorem). If $a_i \leq b_i \leq c_i$ for almost all $i = 1, 2, 3, \dots$, $a = r$-lim l and $a = r$-lim k, then $a = r$-lim l.

Proof. By Definition 2.2.1, $a = r$-lim l means that for any $\varepsilon > 0$, the inequality $| a - a_i | < r + \varepsilon$ is valid for almost all $i = 1, 2, 3, \dots$, and $a = r$-lim k means that for any $\varepsilon > 0$, the inequality $| a - c_i | < r + \varepsilon$ is valid for almost all $i = 1, 2, 3, \dots$. As $a_i \leq c_i$, we have three possibilities: $a_i \leq c_i \leq a$, $a \leq a_i \leq c_i$ or $a_i \leq a \leq c_i$ where i is any number for which the inequality $a_i \leq c_i$ holds.

In the first case, $| a - b_i | \leq | a - a_i | < r + \varepsilon$ for almost all $i = 1, 2, 3, \dots$ because $a_i \leq b_i \leq c_i \leq a$. In the second case, $| a - b_i | \leq | a - c_i | < r + \varepsilon$ for almost all $i = 1, 2, 3, \dots$ because $a \leq a_i \leq b_i \leq c_i$. In the third case, $| a - b_i | \leq | a - a_i | < r + \varepsilon$ when $b_i \leq a$ or $| a - b_i | \leq | a - a_i | < r + \varepsilon$ when $a \leq b_i$. Thus, in all cases, for any $\varepsilon > 0$, $| a - b_i | < r + \varepsilon$ for almost all $i = 1, 2, 3, \dots$ It means by Definition 2.2.1, that $a = r$-lim h.

Theorem is proved.

Let us consider two sequences $l = \{a_i \in \mathbf{R}; i = 1, 2, 3, \dots\}$ and $h = \{b_i \in \mathbf{R}; i = 1, 2, 3, \dots\}$.

Definition 2.2.11 (Phu, 2001). Sequences l and h are called *r-close* to one another if $| a_i - b_i | \leq r$ for all $i = 1, 2, 3, \dots$.

Proposition 2.2.8 (Phu, 2001). We have $a = (q + r)$-lim l if and only if there is a sequence h such that is $a = r$-lim h and l and h are q-close to one another.

Proof. Necessity. Let us assume that $l = \{a_i \in \mathbf{R}; i = 1, 2, 3, \dots\}$ and $a = (q + r)$-lim l. Then we build the sequence $h = \{b_i \in \mathbf{R}; i = 1, 2, 3, \dots\}$ in which $b_i = a_i - q$ when $a \leq a_i$ and $b_i = a_i + q$ when $a > a_i$. By construction, the sequences l and h are r-close to one another. In addition, for any $\varepsilon > 0$, we have $| a - b_i | \leq r + \varepsilon$ for almost all $i \in \omega$ as $| a - a_i | \leq q + r + \varepsilon$ for almost all $i \in \omega$. Thus, $a = r$-lim h.

Sufficiency. Let us assume that $l = \{a_i \in \mathbf{R}; i = 1, 2, 3, \dots\}$, there is a sequence $h = \{b_i \in \mathbf{R}; i = 1, 2, 3, \dots\}$ such that is $a = r$-lim h, and l and h are q-close to one another. Then by Definition 2.2.1, $| a - b_i | \leq r + \varepsilon$ for almost all $i \in \omega$ and $| a - a_i | = | a - b_i + b_i - a_i | \leq | a - b_i | + | b_i - a_i | \leq r + \varepsilon + q$. Consequently, $a = (q + r)$-lim l.

Proposition is proved.

Corollary 2.2.14. A sequence l r-converges if and only if there is a convergent sequence r-close to l.

Convergence of one sequence to another was studied by Kalina in the context of nearness relations. This concept is closely related to the concept of closedness. Here we study convergence of one sequence to another by means of neoclassical analysis.

Let us consider two sequences $l = \{a_i \in \mathbf{R}; i = 1, 2, 3, \dots\}$ and $h = \{b_i \in \mathbf{R}; i = 1, 2, 3, \dots\}$.

Definition 2.2.12. a) The sequence l *converges* to the sequence h if for any $\varepsilon \in \mathbf{R}^{++}$, there is $n \subset \omega$ such that for any $i \geq n$, we have $| a_i - b_i | < \varepsilon$.

b) The sequence l *r-converges* to the sequence h if for any $\varepsilon \in \mathbf{R}^{++}$, there is $n \in \omega$ such that for any $i \geq n$, we have $| a_i - b_i | < 2r + \varepsilon$.

c) The sequence l *fuzzy converges* to the sequence h if it r-converges to h for some $r \in R^+$.

Lemma 2.2.16. A sequence l converges (r-converges) to a sequence h if and only if the sequence h converges (r-converges) to the sequence l.

Lemma 2.2.17. If a sequence l converges (fuzzy converges) to a sequence h and a sequence h converges (fuzzy converges) to a sequence k, then the sequence l converges (fuzzy converges) to the sequence k.

Lemmas 2.2.16 and 2.2.17 imply the following result.

Proposition 2.2.9. Relations of convergence and fuzzy convergence of a sequence to a sequence are equivalence relations.

Remark 2.2.8. Relation of r-convergence of a sequence to a sequence is not transitive in general and thus, it is not an equivalence relation.

Remark 2.2.9. The relation of convergence of a sequence to a sequence is used to define real and complex hypernumbers (Burgin, 1990a; 2002; 2004b).

Proposition 2.2.10. Any r-close sequences r-converge to one another.

Proposition 2.2.11. If $c \leq a_i$ ($c \geq a_i$) for almost all $i \in \omega$ and a sequence $l = \{a_i \in R;\ i = 1, 2, 3, \ldots\}$ converges to a sequence $h = \{b_i \in R;\ i = 1, 2, 3, \ldots\}$, then for any $\varepsilon > 0$, $c + \varepsilon \leq b_i$ ($c + \varepsilon \geq b_i$) for almost all $i \in \omega$.

Let us consider two sequences $l = \{a_i \in R;\ i = 1, 2, 3, \ldots\}$ and $h = \{b_i \in R;\ i = 1, 2, 3, \ldots\}$.

Theorem 2.2.10. If $0 \leq a_i \leq b_i$ for almost all $i \in \omega$ and $0 \leq b = r\text{-lim } h$, then l is $\frac{1}{2}(b + r)$-convergent and $b = u\text{-lim } l$ where $u = \max \{b, r\}$.

Proof. Let $0 \leq b = r\text{-lim } h$. Definition 2.2.1 implies that for any $\varepsilon > 0$, almost all b_i belong to the interval $(-\varepsilon, b + r + \varepsilon) = (\frac{1}{2}(b + r) - \frac{1}{2}(b + r) - \varepsilon, \frac{1}{2}(b + r) + \frac{1}{2}(b + r) + \varepsilon)$ As $0 \leq a_i \leq b_i$ for almost all $i \in \omega$, we have that almost all a_i belong to the interval $(\frac{1}{2}(b + r) - \varepsilon, \frac{1}{2}(b + r) + \frac{1}{2}(b + r) + \varepsilon)$. By Definition 2.2.1, it means that $\frac{1}{2}(b + r) = \frac{1}{2}(b + r)\text{-lim } l$ and thus, the sequence l is $\frac{1}{2}(b + r)$-convergent.

If $u = \max \{b, r\}$, then $u \geq \frac{1}{2}(b + r)$. Consequently, the interval $(\frac{1}{2}(b + r) - \frac{1}{2}(b + r) - \varepsilon, \frac{1}{2}(b + r) + \frac{1}{2}(b + r) + \varepsilon)$ is a subset of the interval $(b - u - \varepsilon, b + u + \varepsilon)$. Thus, almost all a_i belong to the interval $(b - u - \varepsilon, b + u + \varepsilon)$. By Definition 2.2.1, it means that $b = u\text{-lim } l$.

Theorem is proved.

Remark 2.2.8. The estimate for u in Theorem 2.2.10 cannot be improved as the following example demonstrates.

Example 2.2.12. Let us consider sequences $l = \{ 1/n;\ n = 1, 2, 3, \ldots\}$ and $h = \{2 + 1/n;\ n = 1, 2, 3, \ldots\}$. Both sequences consist of positive elements and all elements from the second sequence are larger than all corresponding elements from the first sequence. Thus, all conditions from Theorem 2.2.10 are valid for these sequences and number 2. Namely, $2 = 0\text{-lim } h$. Here $u = \max \{b, r\} = 2$. According to Theorem 2.2.10, $2 = 2\text{-lim } l$, but for any number r smaller than 2, it is not true that $2 = r\text{-lim } l$.

Remark 2.2.9. The estimate for fuzzy convergence of l in Theorem 2.2.10 cannot be improved as the following example demonstrates.

Example 2.2.13. Let us consider sequences $l = \{ 2 + 1/n + 2\cdot (-1)^n;\ n = 1, 2, 3, \ldots\}$ and $h = \{3 + 1/(n - 1) + (-1)^n;\ n = 1, 2, 3, \ldots\}$. Both sequences consist of positive elements and all elements from the second sequence are larger than all corresponding elements from the first sequence. Thus, all conditions from Theorem 2.2.9 are valid for these sequences and numbers

2 and 3. Namely, 3 = 1-lim h. Thus, 2 = ½(1 + 3) and 2 = 2-lim l. However, l does not r-converge for any number $r < 2$.

Remark 2.2.10. It is possible that $0 \leq a_i \leq b_i$ for all $i = 1, 2, 3, \dots$, the sequence h r-converges, while the sequence l r-diverges as the following example demonstrates.

Example 2.2.14. Let us consider sequences $l = \{ 3 + (-1)^n; n = 1, 2, 3, \dots\}$ and $h = \{5 + 1/n ; n = 1, 2, 3, \dots\}$. Then the sequence h 0-converges, while the sequence l 0-diverges and only 1-converges.

Let us consider a sequence $l = \{a_i \in \boldsymbol{R}; i = 1, 2, 3, \dots\}$, the set CS($l$) of all converging subsequences of l, and the set L(l) of all their limits. According to (Burgin, 2002), L(l) is the spectrum of the sequence l.

The set EL(l) of extended limits of l consists of the set L(l), plus the element ∞ when there is a subsequence h of l that converges to ∞, and/or plus of the element $-\infty$ when there is a subsequence g of l that converges to $-\infty$. When EL(l) contains, at least, one infinite element, the sequence l is unbounded and the set EL(l) is unbounded.

Theorem 2.2.11. a) A sequence l fuzzy converges if and only if the set EL(l) is bounded.

b) $a = r$-lim l if and only if EL(l) = L(l) and for any c from L(l), we have $| a - c | \leq r$.

c) If the set EL(l) is bounded, then l is k-convergent where k = ½ (sup L(l) - inf L(l)) and l is q-divergent for any $q < k$.

Proof. Let us assume that a sequence $l = \{a_i \in \boldsymbol{R}; i = 1, 2, 3, \dots\}$ is fuzzy converging. Then there are numbers a and r such that $a = r$-lim l. We show that for any c from L(l), we have $| a - c | \leq r$. Indeed, if $c \in$ L(l) and $| a - c | > r$, then the interval $[a - r, a + r]$ and the point c have neighborhoods $A = (a - r - \varepsilon, a + r + \varepsilon)$ and $B = (c - \varepsilon, c + \varepsilon)$ that do not intersect. By Definition 2.2.1, almost all elements a_i from l belong to A. At the same time, as c = lim h for some subsequence h of the sequence l, infinitely many elements a_i from l belong to B. As A and B do not intersect, this is a contradiction. Thus, we have $| a - c | \leq r$ according to the statement (b). Consequently, the set L(l) is bounded as it is demanded in the statement (c).

Necessity of conditions in the statements (b) and (c) is proved.

If the set EL(l) is bounded for a sequence $l = \{a_i \in \boldsymbol{R}; i = 1, 2, 3, \dots\}$, then we can take numbers u = sup L(l) and v = inf L(l). By construction of the set L(l), it is a subset of the interval $[v, u]$. Taking the midpoint a of this interval, we show that $a = k$-lim l where k = ½ $(u - v)$.

We prove this by contradiction. If $a = k$-lim l, then for any $\varepsilon > 0$, almost all elements a_i from l belong to the interval $(v - \varepsilon, u + \varepsilon)$ as $| a - a_i | < k + \varepsilon$ for almost all $i = 1, 2, 3, \dots$. If a is not a k-limit of the sequence l, then for some $\varepsilon > 0$, there are infinitely many elements a_i from l that do not belong to the interval $(v - \varepsilon, u + \varepsilon)$. Let us denote the set of all these elements a_i by K.

As the set EL(l) is bounded, the sequence l is also bounded, i.e., all its elements belong to some interval $[x, y]$. Consequently, the set K is contained in the union $H = [x, v - \varepsilon] \cup [u + \varepsilon, y]$.

By Theorem 2.1.7, any infinite bounded set of real numbers contains a converging sequence. Thus, K contains a converging sequence h. As the set H is closed, the limit d of h belongs to H. As a result, d does not belong to the interval $[v, u]$ as H and this interval do not intersect.

However, by construction, $[v, u]$ contains EL(l) = L(l) and d belongs to L(l) because h is a subsequence of l. This contradiction shows that $a = k$-lim l and l is a k-convergent sequence.

This proves the first part of the statement (c) and sufficiency in the statement (a).

Let us consider some number $q < k$. Then for any number d, we have either $| v - d | > q$ or $| u - d | > q$. Let us consider the first case. Then there is $\varepsilon > 0$ such that $| v - d | > q + 4\varepsilon$. As $v = \inf$ L(l), there is a point c such that $c = \lim h$ for some subsequence h of the sequence l, and $| v - c | < \varepsilon$. As $c = \lim h$, there infinitely many elements a_i from l for which $| c - a_i | < \varepsilon$.

Thus for any such element a_i, we have

$$| v - d | = | v - c + c - a_i + a_i - d | \leq$$
$$| v - c | + | c - a_i | + | a_i - d | < \varepsilon + \varepsilon + | a_i - d |$$

and consequently,

$$| a_i - d | > | v - d | - 2\varepsilon > q + 4\varepsilon - 2\varepsilon = q + 2\varepsilon.$$

In other words, for this $\varepsilon > 0$, there are there infinitely many elements a_i from l for which $| a_i - d | > q + 2$. This means that d cannot be a q-limit of l.

The case when $| u - d | > q$ is considered in a similar way.

This concludes the proof of the statement (c).

Now let us assume that for any number $c \in$ L(l) and some number a, we have $| a - c | \leq r$. By contradiction, we prove that $a = r$-lim l. Indeed, if a is not an r-limit of the sequence l, then for some $\varepsilon > 0$, there are infinitely many elements a_i from l that do not belong to the interval $(a - r - \varepsilon, a + r + \varepsilon)$. Let us denote the set of all these elements a_i by D.

As the set EL(l) is bounded, the sequence l and its subset K are bounded. By Theorem 2.1.7, any infinite bounded set of real numbers contains a converging sequence. Thus, D contains a converging sequence h. If $d = \lim h$, then d belongs to L(l) but does not belong to the interval $(a - r - \varepsilon, a + r + \varepsilon)$. At the same time, the inequality $| a - c | \leq r$ valid for all $c \in$ L(l) provides that all elements from L(l) belong to this interval.

This contradiction shows that $a = r$-lim l and l is an r-convergent sequence.

Theorem is proved.

Fuzzy limits are essential for the development of principal structures in neoclassical analysis. At the same time, fuzzy limits find various applications beyond neoclassical analysis. The following example explains why we need fuzzy limits in discrete dynamics.

Let us consider the following situation. We can compute with the precision of 100 digits after the decimal point. We want to analyze results of experiments in which measurements have the precision equal to10 digits after the decimal point. Thus, both sets, computational data and experimental data, are discrete.

Analyzing experimental data, it is natural to ask the question whether they converge or not. However, the classical calculus gives a very imprecise answer to this question because a sequence in a discrete set converges if and only if it becomes constant after some point. However, all results in measurements are, as a rule, approximate, i.e., obtained only with some finite precision. Consequently, it is impossible to know for sure when the sequence of experimental data stabilizes and whether it stabilizes at all.

The theory of fuzzy limits allows engineers and applied mathematicians to answer the question whether a sequence or a series of experimental data converges or not. Qualified engineers and experts in applied mathematics know that it is not reasonable to perform computations with higher precision than the precision of measurements (cf., for example, (Krylov, 1979)). For instance, this means that it is natural to assume that sequences that have an r-limit in the sense of neoclassical analysis with a sufficiently small r (e.g., 10^{-10}-limit) are converging. Such a fuzzy limit means that the data are grouped in a very small interval (e.g., with the length $2 \cdot 10^{-10}$ m). Thus, it is reasonable, for example, to consider 10^{-10}-limits in the set of all numbers for which computation is possible and to check if the sequence of experimental data has such a limit.

Besides, the majority of models of real phenomena (in physics, chemistry, biology, economics, etc.) are build based on real and complex numbers. However, when it comes to computation, only computable real and complex numbers have to be considered (cf., for example, (McCauley, 1997)). Moreover, computable real and complex numbers are, as a rule, only theoretically computable, e.g., by such a computational model as Turing machine. In practice of computation, computers work only with rational numbers represented by binary sequences. This shows that methods of classical calculus, in particular, of classical limits, do not work under such restrictions (cf., also Section 1.2).

At the same time, in neoclassical analysis, it is not necessary to employ irrational numbers as limits and consequently, as values of derivatives and integrals. The following theorem demonstrates that it is always possible to take some rational point as a fuzzy limit that gives a necessary precision for sequences that come from measurement and computation.

Theorem 2.2.12. If a sequence of real numbers $l = \{a_i \in \boldsymbol{R};\ i = 1, 2, 3, \dots\}$ has an r-limit a with $r > 0$, then for any $\alpha > 0$, this sequence has a rational $(r+\alpha)$-limit b.

Proof. Let us take a rational number b such that $|\ a - b\ | \leq \alpha$. Properties of real numbers show that we can do this. Then by Definition 2.2.1, for any $\alpha > 0$, we have $|\ a - a_i\ | \leq r + \varepsilon$ for almost all $i \in \omega$ and

$$|\ b - a_i\ | == |\ b - a + a - a_i\ | \leq |\ b - a\ | + |\ a - a_i\ | \leq r + \varepsilon + \alpha.$$

Consequently, $b = (r+\alpha)\text{-lim } l$.

Theorem is proved.

For computation, it is important not only that a number that we want to compute is in general rational but also how many digits this number contains in general and after the decimal point in particular. The reason is that even the best computers have restrictions on the number of digits in numbers that these computers represent and process. The following theorems give answers to these questions.

Theorem 2.2.13. If a sequence of real numbers $l = \{a_i \in \boldsymbol{R};\ i = 1, 2, 3, \dots\}$ has an r-limit a, then this sequence has a rational $(r + 2^{-p-1})$-limit b that its binary representation has less than $p + 1$ binary digits after the point.

Proof. The set of all rational numbers that have less than $p + 1$ binary digits after the point form 2^{-p-1}-net on the real line $\boldsymbol{R}$. That is, for any point a from $\boldsymbol{R}$, there is a rational point b such that the distance between a and b is less than or equal to 2^{-p-1} and the binary representation of b has less than $p + 1$ binary digits after the point. By Lemma 2.2.5, $b = q\text{-lim } l$ where $q = r + 2^{-p-1}$.

Theorem is proved.

Theorem 2.2.13 implies the following result.

Theorem 2.2.14. If a sequence of real numbers $l = \{a_i \in \mathbf{R};\ i = 1, 2, 3, \dots\}$ has an r-limit a where $a < 2^q$ and $2^{-p} > r > 0$, then this sequence has a rational r-limit b a binary representation of which has not more than $p + q + 1$ binary digits.

Thus, Theorems 2.2.12 - 2.2.14 demonstrate that instead of using irrational numbers as limits and fuzzy limits, it is possible to have rational fuzzy limits that have a good precision and are in the range of the computing device (e.g., computer, calculator, etc.) operations.

2.3. Limits and Fuzzy Limits of Functions

In order to be able to set a limit to thought,
we should have to find both sides of the limit thinkable
(i.e. we should have to be able to think what cannot be thought).
Ludwig Wittgenstein (1889-1951)

We begin with the classical concept of the limit of a function at a point. The first definition we give is a reduction of the limit construction for a function to the limit construction for a sequence.

Let us assume that $r \in \mathbf{R}^+$ and $f: \mathbf{R} \to \mathbf{R}$ is a partial function.

Definition 2.3.1. a) A number b is called the *limit of a function* $f(x)$ at a point $a \in \mathbf{R}$ (it is denoted by $b = \lim_{x\to a} f(x)$) if for any sequence $l = \{ a_i ;\ a_i \in \text{Dom} f;\ i = 1, 2, 3, \dots ,\ a_i \neq a\}$, the condition $a = \lim l$ implies $b = \lim_{i\to\infty} f(a_i)$.

b) a function $f(x)$ *converges* at a point $a \in \mathbf{R}$ if it has the limit at this point.

Example 2.3.1. $9 = \lim_{x\to 3} x^2$, $0 = \lim_{x\to 0} |x|$, and $1 = \lim_{x\to 3} \cos x$.

However, the most popular is the direct definition, or, the so-called, (ε, δ)-definition of the classical concept of the limit of a function at a point. Namely, we have:

Definition 2.3.2. A number b is called the *limit of a function* $f(x)$ at a point $a \in \mathbf{R}$ if for any $\varepsilon > 0$, there is $\delta > 0$ such that the inequality $| x - a | \leq \delta$ implies the inequality $|f(x) - b | \leq \varepsilon$.

Proposition 2.3.1. Definitions 2.3.1 and 2.3.2 define the same concept for all points in $\mathbf{R}$, i.e., b is the limit of a function $f(x)$ at a point a according to Definition 2.3.1 if and only if b is the limit of the function $f(x)$ at a point a according to Definition 2.3.2.

Proof. (a) Let us assume that b is the limit of a function $f(x)$ at a point a according to Definition 2.3.1, but the condition from Definition 2.3.2 is not true. That is, there is some $\varepsilon > 0$ such that for any $\delta > 0$, there is a number x for which both inequalities $| x - a | \leq \delta$ and $| f(x) - b | > \varepsilon$ are true.

Let us consider the sequence $h = \{1/i\ ;\ i = 1, 2, 3, \dots\}$. Taking $\delta = 1/i$, we can find numbers a_i such that $| a_i - a | \leq 1/i$ and the inequality $| f(a_i) - b | > \varepsilon$ is true for each $i = 1, 2, 3, \dots\}$. Then the sequence $l = \{ a_i ;\ i = 1, 2, 3, \dots\}$ converges to a, while the sequence $k = \{ f(a_i)\ ;\ i = 1, 2, 3, \dots\}$ does not converge to b. This contradicts Definition 2.3.1 and shows that Definition 2.3.1 implies Definition 2.3.2.

(b) Let us assume that b is the limit of a function $f(x)$ at a point a according to Definition 2.3.2, but the condition from Definition 2.3.1 is not true.

Let us take some sequence $l = \{ a_i ; i = 1, 2, 3, \dots\}$ that converges to a. It means that for any $\delta > 0$, there is a number $m > 0$ such that the inequality $i > m$ implies the inequality $| a_i - a | \leq \delta$. Taking some $\varepsilon > 0$, we can find $\delta > 0$ such that the inequality $| x - a | \leq \delta$ implies the inequality $| f(x) - b | \leq \varepsilon$. Consequently, we can find a number $m > 0$ such that the inequality $i > m$ implies the inequality $| f(a_i) - b | \leq \varepsilon$.

As l is an arbitrary sequence that converges to a, Definition 2.3.2 implies Definition 2.3.1.

Proposition is proved.

It is also possible to characterize the limit of a function utilizing limits of sets of sequences (cf. Section 2.6).

Let us consider the set $E_{f,a}$ of all sequences $h = \{ f(a_i); i = 1, 2, 3, \dots \}$ such that the corresponding sequence $l = \{ a_i ; i = 1, 2, 3, \dots \}$ converges to a.

Proposition 2.3.2. $b = \lim_{x\to a} f(x))$ if and only if $b = \lim E_{f,a}$.

Definition 2.3.1 and Theorem 2.1.1 imply such a classical result as the following theorem.

Theorem 2.3.1. A limit of a function at a point is unique (if this limit exists).

In our study of fuzzy limits of functions, we, at first, develop and explore a construction of a fuzzy limit of a function that is similar to fuzzy limits of sequences because it is only one level of generality higher than the classical concept of the limit of a function. This construction allows one to model and study continuous in some, more general than traditional, sense mappings in discrete spaces.

Let us assume that $r \in \boldsymbol{R}^+$ and $f: \boldsymbol{R} \to \boldsymbol{R}$ is a partial function.

Definition 2.3.3. a) A number b is called an *r-limit of a function* $f(x)$ at a point $a \in \boldsymbol{R}$ (it is denoted by $b = r\text{-}\lim_{x\to a} f(x))$ if for any sequence $l = \{ a_i ; a_i \in \text{Dom} f; i = 1, 2, 3, \dots , a_i \neq a\}$, the condition $a = \lim l$ implies $b = r\text{-}\lim_{i\to\infty} f(a_i)$.

b) a function $f(x)$ *r-converges* at a point $a \in \boldsymbol{R}$ if it has an r-limit at this point.

Example 2.3.2. $1.5 = 0.5\text{-}\lim_{x\to 1} x^2$, $0.5 = 0.3\text{-}\lim_{x\to 0} |x|$, and $0 = 1\text{-}\lim_{x\to 3} \cos x$.

Example 2.3.3. The function $\sin (1/x)$ does not have the limit at the point 0. However, $0.5 = 0.5\text{-}\lim_{x\to 0} \sin (1/x)$ and $0 = 1\text{-}\lim_{x\to 0} \sin (1/x)$.

Remark 2.3.1. It is possible to define an r-limit of a function $f(x)$ at a point $a \in \boldsymbol{R}$ independently of the concept of an r-limit of a sequence (cf. Theorem 2.3.2). This allows one to get the latter concept as a particular case of the former concept. Really, a sequence with values in $\boldsymbol{R}$ is a partial function $f: \boldsymbol{R} \to \boldsymbol{R}$ that is defined only for natural numbers, i.e., $f: \boldsymbol{N} \to \boldsymbol{R}$. Then an r-limit of a sequence $l = \{a_i \in \boldsymbol{R}; i = 1, 2, 3, \dots \}$ is an r-limit of the function $f(x)$ at the point ∞.

Remark 2.3.2. The concept of an r-limit of a function allows one to develop a theory of limits for functions that take values in discrete sets.

Lemma 2.2.1 and Definition 2.3.3 imply the following result.

Lemma 2.3.1. If $\text{Dom} f = \boldsymbol{R}$, then $b = 0\text{-}\lim_{x\to a} f(x)$ if and only if $b = \lim_{x\to a} f(x)$ in the classical sense.

This result demonstrates that the concept of an r-limit of a function is a natural extension of the concept of conventional limit of a function. However, the concept of an r-limit actually extends the conventional construction of a limit (cf. Example 2.3.1).

Remark 2.3.3. The condition Dom $f = \mathbf{R}$ is essential in Lemma 2.3.1 because if Dom f is a discrete set, then all sequences converging to any element from Dom f stabilize after some number of its elements. This results in the property that any function $f(x)$ defined on a discrete set has the limit at any point of this set, which is equal to value of $f(x)$, and thus, $f(x)$ is continuous.

Lemma 2.3.2. If $b = r\text{-lim}_{x\to a} f(x)$, then $b = q\text{-lim}_{x\to a} f(x)$ for any $q > r$.

Theorem 2.3.2. a) The statement $b = r\text{-lim}_{x\to a} f(x)$ is true if and only if for any open neighborhood Ob of b that contains the interval $[b - r, b + r]$ there is a neighborhood Oa of a such that $f(\text{O}a \cap \text{Dom} f) \subseteq \text{O}b$.

b) The statement $b = r\text{-lim}_{x\to a} f(x)$ is true if and only if for any $\varepsilon \in \mathbf{R}^{++}$ there is $\delta > 0$ such that for all x, the inequality $|\, x - a \,| \leq \delta$ implies the inequality $|\, f(x) - b \,| \leq r + \varepsilon$.

Proof. Necessity. Let $b = r\text{-lim}_{x\to a} f(x)$ and Ob is an open neighborhood of b that contains the interval $[b - r, b + r]$. Let us suppose that in any neighborhood Oa of a, there is a point $a_i \neq a$ such that $f(a_i)$ does not belong to Ob. We can take a sequence of neighborhoods $\text{O}_i a$ such that $\text{O}_i a \subseteq \text{O}_{i\text{-}1} a$ for all $i = 2, 3, \dots$, in any neighborhood $\text{O}_i a$, there is a point $a_i \neq a$ such that $f(a_i)$ does not belong to Ob, and the intersection $\bigcap \text{O}_i a = \{a\}$. Then we have a sequence

$$l = \{a_i \in \mathbf{R};\ i = 1, 2, 3, \dots\ a_i \neq a\},$$

for which the condition $a = \lim l$ does not imply $b = r\text{-lim}_{x\to a} f(a_i)$. This contradicts the initial condition that $b = r\text{-lim}_{x\to a} f(x)$ and concludes the proof of necessity.

Sufficiency. Let us assume that for any open neighborhood Ob of b that contains the interval $[b - r, b + r]$, there is a neighborhood Oa of a such that $f(\text{O}a) \subseteq \text{O}b$ and $l = \{a_i \in \mathbf{R};\ i = 1, 2, 3, \dots;\ a_i \neq a\}$ is a sequence such that $a = \lim l$. Then almost all elements of l belong to Oa. Consequently, almost all elements $f(a_i)$ belong to Ob. By the definition of r-limit, we have $b = r\text{-lim}_{i\to\infty} f(a_i)$. As the sequence l is chosen arbitrarily, by Definition 2.3.1, $b = r\text{-lim}_{x\to a} f(x)$. This concludes the proof of the part a) of the theorem.

Proof of the part b) is similar to the proof of Proposition 2.3.1.

Theorem is proved.

Theorem 2.2.1 implies the following result.

Theorem 2.3.3. If $b = r\text{-lim}_{x\to a} f(x)$ and $b > d + r$, then there is a neighborhood Oa of a such that $f(x) > d$ for all x from O$a \cap$ Dom f.

Corollary 2.3.1. If $b = r\text{-lim}_{x\to a} f(x)$ and $b > d + r$, then for any sequence $l = \{a_i \in \text{Dom} f\ ;\ i = 1, 2, 3, \dots , a_i \neq a \}$ with $a = \lim l$, we have $f(a_i) > d$ for almost all a_i from l.

Corollary 2.3.2. If $f(x) \leq q$ for all x from some neighborhood Oa of a and $a = r\text{-lim}_{x\to a} f(x)$, then $a \leq q + r$.

These results allow us to obtain important properties of conventional limits of functions that are studied in courses of calculus.

Corollary 2.3.3. If $a = \lim_{x \to a} f(x)$ and $a > b$, then $f(x) > b$ for all x from some neighborhood Oa of a.

Corollary 2.3.4. If $a = \lim_{x \to a} f(x)$ and $a > 0$, then $f(x) > 0$ for all x from some neighborhood Oa of a.

Corollary 2.3.5. If $f(x) \leq q$ for all x from some neighborhood Oa of a and $a = \lim_{x \to a} f(x)$, then $a \leq q$.

Definition 2.3.4. a) A number a is called a *fuzzy limit* of a function $f(x)$ at a point $a \in \boldsymbol{R}$ if it is an r-limit of a function f at a point a for some $r \in \boldsymbol{R}^+$.

b) a function $f(x)$ *fuzzy converges* at a point $a \in \boldsymbol{R}$ if it has a fuzzy limit at this point.

Example 2.3.4. Let us consider the following function $f(x)$

$$f(x) = \begin{cases} 1 + 1/i \text{ when } x = 1 - 1/i \text{ for } i = 1, 2, 3, \ldots ; \\ 2 - 1/i \text{ when } x = 1 + 1/2i \text{ for } i = 1, 2, 3, \ldots ; \\ 1 + (-1)^i \text{ when } x = 1 + 1/(2i+1) \text{ for } i = 1, 2, 3, \ldots ; \\ x \text{ otherwise.} \end{cases}$$

The function $f(x)$ does not have the conventional limit at the point 1 but has different fuzzy limits at this point. For instance, 1 is a 1-limit of $f(x)$, while 2, 0, 1.5, 1.7, and 0.5 are 2-limits of $f(x)$.

Thus, we see that many functions that do not have the conventional limit at some point have lots of fuzzy limits.

It is possible to define a measure $\mu(z = \lim_{x \to a} f(x))$ of convergence of a function $f(x)$ at a point a to points from $\boldsymbol{R}$

$$\mu(z = \lim_{x \to a} f(x)) = \inf \{r ; z = r\text{-}\lim_{x \to a} f(x)\}$$

Remark 2.3.4. This measure defines the normal fuzzy set $\mathrm{Lim}_a f = [\mathrm{L}, \mu(z = \lim_{x \to a} f(x))]$ of fuzzy limits of the function $f(x)$ at the point a.

Remark 2.3.5. The property of fuzzy limits of a function at a point implies that for a bounded function $f(x)$ at the point a (cf. Appendix B) all real numbers become its fuzzy limits at this points. However, the advantage of our approach is that we classify all these fuzzy limits by the measure of convergence. Thus, it is not reasonable to consider the set of all fuzzy limits of a bounded function $f(x)$ at a point a, while the fuzzy set $\mathrm{Lim}_a f$ of all fuzzy limits of a bounded function $f(x)$ at a point a gives a lot of information about this function.

Definition 2.3.5. A number b is called a *weak* or *partial* r-limit of a function $f(x)$ at a point $a \in \boldsymbol{R}$ (it is denoted by $b = r\text{-wlim}_{x \to a} f(x)$) if there is a sequence $l = \{a_i \in \boldsymbol{R}; i = 1, 2, 3, \ldots ; a_i \neq a\}$ such that $a = \lim l$ and $b = r\text{-}\lim_{i \to \infty} f(a_i)$.

Lemma 2.3.3. If $b = r\text{-wlim}_{x \to a} f(x)$, then $b = q\text{-wlim}_{x \to a} f(x)$ for any $q > r$.

Definition 2.3.6. A number a is called a *weak fuzzy limit* of a function $f(x)$ at a point $a \in \boldsymbol{R}$ if it is a weak r-limit of a function $f(x)$ at the point a for some $r \in \boldsymbol{R}^+$.

The concept of a weak r-limit of a function gives a new concept for the classical case.

Definition 2.3.7. A number b is called a *weak* or *partial limit* of a function $f(x)$ at a point $a \in \boldsymbol{R}$ (it is denoted by $b = \mathrm{wlim}_{x \to a} f(x)$) if there is a sequence $l = \{a_i \in \mathrm{Dom} f ; i = 1, 2, 3, \ldots ; a_i \neq a\}$ such that $a = \lim l$ and $b = \lim_{i \to \infty} f(a_i)$.

In the classical mathematical analysis, some cases of weak limits (such as $\underline{\lim}$ = lim inf or $\overline{\lim}$ = lim sup) are considered. A general case of partial limits is treated in (Randolph, 1968) where weak limits, i.e., the same as weak 0-limits by Lemma 2.3.1, are called subsequential limits.

Remark 2.3.6. In the theory of hypernumbers and extrafunctions, weak limits are used to build the spectrum of a hypernumber (Burgin, 2002).

Lemma 2.3.4. If Dom $f = \boldsymbol{R}$, then the number $b = \text{wlim}_{x\to a} f(x)$ if and only if $b = 0\text{-wlim}_{x\to a} f(x)$.

Important special cases of partial limits are one-sided limits.

Definition 2.3.8. A number b is called a *left* (*right*) *limit* or *limit from the left* (*right*) or *left-hand* (*right-hand*) *limit* of a function $f(x)$ at a point a if for any $\varepsilon \in \boldsymbol{R}^{++}$ there is $\delta > 0$ such that for all $x < a$ ($x > a$), the inequality $| x - a | \leq \delta$ implies the inequality $| f(x) - b | \leq \varepsilon$.

A right limit b of $f(x)$ at a point a is denoted by $b = \lim_{x\to a+} f(x)$ or $\lim_{x\to a+} f(x) = b$ and a left limit c of $f(x)$ at a point a is denoted by $c = \lim_{x\to a-} f(x)$ or $\lim_{x\to a-} f(x) = c$.

There are sequences that do not converge but have both right and left limits. For instance, the sequence $l = \{ (-1)^i ; i = 1, 2, 3, \dots\}$ does not have a limit but its right limit is 1 and its left limit is -1.

Theorem 2.1.2 implies the following result.

Theorem 2.3.4. A right (left) limit of a function at a point is unique (if this limit exists).

Proposition 2.3.3. A function $f(x)$ has a limit at a point a if and only if $f(x)$ has both right and left limits at the point a and these limits coincide.

In a similar way, we define left and right fuzzy limits.

Definition 2.3.9. A number a is called a *left* (*right*) *r-limit* or *r-limit from the left* (*right*) of a function $f(x)$ at a point a if for any $\varepsilon \in \boldsymbol{R}^{++}$ there is $\delta > 0$ such that for all $x < a$ ($x > a$), the inequality $| x - a | \leq \delta$ implies the inequality $| f(x) - b | \leq r + \varepsilon$.

A right r-limit b of $f(x)$ at a point a is denoted by $b = r\text{-lim}_{x\to a+} f(x)$ and a left r-limit c of $f(x)$ at a point a is denoted by $c = r\text{-lim}_{x\to a-} f(x)$.

There are sequences that do not converge but have both right and left limits. For instance, the sequence $l = \{ (-1)^i ; i = 1, 2, 3, \dots \}$ does not have a limit but its right limit is 1 and its left limit is -1.

Proposition 2.3.4. If b is a right r-limit and left q-limit of a function $f(x)$ at a point a, then b a p-limit of the function $f(x)$ at the point a where $p = \max \{r, q\}$.

Indeed, by Definition 2.3.9, for any $\varepsilon \in \boldsymbol{R}^{++}$ there is $\delta_1 > 0$ such that for all $x > a$, the inequality $| x - a | \leq \delta$ implies the inequality $| f(x) - b | \leq r + \varepsilon$ and there is $\delta_2 > 0$ such that for all $x < a$, the inequality $| x - a | \leq \delta$ implies the inequality $| f(x) - b | \leq q + \varepsilon$. Then taking $\delta = \min \{\delta_1 , \delta_2\}$ and $p = \max \{r, q\}$, we have that the inequality $| x - a | \leq \delta$ implies the inequality $| f(x) - b | \leq p + \varepsilon$ for all x. It means that b a p-limit of the function $f(x)$ at the point a.

for any $\varepsilon \in \boldsymbol{R}^{++}$ there is $\delta > 0$ such that for all $x < a$ ($x > a$), the inequality $| x - a | \leq \delta$ implies the inequality $| f(x) - b | \leq r + \varepsilon$.

Corollary 2.3.6. A point b is both a right r-limit and left r-limit of a function $f(x)$ at a point a if and only if b an r-limit of the function $f(x)$ at the point a.

Definition 2.3.10. A number a is called a *left* (*right*) *fuzzy limit* or *fuzzy limit from the left* (*right*) of a function $f(x)$ at a point $a \in \boldsymbol{R}$ if it is a left (right) r-limit of a function $f(x)$ at the point a for some $r \in \boldsymbol{R}^{+}$.

Proposition 2.3.4 implies the following result.

Corollary 2.3.7. A point b is both a right fuzzy limit and left fuzzy limit of a function $f(x)$ at a point a if and only if b a fuzzy limit of the function $f(x)$ at the point a.

Definition 2.3.11. A number b is called a *left* (*right*) *weak* or *partial r-limit* of a function $f(x)$ at a point $a \in \boldsymbol{R}$ (it is denoted by $b = -r\text{-wlim}_{x\to a} f(x)$ or $b = +r\text{-wlim}_{x\to a} f(x)$, respectively) if there is a sequence $l = \{a_i \in \text{Dom} f;\ i = 1, 2, 3, \ldots;\ a_i < a\}$ (a sequence $h = \{d_i \in \text{Dom} f;\ i = 1, 2, 3, \ldots;\ d_i > a\}$) such that $a = \lim l$ and $b = r\text{-lim}_{i\to\infty} f(a_i)$ ($a = \lim h$ and $b = r\text{-lim}_{i\to\infty} f(d_i)$).

Lemma 2.3.5. If $b = +r\text{-wlim}_{x\to a} f(x)$ ($b = -r\text{-wlim}_{x\to a} f(x)$) , then $b = +q\text{-wlim}_{x\to a} f(x)$ ($b = -q\text{-wlim}_{x\to a} f(x)$) for any $q > r$.

Fuzzy limits and weak fuzzy limits of functions are invariant with respect to a continuous monotone change of the scale of the argument.

Theorem 2.3.5. If $b = r\text{-lim}_{x\to a} f(x)$ ($b = r\text{-wlim}_{x\to a} f(x)$)) and $g(x)$ is a monotone continuous function with the continuous inverse function, then $b = r\text{-lim}_{x\to h(a)} f(g(x))$ ($b = r\text{-wlim}_{x\to h(a)} f(g(x))$) where $h(a) = g^{-1}(x)$.

Proof. (a) Let us assume that $b = r\text{-lim}_{x\to a} f(x)$ and take some sequence $l = \{a_i \in \boldsymbol{R};\ i \in \omega\}$ that satisfies the condition $h(a) = \lim_{i\to\infty} a_i$. As $g(x)$ is a continuous function, we have $a = g(h(a)) = g(g^{-1}(a)) = \lim_{i\to\infty} g(a_i)$. Then by Definition 2.3.1, we have $b = r\text{-lim}_{x\to h(a)} f(g(a_i))$. Thus, $b = r\text{-lim}_{x\to h(a)} f(g(x))$ because l is an arbitrary sequence.

(b) Let us assume that $b = r\text{-lim}_{x\to a} f(x)$ and take such sequence $l = \{a_i \in \boldsymbol{R};\ i = 1, 2, 3, \ldots\}$ that satisfies the condition $h(a) = \lim_{i\to\infty} a_i$. By Definition 2.3.2, such a sequence exists. Then $h(a) = \lim_{i\to\infty} g^{-1}(a_i)$ because $g^{-1}(x)$ is a continuous function. By the definition of an r-limit, this implies that for any $\varepsilon > 0$ there is such a $\delta > 0$ that whenever $|g^{-1}(a_i) - h(a)| \leq \delta$, we have

$$|f(a_i) - a| = |f(g(g^{-1}(a_i))) - a| \leq \varepsilon$$

Consequently, $b = r\text{-wlim}_{x\to h(a)} f(g(x))$.

Theorem is proved.

Corollary 2.3.8. If $b = r\text{-lim}_{x\to a} f(x)$ ($b = r\text{-wlim}_{x\to a} f(x)$)) and $k \in \boldsymbol{R}^+$, then $b = r\text{-lim}_{x\to ha} f(kx)$ ($b = r\text{-wlim}_{x\to ha} f(kx)$)) where $h = k^{-1}$.

Theorem 2.5.1 implies the following result.

Theorem 2.3.6. For an arbitrary number $r \in \boldsymbol{R}^+$, all r-limits at a point a of a locally bounded function $f(x)$ belong to some finite interval, the length of which is equal to $2r$.

Remark 2.3.7. For unbounded functions, this property can be false.

Theorem 2.3.4 and Lemma 2.3.1 imply the following result.

Corollary 2.3.9. (Any course of calculus, cf., for example, Ross, 1996). If the limit of a function at some point exists, then it is unique.

Theorem 2.3.7. Let $b = r\text{-lim}_{x\to a} f(x)$ and $c = q\text{-lim}_{x\to a} g(x)$. Then:

a) $b + c = (r+q)\text{-lim}_{x\to a} (f + g)(x)$;
b) $b - c = (r+q)\text{-lim}_{x\to a} (f - g)(x)$;
c) $kb = (|k|\cdot r)\text{-lim}_{x\to a} (kf)(x)$ for any $k \in \boldsymbol{R}$ where $(kf)(x) = k\cdot f(x)$;
d) $b^2 = r(2|b| + r)\text{-lim}_{x\to a} f^2(x)$.

This theorem is derived as a corollary from a more general Theorem 2.3.13 proved below.

Corollary 2.3.10. (Any course of calculus, cf., for example, (Ross, 1996)). Let $b = \lim_{x\to a} f(x)$ and $c = \lim_{x\to a} g(x)$. Then:

a) $b + a = \lim_{x\to a} (f + g)(x)$;
b) $b - c = \lim_{x\to a} (f - g)(x)$;
c) $ka = \lim_{x\to a} (kf)(x)$ for any $k \in \boldsymbol{R}$.
d) $b^2 = \lim_{x\to a} f^2(x)$.

Theorem 2.3.8. A function $f(x)$ fuzzy converges at a point a if and only if $f(x)$ is bounded at this point.

Proof. Necessity. Let us take a function $f(x)$ and assume that it r-converges at a point a, but not bounded at this point. A function can be unbounded either from above or from below or from both sides. For convenience, we consider the first case. Two others are treated in a similar way.

Let us consider the sequence of closed intervals $\{[a - 1/n, a + 1/n]\ ;\ n = 1, 2, 3, \dots\}$. As $f(x)$ is not bounded at the point a, for any number n, there is a number $c_n \in [a - 1/n, a + 1/n]$ such that $f(c_n) > n$. Thus, we can choose a sequence $l = \{c_i\ ;\ i = 1, 2, 3, \dots\}$ such that $f(c_i) > i$ for all $i = 1, 2, 3, \dots$. The lengths of the intervals $\{[a - 1/n, a + 1/n]$ converge to zero. Consequently, the sequence l converges to a. The function $f(x)$ r-converges at the point a. It means (cf. Definitions 2.3.3 and 2.3.4) that it has an r-limit at the point a and for any sequence $l = \{ a_i\ ;\ a_i \in \text{Dom}\, f;\ i = 1, 2, 3, \dots, a_i \neq a\}$, the condition $a = \lim l$ implies $b = r\text{-}\lim_{i\to\infty} f(a_i)$. This inequality implies that $f(x) < b + 2r$ in some small neighborhood of a. However, this not true as the sequence of $f(c_i)$ tends to infinity. This contradiction implies that $f(x)$ has to be bounded at the point a.

Sufficiency. Let us consider a bounded at a point a function $f(x)$. It means that $f(x)$ is a bounded both above and below at the point a. The first condition means that there are a number M and an interval $[b, c]$ such that a belongs to this interval and $f(x) < M$ for all x from the interval $[b, c]$. The second condition means that there are a number m and an interval $[u, v]$ such that a belongs to this interval and $f(x) > m$ for all x from the interval $[u, v]$. Consequently, in the interval $[p, q]$ where $p = \max \{b, u\}$ and $q = \min \{c, v\}$, the equalities $m < f(x) < M$ hold for all $x \in [p, q]$.

Let us take $r = M - m$. In this case, taking some point b inside the interval (m, M) and any point x inside the interval $[p, q]$, we have $|f(x) - b| < r$. This allows us to conclude that for any $\varepsilon > 0$ there is $\delta > 0$ such that the inequality $|a - x| < \delta$ implies the inequality $|f(x) - f(a)| < r + \varepsilon$. By Definition 2.3.3, $f(x)$ r-converges at the point a.

Theorem is proved.

As in the case of sequences, fuzzy limits of functions are more adequate to real situations and provide means for more relevant models than classical limits. However, to treat functions defined on discrete sets, we need a more advanced concept of a limit.

Let $q, r \in \boldsymbol{R}^+$ and $f: \boldsymbol{R} \to \boldsymbol{R}$ be a partial function.

Definition 2.3.12. A number b is called a *(q, r)-limit* of a function $f(x)$ at a point $a \in \boldsymbol{R}_\infty$ (it is denoted by $b = (q, r)\text{-}\lim_{x\to_q a} f(x)$) if for any sequence $l = \{a_i \in \text{Dom}\, f;\ i = 1, 2, 3, \dots;\ a_i \neq a\}$, the condition $a = q\text{-}\lim l$ implies $b = r\text{-}\lim_{i\to\infty} f(a_i)$.

Example 2.3.5. Let us consider the function $f(x) = x/|x|$. The graph is presented in Figure 2.1. The classical limit $\lim_{x\to 0} (x/|x|)$ does not exist. At the same time, this function has fuzzy and fuzzy fuzzy limits. For instance, $0 = 1\text{-}\lim_{x\to 0} (x/|x|)$, $0 = (0, 1)\text{-}\lim_{x\to 0} (x/|x|)$ and $1 = (1, 2)\text{-}\lim_{x\to_1 0} (x/|x|)$.

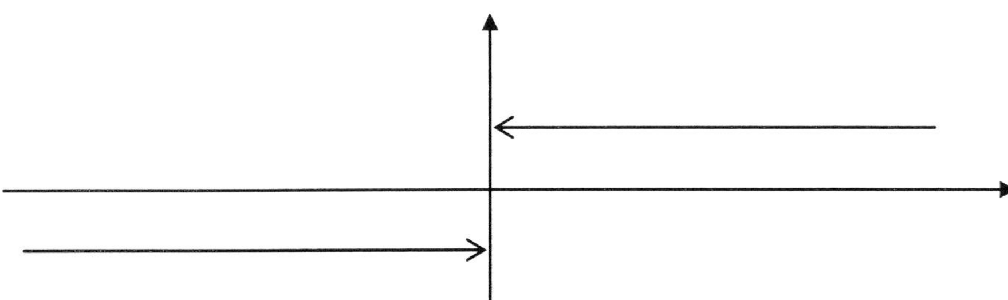

Figure 2.1. The graph of the function $f(x) = x/|x|$.

Remark 2.3.8. It is possible to define a (q, r)-limit of a function $f(x)$ at a point $a \in \boldsymbol{R}$ independently of the concept of an (q, r)-limit of a sequence (cf. Theorem 2.3.7).

Remark 2.3.9. The concept of an (q, r)-limit of a function allows one to develop a theory of limit for functions that are defined and take values in discrete sets.

Lemma 2.3.6. The number $b = r\text{-}\lim_{x\to a} f(x)$ if and only if $b = (0, r)\text{-}\lim_{x\to_0 a} f(x)$.

This result demonstrates that the concept of a (q, r)-limit of a function is a natural extension of the concept of an r-limit of a function.

Lemma 2.3.2 and properties of bounded at a point functions imply the following result.

Lemma 2.3.7. If $b = (q, r)\text{-}\lim_{x\to_q a} f(x)$, then $b = (u, v)\text{-}\lim_{x\to_u a} f(x)$ for any $v \geq r$ and $u \leq q$.

Indeed, let us take some sequence $l = \{a_i \in \text{Dom}\, f;\ i = 1, 2, 3, \ldots ;\ a_i \neq a\}$ such that $a = q\text{-}\lim l$. As $u \leq q$, for any sequence l, the condition $a = u\text{-}\lim l$ implies (by Lemma 2.3.2) the condition $a = q\text{-}\lim l$. If $b = (q, r)\text{-} \lim_{x\to_q a} f(x)$, then by properties of bounded at a point functions, for any sequence $h = \{c_i \in \text{Dom}\, f;\ i = 1, 2, 3, \ldots ;\ c_i \neq a\}$, the condition $a = q\text{-}\lim l$ implies $b = r\text{-}\lim_{i\to\infty} f(c_i)$. In particular, we have $b = r\text{-}\lim_{i\to\infty} f(a_i)$. As $v \geq r$, the condition $b = r\text{-}\lim_{i\to\infty} f(a_i)$ implies (by Lemma 2.3.2) the condition $b = v\text{-}\lim_{i\to\infty} f(a_i)$. As l is an arbitrary sequence of elements from the domain $\text{Dom}\, f$, $b = (u, v)\text{-}\lim_{x\to_u a} f(x)$ by properties of bounded at a point functions.

Theorem 2.3.9. The condition $b = (q, r)\text{-}\lim_{x\to_q a} f(x)$ is valid if and only if for any $x \in \text{Dom}\, f$ and any $\varepsilon \in \boldsymbol{R}^{++}$ there is $\delta \in \boldsymbol{R}^{++}$ such that the inequality $|\, x - a \,| < r + \delta$ implies the inequality $|\, f(x) - b \,| < r + \varepsilon$.

Proof. Necessity. Let $b = (q, r)\text{-}\lim_{x\to_q a} f(x)$ and Ob is an open neighborhood of b that contains the interval $[b - r, b + r]$. Let us suppose that in any neighborhood Oa of a such that Oa contains the interval $[a - q, a + q]$, there is a point $a_i \neq a$ such that $f(a_i)$ does not belong to Ob. We can take a sequence of neighborhoods $O_i a$ such that $O_i a \subseteq O_{i-1} a$ for all $i = 2, 3, \ldots$, in any neighborhood O_ia , there is a point $a_i \neq a$ such that $f(a_i)$ does not belong to Ob, and the intersection $\bigcap O_ia = \{a\}$. Then we have a sequence

$$l = \{a_i \in \mathbf{R};\ i = 1, 2, 3, \ldots , a_i \neq a\},$$

for which the condition $a = q$-lim l does not imply $b = r\text{-lim}_{i\to\infty} f(a_i)$. This contradicts the initial condition that $b = (q, r)\text{-lim}_{x\to_q a} f(x)$ and concludes the proof of necessity.

Sufficiency. Let for any open neighborhood Ob of b that contains the interval $[b - r, b + r]$ there is a neighborhood Oa of a such that Oa contains the interval $[a - q, a + q]$, $f(Oa) \subseteq Ob$, and $l = \{a_i \in \mathbf{R};\ i = 1, 2, 3, \ldots ;\ a_i \neq a\}$ is a sequence such that $a = q$-lim l. Then almost all elements of l belong to Oa. Consequently, almost all elements $f(a_i)$ belong to Ob. By the definition of r-limit, we have $b = r\text{-lim}_{i\to\infty} f(a_i)$. As the sequence l is chosen arbitrarily, by Definition 2.3.12, $b = (q, r)\text{-lim}_{x\to_q a} f(x)$. This concludes the proof of the theorem.

Theorem 2.3.7 and Lemma 2.3.7 imply the following result.

Theorem 2.3.10. If $b = (q, r)\text{-lim}_{x\to_q a} f(x)$ and $b > d + r$, then there is a neighborhood Oa of a such that Oa contains the interval $[a - q, a + q]$ and $f(x) > d$ for all x from $Oa \cap \text{Dom} f$.

Corollary 2.3.11. If $b = (q, r)\text{-lim}_{x\to_q a} f(x)$ and $b > d + r$, then for any sequence $l = \{a_i \in \mathbf{R};\ i = 1, 2, 3, \ldots\ a_i \neq a\}$ with $a = r$-lim l, we have $f(a_i) > d$ for almost all a_i from l.

Corollary 2.3.12. If $f(x) \leq d$ for all x from some neighborhood Oa of a that contains the interval $[a - q, a + q]$ and $b = (q, r)\text{-lim}_{x\to_q a} f(x))$, then $b \leq d + r$.

Definition 2.3.13. a) A number a is called a 2-*fuzzy limit* of a function $f(x)$ at a point $a \in \mathbf{R}$ if it is an (q, r)-limit of a function $f(x)$ at the point a for some $q, r \in \mathbf{R}^+$.

b) A function $f(x)$ 2-*fuzzy converges* at a point $a \in \mathbf{R}$ if it has a 2-fuzzy limit at this point.

Example 2.3.6. Let us consider the function $g(x) = [x]$ where $[x]$ is the largest integer that is less than or equal to x. The classical limit $\text{lim}_{x\to n} [x]$ does not exist for any integer number n. At the same time, this function has fuzzy and fuzzy fuzzy limits. For instance, $2.5 = (½ , ½)\text{-lim}_{x\to_{0.5} 3} [x]$, $5 = (0, 1)\text{- lim}_{x\to_0 5} [x]$, and $3 = (¼ , 0)\text{-lim}_{x\to_{0.25} 3} [x]$. Thus, 2.5 and 3 are 2-fuzzy limits of $g(x)$.

Remark 2.3.10. The concept of a fuzzy limit takes into account fuzziness only in one dimension. Taking the conventional x-y coordinates, this is the y- dimension. The concept of a fuzzy-fuzzy limit takes into account fuzziness in two dimensions. Namely, they are both x- and y-dimensions.

Remark 2.3.11. The concept of an (q, r)-limit of a function allows one to develop a theory of limit for functions that are defined and take values in discrete sets. Let us consider some examples.

Example 2.3.7. Let us consider the function $f(x) = x$ defined for all integers. The classical limit $\text{lim}_{x\to n} x$, as well as any r-limit $r\text{-lim}_{x\to n} x$, becomes meaningless (being automatically equal to n for any integer number n) for discerning continuous, in some sense, functions on integers as x can never come infinitely close to n without being equal to n. At the same time, this function has different fuzzy-fuzzy limits that provide for a broader concept of continuity on discrete sets. For instance, $2.5 = (1, 1.5)\text{-}\lim_{x \to_1 3} [x]$, $5 = (0.5, 1)\text{-}\lim_{x \to_{0.5} 5} [x]$, and $3 = (1,1)\text{-}\lim_{x \to_1 3} [x]$. Thus, 2.5 and 3 are fuzzy-fuzzy limits of $f(x)$.

Remark 2.3.12. Fuzzy limits or r-limits of functions form a base for the concept of fuzzy continuity or r-continuity (cf. Chapter 3). In a similar way, fuzzy-fuzzy limits or (q, r)-limits of functions form a base for the concept of fuzzy-fuzzy continuity or (q, r)-continuity.

There is another way to construct 2-fuzzy limits of functions.

Let $r \in \mathbf{R}^+$, $f: \mathbf{R} \to \mathbf{R}$ be a partial function, and $[u, v]$ be an interval in $\mathbf{R}$.

Definition 2.3.14. A number b is called an *r-limit* of a function $f(x)$ *at an interval* $[u, v]$ (it is denoted by $b = r\text{-}\lim_{x \to [u, v]} f(x))$ if for any point a from the interval $[u, v]$ and any sequence $l = \{a_i \in \text{Dom} f;\ i = 1, 2, 3, \dots\ a_i \neq a\}$ such that $a = \lim l$, we have $b = r\text{-}\lim_{i \to \infty} f(a_i)$.

Example 2.3.8. Number 2 is a 2-limit and number ½ is a ½-limit of the function $f(x) = x$ at the interval $[0, 1]$.

Theorem 2.3.11. For any point $a \in \mathbf{R}$, we have $b = (q, r)\text{-}\lim_{x \to_q a} f(x)$ if and only if $b = r\text{-}\lim_{x \to [u, v]} f(x))$ with $u = a - q$ and $v = a + q$.

Proof. Necessity. Let $b = (q, r)\text{-}\lim_{x \to_q a} f(x)$ for some point a and d be an arbitrary point from the interval $[a - q, a + q]$. If $d = \lim l$ for a sequence $l = \{a_i \in \text{Dom} f;\ i = 1, 2, 3, \dots ;\ a_i \neq a\}$, then results from Section 2.2 imply that $a = q\text{-}\lim l$. This, by Definition 2.3.7, results in $b = r\text{-}\lim_{i \to \infty} f(a_i)$. As the point d is arbitrary in the interval $[a - q, a + q]$, we have, by Definition 2.3.14, that $b = r\text{-}\lim_{x \to [u, v]} f(x))$ with $u = a - q$ and $v = a + q$.

Sufficiency. Let us assume that $b = r\text{-}\lim_{x \to [u, v]} f(x))$ with $u = a - q$ and $v = a + q$ and let us consider a sequence $h = \{c_i \in \text{Dom} f;\ i = 1, 2, 3, \dots ;\ c_i \neq a\}$ such that $a = q\text{-}\lim h$, but $b \neq r\text{-}\lim_{x \to [u, v]} f(x))$. It means that there is $\varepsilon > 0$ such for any $n \in \omega$ there is $c_i \in h$ such that $i > n$ and $| f(c_i) - b | > r + \varepsilon$. Consequently, there is a subsequence $k = \{d_i \in \text{Dom} f;\ i = 1, 2, 3, \dots ;\ d_i \neq a\}$ of the sequence h such that $| f(d_i) - b | > r + \varepsilon$ for all $i = 1, 2, 3, \dots$.

As h is a bounded sequence, k is also a bounded sequence. By Theorem 2.1.7, k has a convergent subsequence $l = \{b_i \in \text{Dom} f;\ i = 1, 2, 3, \dots ;\ b_i \neq a\}$. Let us consider the limit of this subsequence $d = \lim l$. If d does not belong to the interval $[a - q, a + q]$, then (cf. Section 2.1) almost all elements b_i from l do not belong to the interval $[a - q, a + q]$. This contradicts to the assumption that $a = q\text{-}\lim h$ (cf. Definition 2.2.1) and shows that $d \in [a - q, a + q]$.

Thus, by Definition 2.3.14, $b = r\text{-}\lim_{i \to \infty} f(b_i)$, while by our construction of the sequence l, we have $| f(d_i) - b | > r + \varepsilon$ for all $i = 1, 2, 3, \dots$. This contradiction and Definition 2.3.12 show that $b = (q, r)\text{-}\lim_{x \to_q a} f(x)$.

Theorem is proved.

Properties of fuzzy limits of function at intervals are similar to properties of fuzzy limits of function at points. For instance, Lemma 2.3.2 implies the following result.

Lemma 2.3.8. For any interval $[u, v]$, if $b = r\text{-}\lim_{x \to [u, v]} f(x)$, then $b = q\text{-}\lim_{x \to [u, v]} f(x)$ for any $q \geq r$.

Lemma 2.3.9. If $b = r\text{-}\lim_{x \to [u, v]} f(x)$ and $[u, v] \supseteq [h, k]$, then $b = r\text{-}\lim_{x \to [h, k]} f(x)$.

Definition 2.3.15. A number b is called a *weak r-limit* of a function $f(x)$ at an interval $[u, v]$ (it is denoted by $b = r\text{-wlim}_{x \to [u, v]} f(x))$ if there is a point a from the interval $[u, v]$ and any sequence $l = \{a_i \in \text{Dom} f;\ i = 1, 2, 3, \dots\ a_i \neq a\}$ such that $a = \lim l$, we have $b = r\text{-}\lim_{i \to \infty} f(a_i)$.

The concept of an r-limit of a function $f(x)$ at an interval gives new concepts for the classical case.

Definition 2.3.16. A number b is called a *limit* of a function $f(x)$ at an interval $[u, v]$ (it is denoted by $b = \lim_{x \to [u, v]} f(x)$) if for any point a from the interval $[u, v]$ and any sequence $l = \{a_i \in \text{Dom} f;\ i = 1, 2, 3, \dots\ a_i \neq a\}$ with $a = \lim l$, we have $b = \lim_{i \to \infty} f(a_i)$.

However, this construction is not very interesting as the following proposition shows.

Proposition 2.3.5. If a limit a function $f(x)$ at an interval $[u, v]$ exists, then $f(x)$ is almost constant on $[u, v]$, that is, for all but finitely many points from $[u, v]$, the values of $f(x)$ coincide.

Definition 2.3.17. A number b is called a *weak limit* of a function $f(x)$ at an interval $[u, v]$ (it is denoted by $b = \text{wlim}_{x \to [u, v]} f(x)$) if there is a point a from the interval $[u, v]$ such that for any sequence $l = \{a_i \in \text{Dom} f;\ i = 1, 2, 3, \dots\ ;\ a_i \neq a\}$, the condition $a = \lim l$ implies $b = r\text{-}\lim_{i \to \infty} f(a_i)$.

Remark 2.3.13. It is known that a function $f(x)$ is continuous at a point $a \in \mathbf{R}$ if $f(a) = \lim_{x \to a} f(x)$. A similar statement for weak limits at an interval is not true as the following example demonstrates.

Example 2.3.9. Let us consider the following function

$$f(x) = \begin{cases} 1/3 \text{ when } x = 2/3; \\ 2/3 \text{ when } x = 1/3; \\ x \text{ otherwise.} \end{cases}$$

Then $f(\ [0, 1]\ ) = \text{L}(\ \lim_{x \to [0, 1]} f(x))$. However, the function f is not continuous in the interval $[0, 1]$.

Definition 2.3.18. A number b is called a weak (q, r)-limit of a function $f(x)$ at a point $a \in \mathbf{R}$ (it is denoted by $b = (q, r)\text{-wlim}_{x \to_q a} f(x)$) if there is a sequence $l = \{a_i \in \text{Dom} f;\ i = 1, 2, 3, \dots\ a_i \neq a\}$ such that $a = q\text{-}\lim l$ and $b = r\text{-}\lim_{i \to \infty} f(a_i)$.

Lemma 2.3.10. The number $b = r\text{-wlim}_{x \to a} f(x)$ if and only if $b = (0, r)\text{-wlim}_{x \to_0 a} f(x)$.

This result demonstrates that the concept of a weak (q, r)-limit of a function is a natural extension of the concept of a weak r-limit of a function.

Lemma 2.3.11. If $a = (q, r)\text{-wlim}_{x \to_q a} f(x)$, then $a = (u, v)\text{-wlim}_{x \to_u a} f(x)$ for any $v \geq r$ and $u \geq q$.

Proof is similar to the proof of Lemma 2.3.7.

Theorems 2.5.1 and 2.3.9 imply the following result.

Theorem 2.3.12. For an arbitrary interval $[u, v]$, all r-limits at points from the interval $[u, v]$ of a bounded in $[u, v]$ function $f(x)$ belong to some finite interval, the length of which is equal to $v - u + 2r$.

Remark 2.3.13. For unbounded functions, this property can be false.

Theorem 2.3.13. Let $b = (q, r)\text{-}\lim_{x \to_q a} f(x)$ and $c = (h, k)\text{-}\lim_{x \to_h a} g(x)$. Then:

a) $b + c = (u, v)\text{-}\lim_{x \to_u a} (f + g)(x)$ where $v = r + k$ and $u = \min \{q, h\}$;

b) $b - c = (u, v)\text{-}\lim_{x \to_u a} (f - g)(x)$ where $v = r + k$ and $u = \min \{q, h\}$;

c) $kb = (q, |k| \cdot r)\text{-}\lim_{x \to_q a} (kf)(x)$ for any $k \in \mathbf{R}$ where $(kf)(x) = k \cdot f(x)$.

d) $b^2 = (q, r(2|b| + r))\text{-}\lim_{x \to_q a} f^2(x)$.

Proof. a) Let $b = (q, r)\text{-}\lim_{x \to_q a} f(x)$ and $c = (h, k)\text{-}\lim_{x \to_q a} g(x)$. Then by Definition 2.3.7, for any sequence $l = \{b_i \in \text{Dom} f;\ i = 1, 2, 3, \dots\ , b_i \neq a\}$ such that $a = q\text{-}\lim l$, we have $b = r\text{-}$

$\lim_{i\to\infty} f(b_i)$ and for any sequence $t = \{c_i \in \text{Dom } g;\ i = 1, 2, 3, \dots, c_i \neq a\}$ such that $a = h\text{-lim } t$, we have $c = k\text{-lim}_{i\to\infty} f(c_i)$. Let $u = \min \{q, h\}$. Then for any sequence $d = \{d_i \in \text{Dom } f \cap \text{Dom } g;\ i = 1, 2, 3, \dots, d_i \neq a\}$ such that $a = u\text{-lim } d$, we have (by Lemma 2.2.4) $a = q\text{-lim } d$ and $a = h\text{-lim } d$. Thus, by Definition 2.3.7, we have $b = r\text{-lim}_{i\to\infty} f(d_i)$ and $c = k\text{-lim}_{i\to\infty} f(d_i)$. By Theorem 2.2.5, we have $b + c = (r + k)\text{-lim}_{i\to\infty} f(d_i)$. As d is an arbitrary sequence of elements from the domain Dom $f + g$, we have $b + c = (u, v)\text{-lim}_{x\to_u a} (f + g)(x)$ by Definition 2.3.7.

Parts b), c) and d) are proved in a similar way based on corresponding results for fuzzy limits of sequences.

Theorem is proved.

It is possible to deduce Theorem 2.3.7 from Theorem 2.3.13.

Proposition 2.3.6. If $g(x) \leq f(x) \leq h(x)$ in some neighborhood of the point a that contains the interval $[a - q, a + q]$, $b = (q, r)\text{-lim}_{x\to qa}\, g(x)$ and $b = (q, r)\ \lim_{x\to_q a} h(x)$ ($b = (q, r)\text{-wlim}_{x\to_q a}\, g(x)$ and $b = (q, r)\text{-wlim}_{x\to_q a}\, h(x)$), then $b = (q, r)\text{-lim}_{x\to_q a} f(x)$ ($b = (q, r)\text{-wlim}_{x\to_q a} f(x)$).

Corollary 2.3.13. If $g(x) \leq f(x) \leq h(x)$ in some neighborhood of the point a, $b = r\text{-lim}_{x\to a}\, g(x)$ and $b = r\text{-lim}_{x\to a}\, h(x)$ ($b = r\text{-wlim}_{x\to a}\, g(x)$ and $b = r\text{-wlim}_{x\to a}\, h(x)$), then $b = r\text{-lim}_{x\to a} f(x)$ ($b = r\text{-wlim}_{x\to a} f(x)$).

Corollary 2.3.14. If $g(x) \leq f(x) \leq h(x)$ in some neighborhood of the point a, $b = \lim_{x\to a}\, g(x)$ and $b = \lim_{x\to a}\, h(x)$ ($b = \text{wlim}_{x\to a}\, g(x)$ and $b = \text{wlim}_{x\to a}\, h(x)$), then $b = \lim_{x\to a} f(x)$ ($b = \text{wlim}_{x\to a} f(x)$).

Definition 2.3.19. A number b is called a *right* (*left*) *(q, r)-limit* of a function $f(x)$ at a point $a \in \mathbf{R}_\infty$ (it is denoted by $b = (q, r)\text{-lim}_{x\to_q a+} f(x)$ or $b = (q, r)\text{-lim}_{x\to_q a-} f(x)$) if for any sequence $l = \{a_i \in \text{Dom } f;\ i = 1, 2, 3, \dots ;\ a_i > a\}$ ($l = \{a_i \in \text{Dom } f;\ i = 1, 2, 3, \dots ;\ a_i < a\}$), the condition $a = q\text{-lim } l$ implies $b = r\text{-lim}_{i\to\infty} f(a_i)$.

Lemma 2.3.12. If $b = (q, r)\text{-lim}_{x\to_q a+} f(x)$ [$b = (q, r)\text{-lim}_{x\to_q a-} f(x)$], then $b = (u, v)\text{-lim}_{x\to_u a+} f(x)$ [$b = (u, v)\text{-} \lim_{x\to_u a-} f(x)$] for any $v \geq r$ and $u \leq q$.

Proof is similar to the proof of Lemma 2.3.7.

Theorem 2.3.14. Let $b = (q, r)\text{-lim}_{x\to_q a+} f(x)$ and $c = (h, k)\text{-lim}_{x\to_u a+}\, g(x)$ [$b = (q, r)\text{-lim}_{x\to_q a-} f(x)$ and $c = (h, k)\text{-lim}_{x\to_u a-}\, g(x)$]. Then:

a) $b + c = (u, v)\text{-lim}_{x\to_u a+} (f + g)(x)$ [$b + c = (u, v)\text{-lim}_{x\to_u a-} (f + g)(x)$] where $v = r + k$ and $u = \min \{q, h\}$;

b) $b - c = (u, v)\text{-lim}_{x\to_u a+} (f - g)(x)$ [$b - c = (u, v)\text{-lim}_{x\to_u a-} (f - g)(x)$] where $v = r + k$ and $u = \min \{q, h\}$;

c) $kb = (q, |k|\cdot r)\text{-lim}_{x\to_q a+} (kf)(x)$ [$kb = (q, |k|\cdot r)\text{-lim}_{x\to_q a-} (kf)(x)$] for any $k \in \mathbf{R}$ where $(kf)(x) = k\cdot f(x)$.

d) $b^2 = (q, r(2|b| + r))\text{-lim}_{x\to_q a+} f^2(x)$ [$b^2 = (q, r(2|b| + r))\text{-lim}_{x\to_q a-} f^2(x)$].

Proof is similar to the proof of Theorem 2.3.13.

2.4. Conditional Fuzzy Limits of Sequences and Nearness Relations

We were allowed only so much film per picture,
but there was no limit to the creativity.
Helmut Newton (1920-2004)

2.4.1. Conditional Fuzzy Limits of Sequences

The idea of conditional fuzzy convergence is rooted in some fundamental computational problems. For a numerical computation, it is important to prove not just the convergence of an algorithm, but that it converges at a sufficient rate. Thus, in problems of numerical analysis and computations, we need not only to know that a sequence or a series converges, but also to estimate the rate of the convergence.

Let us assume that $r \in \boldsymbol{R}^+$, $l = \{a_i \in \boldsymbol{R};\ i = 1, 2, 3, \dots\}$ is a sequence of real numbers, and B is a condition for (property of) sequences, that is, B is a subset of the set $\boldsymbol{R}^\omega$ of all real sequences.

Definition 2.4.1. a) A number a is called a B-*conditional r-limit* of a sequence l (it is denoted by $a = (\mathrm{B}, r)\text{-}\lim_{i\to\infty} a_i$ or $a = (\mathrm{B}, r)\text{-}\lim\ l$) if:

(1) for any $\varepsilon \in \boldsymbol{R}^{++}$ the inequality $|\ a - a_i\ | < r + \varepsilon$ is valid for almost all a_i , i.e., there is such n that for any $i > n$, we have $|\ a - a_i\ | < r + \varepsilon$;

(2) l satisfies the condition B, i.e., $l \in \mathrm{B}$.

b) a sequence l B-*conditionally r-converges* if it has a B-conditional r-limit.

For $r = 0$, we have new concepts of the standard analysis.

Definition 2.4.2. A number a is called a B-*conditional limit* of a sequence l (it is denoted by $a = \mathrm{B}\text{-}\lim_{i\to\infty} a_i$ or $a = \mathrm{B}\text{-}\lim\ l$) if $a = \lim\ l = \lim_{i\to\infty} a_i$ and all triples (a, a_i , i) with a_i from l satisfy the condition B.

b) a sequence l B-*conditionally converges* if it has a B-conditional fuzzy limit.

Remark 2.4.1. If $a = (\mathrm{B}, r)\text{-}\lim\ l$ for some condition B, then $a = r\text{-}\lim\ l$. When the condition B is empty, B-conditional r-limit coincides with the concept of an r-limit of a sequence.

In a similar way, if $a = \mathrm{B}\text{-}\lim\ l$ for some condition B, then $a = \lim\ l$.

Let us consider some examples in which we define several important types of such conditions B.

Example 2.4.1. a) $\mathrm{B}_{\mathrm{lin}(r,\,k)}$ consists of all sequences $l = \{a_i \in \boldsymbol{R};\ i = 1, 2, 3, \dots\}$ such that $|\ a - a_i\ | < r + k/i$ for some number k and all $i > n$ where n is some natural number. In particular, $\mathrm{B}_{\mathrm{lin}(k)}$ consists of all sequences $l = \{a_i \in \boldsymbol{R};\ i = 1, 2, 3, \dots\}$ such that $|\ a - a_i\ | < k/i$. This condition defines *absolute linear convergence*.

b) $\mathrm{B}_{\mathrm{Lin}(r,\,k)}$ consists of all sequences $l = \{a_i \in \boldsymbol{R};\ i = 1, 2, 3, \dots\}$ such that $|\ a_i - a_{i+1}| < r + k/i$ for some number k and all $i > n$ where n is some natural number. In particular, $\mathrm{B}_{\mathrm{Lin}(k)}$ consists of all sequences $l = \{a_i \in \boldsymbol{R};\ i = 1, 2, 3, \dots\}$ such that $|\ a_i - a_{i+1}| < k/i$. This condition defines *relative linear convergence*.

Example 2.4.2. a) $\mathrm{B}_{\exp(r,\,k)}$ consists of all sequences $l = \{a_i \in \boldsymbol{R};\ i = 1, 2, 3, \dots\}$ such that $|\ a - a_i\ | < r + q^{ki}$ for some number k, number q $(0 < q < 1)$, and all $i > n$ where n is some natural number. In particular, $B_{\exp(k)}$ consists of all sequences $l = \{a_i \in \boldsymbol{R};\ i = 1, 2, 3, \dots\}$ such that $| a - a_i | < q^{ki}$. This condition defines *absolute exponential convergence.*

b) $B_{\mathrm{Exp}(r,\ k)}$ consists of all sequences $l = \{a_i \in \boldsymbol{R};\ i = 1, 2, 3, \dots\}$ such that $| a_i - a_{i+1}| < r + q^{ki}$ for some number k, number q $(0 < q < 1)$, and all $i > n$ where n is some natural number. In particular, $B_{\mathrm{Exp}(k)}$ consists of all sequences $l = \{a_i \in \boldsymbol{R};\ i = 1, 2, 3, \dots\}$ such that $| a_i - a_{i+1}| < q^{ki}$. This condition defines *relative exponential convergence.*

Exponential strict convergence is basic for asymptotic series.

Example 2.4.3. a) $B_{\log(r,\ k)}$ consists of all sequences $l = \{a_i \in \boldsymbol{R};\ i = 1, 2, 3, \dots\}$ such that $| a - a_i | < r + 1/(k(\log i))$ for some number k and all $i > n$ where n is some natural number. In particular, $B_{\log(k)}$ consists of all sequences $l = \{a_i \in \boldsymbol{R};\ i = 1, 2, 3, \dots\}$ such that $| a - a_i | < r + 1/(k(\log i))$. This condition defines *absolute logarithmic convergence.*

b) $B_{\mathrm{Log}(r,\ k)}$ consists of all sequences $l = \{a_i \in \boldsymbol{R};\ i = 1, 2, 3, \dots\}$ such that $| a_i - a_{i+1}| < r + 1/(k(\log i))$ for some number k, and all $i > n$ where n is some natural number. In particular, $B_{\mathrm{Log}(k)}$ consists of all sequences $l = \{a_i \in \boldsymbol{R};\ i = 1, 2, 3, \dots\}$ such that $| a_i - a_{i+1}| < r + 1/(k(\log i))$. This condition defines *relative logarithmic convergence.*

In all previous examples, the condition B is defined by some sequences. We call such condition *sequential* because it is defined for sequences. For example, a sequence l satisfies the condition B if all pairs (a_i, a_{i+1}) with a_i , a_{i+1} from l belong to some set B_0 . We have another example of a condition B only for one sequence, namely, a sequence l satisfies the condition B if there is a number $n > 0$ such that for all $i > n$ each pair (a_i, a_{i+1}) with a_i , a_{i+1} from l belongs to B. One more example is given by the Generalized Cauchy Property: a sequence l satisfies the Generalized Cauchy Property with respect to B if there is a number $n > 0$ such that for all $i, j > n$ each pair (a_i , a_j) with $j > i$ and a_i , a_j from l belongs to B.

However, there are other important types of conditions for convergence, for example, a condition for pairs that consist of a sequence and a number. In this case, B_0 is a subset of the direct product $\boldsymbol{R}^{\omega}\times\boldsymbol{R}$ and condition 2) from Definition 2.4.1.a) is modified to the following condition:

(2°) the pair (l , a) satisfies the condition B, i.e., $(l , a) \in B_0$.

Example 2.4.4. A sequence $l = \{a_i \in \boldsymbol{R};\ i = 1, 2, 3, \dots\}$ and a number a satisfy the condition B if all a_i and a are rational numbers. This conditions reflects the situation with computers, which can operate exactly only with rational numbers.

Let b be some positive real number.

Example 2.4.5. A sequence $l = \{a_i \in \boldsymbol{R};\ i = 1, 2, 3, \dots\}$ and a number a satisfy the condition B if all a_i and a are rational numbers from the interval $[- b, b]$. This condition reflects the situation with computers, which can exactly operate only with rational numbers that have bounded absolute values.

Example 2.4.6. A sequence $l = \{a_i \in \boldsymbol{R};\ i = 1, 2, 3, \dots\}$ and a number a satisfy the condition B if all a_i are less (greater) than or equal to a. This condition defines convergence from the left (from the right).

Remark 2.4.2. Actually it is possible to include condition 1) from Definition 2.4.1 in the condition B. Then a sequence l B-conditionally r-converges if and only if it satisfies the condition B (see Examples 2.4.1 –2.4.3).

Definition 2.4.3. a) A number a is called a B-*conditional fuzzy limit* of a sequence l if it is a B-conditional r-limit of l for some $r \in \boldsymbol{R}^+$.

b) a sequence l *fuzzy* B-*converges* if it has a B-conditional fuzzy limit.

In what follows, we mostly consider conditions B that are applied only to sequences.

Some properties of conditional fuzzy limit of a sequence are similar to the corresponding properties of ordinary fuzzy limits and have similar proofs, while other properties are essentially different.

For example, we have the following result.

Proposition 2.4.1. If a = (B, r)-lim l, then a = (B, q)-lim l for any $q > r$.

Indeed, a = (B, r)-lim l implies that for any $\varepsilon \in \boldsymbol{R}^{++}$, the inequality $| a - a_i | < r + \varepsilon$ is valid for almost all a_i, i.e., there is such n, that for any $i > n$, we have $| a - a_i | < r + \varepsilon$. As $q > r$, we have that for any $\varepsilon \in \boldsymbol{R}^{++}$, the inequality $| a - a_i | < q + \varepsilon$ is valid for almost all a_i. In addition, a = (B, r)-lim l implies that all pairs (a, a_i) satisfy the condition B. Consequently, a = (B, q)-lim l.

Remark 2.4.3. Not all properties of fuzzy limits remain valid for conditional fuzzy limits. For example, for ordinary fuzzy limits, we have the following result.

Let $r \in \boldsymbol{R}^+$, $l = \{a_i \in \boldsymbol{R};\ i = 1, 2, 3, \ldots\}$, $h = \{b_i \in \boldsymbol{R};\ i = 1, 2, 3, \ldots\}$, $k = \{c_i \in \boldsymbol{R};\ i = 1, 2, 3, \ldots\}$, and the sequence l is the disjoint union of the sequences h and k, that is, elements of l are taken from h and k without repetition. For example, if $a_{2i} = b_i$ and $a_{2i+1} = c_i$ for all $i = 1, 2, 3, \ldots\}$, then $l = \{a_i \in \boldsymbol{R};\ i = 1, 2, 3, \ldots\}$ is the disjoint union of h and k. In this case (cf. Section 2.2), $a = r$-lim l if and only if $a = r$-lim h and $a = r$-lim k.

In particular, the limit of a sequence is the limit of any its subsequence. For conditional convergence, this property is usually violated.

Let us take the condition B that is dual to the condition of absolute linear convergence $B_{lin(1)}$: $|b - a_i| > 1/i$ for all $i > n$ where n is some natural number. (cf., example 2.4.1.a). Then the sequence $l = \{1/(i - 1)\ ;\ i = 1, 2, 3, \ldots\}$ B-converges to 0. However, if we take as the sequence h such subsequence of l that consists of all odd elements from l and as the sequence h such subsequence of l that consists of all even elements from l, then neither h nor k B-converges to 0.

At the same time, it is possible to deduce some properties of conditional fuzzy limits from the corresponding properties of ordinary fuzzy limits.

Theorem 2.2.1 implies the following result.

Theorem 2.4.1. If a = (B, r)-lim l and $a > b + r$, then $a_i > b$ for almost all a_i from l.

Theorem 2.5.1 imply the following result.

Theorem 2.4.2. For an arbitrary number $r \in \boldsymbol{R}^+$, all (B, r)-limits of a sequence l belong to some interval, the length of which is equal to $2r$.

Corollary 2.4.1. If a conditional limit of a sequence exists, then it is unique.

Corollary 2.4.2. (Any course of calculus, cf., for example, (Ross, 1996)). If a limit of a sequence exists, then it is unique.

Theorem 2.2.2 implies the following result.

Theorem 2.4.3. A sequence l fuzzy B-converges if and only if it is bounded and satisfies the condition B.

Corollary 2.4.3. (Any course of calculus: cf., for example, (Ross, 1996)). Any convergent sequence is bounded.

Let B and H be some conditions on sequences and $l = \{a_i \in \boldsymbol{R};\ i = 1, 2, 3, \ldots\}$.

Proposition 2.4.2. If the condition B implies the condition H and a = (B, r)-lim l, then a = (H, q)-lim l.

Really, any B-sequence is an H-sequence. Thus, if a = (B, r)-lim l, l is a B-sequence, and consequently, H-sequence. As the first condition from Definition 2.4.1 is satisfied, a = (H, q)-lim l.

Let us consider such condition B that separates all sequences of real numbers into two classes: all sequences that satisfy B and those sequences that do not satisfy B.

Theorem 2.4.4. For an arbitrary number $r \in \mathbf{R}^+$ and an arbitrary sequence l, the set $L_{B,r}(l) = \{a \in \mathbf{R};\ a =$(B, r)-lim $l\}$ of all (B, r)-limits of the sequence l is a convex closed set, i.e., either $L_{B,r}(l) = [a, b]$ for some $a,b \in \mathbf{R}$, or $L_{B,r}(l) = \varnothing$ when l has no (B, r)-limits.

Really, if the sequence l does not satisfy the condition B, then $L_{B,r}(l) = \varnothing$. If the sequence l satisfy the condition B, but does not have r-limits, then $L_{B,r}(l) = \varnothing$. If the sequence l satisfy the condition B and has r-limits, then any r-limit of l is a (B, r)-limit of l. Results from Section 2.5 show the set of all r-limits of l is a convex closed set. So, $L_{B,r}(l) = [a, b]$ for some $a, b \in \mathbf{R}$.

Let us take a condition B that is defined by a ternary relation $B \subseteq \mathbf{R} \times \mathbf{R} \times \mathbf{N}$.

Definition 2.4.4. A condition B is called:

1. *closed under contractions* if the conditions $(x, y, i) \in B$ and $|u - v| \leq |x - y|$ imply $(u, v, i) \in B$;
2. *closed under translations* if $(x, y, i) \in B$ and $t \in \mathbf{R}$ imply $(x + t, y + t, i) \in B$;
3. *closed with respect to addition* (*subtraction* or ***R**-multiplication*) if $(x, y, i) \in B$ and $(u, v, i) \in B$ imply $(x + u, y + v, i) \in B$ (respectively, $(x - u, y - v, i) \in B$ or $(kx, ky, i) \in B$).

Let $l = \{a_i \in \mathbf{R};\ i = 1, 2, 3, \dots\}$ and $h = \{b_i \in \mathbf{R};\ i = 1, 2, 3, \dots\}$. Then their sum $l+h$ is equal to the sequence $\{ a_i + b_i ;\ i = 1, 2, 3, \dots\}$ and their difference $l - h$ is equal to the sequence $\{ a_i - b_i ;\ i = 1, 2, 3, \dots\}$.

Theorem 2.4.5. Let a = (B, r)-lim l and b = (B, q)-lim h. Then:

a. if the condition B is closed with respect to addition, then $a + b$ = (B, $r+q$)-lim($l+h$) ;
b. if the condition B is closed with respect to subtraction, then $a - b$ = (B, $r+q$)-lim($l - h$);
c. if the condition B is closed with respect to **R**-multiplication, then ka = (B, $|k|\cdot r$)-lim (kl) for any $k \in \mathbf{R}$ where $kl = \{ka_i ;\ i = 1, 2, 3, \dots\}$.
d. if the condition B is closed under translations, then a = (B, r)-lim $(+k)l$ for any $k \in \mathbf{R}$ where $(+k)l = \{a_i + k ;\ i = 1, 2, 3, \dots\}$.

Proof. a) If the condition B is closed with respect to addition and the sequences l and h satisfy the condition B, then the sequences $l + h$ also satisfies the condition B. Consequently, any r-limit of $l + h$ is a (B, r)-limit of $l + h$. By Theorem 2.2.5, $a + b$ = ($r+q$)-lim($l+h$). Consequently, $a + b$ = (B, $r+q$)-lim($l+h$).

Proofs for the parts b), c) and d) are similar.

Corollary 2.4.4. Let a = B-lim l and b = B-lim h. Then:

a) if the condition B is closed with respect to addition, then $a + b =$ B-lim$(l+h)$;
b) if the condition B is closed with respect to subtraction, then $a - b =$ B-lim$(l - h)$;
c) if the condition B is closed with respect to $\boldsymbol{R}$-multiplication, then $ka =$ B-lim (kl) for any $k \in \boldsymbol{R}$ where $kl = \{ka_i ; i = 1, 2, 3, \ldots\}$.

Remark 2.4.4. Properties of the condition B are essential for the validity of Theorem 2.4.5 and Corollary 2.4.4. To show this for the part a), let us take the condition B: $| a - a_i | < 1/i$ and the sequence $l = \{1/i; i = 1, 2, 3, \ldots\}$. Then 0 is the limit of l, but $0 + 0 = 0$ is not a B-conditional limit of the sequence $l + l$ because $l + l$ does not satisfy the condition B.

However, if we take the condition $\mathrm{B}_{\mathrm{lin}(k)}$: $| a - a_i | < k/i$, which is closed with respect to addition, then $0 + 0 = 0$ is a $\mathrm{B}_{\mathrm{lin}(k)}$ -conditional limit of the sequence $l + l$, as well as of any sequence nl where n is an arbitrary natural number.

Let us assume that the condition B is closed under contractions and take two sequences $l = \{a_i \in \boldsymbol{R}; i = 1, 2, 3, \ldots\}$ and $h = \{b_i \in \boldsymbol{R}; i = 1, 2, 3, \ldots\}$.

Proposition 2.4.3. If $a =$ (B, r)-lim l and $| a - b_i | \leq | a - a_i|$ for all $i = 1, 2, 3, \ldots$, then $a =$ (B, q)-lim h.

Proof. As the condition B is closed under contractions, the sequence h satisfies the condition B. By the definition of a conditional limit, for any $\varepsilon \in \boldsymbol{R}^{++}$ the inequality $| a - a_i| < r + \varepsilon$ is valid for almost all a_i. Thus, the inequality $| a - b_i | < r + \varepsilon$ is valid for almost all b_i. It means that $a =$ (B, q)-lim h.

Proposition 2.4.3 is proved.

Corollary 2.4.5. If $a =$ B-lim l and $| a - b_i | \leq | a - a_i|$ for all i then $a =$ B-lim h.

Remark 2.4.5. The assumed property of the condition B is essential for the validity of Proposition 2.4.3 and Corollary 2.4.5.

Corollary 2.4.6. If $a = \lim l$ and $| a - b_i | \leq | a - a_i|$ for all i then $a = \lim h$.

2.4.2. Nearness Convergence

Nearness convergence is based on the concept of a nearness relation.

Definition 2.4.5. (Kalina 1997) A function N: $\boldsymbol{R} \times \boldsymbol{R} \rightarrow [0, 1]$ is called a *nearness relation* on $\boldsymbol{R}$ if it satisfies the following conditions:

a. for each $x, y \in \boldsymbol{R}$, $\mathrm{N}(x, y) = 1$ if $x = y$
b. for each $x, y \in \boldsymbol{R}$, $\mathrm{N}(x, y) = \mathrm{N}(y, x)$
c. if $x < y < t < z$, then $\mathrm{N}(y, t) \geq \mathrm{N}(x, z)$
d. for each $x \in \boldsymbol{R}$, we have $\lim_{y \to \pm\infty} \mathrm{N}(x, y) = 0$

If moreover the following condition holds

e. for each $x, y, c \in \boldsymbol{R}$, we have $\mathrm{N}(x, y) = \mathrm{N}(x + c, y + c)$,

then the nearness relation N is called *shift-invariant*.

In what follows, let N be any nearness relation, $l = \{a_i \in \boldsymbol{R}; i = 1, 2, 3, \ldots \}$ be a sequence of real numbers and $\alpha \in (0, 1)$.

Remark 2.4.6. A nearness relation on $\boldsymbol{R}$ is a kind of a fuzzy binary relation on $\boldsymbol{R}$ (Zimmermann, 1991).

Definition 2.4.6. A sequence l is (α, N)-*convergent* if there exists a number a such that the following holds

$$\text{if } 1 - 1/i \geq \alpha, \text{ then } \mathrm{N}(a_i, a) \geq \alpha$$

In this case, the α-nearness limit of the sequence l is defined by the formula

$$(\alpha, \mathrm{N})\text{-lim } l = \{ b \,;\, \mathrm{N}(a_i, b) \geq \alpha \text{ for all } i \geq 1/(1 - \alpha) \}.$$

Lemma 2.4.1. If a sequence l is (α, N)-convergent for some $\alpha > 0$ (N-convergent), then l is bounded.

Definition 2.4.7. A sequence $l = \{a_i \,;\, i = 1, 2, 3, \ldots \}$ is *nearness convergent* with respect to the nearness N (or N-convergent, for short), if there is $\beta < 1$ for which the sequence l is (α, N)-convergent for all $\alpha \geq \beta$. The nearness limit of the N-convergent sequence l is defined by the formula

$$\mathrm{N}\text{-lim } l = \{ b \,;\, \mathrm{N}(b, \liminf_{i\to\infty} a_i) = 1\} \cap \{ b \,;\, \mathrm{N}(b, \limsup_{i\to\infty} a_i) = 1\}$$

In the case when N-lim $l = \{d\}$ (is a one-point set), we write N-lim $l = d$.

Proposition 2.4.4. If N and M are nearness relations and M $\geq$ N, then a sequence l is (α, M)-convergent (M-convergent) whenever the sequence l is (α, N)-convergent (N-convergent).

Remark 2.4.7. Definitions 2.4.5 and 2.4.6, as well as Definitions 2.4.7 and 2.4.8, show the main difference between the nearness limit and fuzzy or conditional fuzzy limit of a sequence. The nearness limit of a sequence is a set valued mapping that corresponds an interval of real numbers to the sequence. At the same time, the fuzzy and conditional fuzzy limits of a sequence are conventional mappings that correspond a number to the sequence as its fuzzy or fuzzy conditional limit, as well as the fuzzy limit set and fuzzy conditional limit set of this sequence. Fuzzy limits are elements of a fuzzy limit set, while conditional fuzzy limits are elements of a fuzzy conditional limit set. However, different numbers may be corresponded in such way to the same sequence. The nearness convergence approach is closer to the conventional calculus and interval calculus (Moore, 1966; Alefeld, and Herberger, 1983) with their actualization of infinite sets. The fuzzy conditional convergence approach is closer to the computational environment, in which only separate values or finite sets of such values are computed.

At the same time, as it is demonstrated below nearness limits and conditional fuzzy limits of sequences are closely related to each other.

Let us consider two sequences $l_1 = \{ a_i \in \boldsymbol{R} \,;\, i = 1, 2, 3, \ldots\}$ and $l_2 = \{ b_i \in \boldsymbol{R} \,;\, i = 1, 2, 3, \ldots\}$.

Definition 2.4.8. The sequences l_1 and l_2 relatively (α, N)-*converge* to each other if and only if each inequality $i \geq 1/(1-\alpha)$ yields $\mathrm{N}(a_i, b_i) \geq \alpha$.

We say that l_1 and l_2 relatively N-converge to each other if and only if there exists a number $\beta < 1$ such that for all α with $1 > \alpha \geq \beta$, the sequences l_1 and l_2 relatively (α, N)-converge to each other.

Lemma 2.4.2. A sequence l is (α, N)-convergent (N-convergent) if and only if there is a constant sequence $h = \{d_i ; i = 1, 2, 3, \ldots\}$ with $d_i = d$ for some $d \in \boldsymbol{R}$ such that l and h (α, N)-converge (N-converge) to each other.

Example 2.4.7. Let us consider the following nearness relation:

$$N_1(x, y) = \max\{ 0, 1 - | x - y |\}$$

Then, the sequence $\{ a_i = 1/i; i = 1, 2, 3, \ldots\}$ is N_1–convergent with

$$N_1\text{-}\lim_{i\to\infty} a_i = 0.$$

However, the sequence $\{ b_i = 2/i ; i = 1, 2, 3, \ldots \}$ is not N_1-convergent. It is easy to show that the sequences $\{ a_i \in \boldsymbol{R} ; i = 1, 2, 3, \ldots \}$ and $\{b_i \in \boldsymbol{R} ; i = 1, 2, 3, \ldots \}$ relatively N_1-converge to each other. Thus, we can see that nearness convergence is not a transitive relation on the set of all real sequences.

Example 2.4.8. Let us take the following nearness relation

$$N_2(x,y) = \begin{cases} 1, \text{ iff } x = y \\ \max\{ 0, ¾ (1 - | x - y |) \} \text{ otherwise} \end{cases}$$

Only such sequences that are constant beginning with some index i are N_2-convergent. Only such sequences, elements of which equal to each other beginning with some index i, relatively N_2-converge to each other. Taking the sequence $\{ a_i ; i = 1, 2, 3, \ldots\}$ from Example 2.4.8, we have

$$(1/3, N_2)\text{-}\lim_{i\to\infty} a_i = [0; 1/3]$$

Sequences $\{ a_i ; i = 1, 2, 3, \ldots \}$ and $\{ b_i; i = 1, 2, 3, \ldots \}$ from Example 2.4.8 relatively (1/3, N_2)- converge to each other.

Example 2.4.9. Let us take the following nearness relation

$$N_3(x,y) = \begin{cases} 1, \text{ if } | x - y | \leq 1 \\ \max \{0, 2 - | x - y |\} \text{ otherwise} \end{cases}$$

Consider the sequence given by $\{ c_i = (-1)^i ; i = 1, 2, 3, \ldots\}$. This sequence N_3-converges and the following condition is true

$$N_3\text{-}\lim_{i\to\infty} c_i = 0.$$

If we consider sequences $\{ a_i ; i = 1, 2, 3, \ldots\}$ and $\{ b_i; i = 1, 2, 3, \ldots\}$ from Example 2.4.8, then they also N_3-converge and there holds

$$N_3\text{-}\lim_{i\to\infty} a_i = N_3\text{-}\lim_{i\to\infty} b_i = [-1;1].$$

Sequences $\{ a_i ; i = 1, 2, 3, \ldots\}$ and $\{ c_i; i = 1, 2, 3, \ldots\}$ do not relatively N_3 –converge to each other. If we take the sequence $\{ d_i ; i = 1, 2, 3, \ldots \}$ given by the formula $d_i = (-1)^i a_i$, then $\{ d_i ; i = 1, 2, 3, \ldots\}$ and $\{c_i ; i = 1, 2, 3, \ldots\}$ relatively N_3 –converge to each other.

Example 2.4.10. Let us take the following nearness relation

$$N_4(x,y) = \begin{cases} 0, & \text{if } x\,y = 0 \text{ and } x \neq y \\ 1, & \text{if } x = y \\ \max\{0; \min\{ x/y , y/x \}\} & \text{otherwise} \end{cases}$$

It is possible to demonstrate that no one of the above defined sequences, $\{ a_i ; i = 1, 2, 3, \ldots\}$, $\{ b_i; i = 1, 2, 3, \ldots\}$, $\{ c_i ; i = 1, 2, 3, \ldots\}$ and $\{ d_i; i = 1, 2, 3, \ldots\}$, is N_4-convergent. If we consider the sequence $\{ e_i ; i = 1, 2, 3, \ldots\}$ in which $e_i = ((i +1)/i)^2$, then this sequence N_4-converges and there holds

$$N_4\text{-}\lim_{i\to\infty} e_i = 1.$$

If we consider sequences $\{ x_i; i = 1, 2, 3, \ldots\}$ and $\{ y_i ; i = 1, 2, 3, \ldots\}$ given by the formulas $x_i = i^2$ and $y_i = (i + 1)^2$, then they relatively N_4-converge to each other.

Remark 2.4.8. An important property of nearness relations is shift invariance considered above. In the Examples 2.4.7 – 2.4.9, the considered nearness relations are shift invariant. The nearness relation N_4 from Example 2.4.10 is not shift-invariant.

Applying Definition of relative N-convergence to the nearness relations N_1, N_2, N_3 and N_4 from Examples 2.4.7 – 2.4.10, we get the following result.

Let us consider two sequences $l_1 = \{x_i ; i = 1, 2, 3, \ldots\}$ and $l_2 = \{y_i ; i = 1, 2, 3, \ldots\}$.

Lemma 2.4.3. a) If l_1 and l_2 N_1-relatively converge to each other, then $\lim_{i\to\infty} | x_i - y_i| = 0$.

b) If l_1 and l_2 N_2-relatively converge to each other, then there is a number $n > 0$ such that for all $i > n$, we have $x_i = y_i$.

c) If l_1 and l_2 N_3-relatively converge to each other, then

$$\limsup_{i\to\infty} | x_i - y_i | \leq 1$$

d) If l_1 and l_2 N_4-relatively converge to each other, and let for some $n > 0$ and for all $i > n$, we have $x_i , y_i \neq 0$, then $\lim_{i\to\infty} x_i /y_i = 1$.

Theorem 2.4.5 shows that conditional fuzzy limits have additivity and uniformity properties. Let us study these properties for nearness limits. If $[a, b]$ is an interval of real numbers, then the interval $[a, b]^* = [-b, -a]$ is the *inversion* of the interval $[a, b]$.

Let us take two sequences of real numbers $l = \{a_i ; i = 1, 2, 3, \dots\}$ and $h = \{b_i ; i = 1, 2, 3, \dots\}$ and a nearness relation N.

Proposition 2.4.5. a) N-lim $l \oplus$ N-lim $h \subseteq \mathrm{N}^+$-lim $(l + h)$,

where N^+ is a nearness relation defined by

$$\text{if } \mathrm{N}(x_1,y_1) \geq \alpha \text{ and } \mathrm{N}(x_2,y_2) \geq \alpha, \text{ then } \mathrm{N}^+(x_1 + x_2, y_1 + y_2) \geq \alpha$$

and $\oplus$ is the Minkowski sum (cf. Appendix B).

b) If $d > 0$, then N^d-lim $dl = d\cdot$N-lim l,
where N^d is a nearness relation defined by the condition

$$\mathrm{N}(x, y) \geq \alpha \text{ if and only if } \mathrm{N}^d(x, y) \geq \alpha$$

c) If $d < 0$, then we have $\mathrm{N}^{|d|}$-lim $dl = |d|\cdot$(N-lim l)*.

Let $\boldsymbol{N}$ be a nonempty system of nearness relations.

Definition 2.4.9. A sequence l is $(\alpha, \boldsymbol{N})$*-convergent* if it is (α, N)-convergent for some N $\in \boldsymbol{N}$.

If a sequence l is $(\alpha, \boldsymbol{N})$-convergent, then

$$(\alpha, \boldsymbol{N})\text{-lim } l = \{ b ; (\exists \mathrm{N} \in \boldsymbol{N})(\forall i \geq 1/(1 - \alpha))(\mathrm{N}(a_i , b) \geq \alpha) \}$$

Definition 2.4.10. A sequence l is $\boldsymbol{N}$*-convergent* if there exists some N $\in \boldsymbol{N}$ such that l is (α, N)-convergent for all $\alpha \geq \beta$, where $\beta < 1$.

The corresponding *nearness limit* of l is defined as

$$\boldsymbol{N}\text{-lim } l = \cup_{\mathrm{N}\in\boldsymbol{N}} \mathrm{N}\text{-lim } l.$$

Proposition 2.4.4 implies the following result.

Corollary 2.4.7. If $\boldsymbol{N}$ and $\boldsymbol{M}$ are nonempty systems of nearness relations and for each N $\in \boldsymbol{N}$ there is some M $\in \boldsymbol{M}$ such that M $\geq$ N, then a sequence l is $(\alpha, \boldsymbol{M})$-convergent ($\boldsymbol{M}$-convergent) whenever the sequence l is $(\alpha, \boldsymbol{N})$-convergent ($\boldsymbol{N}$-convergent).

2.4.3. Connection between the Nearness Convergence and Fuzzy Conditional Convergence

Let us find relations between fuzzy conditional and nearness limits. To do this, we introduce a special class of fuzzy conditional limits, which are defined by sequential conditions B and thus, are called *sequential fuzzy limits*.

Let $h = \{ b_i \in R ; i = 1, 2, 3, \dots \}$ be a non-increasing sequence of positive real numbers.

Definition 2.4.11. Validity of the *unitary sequential condition* B with sB = h means for the sequence $l = \{a_i \in R; i = 1, 2, 3, \dots \}$ and a number a that for some number $n \in \omega$, we have $| a - a_i | \le b_i$ for all $i > n$.

Lemma 2.4.4. If a sequence $l = \{a_i \in R; i = 1, 2, 3, \dots \}$ and a number a satisfy a unitary sequential condition B with a converging sB = h, then there is a converging sequence $k = \{d_i \in R; i = 1, 2, 3, \dots \}$ of positive real numbers such that $| a - a_i | \le d_i$ for all $i = 1, 2, 3, \dots$.

Let H be a class of sequences of real numbers.

Definition 2.4.12. An *existential sequential condition* B with sB = H means for a sequence $l = \{a_i \in R; i = 1, 2, 3, \dots \}$ and a number a such that there is a sequence $h = \{b_i \in R; i = 1, 2, 3, \dots \}$ from H that has all positive elements b_i and for some $n \in \omega$, we have $| a - a_i | \le b_i$ for all $i > n$.

In Examples 2.4.1-3, all conditions B are sequential and all limits are sequential conditional fuzzy limits.

Proposition 2.4.6. If a sequence $l = \{a_i \in R; i = 1, 2, 3, \dots \}$ and a number a satisfy a unitary sequential condition B with the decreasing sequence sB = $\{b_i \in R; i = 1, 2, 3, \dots \}$ of positive real numbers with $\lim_{i\to\infty}\{b_i; i = 1, 2, 3, \dots \} \le r$ and $r \ge 0$, then l B-conditionally r-converges to a, i.e., $a = (\mathrm{B}, r)\text{-}\lim l$.

Indeed, we need to check only the first condition from Definition 2.4.1. By the definition of the conventional limit, the inequality $\lim_{i\to\infty}\{b_i; i = 1, 2, 3, \dots \} \le r$ implies that some $n \in \omega$ such that $b_i \le r$ for all $i > n$. Then $| a - a_i | \le b_i \le r$ for all $i > n$. Thus, $a = (\mathrm{B}, r)\text{-}\lim l$.

Corollary 2.4.8. If a sequence $l = \{a_i \in R; i = 1, 2, 3, \dots \}$ and a number a satisfy a unitary sequential condition B with the decreasing sequence sB = $\{b_i \in R; i = 1, 2, 3, \dots \}$ of positive real numbers with $\lim_{i\to\infty}\{b_i; i = 1, 2, 3, \dots \} = 0$, then l B-conditionally converges to a, i.e., $a = \mathrm{B}\text{-}\lim l$.

Corollary 2.4.9. If a sequence $l = \{a_i ; i = 1, 2, 3, \dots \}$ satisfy some unitary sequential condition B with sB = $\{b_i ; i = 1, 2, 3, \dots \}$, such that $b_i > b_{i+1}$ holds, then there is some number $r \ge 0$ such that l B-conditionally r-converges to a, i.e., $a = (\mathrm{B}, r)\text{-}\lim l$.

Indeed, by the properties of decreasing positive sequences (cf. Section 2.1), the sequence $\{b_i ; i = 1, 2, 3, \dots \}$ has the limit. Then by Proposition 2.4.1, $a = (\mathrm{B}, r)\text{-}\lim l$ where r is the limit of $\{b_i ; i = 1, 2, 3, \dots \}$.

Now, we are going to show the links between the (B, r)-limit and the nearness limit.

Example 2.4.11. Let us have the unitary sequential condition B = $\{ 1+1/i; i = 1, 2, 3, \dots \}$, i.e., the sequence $a_i = (-1)^i(1+1/i)$ satisfies the condition B and $0 = (\mathrm{B}, 1)\text{-}\lim a_i$. Our aim is to construct a shift-invariant nearness relation N such that the sequence $l = \{a_i ; i = 1, 2, 3, \dots \}$ is N-convergent. It is enough to put

$$N(x,y) = \begin{cases} 1 & \text{if } | x - y | \le 1 \\ 0 & \text{if } | x - y | \ge 2 \\ 1 - 1/(i+1) & \text{if } 1 + 1/(i+1) \le | x - y | < 1 + 1/i. \end{cases}$$

The condition B is decreasing, i.e. $b_i = 1+1/i > 1+1/(i+1) = b_{i+1}$ and the following equalities are true:

$$\inf (1+1/i) = 1 \text{ (this means that 1 is the core of N);}$$
$$\sup (1+1/i) = 2 \text{ (this means that N } (x,y) = 0 \text{ if } |x-y| \geq 2).$$

This allows one to conclude that $N(0; a_i) \geq 1 - 1/i$ and this is the third condition in the definition of N.

In a similar way, we can start from a given shift-invariant nearness relation N and find some sequential condition B and some r ($r = \sup\{y;\ N(0, y)=1\}$) such that if a sequence $\{a_i\,;\ i = 1, 2, 3, \ldots\}$ is N-convergent, then it is B-conditionally r-convergent.

Theorem 2.4.6. a) If a sequence $l = \{a_i\,;\ i = 1, 2, 3, \ldots\}$ satisfy some unitary sequential condition B with sB $= \{b_i\,;\ i = 1, 2, 3, \ldots\}$, such that $b_i > b_{i+1}$ holds, then the sequence l is N-convergent for some shift-invariant nearness relation N.

b) Moreover, the nearness relation N can be chosen in such a way that the sequence l B-conditionally r-converges to some number a where a belongs to the N-limit of l, i.e., the following holds

$$(\,|a - a_i| \leq b_i \text{ for all } i = 1, 2, 3, \ldots\,) \Leftrightarrow (\,a \in \text{N-lim}\ l\,).$$

Proof. Let us consider some unitary sequential condition sB $= \{b_i\,;\ i = 1, 2, 3, \ldots\}$. By the properties of decreasing positive sequences (Ribenboim, 1964), the sequence $\{b_i\,;\ i = 1, 2, 3, \ldots\}$ has the limit $a = \lim_{i\to\infty} b_i$.

Taking this limit a, we construct the shift-invariant nearness relation N by the formula

$$N(x, y) = \begin{cases} 1, & \text{if } |x-y| \leq r \\ 0, & \text{if } |x-y| \geq b_1 \\ 1 - 1/(i+1), & \text{if } b_{i+1} \leq |x-y| < b_i \end{cases}$$

Taking this nearness relation N, it is possible to check that

$$(\,|a - a_i| \leq b_i \text{ for all } i = 1, 2, 3, \ldots\,) \Leftrightarrow (\,a \in \text{N-lim}\ l\,).$$

Theorem is proved.

Theorem 2.4.7. a) If a sequence $l = \{a_i \in R;\ i = 1, 2, 3, \ldots\}$ is N-convergent for some nearness relation N, then the sequence l B-conditionally r-converges for some unitary sequential condition B and some number $r \geq 0$.

b) If moreover the nearness relation N is shift-invariant, then the sequence sB can be chosen in such a way that the following equivalence holds:

$$(\,|a - a_i| \leq b_i \text{ for all } i = 1, 2, 3, \ldots) \Leftrightarrow (\,a \in \text{N-lim}\ l\,),$$

where $\{b_i\,;\ i = 1, 2, 3, \ldots\} = $ sB.

Proof. To show the existence of a necessary sequential condition B, it is enough to put $b_i = \sup_i | a - a_i |$, where $a \in$ N-lim l. Lemma 2.4.1 shows that this is a correct definition as all numbers b_i exist. If N is shift-invariant, then we can define the sequential condition B with sB $= \{b_i \in R;\ i = 1, 2, 3, \ldots\}$ by the formula

$$b_i = \sup\{ y > 0;\ \mathrm{N}(0, y) \geq 1 - 1/i \}.$$

Taking this sequential condition B, it is possible to check that

$$(| a - a_i | \leq b_i \text{ for all } i = 1, 2, 3, \ldots) \Leftrightarrow (a \in \text{N-lim } l).$$

Theorem is proved.

Corollary 2.4.10. If $l = \{a_i \in R;\ i = 1, 2, 3, \ldots \}$ is N-convergent for some nearness relation N, then the sequence l satisfies some unitary sequential condition B.

Remark 2.4.9. Theorems 2.4.6 and 2.4.7 show that the concept of conditional convergence is more general than the concept of nearness convergence.

Let H be a class of converging sequences of positive real numbers the limits of which are not larger than some number $r \geq 0$.

Proposition 2.4.7. If a sequence $l = \{a_i \in R;\ i = 1, 2, 3, \ldots \}$ and a number a satisfy an existential sequential condition B such that sB = H, then l B-conditionally r-converges to a, i.e., $a =$ (B, r)-lim l.

Corollary 2.4.11. If a sequence $l = \{a_i \in R;\ i = 1, 2, 3, \ldots \}$ and a number a satisfy an existential sequential condition B such that sB consists of decreasing sequences of real numbers converging to 0, then l B-conditionally converges to a, i.e., $a =$ B-lim l.

Let H be a class of decreasing sequences of positive real numbers.

Corollary 2.4.12. If a sequence $l = \{a_i \in R;\ i = 1, 2, 3, \ldots \}$ and a number a satisfy an existential sequential condition B such that sB = H, then there is some number $r \geq 0$ such that the sequence l B-conditionally r-converges to a, i.e., $a =$ (B, r)-lim l.

Let B be an existential sequential condition with such sB that consists of decreasing sequences.

Proposition 2.4.8. There is a system N of shift-invariant nearness relations such that for any sequence $l = \{a_i;\ i = 1, 2, 3, \ldots \}$, we have

$$a \in N\text{-lim } l \Leftrightarrow \exists\, r \geq 0\ (a = (\mathrm{B}, r)\text{-lim } l) \tag{2.1}$$

Proof. Using the construction from the proof of Proposition 2.4.7, we build a shift-invariant nearness relation N related to each sequence $h = \{b_i;\ i = 1, 2, 3, \ldots \}$ from the set sB. Combining all these relations N in one set, we get the system N of shift-invariant nearness relations for which the formula (2.1) is satisfied.

Corollary 2.4.13. There is a system N of shift-invariant nearness relations such that for any sequence $l = \{a_i;\ i = 1, 2, 3, \ldots \}$ the following holds

$$a \in N\text{-lim } l \Leftrightarrow (\exists\, h = \{b_i;\ i = 1, 2, 3, \ldots \} \in \mathrm{sB})(| a - a_i | \leq b_i).$$

Proposition 2.4.9. If a sequence $l = \{a_i \in R;\ i = 1, 2, 3, \dots\}$ is N-convergent for some system of nearness relations N, then l satisfies some existential sequential condition sB. If moreover all nearness relations $\mathrm{N} \in N$ are shift-invariant, then sB can be chosen in such a way that the following equivalence holds:

$$(\exists\ h = \{b_i\,;\ i = 1, 2, 3, \dots\} \in \mathrm{sB})(\ |\ a - a_i\ | \leq b_i \text{ for all } i = 1, 2, 3, \dots) \Leftrightarrow (\ a \in N\text{-lim}\ l\ ).$$

Proof. In the same way as before, for each relation $\mathrm{N} \in N$ and some $a \in \mathrm{N}$-lim l, we build the sequence $h = \{b_i\,;\ i = 1, 2, 3, \dots\}$ such that $b_i = \sup_i |\ a - a_i|$. If the nearness relation N is shift-invariant, then we can define the sequence $h = \{b_i\,;\ i = 1, 2, 3, \dots\}$ by the formula

$$b_i = \sup\{\ y > 0;\ \mathrm{N}(0, y\ ) \geq 1 - 1/i\ \}.$$

Finally, we get the necessary existential condition B by forming the set sB of all constructed sequences h.

Then it is possible to check that the following equivalence is true

$$(\ \exists\ h = \{b_i\,;\ i = 1, 2, 3, \dots\} \in \mathrm{sB})\ (\ |\ a - a_i\ | \leq b_i \text{ for all } i = 1, 2, 3, \dots) \Leftrightarrow (a \in N\text{-lim}\ l\ )$$

Corollary 2.4.14. There is a system N of shift-invariant nearness relations such that for any sequence $l = \{a_i\,;\ i = 1, 2, 3, \dots\}$, we have

$$a \in N\text{-lim}\ l \Leftrightarrow (\exists\ h = \{b_i\,;\ i = 1, 2, 3, \dots\} \in \mathrm{sB})(\ |\ a - a_i\ | \leq b_i\ ).$$

2.5. Sets and Fuzzy Sets of Fuzzy Limits

The whole object of the Prophets and the Sages was
to declare that a limit is set to human reason
where it must halt.
Moses Maimonides (1135-1204)

Let us denote the set of all finite r-limits of a sequence $l = \{a_i \in \boldsymbol{R};\ i = 1, 2, 3, \dots\}$ by $L_r(l)$, that is,

$$L_r(l) = \{a \in \boldsymbol{R};\ a = r\text{-lim}\ l\}.$$

The set $L_r(l)$ exists only for bounded sequences l. That is why in all cases when $L_r(l)$ is considered, it is assumed that the sequence l is bounded.

We have the following property of $L_r(l)$.

Theorem 2.5.1. For any sequence l and number $r \geq 0$, the set $\mathrm{L}_r(l)$ is a convex closed set, i.e., either $\mathrm{L}_r(l) = [a, b]$ for some $a, b \in \boldsymbol{R}$ with $b - a \leq 2r$, or $\mathrm{L}_r(l) = \varnothing$ when l has no r-limits.

Proof. At first, we prove that if $L_r(l)$ is not empty, it is a convex subset of the set $\boldsymbol{R}$ of the real numbers. Let $c, d \in L_r(l)$, $c < d$ and $e \in [c, d]$. Then it is enough to prove that $e \in L_r(l)$. Since $e \in [c, d]$, there is a number $\lambda \in [0, 1]$ such that $e = \lambda c - (1\text{-}\lambda)d$. As $c, d \in L_r(l)$, then for every $\varepsilon > 0$, there are numbers n_1 and n_2 such that $|a_i - c| < r + \varepsilon$ for all $i \in \omega$ such that $i \geq n_1$ and $|a_i - d| < r + \varepsilon$ for all $i \in \omega$ such that $i \geq n_2$. Let us put $n = \max\{n_1, n_2\}$. Then for all $i \geq n$, we have

$$\begin{aligned} | a_i - e | &= | a_i - \lambda c - (1-\lambda)d| \\ &= | \lambda a_i + (1-\lambda)\, a_i - \lambda c - (1-\lambda)d | \\ &= |(\lambda a_i - \lambda c) + ((1-\lambda)\, a_i - (1-\lambda)d)| \\ &\leq |\lambda a_i - \lambda c| + |(1-\lambda)\, a_i - (1-\lambda)d | \\ &\leq \lambda(r + \varepsilon) + (1-\lambda)(r + \varepsilon) = r + \varepsilon. \end{aligned}$$

So, by the definition of an r-limit, we have $e = r\text{-}\lim l$. In turn, this implies $e \in L_r(l)$, i.e., $L_r(l)$ is a convex subset of the real numbers (cf. Definition 1.2.1).

Now we prove that $L_r(l)$ is a closed subset of the real numbers. Let us take an arbitrary number $\varepsilon \in \boldsymbol{R}^{++}$ and consider a sequence $h = \{ d_i ; i = 1, 2, 3, \ldots\}$ of r-limits of l such that h converges to d. Then by the definition of a limit of a sequence, we can find such point d_j, for which $|d_j - d| < \varepsilon/3$. By the definition of an r-limit of a sequence, for almost all elements a_i from the sequence l, we have $|a_i - d_j| < r + \varepsilon/3$. Thus, for almost all elements a_i of the sequence l, we have

$$|a_i - d | = |a_i - d_j + d_j - d| \leq |a_i - d_j| + |d_j - d| < r + \varepsilon/3 + \varepsilon/3 < r + \varepsilon$$

As ε is an arbitrary positive number, this means that $d = r\text{-}\lim l$. Consequently, $L_r(l)$ is a closed subset of the real numbers.

When the set $L_r(l)$ is not empty, the sequence l has, at least, one r-limit. So, by Theorem 2.2.2, the sequence l is bounded. Then by Theorem 2.1.7, the sequence l has a convergent subsequence and thus, l has partial limits.

Let us show that if c is a partial limit of l, then the distance from c to any r-limit e of l cannot be larger than r. Indeed, if $|c - e| = k > r$, then $d = k - r > 0$ and by the definition of a partial limit, there are infinitely many elements c_j from l such that $|c_j - e| > r + \varepsilon$ for any $\varepsilon < k$. This contradicts the definition of an r-limit and proves that the distance from c to any r-limit e of l cannot be larger than r.

Consequently, the length of the interval $L_r(l)$ cannot be larger than $2r$.

Theorem is proved.

Remark 2.5.1. It is possible that $a = b$ and $L_r(l) = \{ a \}$. Indeed, when $r = 0$, this interval shrinks to one point, and we obtain Theorem 2.1.1 as a direct corollary of Theorem 2.5.1.

Remark 2.5.2. In a general case, the estimate $2r$ from cannot be improved in Theorem 2.5.1, that is, it is possible that $b - a = 2r$ as Proposition 2.5.4 demonstrates.

Lemma 2.2.2 implies the following result.

Lemma 2.5.1. If $r \leq p$, then $L_r(l) \subseteq L_p(l)$ for any sequence l.

Lemma 2.2.5 implies the following result.

Lemma 2.5.2. If h is a subsequence of l, then $L_r(l) \subseteq L_r(h)$ for all $r \in \boldsymbol{R}^{+}$.

Let us consider a bounded sequence $l = \{a_i \in \boldsymbol{R};\, i = 1, 2, 3, \ldots\}$.

Lemma 2.5.3. $a = r\text{-}\lim l$ if and only if the distance of a to all partial limits of l is less than or equal to r, i.e., $|a - d| \leq r$ for any partial limit d of l.

Proof. Necessity. Let $a \in \boldsymbol{R}$, $b = \text{plim } l$ and $|a - b| = q > r$. By the definition of a partial limit (cf. Section 2.1), for any $\varepsilon > 0$, there are infinitely many elements a_i from l such that $|b - a_i| < \varepsilon$.

Let $k = q - p$ and $\varepsilon = k/3$. Then there are infinitely many elements a_i from l such that $|b - a_i| < k/3$. This implies that for infinitely many elements a_i from l, we have

$$|a - b| = |a - a_i + a_i - b| \leq |a - a_i| + |a_i - b|$$

In turn, this implies

$$|a - a_i| \geq |a - b| - |a_i - b| = q - |a_i - b| \geq q - k/3 > r + k/3$$

At the same time, if $a = r\text{-lim } l$, then for almost all, that is, for all but a finite number, elements a_i from l, the inequality $|a - a_i| < r + k/3$ has to be valid. This contradiction shows that a is not an r-limit of l and completes the proof of necessity.

Sufficiency. Let us assume that the distance of a to all partial limits of l is less than or equal to r. Then to prove that $a = r\text{-lim } l$, we take some $\varepsilon > 0$ and consider all elements a_i from l such that

$$|a - a_i| > r + \varepsilon \qquad (2.2)$$

Let us assume that there are infinitely many elements a_i satisfying inequality (2.2). The set of all these elements a_i is bounded as l is a bounded sequence. Thus, by Theorem 2.1.7, there is a converging sequence h that consists of these elements a_i. If d is the limit of this sequence, then by Definition 2.1.3, d is a partial limit of l. It is possible that $d > a$ or $d < a$. For convenience, we consider the first case, as the second case is treated in a similar way.

At the same time, as $d > a$, all elements a_i from l that converge to d are larger than $a + r + \varepsilon$. Consequently (cf. Proposition 2.1.7), we have $d > a + r + \varepsilon/2$. This contradicts to our assumption that the distance of a to all partial limits of l is less than or equal to r. Thus, $|a - a_i| > r + \varepsilon$ for almost all elements a_i from l. As ε is an arbitrary positive number, it means that $a = r\text{-lim } l$.

Lemma is proved.

For any bounded sequence l, it is possible to find the least r such that l r-converges.

Theorem 2.5.2. For any sequence l, if $r_0 = \inf \{ r \,;\, l \text{ has an } r\text{-limit}\}$, then there is a point a_0 such that $a_0 = r_0\text{-lim } l$.

Proof. For any r from the set the set $H(l) = \{ r \,;\, l \text{ has an } r\text{-limit}\}$, we fix some point a_r, which is an r-limit of l. As all r are non-negative numbers, the set $H(l)$ is bounded below. Consequently, $H(l)$ has the infimum r_0. By the definition of an infimum, we can find such a sequence$\{ r_i \,;\, i = 1, 2, 3, \ldots\}$ of numbers r_i from $H(l)$ that converges to r_0. Then the set T of corresponding points a_{r_i} , i.e., a_{r_i} is an r_i-limit of l, will be bounded (cf. Theorem 2.2.4 and Proposition 2.2.4). Consequently, this set T contains a sequence h, which converges to some point a_0. Then we can take such sequence $\{ r_i \,;\, i = 1, 2, 3, \ldots\}$ of numbers r_i that converges to r_0 , for which T is equal to the sequence h. When T contains more points, we eleminate these points from T. Thus, we have a sequence $\{r_i \,;\, i = 1, 2, 3, \ldots\}$ converging to r_0 and a sequence $h = \{ a_{r_i} \,;\, i = 1, 2, 3, \ldots\}$ of r_i-limits of l, which converges to a_0.

Let us take an arbitrary number $\varepsilon \in \boldsymbol{R}^{++}$ and consider the neighborhood $O_t a_0 = \{c \in \boldsymbol{R} ; |a_0 - c| < t\}$ of the point a_0 for which $t = r_0 + \varepsilon$. Then by the definition of a limit of a sequence, we can find such point a_{r_i}, for which $|a_{r_i} - a_0| = d_1 < \varepsilon/3$ and $r_i = r_0 + d_2$ with $d_2 < \varepsilon/3$. By the definition of an r-limit of a sequence, almost all elements of the sequence l belong to the neighborhood $O_p a_{r_i} = \{c \in \boldsymbol{R} ; |a_{r_i} - c| < p\}$ of the point a_{r_i} when $p = r_i + d_3$ with $d_2 < \varepsilon/3$.

Then $O_t a_0 \supseteq O_p a_{r_i}$ because $t = r_0 + \varepsilon = t = r_0 + \varepsilon/3 + \varepsilon/3 + \varepsilon/3 > r_0 + d_1 + d_2 + d_3$. Consequently, almost all elements of the sequence l belong to the neighborhood $O_t a_0$ of the point a_0. As ε is an arbitrary number from $\boldsymbol{R}^{++}$, the point a_r is an r_0-limit of l.

Theorem 2.5.2 is proved.

Corollary 2.5.1. For any sequence l, the set $H(l) = \{ r ; l$ has an r-limit$\}$ is a closed ray.

Really, by Theorem 2.5.3, the set $H(l)$ contains its infimum r_0, and by Lemma 2.2.2, $H(l)$ contains its all numbers that are larger than r_0.

Moreover, we have a stronger result.

Proposition 2.5.1. If $r > r_0 = \inf\{ r ; l$ has an r-limit$\}$, then the interval $L_r(l)$ is not empty and the r_0-limit of l is the midpoint of $L_r(l)$.

Let us consider a sequence $l = \{ a_i \geq 0; i = 1, 2, 3, \ldots \}$.

Lemma 2.5.4. If for any $\varepsilon > 0$, almost all elements a_i from l belong to an interval with the length $2r + \varepsilon$, then l is r-convergent.

Proof. Let us consider the sequence $h = \{ 1/n ; n = 1, 2, 3, \ldots \}$. The condition of the lemma allows us to build a system of intervals $Int = \{ I_n ; I_n \subseteq I_{n-1}, n = 1, 2, 3, \ldots \}$ such that the length of I_n is less than or equal to $2r + 1/n$ and almost all elements a_i from l belong to the interval I_n.

Let us consider the sequence $k = \{ c_n; n = 1, 2, 3, \ldots \}$ where c_n is the midpoint of the interval I_n, $n = 1, 2, 3, \ldots$ As $I_n \subseteq I_{n-1}$ for all $n = 1, 2, 3, \ldots$, the the sequence k is bounded. Consequently (cf. Section 2.1, Theorem 2.1.7), k has a convergent subsequence $h = \{ b_n; n = 1, 2, 3, \ldots \}$

Let $a = \text{plim } h$ and $\varepsilon \in \boldsymbol{R}^{++}$. Then there is a number m such that $|a - b_n| < \varepsilon/3$ and the length $|I_n|$ of the interval I_n is less than $r + \varepsilon/3$ for all $n > m$. By the choice of the intervals I_n, almost all elements a_i from l belong to the interval I_n. Then for any of these elements a_i, we have:

$$\begin{aligned} |a - a_i| &= |a - b_n + b_n - a_i| \\ &\leq |a - b_n| + |b_n - a_i| \\ &< \varepsilon/3 + r + \varepsilon/3 < r + \varepsilon \end{aligned}$$

as b_n is the midpoint of I_n.

Thus, for almost all elements a_i from l, we have $|a - a_i| < r + \varepsilon$. It means that $a = r\text{-lim } l$ and l is r-convergent.

Lemma is proved.

Lemma 2.5.5. If $d_t = \text{plim } l$ for $t = 1, 2, 3, \ldots$ and $a = \lim_{t\to\infty} d_t$, then $a = \text{plim } l$.

Proof. The statement $d_t = \text{plim } l$ means that there is a subsequence $h_t = \{ a_{ti} \in \boldsymbol{R}; i = 1, 2, 3, \ldots\}$ of the sequence l, such that $d_t = \lim h_t$. This allows one to find a sequence $k = \{ d_t \in \boldsymbol{R};$

$|a - d_i| < 1/2i$; $i = 1, 2, 3, \ldots\}$. In each sequence h_i, which converges to d_i, there is an element $a_{t(i), i}$ such that $|a_{t(i), i} - d_i| < 1/2i$. Then we have

$$|a - a_{t(i), i}| = |a - d_i + d_i - a_{t(i), i}| \leq |a - d_i| + |a_{t(i), i} - d_i| < 1/i$$

As i tends to infinity, the sequence $h = \{ a_{t(i), i} \in \boldsymbol{R}; i = 1, 2, 3, \ldots\}$ converges to a. As h is a subsequence of l, we have $a = \text{plim } l$.

Lemma is proved.

Remark 2.5.3. By Theorem 2.5.2, the r_0-limit of l is unique for any sequence l and belongs to any non-empty set $L_r(l)$.

Let us consider a sequence l.

Definition 2.5.1. The number

$$r_0 = \inf\{r ; a = r\text{-lim } l\}$$

is called the *defect of convergence* of l to the number a and is denoted by $\delta(a = \lim l) = r_0$. When such a number r_0 does not exist, we put

$$\delta(a = \lim l) = \infty.$$

Lemma 2.5.3 implies the following results.

Lemma 2.5.6. $\delta(a = \lim l) = 0.5\cdot(b - c)$ where c is the smallest and b is the biggest of the partial limits of l.

Proposition 2.5.2. $L_p(l) = [b, c]$ with $p = \delta(a = \lim l)$ if and only if c is the smallest and b is the biggest of the partial limits of l.

As a consequence, we have the following result.

Lemma 2.5.7. For any subsequence h of a sequence l, we have $\delta(a = \lim h) \leq \delta(a = \lim l)$.

Let $L_r(l) = [c, b]$.

Proposition 2.5.3. a) $0 \leq b - c \leq 2r$;

b) the sequence l has an $r - 0.5\cdot(b - c)$-limit;

c) $b - c = |2r - (v - u)|$ where u is the smallest and v is the largest of the partial limits of l.

Proof. (a) follows from Theorem 2.5.1.

(b) Let a be a midpoint of the interval $[b, c]$. As $b = r$-lim l, we come to conclusion that the distance $\mathbf{d}(b, u) = |b - u| \leq r$ for any partial limit u of l. Thus, $\mathbf{d}(a, u) \leq r - 0.5\cdot(b - c)$ for any partial limit u of l such that $u \leq a$. As $c = r$-lim l, we have that the distance $\mathbf{d}(c, u) \leq r$ for any partial limit u of l. Thus, $\mathbf{d}(a, u) \leq r - 0.5\cdot(b - c)$ for any partial limit u of l such that $u \geq a$. Thus, $\mathbf{d}(a, u) \leq r - 0.5\cdot(b - c)$ for any partial limit u of l.

Then by Lemma 2.5.6, $a = (r - 0.5\cdot(b - c))$-lim l. This concludes the proof of part (b).

(c) Let u be the smallest and v be the largest of the partial limits of l. Then by Lemma 4, $b = u + r$ and $c = v - r$. This gives us the following inequalities:

$$b - c = |u + r - (v - r)| = |u + r - v + r)|$$
$$=| u - v + 2r)| = |2r - (v - u)|$$

Proposition 2.5.3 is proved.

It implies the following result.

Let $L_r(l) = [c, b]$.

Proposition 2.5.4. The length of the interval $L_r(l)$ is equal to $2r$ if and only if l is converging.

Indeed, if l is convergent, then it has a limit (say a). Then any point e such that $|e - a| \leq r$ is an r-limit of l by Lemma 2.5.3. Thus, the length of the interval $L_r(l)$ is equal to $2r$.

Assume that l is not convergent, then it has, at least, two partial limits (say c and d) and $|c - d| > k$ for some $k > 0$. As $|c - d| \leq |v - u|$ where u is the smallest and v is the largest of the partial limits of l, by Proposition 2.5.3, we have $b - c = | 2r - |v - u|| \leq | 2r - |c - d|| < 2r - k$. Thus, the length of the interval $L_r(l)$ is less than $2r$.

Proposition 2.5.4 is proved.

Proposition 2.5.3 also implies the following result.

Proposition 2.5.5. The interval $L_r(l)$ is a single point if and only if $r = \inf\{q ; a = q\text{-lim } l\}$.

Theorem 2.2.2 implies the following result.

Proposition 2.5.6. $\delta(a = \lim l)$ is a number, i.e., it is not equal to ∞, if and only if the sequence l is bounded.

Proposition 2.5.7. If $q = \delta(a = \lim l)$, then $a = q\text{-lim } l$.

Proof. To prove that $a = q\text{-lim } l$, it is sufficient to show that for any $\varepsilon > 0$, there is a number n such that $i > n$ implies $| a - a_i | < q + \varepsilon$. As $q = \inf \{r; a = r\text{-lim } l\}$, there is a number r such that $a = r\text{-lim } l$ and $| r - q | < \varepsilon/2$. As $a = r\text{-lim } l$, there is a number n such that $i > n$ implies $| a - a_i | < r + \varepsilon/2 < (q + \varepsilon/2) + \varepsilon/2 = q + \varepsilon$. Thus, $a = q\text{-lim } l$.

Proposition is proved.

Lemma 2.5.6 implies the following results.

Lemma 2.5.8. $\delta(a = \lim l) = 0$ if and only if $a = \lim l$.

Let $l = \{a_i \in \boldsymbol{R}; i = 1, 2, 3, \ldots\}$ and $h = \{b_i \in \boldsymbol{R}; i = 1, 2, 3, \ldots\}$ be bounded sequences.

Proposition 2.5.8. If $a_i \leq b_i$ for almost all $i \in \omega$, then for any partial limit $d = \text{plim } l$, there is a partial limit $c = \text{plim } h$, such that $d \leq c$, and for any partial limit $u = \text{plim } h$, there is a partial limit $v = \text{plim } l$, such that $v \leq u$.

Proof. Let $d = \text{plim } l$. Then there is a subsequence $l_0 = \{a_{ij} \in \boldsymbol{R}; i = 1, 2, 3, \ldots\}$ of the sequence l such that $d = \lim l_0$. Taking the corresponding to a_{ij} elements b_{ij} from h, we get a subsequence $h_0 = \{b_{ij} \in \boldsymbol{R}; i = 1, 2, 3, \ldots\}$ of the sequence h. As h is bounded, its subsequence h_0 is also bounded. Thus (cf. Section 1.1), the sequence h_0 has a converging subsequence h_1.

Let $d = \lim h_1$. As all elements from h_1 are larger than or equal to corresponding elements from l_0, we have (Theorem 2.1.3) $d \leq c$. Thus, the first part of Proposition 2.5.9 is proved.

Proof of the second part is similar.

Proposition 2.5.8 implies the following result.

Proposition 2.5.9. $Lr_o(l) = [b, c]$ with $r_o = \delta(a = \lim l)$ if and only if c is the smallest and b is the biggest of the partial limits of l.

Proposition 2.5.10. If $l = \{a_i \in \boldsymbol{R}; i = 1, 2, 3, \ldots\}$, $h = \{b_i \in \boldsymbol{R}; i = 1, 2, 3, \ldots\}$, $L_r(l) = [a, c]$, $L_r(h) = [b, d]$, and $a_i \leq b_i$ for almost all $i \in \omega$, then $a \leq b$ and $c \leq d$.

Proof. If $a_i < b_i$ for almost all $i \in \omega$, then the least partial limit u_l of l is less than or equal to the least partial limit u_h of h, and the largest partial limit v_l of l is less than or equal to the largest partial limit v_h of h. As it is demonstrated in Proposition 2.5.3, $a = u_l + r$, $c = v_l - r$, $b = u_h + r$, and $c = v_h - r$. Consequently, we have $a \leq b$ and $c \leq d$.

Proposition is proved.

Corollary 2.5.2. (cf., for example, (Ribenboim, 1964) and Section 2.1). If $a_i \leq b_i$ for almost all $i = 1, 2, 3, \ldots$, $a = \lim a_i$, and $b = \lim b_i$, then $a \leq b$.

Let $l = \{a_i \in \boldsymbol{R}; i = 1, 2, 3, \ldots\}$, $h = \{b_i \in \boldsymbol{R}; i = 1, 2, 3, \ldots\}$, $k = \{c_i \in \boldsymbol{R}; i = 1, 2, 3, \ldots\}$, and $a_i \leq b_i \leq c_i$ for almost all $i \in \omega$.

Proposition 2.5.11. If $L_r(l) = [a, u]$, $L_r(h) = [b, v]$, and $L_r(k) = [c, w]$, then:

a) $L_r(h) \subseteq [a, w]$;
b) $L_r(l) = L_r(k) \neq \varnothing$ *implies* $L_r(h) = L_r(l)$ and thus, $L_r(h) \neq \varnothing$.

Proof. The statement (a) follows from Proposition 2.5.9.

(b) $L_r(l) = L_r(k) \neq \varnothing$ means that $[a, u] = [c, w]$. Consequently, $a = c$ and $u = w$. At the same time, by Proposition 2.5.9, we have $a \leq b \leq c$ and $u \leq v \leq w$. Thus, $a = b$, $u = v$, and $L_r(h) = [b, v] = [a, u] = L_r(l) \neq \varnothing$.

Proposition is proved.

As a direct corollary, Proposition 2.5.11 gives such a classical result as the Squeeze Theorem (cf. (Ribenboim, 1964; Fihtengoltz, 1955; Randolph, 1968; Goldstein, *et al*, 1987; Shenk, 1979) and Section 1.3). The Squeeze Theorem states that if both sequences l and k converge to the same limit a, then h also converges and $\lim h = a$.

Definition 2.5.2. The quantity

$$\frac{1}{1 + \delta(a = \lim l)}$$

is called the *measure of convergence* of l to the number a and is denoted by $\mu(a = \lim l)$.

The measure of convergence of l taken as a functions on real numbers defines the fuzzy set $L_{lim}(l) = (\boldsymbol{R}, \mu(a = \lim l), [0, 1])$, which is called the *complete fuzzy limit* of the sequence l.

Note that a fuzzy limit of a sequence is a number, while the complete fuzzy limit of a sequence is a fuzzy set.

Example 2.5.1. Let us consider the complete fuzzy limit $L_{lim}(l)$ of the sequence $l = \{ 1, -1, 1, -1, 1, -1, 1, \ldots \}$. For this sequence and a real number a, we have

$$\mu(a = \lim l) = 1/(1½ + | ½ - a|)$$

An important property of complete fuzzy limit is described in the following result.

Theorem 2.5.3. The complete fuzzy limit $L_{lim}(l)$ of a sequence l is a convex fuzzy set.

Proof. Let $c, d \in L_{lim}(l)$, $c < d$ and $a \in [c, d]$. Then it is enough to prove that $\mu(a = \lim l) = \mu((\lambda c + (1 - \lambda)d) = \lim l) \geq \min\{\mu(c = \lim l), \mu(d = \lim l)\}$. This is equivalent to the inequality $\delta(a = \lim l) = \delta((\lambda c + (1 - \lambda)d) = \lim l) \leq \max\{\delta(c = \lim l), \delta(d = \lim l)\}$.

Let us assume, for convenience, that $q = \delta(c = \lim l) \geq r = \delta(d = \lim l)\}$. Then by Lemma 3.4, $d = q\text{-}\lim l$. Then by Theorem 3.3, $d = q\text{-}\lim l$ as the set $L_{lim}(l)$ is convex. Thus, $\delta(a = \lim l) \leq q = \max\{\delta(c = \lim l), \delta(d = \lim l)\}$.

Theorem is proved.

Theorem 2.5.4. The complete fuzzy limit $L_{lim}(l)$ of a sequence l is a normal fuzzy set if and only if l converges.

Indeed, $\mu(a = \lim l) = 1$ if and only if $\delta(a = \lim l) = 0$, while by Lemma 2.5.3, $\delta(a = \lim l) = 0$ if and only if $a = \lim l$.

2.6. FUZZY LIMITS OF SETS OF SEQUENCES

Everything has its limit –
iron ore cannot be educated into gold.
Mark Twain (1835-1910)

Neoclassical analysis makes possible not only to extend ordinary concepts obtaining new results for classical structures, but also provides for elaboration of new useful concepts. One of such concepts is given in the definition of fuzzy limits of sets of sequences. A fuzzy limit of a function at some point is an example of a fuzzy limit of a set of sequences.

Let E = $\{ l_j ; j \in J \}$ be a set of sequences l_j of real numbers.

Definition 2.6.1. A number a is called an *r-limit of a set* E = $\{ l_j ; j \in J \}$ (it is denoted by $a = r$-lim E) if a is an r-limit of each sequence l_i from E.

Remark 2.6.1. If E has a 0-limit a, then this 0-limit is unique and all sequences from E converge to a. In contrast to this, sequences from a given set may have different limits but a common fuzzy limit. It is demonstrated by the following example.

Example 2.6.1. Let us consider the set E = $\{ \{1/2^n; n = 1, 2,\ldots\}, \{1 + 1/3^n; n = 1, 2,\ldots\}, \{2 + 1/5^n; n = 1, 2,\ldots\} \}$. It consists of three sequences and has a 1-limit 1, but it does not have a limit because the first sequence converges to 0, the second sequence converges to 1, and the third sequence converges to 2.

Lemma 2.2.1 implies the following result.

Proposition 2.6.1. If $a = r$-lim E, then $a = q$-lim E for any $q > r$.

From Theorem 2.2.1, we obtain the following result.

Proposition 2.6.2. If $a = r$-lim E and $a > b + r$, then for each sequence $l_j = \{ a_{ij} ; i = 1, 2, 3, \ldots\}$ from the set E, we have $a_{ij} > b$ for almost all a_{ij} from l_j.

Lemma 2.6.1. If $d = \lim_{n\to\infty} d_n$ and $d_n = r$-lim E for any $n = 1, 2, 3, \ldots$, then $d = r$-lim E.

Proof. Let $l = \{ a_i \in \boldsymbol{R}; i = 1, 2, 3, \ldots\} \in$ E. Then by Definition 2.6.1, $d_n = r$-lim l for any $n = 1, 2, 3, \ldots$. Thus, given $\varepsilon > 0$, we can take a number m_n such that $| a_i - d_n | < r + \varepsilon/3$ when $i > m_n$. At the same time, as $d = \lim d_n$, given $\varepsilon > 0$, we can take a number p such that $| d - d_n | < \varepsilon/3$ when $n > p$. Consequently, $| a_i - d | = | a_i - d_n + d_n - d | \leq | a_i - d_n | + | d_n - d | < r + \varepsilon/3 + \varepsilon/3 < r + \varepsilon$ when $n > p$ and $i > m_n$. By Definition 2.2.1, $d = r$-lim l. As l is an arbitrary sequence from E, $d = r$-lim E.

Lemma is proved.

Proposition 2.6.3. For any set E of sequences, if $r_0 = \inf \{ r ;$ E has an r-limit$\}$, then there is a point a_0 such that $a_0 = r_0$-lim E.

Proof is similar to the proof of Theorem 2.5.2.

Remark 2.6.2. By Proposition 2.2.3, the r_0-lim E is unique for any set E of sequences.

Propositions 2.6.3 and 2.6.1 imply the following result.

Corollary 2.6.1. For any set E of sequences, the set H(E) = { r ; E has an r-limit} is empty or H(E) is a closed ray.

Let us consider a set of sequences E.

Definition 2.6.2. The number

$$r_0 = \inf\{r;\ a = r\text{-lim E}\}$$

(if it exists) is called the *defect* $\delta(a = \lim \text{E})$ *of convergence* of E to the number a. When such a number r_0 does not exist, we put $\delta(a = \lim \text{E}) = \infty$.

Proposition 2.6.4. $\delta(a = \lim \text{E})$ is a number if and only if the set of all partial limits of sequences from E is bounded.

Remark 2.6.3. However, it is not true, in general, that if $\delta(a = \lim \text{E})$ is a number, then the set $\cup$E of all elements of sequences from E is bounded. It is also not true, in general, that $\delta(a = \lim \text{E})$ is a number if all sequences from the set E are bounded. This is demonstrated by the followng examples.

Example 2.6.2. Let us consider the set E = { l_n ; n = 1, 2,…} where $l_n = \{n/i;\ i = 1, 2,\dots\}$. The set E has a 0-limit 0 because all sequences l_n converge to 0. However, the set $\cup$E is unbounded as all natural numbers belong to sequences from E.

Example 2.6.3. Let us consider the set E = { l_n ; n = 1, 2,…} where $l_n = \{n - 1/i;\ i = 1, 2,\dots\}$. All sequences from the set E are bounded. However, $\delta(a = \lim \text{E}) = \infty$ as partial limits of sequences from E tend to infinity.

Definitions imply two following results.

Lemma 2.6.2. For any subset D of a set E, we have $\delta(a = \lim \text{D}) \leq \delta(a = \lim \text{E})$.

Lemma 2.6.3. $\delta(a = \lim l) = 0$ if and only if a = 0-lim E.

Since an intersection of closed intervals is a closed interval (cf., Appendix and (Kuratowski, 1966; 1968)), Theorem 2.5.1 implies the following result.

Theorem 2.6.1. For any set E of sequences and any number $r \geq 0$, the set $\text{L}_r(\text{E}) = \{a \in \boldsymbol{R};\ a = r\text{-lim E}\}$ is a convex closed set, i.e., either $\text{L}_r(\text{E}) = [a, b]$ for some $a, b \in \boldsymbol{R}$ with $b - a \leq 2r$, or $\text{L}_r(\text{E}) = \varnothing$ when l has no r-limits.

Example 2.6.2. For the set E from Example 2.6.1, we have: $\text{L}_0(\text{E})$ is empty, $\text{L}_1(\text{E})$ consists of one point 1, and $\text{L}_2(\text{E}) = [0, 2]$.

Remark 2.6.4. When all sequences from a given arbitrary set E converge to the same point, the length of $\text{L}_r(\text{E})$ is exactly equal to $2r$. Thus Corollary 2.2.18 gives the exact boundary for the length of $\text{L}_r(\text{E})$.

Let E = { l_j ; $j \in J$ } and D = { h_t ; $t \in J$ } be two sets of sequences of real numbers. In a natural way, it is possible to define arithmetical operation for these sets. Namely,

$$\text{E} + \text{D} = \{ l_j + h_t ;\ j, t \in \omega,\ l_j \in \text{E},\ h_t \in \text{D}\},$$
$$\text{E} - \text{D} = \{ l_j - h_t ;\ j, t \in J,\ l_j \in \text{E},\ h_t \in \text{D}\},$$

and

$k\text{E} = \{ kl_j ; j \in J, l_j \in \text{E} \}$ for a real number $k \in \boldsymbol{R}$.

Theorem 2.6.2. If $a = r$-lim E and $b = q$-lim D, then:

a) $a + b = (r + q)$-lim (E + D) ;
b) $a - b = (r + q)$-lim (E - D) ;
c) $ka = |k| \cdot r$-lim (kE) for any $k \in \boldsymbol{R}$.

Proof. By Definition 2.6.1, a is an r-limit of any sequence l from E and b is an r-limit of any sequence h from D. By Theorem 2.2.5, $a + b = (r + q)$-lim $(l + h)$, $a - b = (r + q)$-lim $(l - h)$ and $ka = |k| \cdot r$-lim (kl). As l is an arbitrary sequence from E and h is an arbitrary sequence from D, this implies the necessary properties of fuzzy limits, i.e., $a + b = (r + q)$-lim (E + D), $a - b = (r + q)$-lim (E - D), and $ka = |k| \cdot r$-lim (kE) for any $k \in \boldsymbol{R}$.

Theorem is proved.

Remark 2.6.5. Although many properties of fuzzy limits of sets of sequences are the same or, at least, very similar to properties of fuzzy limits of sequences, not all properties of fuzzy limits of sets of sequences are true for sets of sequences. For instance, the result of Lemma 2.5.5 can be invalid for sets of sequences.

Definition 2.6.3. The number d is called a *partial limit* (*partial r-limit*) of E if d is a partial limit (partial r-limit) of some sequence l from E, i.e., d = plim E ($d = r$-plim E) if there is $l \in$ E such that d = plim l ($d = r$-plim l).

Lemma 2.5.3 implies the following result.

Proposition 2.6.5. $a = r$-lim E if and only if for any partial limit d of E, the distance $|a - d| \leq r$.

Theorem 2.6.3 (The Reduction Theorem). If the set E of sequences is finite or countable, then there is a sequence $l = \{ a_i \in \boldsymbol{R}; i = 1, 2, 3, \ldots \}$ such that for any $a \in \boldsymbol{R}$ and any $r \in \boldsymbol{R}^+$, we have $a = r$-lim E if and only if $a = r$-lim l.

Proof. There are two cases: either E does not have r-limits or E has r-limits. In the first case, we can take any sequence l converging to ∞, e.g., $l = \{ i; i = 1, 2, 3, \ldots \}$. Then the condition of the theorem is true for l because both E and l do not have r-limits.

In the second case, let us take some point $a = r$-lim E. Then by Proposition 2.6.5, $|a - d| \leq r$ for any partial limit d of E. Consequently, the set Plim E of all partial limits of E is bounded. This allows us to take numbers u = sup Plim E and v = inf Plim E and build two sequences: $h = \{ u_i \in \boldsymbol{R}; i = 1, 2, 3, \ldots \}$ converges to u and $h = \{ v_i \in \boldsymbol{R}; i = 1, 2, 3, \ldots \}$ converges to v.

Let us define the sequence $l = \{ b_i ; i = 1, 2, 3, \ldots \}$by the following rule: $b_{2i} = u_i$ and $b_{2i-1} = v_i$. Then u = plim l and v = plim l. Let us assume that $a = r$-lim l. Then by Lemma 2.5.3, $|a - u| \leq r$ and $|a - v| \leq r$. Consequently, $|a - d| \leq r$ for any partial limit d of E. Thus, by Proposition 2.6.5, $a = r$-lim E.

Now let us assume that $a = r$-lim E. Then by Proposition 2.6.5, $|a - d| \leq r$ for any partial limit d of E. As u = sup Plim E and v = inf Plim E, we have $|a - u| \leq r$ and $|a - v| \leq r$. Thus, by Lemma 2.5.3, $a = r$-lim l.

Theorem is proved.

Definition 2.6.4. An interval $I = \lfloor a, b \rfloor$ is called a *uniformity interval* for a set $\text{E} = \{ l_j ; j \in J \}$ (it is denoted by I = uni E) if for any sequence l_i from E there is a number n_j such that for any $i > n_j$ elements a_{ij} belong to I.

Definition 2.6.4 implies the following result.

Lemma 2.6.4. If $[a, b]$ is a uniformity interval for a set E and $[a, b] \subseteq [c, d]$, then $[c, d]$ is a uniformity interval for a set E.

Proposition 2.6.4. If $I = [a, b]$ is a uniformity interval for a set E of sequences, then the midpoint c of I is an r-limit of E where $r = ½ (b - a)$.

Indeed, in the case stated in this proposition, for any sequence $l = \{ a_i \in \boldsymbol{R}; i = 1, 2, 3, \ldots\}$ from E and any $\varepsilon > 0$, there is a number n such that for any $i > n$ elements $| a - a_i | < r + \varepsilon$. Consequently, by Definition 1, $c = r\text{-lim E}$.

Definition 2.6.4 also implies the following result.

Proposition 2.6.5. If $d = r\text{-lim E}$, then for any positive number k, $I = [d - r - k, d + r + k]$ is a uniformity interval of E.

2.7. Series and Fuzzy Summability

How individuals of the same species surpass each other
in these sensations and in other bodily faculties is
universally known, but there is a limit to them,
and their power cannot extend
to every distance or to every degree.
Moses Maimonides (1135-1204)

When operation of addition is applied to several numbers, the ordered set of these numbers is called a series and the result of operation is called the sum of this series. For instance, if we have the finite series $1 + 2 + 3 + 4 + 5 + 6 + 7$, its sum is equal to 28. However, mathematicians found a necessity to work not only with finite series but also with infinite series and to find their sums.

Infinite series appear even in the elementary mathematics: an example is given by infinite decimal fractions, the value of which is given by the sum of a corresponding series, e.g., $0.333\ldots = (3/10) + (3/100) + (3/1000) + \ldots$ or $e = 1 + (1/1!) + (1/2!) + (1/3!) + \ldots$

It is possible to find calculations with finite series in mathematics of ancient Egypt and Babilonia in the second millennium B.C.E. Infinite series appeared in mathematics of ancient Greece. Archimedes (the 3rd century B.C.E.) used infinite series to find the area of parabolic segment.

However, only starting with the development of the calculus, the theory of infinite series was developed. Mathematicians studied representations of functions by infinite series and applied these results to integration, numerical computation of function values, and solving differential equations. Different mathematicians, such as Leibniz, Taylor, Euler, Gauss, Cauchy, Abel and many others, contributed to the development of the theory of series. Now series are used in many areas of mathematics (calculus, differential equations, theory of distributions, numerical integration, function extrapolation, etc.), as well as in different scientific fields. For instance, in physics series are utilized for computation of values of physical quantities by summation of series.

Summation of series is defined by utilizing limits of sequences. The sum of a series is defined as the limit of the sequence of its partial sums. This reduces many problems for series to problems with sequences. For instance, many series do not have the sum because, as we know from any course of calculus, many sequences do not have limits.

The theory of fuzzy limits extends possibilities and applications of convergence operations and we can apply the theory of fuzzy limits to derive a theory of fuzzy summability for series.

2.7.1. General Concepts

In calculus, as a rule (cf., for example, (Ross, 1996) or (Stuart, 1991)), a distinction between a series and its sum is not made. However, there is a necessity to make such a distinction because there are different definitions of a sum of a series. Some authors provide a more rigorous approach to series and their summability. For example, Diedonne (1960, Ch. 5) defines a series as a pair of sequences ($\{ a_i ; i = 1, 2, \dots \}$, $\{ s_i ; i = 1, 2, \dots \}$) where $s_i = a_1 + a_2 + \dots + a_i$ for all $i = 1, 2, \dots$

To achieve more flexibility and generality, here we give a different definition.

Definition 2.7.1. a) A *finite series* is a formal expression of the form $\Sigma_{i=1}^{n} a_i$ or formal expression $a_1 + a_2 + \dots + a_n$ where $\{ a_i ; i = 1, 2, \dots , n \}$ is a sequence of numbers.

b) An *infinite series* is a formal expression of the form $\Sigma_{i=1}^{\infty} a_i$ or formal expression $a_1 + a_2 + \dots + a_n + \dots$ where $\{ a_i ; i = 1, 2, \dots \}$ is an infinite sequence of numbers.

This definition is more general than given by Diedonne (1960) as it includes formal series.

The most popular example are:

- an arithmetic series $\Sigma_{i=1}^{\infty} a_i$ where $a_1 = a$, $a_2 = a + d$, $a_3 = a + 2d$, ... , $a_i = a + d(i - 1)$, ... where a an d are arbitrary numbers;
- a geometric series $\Sigma_{i=1}^{\infty} a_i$ where $a_1 = a$, $a_2 = ar$, $a_2 = ar^2$, ... , $a_i = ar^{i-1}$, ... where a an r are arbitrary numbers;

There are no mathematical problems with summation of finite series. So, we study here only infinite series and call them simply series. Summation of infinite series involves their partial sums. Namely, if $\Sigma_{i=1}^{\infty} a_i$ is a series, then the result of adding its first n elements $S_n = a_1 + a_2 + \dots + a_n$ is called its n-th *partial sum*.

To study properties of summation, we introduce operations with series. For instance, arithmetical operations with real numbers induce corresponding operations with series. Namely, for any series $\Sigma_{i=1}^{\infty} a_i$ and $\Sigma_{i=1}^{\infty} b_i$ and a real number c, we have:

a) Addition: $\Sigma_{i=1}^{\infty} a_i + \Sigma_{i=1}^{\infty} b_i = \Sigma_{i=1}^{\infty} (a_i + b_i)$;
b) Subtraction: $\Sigma_{i=1}^{\infty} a_i - \Sigma_{i=1}^{\infty} b_i = \Sigma_{i=1}^{\infty} (a_i - b_i)$;
c) Multiplication: $(\Sigma_{i=1}^{\infty} a_i) \cdot (\Sigma_{i=1}^{\infty} b_i) = \Sigma_{i=1}^{\infty} (a_i \cdot b_i)$;
d) Multiplication by a number: $c \cdot (\Sigma_{i=1}^{\infty} b_i) = \Sigma_{i=1}^{\infty} (c \cdot b_i)$;

By the definition, all laws of operations with real numbers (commutativity of addition, associativity of addition, commutativity of multiplication, associativity of multiplication, and distributivity) are valid for corresponding operations with series. Thus, series form a linear algebra.

In a natural way, we define a *subseries*. Informally, a subseries of a series S is a series elements of which are a part of elements from S. Formally, a series $\Sigma_{i=1}^{\infty}a_i$ is a *subseries* of a series $\Sigma_{i=1}^{\infty}b_i$ if there is an injection $\boldsymbol{j}: \boldsymbol{N} \to \boldsymbol{N}$ such that $a_i = b_{j(i)}$ for all $i = 1, 2, \dots$.

A given injection $\boldsymbol{j}$ of natural numbers corresponds a subseries $\Sigma_{i=1}^{\infty}a_i$ of a series $\Sigma_{i=1}^{\infty}b_i$ to $\Sigma_{i=1}^{\infty}b_i$. This determines a unary operation on the set of all series. We denote this operation by $\Sigma_{i=1}^{\infty}a_i = \boldsymbol{j}(\Sigma_{i=1}^{\infty}b_i)$. This operation commutes with arithmetical operations. Namely, for any series $\Sigma_{i=1}^{\infty}a_i$ and $\Sigma_{i=1}^{\infty}b_i$ and any injection $\boldsymbol{j}: \boldsymbol{N} \to \boldsymbol{N}$, we have:

1) $\boldsymbol{j}(\Sigma_{i=1}^{\infty}a_i) + \boldsymbol{j}(\Sigma_{i=1}^{\infty}b_i) = \boldsymbol{j}(\Sigma_{i=1}^{\infty}a_i + \Sigma_{i=1}^{\infty}b_i)$;
2) $\boldsymbol{j}(\Sigma_{i=1}^{\infty}a_i) - \boldsymbol{j}(\Sigma_{i=1}^{\infty}b_i) = \boldsymbol{j}(\Sigma_{i=1}^{\infty}a_i - \Sigma_{i=1}^{\infty}b_i)$;
3) $\boldsymbol{j}(\Sigma_{i=1}^{\infty}a_i) \cdot \boldsymbol{j}(\Sigma_{i=1}^{\infty}b_i) = \boldsymbol{j}(\Sigma_{i=1}^{\infty}a_i \cdot \Sigma_{i=1}^{\infty}b_i)$, and
4) $c \cdot \boldsymbol{j}(\Sigma_{i=1}^{\infty}b_i) = \boldsymbol{j}(c \cdot \Sigma_{i=1}^{\infty}b_i)$,

i.e., $\boldsymbol{j}$ is an endomorphism of the linear algebra of all series.

A dual to subseries construction is called a *quotient series*. While the definition of a subseries is similar to the definition of a subsequence, to introduce a quotient series, we need some extra constructions.

A projection $\boldsymbol{p}: \boldsymbol{N} \to \boldsymbol{N}$ is called:

a) *cofinite* if the set $\boldsymbol{p}^{-1}(i)$ is finite for all for all $i = 1, 2, 3, \dots$;
b) *interval* if $\boldsymbol{p}^{-1}(i)$ is a finite interval, i.e., has the form $[k, h]$, for all for all $i = 1, 2, 3, \dots$
c) *monotone* if $i < j$ implies $\boldsymbol{p}(i) < \boldsymbol{p}(j)$ for all for all $i, j = 1, 2, 3, \dots$

Informally, a quotient series is obtained by adding some of the elements of the initial series. Formally, a series $\Sigma_{i=1}^{\infty}a_i$ is a (*monotone*) *quotient series* of a series $\Sigma_{i=1}^{\infty}b_i$ if there is an (monotone) interval projection $\boldsymbol{p}: \boldsymbol{N} \to \boldsymbol{N}$ such that $a_i = \Sigma_{j\in \boldsymbol{p}^{-1}(i)}\, b_j$ for all $i = 1, 2, 3, \dots$.

A given cofinite projection $\boldsymbol{p}$ of natural numbers corresponds to a series $\Sigma_{i=1}^{\infty}b_i$ its quotient series $\Sigma_{i=1}^{\infty}a_i$. This determines a unary operation on the set of all series. We denote this operation by $\Sigma_{i=1}^{\infty}a_i = \boldsymbol{p}(\Sigma_{i=1}^{\infty}b_i)$. This operation also commutes with arithmetical operations. Namely, for any series $\Sigma_{i=1}^{\infty}a_i$ and $\Sigma_{i=1}^{\infty}b_i$ and any projection $\boldsymbol{p}: \boldsymbol{N} \to \boldsymbol{N}$, we have:

1) $\boldsymbol{p}(\Sigma_{i=1}^{\infty}a_i) + \boldsymbol{p}(\Sigma_{i=1}^{\infty}b_i) = \boldsymbol{p}(\Sigma_{i=1}^{\infty}a_i + \Sigma_{i=1}^{\infty}b_i)$;
2) $\boldsymbol{p}(\Sigma_{i=1}^{\infty}a_i) - \boldsymbol{p}(\Sigma_{i=1}^{\infty}b_i) = \boldsymbol{p}(\Sigma_{i=1}^{\infty}a_i - \Sigma_{i=1}^{\infty}b_i)$;
3) $c \cdot \boldsymbol{p}(\Sigma_{i=1}^{\infty}b_i) = \boldsymbol{p}(c \cdot \Sigma_{i=1}^{\infty}b_i)$,

i.e., $\boldsymbol{p}$ is an endomorphism of the linear space of all series.

However, $\boldsymbol{p}$ is not an endomorphism of the linear algebra of all series because $\boldsymbol{p}$ does not commute with multiplication.

Example 2.7.1. Let $\Sigma_{i=1}^{\infty}b_i = 1 + ½ + ¼ + 1/8 + \dots$, i.e., $a_i = 1/2^{i-1}$, and $\boldsymbol{p}(2n) = \boldsymbol{p}(2n - 1) = n$. The projection $\boldsymbol{p}$ determines the monotone quotient series $\Sigma_{i=1}^{\infty}a_i = \boldsymbol{p}(\Sigma_{i=1}^{\infty}b_i) = 1½ + 3/8 + 3/32 + \dots$.

One more operation is permutation. Namely, a series $\Sigma_{i=1}^{\infty}a_i$ is a *permutation* of a series $\Sigma_{i=1}^{\infty}b_i$ if there is a permutation (bijection) $\boldsymbol{p}$: $\boldsymbol{N} \to \boldsymbol{N}$ such that $a_i = b_{p(i)}$ for all $i = 1, 2, \ldots$.

Example 2.7.2. Let $\Sigma_{i=1}^{\infty}b_i = 1 + 2 + 3 + 4 + 5 + \ldots$, i.e., $a_i = i$, and $\boldsymbol{p}(2n) = 2n - 1$, $\boldsymbol{p}(2n - 1) = 2n$. The permutation $\boldsymbol{p}$ determines the series $\Sigma_{i=1}^{\infty}a_i = \boldsymbol{p}(\Sigma_{i=1}^{\infty}b_i) = 2 + 1 + 4 + 3 + 6 + 5 + \ldots$.

2.7.2. Summation of Series

To sum infinite series, we use such a topological concept as convergence.

Definition 2.7.2. The *topological sum* denoted by $top\Sigma_{i=1}^{\infty}a_i$ or simply, *sum* denoted by $sum\Sigma_{i=1}^{\infty}a_i$ of a series $\Sigma_{i=1}^{\infty}a_i$ is the limit of the sequence $\{ S_n ; n = 1, 2, \ldots \}$.

This is the classical definition of series summation.

Definition 2.7.3. A series $\Sigma_{i=1}^{\infty}a_i$ *converges* or is called *convergent* (*summable*) if $sum\Sigma_{i=1}^{\infty}a_i$ exists. Otherwise, this series *diverges*.

Example 2.7.3. Let $a \neq 0$ and q be any real number. Then $\Sigma_{i=1}^{\infty}aq^i$ is called a geometric series. It is possible to show (cf., for example, (Stewart, 2003)) that the partial sum of the geometric series $\Sigma_{i=1}^{\infty}aq^i$ is equal to

$$S_n = (a(1 - q^n))/(1 - q)$$

Consequently, when $-1 < q < 1$, the sum of the geometric series exists and is equal to

$$sum\Sigma_{i=1}^{\infty} aq^i = \{ S_n ; n = 1, 2, \ldots \} = (a(1 - q^n))/(1 - q) = a/(1 - q)$$

Thus, when $-1 < q < 1$, the geometric series is convergent.

When $-1 \geq q$ or $q \geq 1$, the sum of the geometric series does not exist, i.e., the series is divergent.

Remark 2.7.1. In the theory of hypernumbers, there is also another, more general construction for the series summation (Burgin, 2005b). It is called the analytical sum of a series and allows summation of all series of real numbers.

There are many properties and criteria that indicate when a series has the sum in the classical sense (cf., for example, (Jolley, 1961; Davis, 1962; Bromwich and MacRobert, 1991; Sofo, 2003)).

Proposition 2.7.1. For any series $\Sigma_{i=1}^{\infty}a_i$, if its sum $sum\Sigma_{i=1}^{\infty}a_i$ exists, then it coincides with the sum of any its monotone cofinite quotient series.

Indeed, the classical limit of a sequence coincides with the classical limit of any its subsequence (cf., for example, (Ross, 1996)). If a series $\Sigma_{i=1}^{\infty}a_i$ is a monotone quotient series of a series $\Sigma_{i=1}^{\infty}b_i$, then the sequence $\{ S_n ; n = 1, 2, \ldots \}$ of the partial sums of the series $\Sigma_{i=1}^{\infty}a_i$ is a subsequence of the sequence $\{ S_n^o; n = 1, 2, \ldots \}$ of the partial sums of the series $\Sigma_{i=1}^{\infty}b_i$. Consequently, $sum\Sigma_{i=1}^{\infty}a_i = \lim S_n = \lim S_n^o = sum\Sigma_{i=1}^{\infty}b_i$.

The Riemann series theorem (cf. Theorem 2.7.3) shows that for non-monotone quotient series the result of Proposition 2.7.1 is not true.

Theorem 2.7.1 (the **Cauchy Criterion for series**). A series $\Sigma_{i=1}^{\infty} a_i$ is summable if and only if for any for any $\varepsilon > 0$ there is such n that for any $m > n$, we have

$$| a_n + a_{n+1} + \ldots + a_m | < \varepsilon \qquad (2.3)$$

or, in other terms,

$$\lim_{n, m \to \infty} | a_n + a_{n+1} + \ldots + a_m | = 0$$

Proof. Necessity. Let the series $\Sigma_{i=1}^{\infty} a_i$ be summable in $\boldsymbol{R}$. It means that the sequence $\{ S_n ; n = 1, 2, 3, \ldots \}$ of the partial sums of the series $\Sigma_{i=1}^{\infty} a_i$ converges. By the Cauchy Criterion for sequences (cf. Theorem 2.1.10), for any $\varepsilon > 0$ there is such $n \in \omega$ that for any $i, j \geq n$, we have $| S_j - S_i | < \varepsilon$. Taking $j = m$ and $i = n$, we obtain $| S_m - S_n | = | a_n + a_{n+1} + \ldots + a_m | < \varepsilon$. Thus,

$$\lim_{n, m \to \infty} a_n + a_{n+1} + \ldots + a_m = 0$$

Sufficiency. Let for any for any $\varepsilon > 0$, there is such n that for any $m > n$, we have $| a_n + a_{n+1} + \ldots + a_m | < \varepsilon$. As $| S_m - S_n | = | a_n + a_{n+1} + \ldots + a_m |$, it means that by the Cauchy Criterion for sequences (cf. Theorem 2.1.10), the sequence $\{ S_n ; n = 1, 2, 3, \ldots \}$ converges. Definition 2.7.3, implies that the series $\Sigma_{i=1}^{\infty} a_i$ is summable in $\boldsymbol{R}$.

Theorem is proved.

For the case when $m = n + 1$, we have the following result.

Corollary 2.7.1. If a series $\Sigma_{i=1}^{\infty} a_i$ is summable, then the absolute value of its elements tends to 0.

Corollary 2.7.2. A series $\Sigma_{i=1}^{\infty} a_i$ is summable if and only for any natural number n, the series $\Sigma_{i=p}^{\infty} a_i$ is summable.

Indeed, the sequence $\{ S_m^{o} ; m = 1, 2, 3, \ldots \}$ of the partial sums of the series $\Sigma_{i=p}^{\infty} a_i$ is obtained by subtracting one and the same number from each member of the sequence $\{ S_n ; n = 1, 2, 3, \ldots \}$ of the partial sums of the series $\Sigma_{i=1}^{\infty} a_i$.

Corollary 2.7.3. If the absolute value of a_i does not converge to 0, then the series $\Sigma_{i=1}^{\infty} a_i$ is divergent.

Proposition 2.7.2. If a series $\Sigma_{i=1}^{\infty} a_i$ is convergent, then its partial sums are bounded.

Indeed, the sum of a series is the limit of the sequence of its partial sums. Thus, Proposition 2.7.2 is a direct consequence of Theorem 2.1.6 that states that any convergent sequence l is bounded.

Theorem 2.1.4 implies the following result.

Theorem 2.7.2. (Any course of calculus)**.** If series $\Sigma_{i=1}^{\infty} a_i$ and $\Sigma_{i=1}^{\infty} b_i$ are convergent, then:

a) the series $\sum_{i=1}^{\infty} (a_i + b_i)$ is convergent and

$$sum\sum_{i=1}^{\infty} a_i + sum\sum_{i=1}^{\infty} b_i = sum\ (\sum_{i=1}^{\infty} a_i + \sum_{i=1}^{\infty} b_i).$$

b) the series $c{\cdot}\sum_{i=1}^{\infty} b_i$ is convergent and for any real number c, we have

$$c\cdot(\ sum\sum_{i=1}^{\infty}b_i\ ) = sum(c\cdot\sum_{i=1}^{\infty}b_i\ )$$

c) the series $\sum_{i=1}^{\infty}(a_i - b_i)$ is convergent and

$$sum\sum_{i=1}^{\infty}a_i - sum\ \sum_{i=1}^{\infty}b_i = sum(\sum_{i=1}^{\infty}a_i - \sum_{i=1}^{\infty}b_i\ ).$$

One more important concept used in any course of calculus is absolute convergence.

Definition 2.7.4. A series $\Sigma_{i=1}^{\infty}a_i$ is called *absolutely convergent* (*absolutely summable*) if the sum $sum\Sigma_{i=1}^{\infty} |\ a_i\ |$ exists.

Proposition 2.7.3 (the Cauchy Theorem). Any absolutely convergent series $\Sigma_{i=1}^{\infty}a_i$ is convergent.

Proof. Comparing series $S = \Sigma_{i=1}^{\infty}a_i$ and $S^{o} = \Sigma_{i=1}^{\infty} |\ a_i\ |$, we see that the following inequality for partial sums of these series follows from the properties of metric

$$|\ S_n - S_m\ | = |\ a_{m+1} + a_{m+2} + a_{m+3} + \ldots + a_m\ | \leq$$
$$|\ a_{m+1}\ | + |\ a_{m+2}\ | + |\ a_{m+3}\ | + \ldots + |\ a_m\ | = |\ S_n^{o} - S_m^{o}\ |$$

Thus, by the Cauchy criterion (Theorem 2.1.10), if the sequence $\{\ S_n^{o}\ ;\ n = 1, 2, 3, \ldots\ \}$ converges, then the sequence $\{\ S_n\ ;\ n = 1, 2, 3, \ldots\ \}$ also converges. Consequently, summability of $\Sigma_{i=1}^{\infty} |a_i\ |$ implies summability of $\Sigma_{i=1}^{\infty}a_i$.

Proposition is proved.

Definition 2.7.5. A series $\Sigma_{i=1}^{\infty}a_i$ is called *conditionally convergent* (*conditionally summable*) if it is convergent but not absolutely convergent.

Results from (Diedonne, 1960; Ch. 5) imply that summation is associative in the following sense.

Proposition 2.7.4. For any absolutely convergent series, its sum coincides with the sum of any its quotient series.

At the same time, the Riemann series theorem shows that for conditionally convergent series the result of Proposition 2.7.4 is not true.

Theorem 2.7.3 (The Riemann Series Theorem). For any conditionally convergent series $\Sigma_{i=1}^{\infty}a_i$ and any real number a there is a permutation $\Sigma_{i=1}^{\infty}b_i$ of the series $\Sigma_{i=1}^{\infty}a_i$ such that $sum\Sigma_{i=1}^{\infty}b_i = a$, there is a permutation $\Sigma_{i=1}^{\infty}c_i$ of the series $\Sigma_{i=1}^{\infty}a_i$ such that $sum\Sigma_{i=1}^{\infty}c_i = \infty$, that is, the sequence of its partial sums tends to positive infinity, and there is a permutation $\Sigma_{i=1}^{\infty}d_i$ of the series $\Sigma_{i=1}^{\infty}a_i$ such that $sum\Sigma_{i=1}^{\infty}d_i = -\infty$, that is, the sequence of its partial sums tends to negative infinity.

Proof. Let a series $\Sigma_{i=1}^{\infty}a_i$ converges in $\boldsymbol{R}$ but does not absolutely converge in $\boldsymbol{R}$ and a is a real number. As the series $\Sigma_{i=1}^{\infty}a_i$ does not absolutely converge, it has positive members and negative members. Taking the series $\Sigma_{i=1}^{\infty}u_i$ that consists of all positive members from the series $\Sigma_{i=1}^{\infty}a_i$, we see that $\Sigma_{i=1}^{\infty}u_i$ tends to infinity. Taking the series $\Sigma_{i=1}^{\infty}v_i$ that consists of all negative members from the series $\Sigma_{i=1}^{\infty}a_i$, we see that $\Sigma_{i=1}^{\infty}v_i$ tends to negative infinity. Indeed, if both series $\Sigma_{i=1}^{\infty}u_i$ and $\Sigma_{i=1}^{\infty}v_i$ converge, then by the Cauchy criterion (cf. Proposition 2.7.2), $\Sigma_{i=1}^{\infty}a_i$ also converges. If both series $\Sigma_{i=1}^{\infty}u_i$ and $\Sigma_{i=1}^{\infty}v_i$ are bounded, then by Theorem 2.1.8, they both converge as their partial sums form a monotone bounded sequence. If either $\Sigma_{i=1}^{\infty}u_i$ or $\Sigma_{i=1}^{\infty}v_i$ is bounded and the other series is unbounded, then $\Sigma_{i=1}^{\infty}a_i$ diverges.

At first, let us consider the case when $a \geq 0$. As the series $\Sigma_{i=1}^{\infty} u_i$ tends to infinity and consists only of positive numbers, either $u_1 \geq a$ or we can find a number n_1 such that

$$u_1 + \ldots + u_{n_1} = A_1 \geq a$$

and

$$u_1 + \ldots + u_{n_1-1} \leq a.$$

In the first case, we put $u_1 = A_1$ and $n_1 = 1$. As the series $\Sigma_{i=1}^{\infty} v_i$ tends to negative infinity and consists only of negative numbers, either $A_1 + v_1 \leq a$ or we can find a number n_2 such that

$$A_1 + v_1 + \ldots + v_{n_2} = A_2 \leq a$$

and

$$A_1 + v_1 + \ldots + v_{n_2-1} \geq a.$$

In the first case, we put $A_1 + v_1 = A_2$ and $n_2 = 1$. As the series $\Sigma_{i=1}^{\infty} u_i$ tends to infinity and consists only of positive numbers, we can find a number n_3 such that

$$A_2 + u_{n_1+1} + \ldots + u_{n_3} = A_3 \geq a$$

and

$$A_2 + u_{n_1+1} + \ldots + u_{n_3-1} \leq a.$$

As the series $\Sigma_{i=1}^{\infty} v_i$ tends to negative infinity and consists only of negative numbers, we can find a number n_4 such that

$$A_3 + v_{n_2+1} + \ldots + v_{n_4} = A_4 \leq a$$

and

$$A_3 + v_{n_2+1} + \ldots + v_{n_4-1} \geq a.$$

We continue this process, obtaining the sequence $\{ A_n ; n = 1, 2, 3, \ldots \}$. In this sequence, all odd members are larger than or equal to a and all even members are less than or equal to a.

The distance $| A_k - a |$ is less than $| u_{n_k} - u_{n_{k-1}} |$ for odd k and less than $| v_{n_k} - v_{n_{k-1}} |$ for even k. By Corollary 2.7.1, both absolute values $| u_n |$ and $| v_n |$ tend to 0. Consequently, both absolute values $| u_{n_k} - u_{n_{k-1}}|$ for odd k and $| v_{n_k} - v_{n_{k-1}} |$ for odd k tend to 0. Thus, the sequence $\{ A_n ; n = 1, 2, 3, \ldots \}$ converges to a.

Let us consider the series

$$\Sigma_{i=1}^{\infty} b_i = u_1 + \ldots + u_{n_1} + v_1 + \ldots + v_{n_2} + u_{n_1+1} + \ldots + u_{n_3} + v_{n_2+1} + \ldots + v_{n_4} + \ldots$$

It is a permutation of the series $\Sigma_{i=1}^{\infty} a_i$. The sequence $\{ S_n^{o} ; n = 1, 2, \ldots \}$ of the partial sums of the series $\Sigma_{i=1}^{\infty} b_i$ has a property that the distance between its elements and the number a tends to zero because any its element lies between two elements from the sequence

$\{ A_n ; n = 1, 2, 3, \dots \}$, and the sequence $\{ A_n ; n = 1, 2, 3, \dots \}$ converges to a. Thus, the sequence $\{ S_n^{o}; n = 1, 2, \dots \}$ converges to a and $sum\Sigma_{i=1}^{\infty} b_i = a$.

The case when $a < 0$ is treated in a similar way.

Now we have to prove that there is a permutation $\Sigma_{i=1}^{\infty} c_i$ of the series $\Sigma_{i=1}^{\infty} a_i$ such that $sum\Sigma_{i=1}^{\infty} c_i = \infty$, that is, the sequence of its partial sums tends to positive infinity, and there is a permutation $\Sigma_{i=1}^{\infty} d_i$ of the series $\Sigma_{i=1}^{\infty} a_i$ such that $sum\Sigma_{i=1}^{\infty} d_i = - \infty$, that is, the sequence of its partial sums tends to negative infinity.

At first, we consider the first case when $sum\Sigma_{i=1}^{\infty} c_i = \infty$. As the series $\Sigma_{i=1}^{\infty} u_i$ tends to infinity and consists only of positive numbers, we can find numbers $n_1 , n_2 , \dots , n_k , \dots$ such that

$$u_1 + \dots + u_{n_1} > 10 + | v_1 | , u_{n_1+1} + \dots + u_{n_3} > 20 + | v_2 | , \dots ,$$
$$u_{n_k} + \dots + u_{n_{k+1}} > 10k + | v_k | , \dots$$

We put

$$\Sigma_{i=1}^{\infty} c_i = u_1 + \dots + u_{n_1} + v_1 + u_{n_1+1} + \dots + u_{n_3} + v_1 + \dots + u_{n_k} + \dots + u_{n_{k+1}} + v_k + \dots .$$

The series $\Sigma_{i=1}^{\infty} c_i$ is a permutation of the series $\Sigma_{i=1}^{\infty} a_i$ and by construction the sequence of its partial sums tends to positive infinity.

The case when $sum\Sigma_{i=1}^{\infty} d_i = - \infty$ is treated in a similar way.

Theorem is proved.

One more type of summability is commutative convergence of series.

Definition 2.7.6 (Diedonne, 1960). A series $\Sigma_{i=1}^{\infty} a_i$ is called *commutatively convergent* (*summable*) if $sum\Sigma_{i=1}^{\infty} a_i = sum\Sigma_{i=1}^{\infty} b_i$ for any permutation $\Sigma_{i=1}^{\infty} b_i$ of the series $\Sigma_{i=1}^{\infty} a_i$.

Theorem 2.7.4. A series is commutatively convergent if and only if it is absolutely convergent.

Indeed, Proposition 2.7.4 implies that for an absolutely convergent series, its sum is unique. Consequently, absolute convergence is sufficient for commutative convergence.

At the same time, by Theorem 2.7.3, if a series is not absolutely convergent, then either it is divergent and thus, not commutatively convergent, or it is conditionally convergent and by Theorem 2.7.3, not commutatively convergent. Consequently, absolute convergence is necessary for commutative convergence.

2.7.3. Fuzzy Summation of Series

As series summation is defined by a limit procedure, fuzzy limits allow us to introduce fuzzy summation of series. Let $r \in \boldsymbol{R}^{+}$.

Definition 2.7.7. a) A number a is an *r-sum* (denoted by r-sum$\Sigma_{i=1}^{\infty} a_i$) of a series $\Sigma_{i=1}^{\infty} a_i$ if $a = r$-lim$\{ S_n ; n = 1, 2, 3, \dots \}$.

b) A number a is a *fuzzy sum* of a series $\Sigma_{i=1}^{\infty} a_i$ if a is an r-sum of this series for some number r.

Remark 2.7.2. The construction of conditional fuzzy limits of sequences allows us to define in a classical way conditional fuzzy sums of series as conditional fuzzy limits of sequences of partial sums of these series. Such sums of series are more relevant to the computational practice where many solutions of differential equations are built in a form of series and where, due to the objective limitations of computers, is impossible to compute exact values of such sums. Fuzzification of model for numerical computation makes this model more adequate. Obtained in Section 2.4 properties of conditional fuzzy limits of sequences allows us to derive similar properties of conditional fuzzy sums of series.

Definition 2.7.8. a) A series $\Sigma_{i=1}^{\infty} a_i$ *r-converges* or is called *r-convergent* (*r-summable*) if there is a number $a = r\text{-sum}\Sigma_{i=1}^{\infty} a_i$. Otherwise, this series *r-diverges* or is *r-divergent.*

b) A series $\Sigma_{i=1}^{\infty} a_i$ *fuzzy converges* or is called *fuzzy convergent* (*fuzzy summable*) if this series r-converges for some number r. Otherwise, this series *fuzzy diverges* or is called *fuzzy divergent*

Thus, the number r plays the role of a measure for series converegnce and divergence.

For series with almost all positive (or almost all negative) elements, Proposition 2.2.4 implies that r-summability coincides with the classical summability of series. However, we know (cf. Chapter 1) that we cannot add one by one all elements of the infinite series even with the best computer. So, when a series converges to an irrational number, we need to take an approximate sum of this series. Thus, the necessity to use approximate limits of partial sums (e.g., when the exact limit is an irrational number) makes reasonable to study fuzzy summability of series with almost all positive (or almost all negative) elements, as well as relations between sums of such series.

At the same time, for series that have infinitely many both positive and negative elements, fuzzy summability extends the scope of summable series and ground some operations with conventionally divergent series.

We have seen in Section 2.7.1 that if a series $\Sigma_{i=1}^{\infty} a_i$ is convergent, then its partial sums are bounded. For fuzzy summability, it is possible to obtain not only a necessary condition of summability but also a complete criterion.

Lemma 2.2.7 and Theorem 2.2.5 imply the following result.

Lemma 2.7.1. A series $\Sigma_{i=1}^{\infty} a_i$ is r-summable if and only is for any natural number n, the series $\Sigma_{i=n}^{\infty} a_i$ is r-summable.

Corollary 2.7.4. A series $\Sigma_{i=1}^{\infty} a_i$ is summable if and only is for any natural number n, the series $\Sigma_{i=n}^{\infty} a_i$ is summable.

Theorem 2.7.5. A series $\Sigma_{i=1}^{\infty} a_i$ is fuzzy convergent if and only if its partial sums are bounded.

Indeed, the sum of a series is the limit of its partial sums. At the same time, by Theorem 2.2.2, a sequence l fuzzy converges if and only if it is bounded.

Let $r \in \boldsymbol{R}^+$.

Lemma 2.7.2. A series with non-negative elements is r-convergent if and only if it is convergent.

Indeed, in this case, the sequence of partial sums of this series is monotone and Lemma 2.7.2 is a direct consequence of Proposition 2.2.5.

Theorem 2.7.6 (the Fuzzy Cauchy Criterion for series). A series $\Sigma_{i=1}^{\infty} a_i$ is r-summable if and only if for any for any $\varepsilon > 0$, there is such n that for any $m > n$, we have

$$| a_n + a_{n+1} + \ldots + a_m | < 2r + \varepsilon$$

or, in other terms,

$$0 = 2r\text{-lim}_{n, m\to\infty} | a_n + a_{n+1} + \ldots + a_m |$$

Proof. Necessity. Let the series $\Sigma_{i=1}^{\infty} a_i$ be r-summable in $\boldsymbol{R}$. It means that the sequence $\{ S_n ; n = 1, 2, 3, \ldots \}$ of the partial sums of the series $\Sigma_{i=1}^{\infty} a_i$ r-converges. By the Generalized Cauchy Criterion for sequences (cf. Theorem 2.2.7), for any $\varepsilon > 0$ there is such $n \in \boldsymbol{N}$ that for any $i, j \geq n$, we have $| S_j - S_i | < 2r + \varepsilon$. Taking $j = m$ and $i = n$, we obtain $| S_m - S_n | = | a_n + a_{n+1} + \ldots + a_m | < 2r + \varepsilon$. Thus,

$$0 = 2r\text{-lim}_{n, m\to\infty} a_n + a_{n+1} + \ldots + a_m$$

Sufficiency. Let for any for any $\varepsilon > 0$ there is such n that for any $m > n$, we have $| a_n + a_{n+1} + \ldots + a_m | < 2r + \varepsilon$. As $| S_m - S_n | = | a_n + a_{n+1} + \ldots + a_m |$, it means that by the Generalized Cauchy Criterion for sequences (cf. Theorem 2.2.7), the sequence $\{ S_n ; n = 1, 2, 3, \ldots \}$ is r-fundamental and thus, r-converges. Definition 2.7.6, implies that the series $\Sigma_{i=1}^{\infty} a_i$ is r-summable in $\boldsymbol{R}$.

Theorem is proved.

For the case when $m = n + 1$, this gives us the following result.

Corollary 2.7.5. If a series $\Sigma_{i=1}^{\infty} a_i$ is r-summable, then $0 = 2r\text{-lim}_{i\to\infty} | a_i |$.

Remark 2.7.3. It is impossible to improve the estimate of fuzzy convergence in Theorem 2.7.6 and Corollary 2.7.5 as the following example demonstrates.

Example 2.7.4. Let us consider the sequence $l = \{ (-1)^n; n = 1, 2, 3, \ldots \}$ and the series $\Sigma_{n=1}^{\infty} (-1)^n$. Definition 2.2.1 implies that $0 = 1\text{-lim } l$ and it is demonstrated at the end of this section, $\frac{1}{2}\text{-sum}\Sigma_{n=1}^{\infty} (-1)^n = \frac{1}{2}$.

Corollary 2.7.6. If the sequence of absolute values of a_i does not $2r$-converge to 0, then the series $\Sigma_{i=1}^{\infty} a_i$ is r-divergent.

A direct corollary is such classical result as the Cauchy Criterion for series (Theorem 2.7.2).

Mathematicians found many other conditions that allow us to determine whether a given series is convergent or divergent. Here, we find conditions for fuzzy convergence and divergence. These conditions are called tests for convergence/divergence and respectively, tests for fuzzy convergence/divergence. At first, we consider series in which almost all members are non-negative, i.e., larger than or equal to zero. For such series Lemma 2.7.2 shows that r-convergence coincides with conventional convergence for any number r. However, in many situations when data come for measurement or computation, it is impossible to get the exact value of the series sum. We can have only some approximations. For instance, when in theory the sum of a series is equal to π or e, practical summation gives only some rational number close to the theoretical sum. Consequently, in practice, we have fuzzy summability in the majority of cases. That is why, we formulate here results for series with non-negative members in a more general form of fuzzy summation and derive the corresponding classical results as corollaries, although both forms are equivalent. When series are arbitrary fuzzy summability is more powerful than conventional summability. Besides, even for series with non-negative members, fuzzy context allows one to find more general conditions for convergence than those that are suggested by classical results. Classical tests for convergence/divergence become direct corollaries of the corresponding tests for fuzzy convergence/divergence when $a = 0$.

In comparison tests, we compare a given series with a series that is known to be r-convergent or r-divergent.

Theorem 2.7.7 (Fuzzy Comparison Test). If for series $\Sigma_{i=1}^{\infty} a_i$ and $\Sigma_{i=1}^{\infty} b_i$, we have $0 \leq a_i \leq b_i$ for almost all $i \in \omega$, then if the series $\Sigma_{i=1}^{\infty} b_i$ r-converges for some r, then the series $\Sigma_{i=1}^{\infty} a_i$ r-converges, and if the series $\Sigma_{i=1}^{\infty} a_i$ r-diverges for some r, then the series $\Sigma_{i=1}^{\infty} b_i$ r-diverges.

Proof. Lemma 2.7.1 allows us to assume that the condition of the theorem is true for all $i \in \omega$. As all elements from both series are positive, sequences of partial sums for both series are monotone as all their elements are non-negative. Then by Proposition 2.2.5, these sequences r-converge for some $r > 0$ if and only if they converge in the classical sense. Thus, the statement of Theorem 2.7.7 is a consequence of Theorem 2.1.3. Namely, if the series $\Sigma_{i=1}^{\infty} b_i$ r-converges for some r, then this series converges. Then by Theorem 2.1.3, the series $\Sigma_{i=1}^{\infty} a_i$ converges and by Lemmas 2.2.1 and 2.2.4, it r-converges for any r.

Conversely, if the series $\Sigma_{i=1}^{\infty} a_i$ r-diverges for some r, then it diverges. Then by Theorem 2.1.3, the series $\Sigma_{i=1}^{\infty} b_i$ diverges and by Lemmas 2.2.1 and 2.2.4, it r-diverges for any r.

Theorem is proved.

Theorem 2.7.7 implies the following classical result.

Corollary 2.7.7 (Comparison Test). If for series $\Sigma_{i=1}^{\infty} a_i$ and $\Sigma_{i=1}^{\infty} b_i$, we have $0 \leq a_i \leq b_i$ for almost all $i \in \omega$, then if the series $\Sigma_{i=1}^{\infty} b_i$ converges, then the series $\Sigma_{i=1}^{\infty} a_i$ converges, and if the series $\Sigma_{i=1}^{\infty} a_i$ diverges, then the series $\Sigma_{i=1}^{\infty} b_i$ diverges.

Theorem 2.7.7 allows us to obtain one more summability test.

Theorem 2.7.8 (Fuzzy Ratio Comparison Test 1). If for series $\Sigma_{i=1}^{\infty} a_i$ and $\Sigma_{i=1}^{\infty} b_i$, we have a_i , b_i, > 0 and $0 < a_i/b_i < c$ for almost all $i \in \omega$ and some $c > 0$, then if the series $\Sigma_{i=1}^{\infty} b_i$ r-converges for some r, then the series $\Sigma_{i=1}^{\infty} a_i$ r-converges, and if the series $\Sigma_{i=1}^{\infty} a_i$ r-diverges for some r, then the series $\Sigma_{i=1}^{\infty} b_i$ r-diverges.

Indeed, for series with non-negative elements, fuzzy convergence coinsides with the conventional conventional convergence. Besides, the inequality $a_i/b_i < c$ valid for almost all $i \in \omega$ implies that $0 \leq a_i \leq c \cdot b_i$ for almost all $i \in \omega$. Then Theorem 2.7.8 is a consequence of Theorem 2.7.7 as the series $\Sigma_{i=1}^{\infty} b_i$ converges if and only if the series $\Sigma_{i=1}^{\infty} cb_i$ converges.

Theorem 2.7.8 implies the following classical result.

Corollary 2.7.8 (Ratio Comparison Test 1). If for series $\Sigma_{i=1}^{\infty} a_i$ and $\Sigma_{i=1}^{\infty} b_i$, we have a_i , b_i, > 0 and $0 < a_i/b_i < c$ for almost all $i \in \omega$ and some $c > 0$, then if the series $\Sigma_{i=1}^{\infty} b_i$ converges, then the series $\Sigma_{i=1}^{\infty} a_i$ converges, and if the series $\Sigma_{i=1}^{\infty} a_i$ diverges, then the series $\Sigma_{i=1}^{\infty} b_i$ diverges.

Similar conditions for fuzzy convergence and divergence are provided by the following result.

Theorem 2.7.9 (Fuzzy Ratio Comparison Test 2). If for series $\Sigma_{i=1}^{\infty} a_i$ and $\Sigma_{i=1}^{\infty} b_i$, we have a_i , b_i, > 0 and

$$a_{i+1}/a_i \leq b_{i+1}/b_i \qquad (2.4)$$

for almost all $i \in \omega$, then if the series $\Sigma_{i=1}^{\infty} b_i$ r-converges, then the series $\Sigma_{i=1}^{\infty} a_i$ r-converges, and if the series $\Sigma_{i=1}^{\infty} a_i$ r-diverges, then the series $\Sigma_{i=1}^{\infty} b_i$ r-diverges.

Proof. It is possible to assume (cf. Lemma 2.7.1) that inequality (2.4) is true for all $i \in \omega$. Then we have

$$a_2/a_1 \leq b_2/b_1 \text{ , } a_3/a_2 \leq b_3/b_2 \text{ , } \dots \text{ , } a_i/a_{i-1} \leq b_i/b_{i-1}$$

If we multiple respectively left and right sides of these inequalities, we have

$$a_i/a_1 \leq b_i/b_1 \text{ or } a_i/b_i \leq a_1/b_1$$

for all $i = 1, 2, 3, \dots$. Taking $a_1/b_1 = c$, we obtain the condition from Theorem 2.7.8. Thus, Theorem 2.7.9 is a consequence of Theorem 2.7.8.

This gives us the following classical test of series convergence.

Corollary 2.7.9 (Ratio Comparison Test 2). If for series $\Sigma_{i=1}^{\infty} a_i$ and $\Sigma_{i=1}^{\infty} b_i$, we have a_i , b_i, > 0 and $a_{i+1}/a_i \leq b_{i+1}/b_i$ for almost all $i \in \omega$, then if the series $\Sigma_{i=1}^{\infty} b_i$ converges, then the series $\Sigma_{i=1}^{\infty} a_i$ converges, and if the series $\Sigma_{i=1}^{\infty} a_i$ diverges, then the series $\Sigma_{i=1}^{\infty} b_i$ diverges.

Theorem 2.7.10 (Fuzzy Ratio Limit Comparison Test). a) If for series $\Sigma_{i=1}^{\infty} a_i$ and $\Sigma_{i=1}^{\infty} b_i$, we have a_i , b_i, > 0, $d = r\text{-lim}_{i\to\infty} b_i/a_i$ and $d - r > 0$, then the series $\Sigma_{i=1}^{\infty} b_i$ converges if and only if the series $\Sigma_{i=1}^{\infty} a_i$ converges.

b) If a_i , b_i, > 0, $d = r\text{-lim}_{i\to\infty} | b_i/a_i |$ and $d - r \geq 0$, then if the series $\Sigma_{i=1}^{\infty} a_i$ converges, then the series $\Sigma_{i=1}^{\infty} b_i$ converges.

c) If a_i , b_i, > 0, $d = r\text{-lim}_{i\to\infty} | a_i/b_i |$ and $d - r \geq 0$, then if the series $\Sigma_{i=1}^{\infty} a_i$ diverges, then the series $\Sigma_{i=1}^{\infty} b_i$ diverges.

Proof. (a) If $d = r\text{-lim}_{i\to\infty} b_i/a_i$, then for any $\varepsilon > 0$, almost all elements a_i and b_i satisfy the inequality

$$| d - b_i/a_i | < r + \varepsilon$$

This inequality implies

$$-(r + \varepsilon) < d - b_i/a_i < r + \varepsilon \tag{2.5}$$

From the right part of the inequality (2.5), we have

$$b_i/a_i > d - r - \varepsilon$$

As $d - r > 0$, we can take an $\varepsilon > 0$ such that $d - r - \varepsilon = k > 0$. Thus, $b_i/a_i > k$ and $b_i > ka_i$ for almost all elements a_i and b_i. As $k > 0$, the series $\Sigma_{i=1}^{\infty} a_i$ converges if and only if the series $\Sigma_{i=1}^{\infty} ka_i$ converges.

Then by Theorem 2.7.2, if the series$\Sigma_{i=1}^{\infty} b_i$ converges, then the series $\Sigma_{i=1}^{\infty} ka_i$ converges. Consequently, if the series$\Sigma_{i=1}^{\infty} b_i$ converges, then the series $\Sigma_{i=1}^{\infty} a_i$ converges.

At the same time, from the left part of the inequality (2.5), we have

$$b_i/a_i < d + r + \varepsilon$$

As $d - r > 0$, we can take an $\varepsilon > 0$ such that $d - r - \varepsilon = k > 0$. Thus, $b_i/a_i < h$ where $h = d + r + \varepsilon > 0$ and $b_i < ha_i$ for almost all elements a_i and b_i. As $h > 0$, the series $\Sigma_{i=1}^{\infty} a_i$ converges if and only if the series $\Sigma_{i=1}^{\infty} ha_i$ converges.

Then by Theorem 2.7.2, if the series $\Sigma_{i=1}^{\infty} ha_i$ converges, then the series $\Sigma_{i=1}^{\infty} b_i$ converges. Consequently, if the series $\Sigma_{i=1}^{\infty} a_i$ converges, then the series $\Sigma_{i=1}^{\infty} b_i$ converges.

This allows us to conclude that the series $\Sigma_{i=1}^{\infty} b_i$ converges if and only if the series $\Sigma_{i=1}^{\infty} a_i$ converges.

Part (a) of the theorem is proved.

(b) When $d - r > 0$, part (b) is a consequence of the part (a). So, we need to consider the case when $d - r = 0$.

From the inequality (2.5), we have that for any $\varepsilon > 0$, almost all elements a_i and b_i satisfy the inequality

$$- \varepsilon < b_i/a_i < \varepsilon$$

By the definition of the limit of a sequence, we have

$$\lim_{i \to \infty} b_i/a_i = 0$$

Then almost all elements a_i and b_i satisfy the inequality

$$b_i \leq \varepsilon a_i$$

By the same argument as in in the proof of the part (a), convergence of the series $\Sigma_{i=1}^{\infty} a_i$ implies convergence of the series $\Sigma_{i=1}^{\infty} b_i$

(c) When $d - r > 0$, part (c) is also a consequence of the part (a). So, we need to consider the case when $d - r = 0$.

From the inequality (2.5), we have that for any $\varepsilon > 0$, almost all elements a_i and b_i satisfy the inequality

$$- \varepsilon < a_i/b_i < \varepsilon$$

By the definition of the limit of a sequence, we have

$$\lim_{i \to \infty} a_i/b_i = 0$$

Then almost all elements a_i and b_i satisfy the inequality

$$a_i \leq \varepsilon b_i$$

By the same argument as in in the proof of the part (a), divergence of the series $\Sigma_{i=1}^{\infty} a_i$ implies divergence of the series $\Sigma_{i=1}^{\infty} b_i$.

Theorem is proved.

This gives us the following classical test of series convergence.

Corollary 2.7.10 (Ratio Limit Comparison Test). a) If for series $\Sigma_{i=1}^{\infty} a_i$ and $\Sigma_{i=1}^{\infty} b_i$, we have a_i, b_i, > 0, $d = \lim_{i\to\infty} b_i/a_i$ and $d > 0$, then the series $\Sigma_{i=1}^{\infty} b_i$ converges if and only if the series $\Sigma_{i=1}^{\infty} a_i$ converges.

b) If a_i, b_i, > 0, $0 = \lim_{i\to\infty} b_i/a_i$, then if the series $\Sigma_{i=1}^{\infty} a_i$ converges, then the series $\Sigma_{i=1}^{\infty} b_i$ converges.

c) If a_i, b_i, > 0, $0 = r\text{-}\lim_{i\to\infty} a_i/b_i$, then if the series $\Sigma_{i=1}^{\infty} a_i$ diverges, then the series $\Sigma_{i=1}^{\infty} b_i$ diverges.

Theorem 2.7.11 (Fuzzy Ratio Test). If for a series $\Sigma_{i=1}^{\infty} a_i$, we have $a_i > 0$ and $a_{i+1}/a_i \le q < 1$ for almost all $i \in \omega$, then the series $\Sigma_{i=1}^{\infty} a_i$ r-converges for all $r \ge 0$, and if $a_{i+1}/a_i \ge 1$ for almost all $i \in \omega$, then the series $\Sigma_{i=1}^{\infty} a_i$ r-diverges for all $r \ge 0$.

Proof. Let us consider the geometric series $\Sigma_{i=1}^{\infty} q^i$. Then the condition $a_{i+1}/a_i \le q$ gives us $a_{i+1}/a_i \le q^{i+1}/ q^i$. At the same time, when $0 < q < 1$, the geometric series is convergent (cf. Example 2.7.2). Consequently, by Theorem 2.7.9, the series $\Sigma_{i=1}^{\infty} a_i$ r-converges.

When $a_{i+1}/a_i \ge 1$, we have $a_{i+1}/a_i \ge q^{i+1}/ q^i$ with $q \ge 1$. At the same time, when $q \ge 1$, the geometric series is divergent (cf. Example 2.7.2). Consequently, by Theorem 2.7.9, the series $\Sigma_{i=1}^{\infty} a_i$ r-diverges.

Theorem is proved.

This gives us the following classical test of series convergence.

Corollary 2.7.11 (Ratio Test or d'Alembert Criterion). If for a series $\Sigma_{i=1}^{\infty} a_i$, we have $a_i > 0$ and $a_{i+1}/a_i \le q < 1$ for almost all $i \in \omega$, then the series $\Sigma_{i=1}^{\infty} a_i$ converges, and if $a_{i+1}/a_i \ge 1$ for almost all $i \in \omega$, then the series $\Sigma_{i=1}^{\infty} a_i$ diverges.

Theorem 2.7.12 (Fuzzy Ratio Limit Test). If for a series $\Sigma_{i=1}^{\infty} a_i$, we have $a_i > 0$ and $d = r\text{-}\lim_{i\to\infty} | a_{i+1}/a_i |$, then if $d + r < 1$, then the series $\Sigma_{i=1}^{\infty} a_i$ r-converges, and if $d - r \ge 1$, then the series $\Sigma_{i=1}^{\infty} a_i$ r-diverges.

Indeed, $d + r < 1$ implies that $d + r + k < 1$ for some $k > 0$. Taking $\varepsilon < k$ and $0 < \varepsilon$, we have

$$a_{i+1}/a_i < d + r + \varepsilon < d + r + k < 1$$

for almost all $i \in \omega$.

Thus, by Theorem 2.7.11, the series $\Sigma_{i=1}^{\infty} a_i$ r-converges.

In the second case, $d - r > 1$ implies that $d + r - k > 1$ for some $k > 0$. Taking $\varepsilon < k$ and $0 < \varepsilon$, we have

$$1 < d + r + k < d - r - \varepsilon < a_{i+1}/a_i$$

for almost all $i \in \omega$.

Thus, by Theorem 2.7.11, the series $\Sigma_{i=1}^{\infty} a_i$ r-diverges.

Theorem is proved.

This gives us the following classical test of series convergence.

Corollary 2.7.12 (Ratio Limit Test). If for a series $\Sigma_{i=1}^{\infty} a_i$, we have $a_i > 0$ and $d = r\text{-}\lim_{i\to\infty} | a_{i+1}/a_i |$, then if $d - r < 1$, then the series $\Sigma_{i=1}^{\infty} a_i$ converges, and if $d - r \geq 1$, then the series $\Sigma_{i=1}^{\infty} a_i$ diverges.

Theorem 2.7.13 (Fuzzy Root Test). If for a series $\Sigma_{i=1}^{\infty} a_i$, we have $a_i > 0$ and $\sqrt[i]{a_i} \leq q < 1$ for almost all $i \in \omega$, then the series $\Sigma_{i=1}^{\infty} a_i$ r-converges, and if and $\sqrt[i]{a_i} \geq 1$ for almost all $i \in \omega$, then the series $\Sigma_{i=1}^{\infty} a_i$ r-diverges.

Indeed, the condition

$$\sqrt[i]{a_i} \leq q < 1$$

implies the condition

$$a_i \leq q^i$$

This reduces convergence of the series $\Sigma_{i=1}^{\infty} a_i$ convergence of the geometric series as in Theorem 2.7.11.

When $\sqrt[i]{a_i} \geq 1$, we also reduces divergence of the series $\Sigma_{i=1}^{\infty} a_i$ to divergence of the geometric series as it is done in Theorem 2.7.11.

Theorem 2.7.13 gives us the following classical test of series convergence.

Corollary 2.7.13 (Root Test or Cauchy Root Criterion). If for a series $\Sigma_{i=1}^{\infty} a_i$, we have $a_i > 0$ and $\sqrt[i]{a_i} \leq q < 1$ for almost all $i \in \omega$, then the series $\Sigma_{i=1}^{\infty} a_i$ converges, and if and $\sqrt[i]{a_i} \geq 1$ for almost all $i \in \omega$, then the series $\Sigma_{i=1}^{\infty} a_i$ diverges.

Theorem 2.7.14 (Fuzzy Harmonic Test). If for a series $\Sigma_{i=1}^{\infty} a_i$, we have $a_i > 0$ and $i(1 - a_{i+1}/a_i) \geq q > 1$ for almost all $i \in \omega$, then the series $\Sigma_{i=1}^{\infty} a_i$ r-converges, and if $i(1 - a_{i+1}/a_i) \leq 1$ for almost all $i \in \omega$, then the series $\Sigma_{i=1}^{\infty} a_i$ r-diverges.

Indeed, the condition

$$i(1 - a_{i+1}/a_i) \geq q > 1$$

implies the condition

$$1 - a_{i+1}/a_i \geq q/i$$

Consequently, we have

$$a_{i+1}/a_i \leq 1 - q/i < 1$$

This reduces Theorem 2.7.14 to Theorem 2.7.11.

It is possible to to do similar reduction in the case when $i(1 - a_{i+1}/a_i) \leq 1$.

Theorem 2.7.14 gives us the following classical test of series convergence.

Corollary 2.7.14 (Ratio Harmonic Test or Raabe Criterion). If for a series $\Sigma_{i=1}^{\infty} a_i$, we have $a_i > 0$ and $i(1 - a_{i+1}/a_i) \geq q > 1$ for almost all $i \in \omega$, then the series $\Sigma_{i=1}^{\infty} a_i$ converges, and if $i(1 - a_{i+1}/a_i) \leq 1$ for almost all $i \in \omega$, then the series $\Sigma_{i=1}^{\infty} a_i$ diverges.

Theorem 2.7.15 (Fuzzy Harmonic Limit Test). If for a series $\Sigma_{i=1}^{\infty} a_i$, we have $a_i > 0$ and $d = r\text{-}\lim_{i\to\infty} i(1 - a_{i+1}/a_i)$, then if $d - r < 1$, then the series $\Sigma_{i=1}^{\infty} a_i$ r-converges, and if $d - r \geq 1$, then the series $\Sigma_{i=1}^{\infty} a_i$ r-diverges.

Proof is similar to the proof of Theorem 2.7.12.

This gives us the following classical test of series convergence.

Corollary 2.7.15 (Ratio Harmonic Limit Test or Raabe Limit Criterion). If for a series $\Sigma_{i=1}^{\infty} a_i$, we have $a_i > 0$ and $d = r\text{-}\lim_{i\to\infty} i(1 - a_{i+1}/a_i)$, then if $d - r < 1$, then the series $\Sigma_{i=1}^{\infty} a_i$ converges, and if $d - r \geq 1$, then the series $\Sigma_{i=1}^{\infty} a_i$ diverges.

Now we find a test for alternating series fuzzy convergence. In an alternating series, signs of elements are changing from plus to minus and from minus to plus, i.e., each member has an opposite sign in comparison with the previous and next members.

Theorem 2.7.16 (Fuzzy Alternating Series Condition). If for an alternating series $\Sigma_{i=1}^{\infty} a_i$, the sequence $l = \{ | a_i | \geq 0; i = 1, 2, 3, \ldots \}$ is monotone decreasing and r-converges to 0, then the series $\Sigma_{i=1}^{\infty} a_i$ $(½)r$-converges.

Proof. For an alternating series, we have $S_n - S_{n-1} = a_i$. As the sequence $l = \{ |a_i| \geq 0; i = 1, 2, 3, \ldots \}$ is monotone decreasing, we have either $S_{n-1} \leq S_{n+1} \leq S_n$ or $S_n \leq S_{n+1} \leq S_{n-1}$. In addition, r-convergence of l to 0 for any $\varepsilon > 0$, there is a number m such that for all $n \geq m$, we have $|a_i| < r + \varepsilon$. For an alternating series, we have $S_n - S_{n-1} = a_i$. Thus, for all $n \geq m$, all partial sums S_n belong to the interval with the endpoints S_m and S_{m+1} and the length $r + \varepsilon$. By Lemma 2.5.4, the sequence of partial sums $(½)r$-converges, and consequently, the series $\Sigma_{i=1}^{\infty} a_i$ $(½)r$-converges.

Theorem is proved.

If a sequence converges to 0, then it r-converges to 0 for any $r > 0$. Thus, as a direct consequence of Theorem 2.7.14, we obtain the following classical result

Corollary 2.7.16 (the Leibniz Theorem (cf., for example, (Fihtengoltz, 1955)). If for a series $\Sigma_{i=1}^{\infty} (-1)^i a_i$, the sequence $l = \{ a_i \geq 0; i = 1, 2, 3, \ldots \}$ is monotone decreasing and converges to 0, then the series $\Sigma_{i=1}^{\infty} a_i$ converges.

Remark 2.7.4. For alternating series, r-convergence does not coincide with convergence. For instance (cf. Example 2.7.4), the series $\Sigma_{i=1}^{\infty}(-1)^i$ diverges in the classical sense, but it ½-converges. Thus, Theorem 2.7.16 is more encompassing than the Leibnitz Theorem.

One more useful concept is absolute fuzzy convergence.

Definition 2.7.9. A series $\Sigma_{i=1}^{\infty} a_i$ is called *absolutely r-convergent* (*absolutely r-summable*) if the fuzzy sum $r\text{-}sum\Sigma_{i=1}^{\infty} | a_i |$ exists.

Lemma 2.7.2 implies the following result.

Corollary 2.7.17. A series $\Sigma_{i=1}^{\infty} a_i$ is absolutely r-convergent if and only if it is absolutely convergent.

Theorem 2.7.17. Any absolutely r-convergent series is r-convergent.

Proof. Let us take an absolutely r-convergent series $\Sigma_{i=1}^{\infty} a_i$. By Corollary 2.7.17, the series $\Sigma_{i=1}^{\infty} a_i$ is absolutely convergent. Then by Proposition 2.7.4, $\Sigma_{i-1}^{\infty} a_i$ converges. Then by Lemma 2.2.1, the series $\Sigma_{i=1}^{\infty} a_i$ is 0-convergent. Then by Lemma 2.2.4, the series $\Sigma_{i=1}^{\infty} a_i$ is r-convergent for any $r > 0$.

Theorem is proved.

A direct corollary is such classical result as the Cauchy Theorem (Proposition 2.7.4).

Definition 2.7.10. A series $\Sigma_{i=1}{}^{\infty} a_i$ is called *conditionally r-convergent* (*conditionally r-summable*) if it is r-convergent but not absolutely r-convergent.

Now we find tests of convergence and divergence for arbitrary series. In this case, fuzzy convergence can be essentially different from conventional convergence.

Example 2.7.5. Let us consider the series $\Sigma_{i=1}{}^{\infty}(-1)^{[(i-1)/2]} = 1 + 1 - 1 - 1 + 1 + 1 - 1 - 1 + \ldots$. Here if $a \in \boldsymbol{R}$, then $[a]$ is the integral part of a, which is equal to the largest integer number that is less than a. Its partial sums are $S_1 = 1, S_2 = 2, S_3 = 1, S_4 = 0, S_5 = 1, S_6 = 2, S_7 = 1, \ldots$. It diverges in the classical sense. However, it 1-converges and has 1-sum equal to 1.

Theorem 2.7.18 (General Fuzzy Ratio Test). If for a series $\Sigma_{i=1}{}^{\infty} a_i$, we have $d = r\text{-}\lim_{i\to\infty} | a_{i+1}| / | a_i |$ for some number d, then the inequality $d + r < 1$ implies that the series $\Sigma_{i=1}{}^{\infty} a_i$ absolutely converges (and therefore, converges), and the inequality $d + r > 1$ implies that the series $\Sigma_{i=1}{}^{\infty} a_i$ diverges.

This gives us the following classical test of series convergence.

Corollary 2.7.18 (General Ratio Test). If for a series $\Sigma_{i=1}{}^{\infty} a_i$, we have $d = \lim_{i\to\infty} | a_{i+1}| / | a_i |$ for some number d, then the inequality $d < 1$ implies that the series $\Sigma_{i=1}{}^{\infty} a_i$ converges, and the inequality $d > 1$ implies that the series $\Sigma_{i=1}{}^{\infty} a_i$ diverges.

Theorem 2.7.19 (Fuzzy Root Test). If for a series $\Sigma_{i=1}{}^{\infty} a_i$, we have $d = r\text{-}\lim_{i\to\infty} \sqrt[i]{| a_i |}$ for some number d, then the inequality $d + r < 1$ implies that the series $\Sigma_{i=1}{}^{\infty} a_i$ absolutely converges (and therefore, converges), and the inequality $d + r > 1$ implies that the series $\Sigma_{i=1}{}^{\infty} a_i$ diverges.

This gives us the following classical test of series convergence.

Corollary 2.7.19 (Root Test). If for a series $\Sigma_{i=1}{}^{\infty} a_i$, we have $d = \lim_{i\to\infty} \sqrt[i]{| a_i |}$ for some number d, then the inequality $d < 1$ implies that the series $\Sigma_{i=1}{}^{\infty} a_i$ converges, and the inequality $d > 1$ implies that the series $\Sigma_{i=1}{}^{\infty} a_i$ diverges.

Theorem 2.7.20 (Fuzzy *n*-th Term Test). If for a series $\Sigma_{i=1}{}^{\infty} a_i$, $\lim_{i\to\infty} a_i \neq 0$, then the series fuzzy diverges.

Theorem 2.7.20 gives us the following classical test of series convergence.

Corollary 2.7.20 (*n*-th Term Test). If for a series $\Sigma_{i=1}{}^{\infty} a_i$, $\lim_{i\to\infty} a_i \neq 0$, then the series diverges.

Fuzzy summability extends the power of summability techniques and allows one, for example, to make mathematically meaningful some operations that originators of the calculus performed with classically divergent series. As we know (cf., for example, (Sandifer, 2006)), in the 1700's, many mathematicians thought that with enough brilliance and enough work they could sum any series. For instance, such great mathematicians as Leibniz, Daniel Bernoulli and Euler reasoned that the series $S = 1 - 1 + 1 - 1 + 1 - 1 + \ldots$ is summable and its sum is equal to ½. They based their assumption on the following reasoning.

Take the series $S_0 = 0 + 1 - 1 + 1 - 1 + 1 - 1 + \ldots$ and add it to S by the rule

$$\Sigma_{i=1}{}^{\infty} a_i + \Sigma_{i=1}{}^{\infty} b_i = \Sigma_{i=1}{}^{\infty}(a_i + b_i)$$

The result is equal to 1. At the same time, as the sum s, which was not formalized at the times of Leibniz, Bernoulli, and Euler, of the series S is equal to the sum of the series S_0, it gives the equality $2s = 1$. Consequently, $s = ½$.

Now it is generally assumed that this series is divergent. However, fuzzy summability allows one to make a rigorous statement about summability of this series. Namely, we show that ½ is a fuzzy limit of S with the least parameter r of convergence.

Indeed, let us consider the sequence $l = \{ S_n ; n = 1, 2, \dots \}$ of partial sums of the series S. Here $S_1 = 1, S_2 = 0, S_3 = 1, S_4 = 0, S_5 = 1, \dots$. Then the set L(l) of all limits of subsequences of l consists of two elements 1 and 0. By Definition 2.2.1, ½ is a ½ -limit of the sequence l and l is q-divergent for any $q < ½$.

2.8. Statistical and Statistical Fuzzy Convergence

Statistics: The only science that enables different experts
using the same figures to draw different conclusions.
Evan Esar (1899 - 1995)

The idea of statistical convergence goes back to the first edition (published in Warsaw in 1935) of the monograph of Zygmund (1979). Formally the concept of statistical convergence was introduced by Steinhaus (1951) and Fast (1951) and later reintroduced by Schoenberg (1959).

Statistical convergence, while introduced over nearly fifty years ago, has only recently become an area of active research. Different mathematicians studied properties of statistical convergence and applied this concept in various areas such as measure theory (Miller, 1995), trigonometric series (Zygmund, 1979), approximation theory (Duman, et al, 2003), locally convex spaces (Maddox, 1988), finitely additive set functions (Connor and Kline, 1996), in the study of subsets of the Stone-Čhech compactification of the set of natural numbers (Connor and Swardson, 1993), and Banach spaces (Connor, et al, 2000).

Here we extend the concept of statistical convergence by means of neoclassical analysis. At first, we consider statistical convergence. Then we study relations between statistical convergence, ergodic systems, and convergence of statistical characteristics such as the mean (average), and standard deviation. Described here relations between statistical convergence and convergence of statistical characteristics (such as mean and standard deviation) explain why this kind of convergence is called statistical.

The next natural step is introduction of a new type of fuzzy convergence, the concept of statistical fuzzy convergence. Different properties of statistical fuzzy convergence are obtained. Then relations between statistical fuzzy convergence and fuzzy convergence of statistical characteristics such as the mean (average) and standard deviation are explicated.

2.8.1. Statistical Convergence

Let us take a subset K of the set $\mathbf{N}$ of all natural numbers and define $K_n = \{k \in K; k \le n\}$. As always, $|X|$ is the number of elements in (cardinality of) a set X.

Definition 2.8.1. The *asymptotic density* $d(K)$ of the set K is equal to

$$\lim_{n\to\infty} (1/n)\ |K_n|$$

whenever the limit exists.

By this definition, $d(K) = 0$ for any finite set $K \subseteq \mathbf{N}$.

Lemma 2.8.1. a) For any $K, H \subseteq \mathbf{N}$, we have

$$d(K \cup H) \leq d(K) + d(H),$$

$$d(K \cap H) \leq \min\{\ d(K), d(H)\ \}$$

b) If $K \cap H = \emptyset$, then

$$d(K \cup H) = d(K) + d(H)$$

c) If $K \subseteq H$, then

$$d(K) \leq d(H)$$

and

$$d(H \setminus K) = d(H) - d(K)$$

d) If K is a cofinite subset of H, i.e., the difference $H \setminus K$ is finite, then

$$d(K) = d(H)$$

If $l = \{a_i\ ;\ i = 1, 2, 3,\ldots\}$ is a sequence of real numbers, a is a real number, then $L_\varepsilon(l, a) = \{i \in \mathbf{N};\ |\ a_i - a| \geq \varepsilon\ \}$.

Definition 2.8.2. The *asymptotic density*, or simply, *density* $d(l)$ of the sequence l with respect to a and ε is equal to $d(L_\varepsilon(l, a))$.

Asymptotic density allows us to define statistical convergence.

Definition 2.8.3. A sequence $l = \{a_i\ ;\ i = 1, 2, 3, \ldots\}$ is *statistically convergent* to a if $d(L_\varepsilon(l, a)) = 0$ for every $\varepsilon > 0$. The number (point) a is called the *statistical limit* of l and is denoted by $a =$ *stat*-lim l.

As in the case of conventional limits, we have the following result.

Lemma 2.8.2. The *statistical limit* of l is unique.

Lemma 2.8.3. All convergent sequences are statistically convergent.

Indeed, as any finite subset of $\mathbf{N}$ has zero density, conditions from Definition 2.8.3 are true for any convergent sequence.

However, the converse of Lemma 2.8.3 is not true as also the following example demonstrates.

Example 2.8.1. Let us consider the sequence $l = \{a_i\ ;\ i = 1, 2, 3, \ldots\}$ whose terms are

$$a_i = \begin{cases} i \text{ when } i = n^2 \text{ for all } n = 1, 2, 3, \dots \\ 1/i \text{ otherwise.} \end{cases}$$

It is easy to see that the sequence l is divergent in the ordinary sense, while 0 is the statistical limit of l since $d(K) = 0$ where $K = \{n^2 \text{ for all } n = 1, 2, 3, \dots\}$.

Not all properties of convergent sequences are true for statistical convergence. For instance, it is known that a subsequence of a convergent sequence is convergent. However, for statistical convergence this is not true. Indeed, the sequence $h = \{i\ ;\ i = 1, 2, 3, \dots\}$ is a subsequence of the statistically convergent sequence l from Example 2.8.1. At the same time, h is statistically divergent.

However, if we consider dense subsequences of fuzzy convergent sequences, it is possible to prove the corresponding result.

Definition 2.8.4. A subset K of the set $\boldsymbol{N}$ is called *statistically dense* if $d(K) = 1$.

Example 2.8.2. The set $\{ i \neq n^2\ ;\ i = 1, 2, 3, \dots;\ n = 1, 2, 3, \dots\}$ is statistically dense, while the set $\{ 3i;\ i = 1, 2, 3, \dots\}$ is not.

Lemma 2.8.4. a) A statistically dense subset of a statistically dense set is a statistically dense set.

b) The intersection and union of two statistically dense sets are statistically dense sets.

Definition 2.8.5. A subsequence h of the sequence l is called *statistically dense* in l if the set of all indices of elements from h is statistically dense.

Lemma 2.8.4 implies the following result..

Corollary 2.8.1. a) A statistically dense subsequence of a statistically dense subsequence of l is a statistically dense subsequence of l.

b) The intersection and union of two statistically dense subsequences are statistically dense subsequences.

Theorem 2.8.1. A sequence l is statistically convergent if and only if any statistically dense subsequence of l is statistically convergent.

Proof. Necessity. Let us take a statistically convergent sequence $l = \{a_i\ ;\ i = 1, 2, 3, \dots\}$ and a statistically dense subsequence $h = \{b_k\ ;\ k = 1, 2, 3, \dots\}$ of l. Let us also assume that h statistically diverges. Then we show that l is also statistically divergent.

Indeed, for any real number a, there is some $\varepsilon > 0$ such that $d(L_\varepsilon(h, a)) = \lim_{n\to\infty} (1/n)$ $|H_{n,\varepsilon}(a)| = d > 0$ for some $d \in (0, 1)$, where $H_{n,\varepsilon}(a) = \{k \leq n;\ |b_k - a| > \varepsilon\}$. Let us put $k_n = |\ \{ a_i\ ;\ a_i \in l \setminus h \}|$. As h is a statistically dense subsequence of l, we have $\lim_{n\to\infty} (k_n /n) = 0$. Then

$$\begin{gathered} d(L_\varepsilon(l, a)) = \lim_{n\to\infty} (1/n)\, |\{i \in \boldsymbol{N};\ |\, a_i - a| \geq \varepsilon \}| \geq \\ \lim_{n\to\infty} (1/n)\, |\{i \in \boldsymbol{N};\ |\, a_i - a| \geq \varepsilon \text{ and } a_i \in h \}| = \\ \lim_{n\to\infty} (1/(n + k_n))\, |H_{n,\varepsilon}(a)| = \\ \lim_{n\to\infty} (1/n)(1/(1+ (k_n/n)))\, |H_{n,\varepsilon}(a)| = \\ \lim_{n\to\infty} (1/n)\, |H_{n,\varepsilon}(a)| = d \end{gathered}$$

Thus, l is also statistically divergent.

Sufficiency follows from the fact that l is a statistically dense subsequence of itself.

Theorem is proved.

Corollary 2.8.2. A statistically dense subsequence of a statistically convergent sequence is statistically convergent.

Let $l = \{a_i \in \boldsymbol{R}; i = 1, 2, 3, \ldots\}$and $h = \{b_i \in \boldsymbol{R}; i = 1, 2, 3, \ldots\}$. Then their sum $l + h$ is equal to the sequence $\{a_i + b_i; i = 1, 2, 3, \ldots\}$ their difference $l - h$ is equal to the sequence $\{a_i - b_i; i = 1, 2, 3, \ldots\}$, and $kl = \{ka_i ; i = 1, 2, 3, \ldots \}$ for any $k\in\boldsymbol{R}$. Lemma 2.8.4 allows us to prove the following result.

Theorem 2.8.2 (Šalat, 1980). If $b = stat\text{-}\lim l$ and $c = stat\text{-}\lim h$, then:

(a) $a + b = stat\text{-}\lim (l+h)$;
(b) $a - b = stat\text{-}\lim (l - h)$;
(c) $ka = stat\text{-}\lim (kl)$ for any $k\in\boldsymbol{R}$.

We do not give a proof of this theorem as it is a direct corollary of Theorem 2.8.9.

2.8.2. Convergence in Statistics and Statistical Convergence

Statistics is concerned with the collection and analysis of data and with making estimations and predictions from the data. Typically two branches of statistics are discerned: descriptive and inferential. Inferential statistics is usually used for two tasks: to estimate properties of a population given sample characteristics and to predict properties of a system given its past and current properties. To do this, specific statistical constructions were invented. The most popular and useful of them are the average or mean (or more exactly, arithmetic mean) μ and standard deviation σ (variance σ^2).

To make predictions for future, statistics accumulates data for some period of time. To know about the whole population, samples are used. Normally such inferences (for future or for population) are based on some assumptions on limit processes and their convergence. Iterative processes are used widely in statistics. For instance, the empirical approach to probability is based on the law (or better to say, conjecture) of big numbers, which states that a procedure repeated again and again, the relative frequency probability tends to approach the actual probability. The foundation for estimating population parameters and hypothesis testing is formed by the central limit theorem, which tells us how sample means change when the sample size grows. In experiments, scientists measure how statistical characteristics (e.g., means or standard deviations) converge (cf., for example, Harris, and Chiang, 1999; Moran and Lienhard, 1997)).

Convergence of means/averages and standard deviations have been studied by many authors and applied to different problems (cf., for example, (Dunford and J. Schwartz, 1955; Vapnik and Chervonenkis, 1981; Leibman, 2002; 2005; Host and Kra, 2005; 2005a; Frantzikinakis and Kra, 2005; 2005a)).

Convergence of statistical characteristics such as the average/mean and standard deviation are related to statistical convergence.

To each sequence $l = \{a_i ; i = 1, 2, 3, \ldots\}$ of real numbers, it is possible to correspond a new sequence $\mu(l) = \{\mu_n = (1/n) \sum_{i=1}^{n} a_i ; n = 1, 2, 3, \ldots\}$ of its partial averages (means). Here a partial average of l is equal to $\mu_n = (1/n) \sum_{i=1}^{n} a_i$.

Sequences of partial averages/means play an important role in the theory of ergodic systems (cf., for example, (Billingsley, 1965)). Indeed, the definition of an ergodic system is based on the concept of the "time average" of the values of some appropriate function g arguments for which are dynamic transformations T of a point x from the manifold of the dynamical system. This average is given by the formula

$$\hat{g}(x) = \lim (1/n) \Sigma_{k=1}^{n-1} g(T^k x).$$

In other words, the dynamic average is the limit of the partial averages/means of the sequence $\{ T^k x ; k = 1, 2, 3, \dots \}$.

Let $l = \{a_i ; i = 1, 2, 3, \dots\}$ be a bounded sequence, i.e., there is a number m such that $|a_i| < m$ for all $i \in N$. This condition is usually true for all sequences generated by measurements or computations, i.e., for all sequences of data that come from real life.

Theorem 2.8.3. If $a = stat\text{-}\lim l$, then $a = \lim \mu(l)$.

Proof. Since $a = stat\text{-}\lim l$, for every $\varepsilon > 0$, we have

$$\lim_{n\to\infty} (1/n) \left|\{i \le n, i \in N; |a_i - a| \ge \varepsilon\}\right| = 0 \tag{2.6}$$

As $|a_i| < m$ for all $i \in N$, there is a number k such that $|a_i - a| < k$ for all $i \in N$. Namely, $| a_i - a| \le | a_i | + | a| \le m + | a| = k$. Taking the set $L_{n,\varepsilon} (a) = \{i \le n, i \in N; | a_i - a| \ge \varepsilon\}$, denoting $|L_{n,\varepsilon} (a)|$ by u_n , and using the hypothesis $|a_i| < m$ for all $i \in N$, we have the following system of inequalities:

$$\begin{aligned} |\mu_n - a| &= |(1/n) \Sigma_{i=1}^{n} a_i - a| \\ &\le (1/n) \Sigma_{i=1}^{n} | a_i - a| \\ &\le (1/n) \{ku_n + (n - u_n)\varepsilon\} \\ &\le (1/n) (ku_n + n\varepsilon) \\ &= \varepsilon + k (u_n/n). \end{aligned}$$

From the equality (2.6), we get, for sufficiently large n, the inequality $|\mu_n - a| < \varepsilon + k\varepsilon$. Thus, $a = \lim \mu(l)$.

Theorem is proved.

Remark 2.8.1. However, convergence of the partial averages/means of a sequence does not imply statistical convergence of this sequence as the following example demonstrates.

Example 2.8.3. Let us take the sequence $l = \{a_i ; i = 1,2,3,\dots\}$ with terms $a_i = (-1)^i \sqrt{i}$. This sequence is statistically divergent although $\lim \mu(l) = 0$.

Taking a sequence $l = \{a_i ; i = 1, 2, 3,\dots\}$ of real numbers, it is possible to construct not only the sequence $\mu(l) = \{ \mu_n = (1/n) \Sigma_{i=1}^{n} a_i ; n = 1, 2, 3,\dots\}$ of its partial averages (means) but also the sequences $\sigma(l) = \{\sigma_n = ((1/n) \Sigma_{i=1}^{n} (a_i - \mu_n)^2)^{1/2} ; n = 1, 2, 3, \dots\}$ of its partial standard deviations σ_n and $\sigma^2(l) = \{\sigma_n^2 = (1/n) \Sigma_{i=1}^{n} (a_i - \mu_n)^2 ; n = 1, 2, 3, \dots\}$ of its partial variances σ_n^2.

Theorem 2.8.4. If $a = stat\text{-}\lim l$ and $| a_i | < m$ for all $i \in N$, then $\lim \sigma(l) = 0$.

Proof. We will show that $\lim \sigma^2(l) = 0$. By the definition, $\sigma_n^2 = (1/n) \sum_{i=1}^n (a_i - \mu_n)^2 = (1/n) \Sigma_{i=1}^n (a_i)^2 - \mu_n^2$. Thus, $\lim \sigma^2(l) = \lim_{n\to\infty} (1/n) \Sigma_{i=1}^n (a_i)^2 - \lim_{n\to\infty} \mu_n^2$.

If $| a_i| < m$ for all $i\in \boldsymbol{N}$, then there is a number p such that $| a_i^2 - a^2| < p$ for all $i\in \boldsymbol{N}$. Namely, $| a_i^2 - a^2| \leq | a_i |^2 + | a|^2 < m^2 + | a|^2 < m^2 + | a|^2 + m + | a| = p$. Let us consider the absolute value of the difference $\mu_n^2 - (1/n) \sum_{i=1}^n (a_i)^2 = \sigma^2_n$. Taking the set $L_{n,\varepsilon} (a) = \{i \leq n, i \in \boldsymbol{N}; | a_i - a| \geq \varepsilon \}$, denoting $|L_{n,\varepsilon} (a)|$ by u_n, and using the hypothesis $|a_i| < m$ for all $i\in \boldsymbol{N}$, we have the following system of inequalities:

$$
\begin{aligned}
|\sigma^2_n | &= | (1/n) \sum_{i=1}^n (a_i)^2 - \mu_n^2 | \\
&= | (1/n) \sum_{i=1}^n (a_i^2 - a^2) - (\mu_n^2 - a^2)| \\
&\leq (1/n) \sum_{i=1}^n | a_i^2 - a^2| + |\mu_n^2 - a^2| \\
&< (p/n) \sum_{i=1}^n |a_i - a| + |\mu_n^2 - a^2| \\
&< (p/n) (u_n + (n - u_n)\varepsilon) + |\mu_n^2 - a^2| \\
&< (p/n) (u_n + n\varepsilon) + |\mu_n^2 - a^2| \\
&= p (u_n /n) + \varepsilon p + |\mu_n^2 - a^2|
\end{aligned}
$$

as $|a_i^2 - a^2| \leq |a_i - a|\ |a_i + a| < |a_i - a| \cdot p$. By Theorem 2.8.3, we have $a = \lim \mu(l)$, which guarantees that $\lim \mu_n^2 = a^2$. Formula (2.6) implies that $\lim (u_n /n) = 0$. Since $\varepsilon > 0$ was arbitrary, the right hand side of the above inequality tends to zero as $n \to \infty$. Therefore, we have $\lim \sigma(l) = 0$.

Theorem is proved.

Corollary 2.8.3. If $a =$ *stat*-lim l and $| a_i | < m$ for all $i\in \boldsymbol{N}$, then $\lim \sigma^2(l) = 0$.

Theorem 2.8.5. A sequence l is statistically convergent if its sequence of partial averages $\mu(l)$ converges and $a_i \leq \lim \mu(l)$ (or $a_i \geq \lim \mu(l)$) for all $i = 1, 2, 3, \ldots$.

Proof. Let us assume that $a = \lim \mu(l)$, $a_i \leq \lim \mu(l)$ and take some $\varepsilon > 0$, the set $L_{n,\varepsilon}(a) = \{i \leq n, i \in \boldsymbol{N}; |a_i - a| \geq \varepsilon\}$, and denote $|L_{n,\varepsilon} (a)|$ by u_n. Then we have

$$
\begin{aligned}
| a - \mu_n | &= | a - (1/n) \sum_{i=1}^n a_i | \\
&= |(1/n) \sum_{i=1}^n (a - a_i)| \\
&= (1/n) \sum_{i=1}^n (a - a_i) \\
&\geq (1/n) \sum_{| a_i - a | \geq \varepsilon} (a - a_i) \geq (u_n/n)\varepsilon
\end{aligned}
$$

Consequently, $\lim_{n\to\infty} | a - \mu_n| \geq \lim_{n\to\infty} (u_n/n)\, \varepsilon$. As $\lim_{n\to\infty} | a - \mu_n| =0$ and ε is a fixed number, we have $\lim_{n\to\infty} (1/n) |\{i \leq n, i \in \boldsymbol{N}; | a_i - a| \geq \varepsilon \}| = 0$, i.e., $a =$ *stat*-lim l.

The case when $a_i \geq \lim \mu(l)$) for all $i = 1, 2, 3, \ldots$ is considered in a similar way.

Theorem is proved as ε is an arbitrary positive number.

Let $l = \{a_i ; i = 1,2,3,\ldots\}$ be a bounded sequence, i.e., there is a number m such that $|a_i| < m$ for all $i\in\boldsymbol{N}$.

Theorem 2.8.6. A sequence l is statistically convergent if and only if its sequence of partial averages $\mu(l)$ converges and its sequence of partial standard deviations $\sigma(l)$ converges to 0.

Proof. Necessity follows from Theorems 2.8.3 and 2.8.4.

Sufficiency. Let us assume that $a = \lim \mu(l)$, $\lim \sigma(l) = 0$, and take some $\varepsilon > 0$. This implies that for any $\lambda > 0$, there is a number n such that $\lambda > | a - \mu_n|$. Then taking such n that implies the inequality $\varepsilon > \lambda$, we have

$$\sigma_n^{\,2} = (1/n) \sum_{i=1}^{n} (a_i - \mu_n)^2$$
$$\geq (1/n) \sum_{i=1}^{n} \{(a_i - \mu_n)^2 ;\ |a_i - a| \geq \varepsilon\}$$
$$= (1/n) \sum_{i=1}^{n} \{((a_i - a) + (a - \mu_n))^2 ;\ |a_i - a| \geq \varepsilon\}$$
$$> (1/n) \sum_{i=1}^{n} \{((a_i - a) \pm \lambda)^2 ;\ |a_i - a| \geq \varepsilon\} \tag{2.7}$$

$$= (1/n) \sum_{i=1}^{n} \{((a_i - a)^2 \pm 2\lambda(a_i - a) + \lambda^2);\ |a_i - a| \geq \varepsilon\}$$
$$= (1/n) \sum_{i=1}^{n} \{(a_i - \mu_n)^2;\ |a_i - a| \geq \varepsilon\} \pm 2\lambda\, (1/n) \sum_{i=1}^{n}\{(a_i - a);\ |a_i - a| \geq \varepsilon\} + \lambda^2 \tag{2.8}$$

as $(a_i - \mu_n) = (a_i - a) + (a - \mu_n)$ and we take $+ \lambda$ or $- \lambda$ in the expression (2.7) according to the following rules:

1) if $(a_i - a) \geq 0$ and $(a - \mu_n) \geq 0$, then $(a_i - a) + (a - \mu_n) \geq (a_i - a) > (a_i - a) - \lambda$, and we take $- \lambda$;
2) if $(a_i - a) \geq 0$ and $(a - \mu_n) \leq 0$, then $(a_i - a) + (a - \mu_n) \geq (a_i - a) - |a - \mu_n| > (a_i - a) - \lambda$, and we take $- \lambda$;
3) if $(a_i - a) \leq 0$ and $(a - \mu_n) \geq 0$, then $|(a_i - a) + (a - \mu_n)| = |(a - a_i) - (a - \mu_n)| > |(a - a_i) -\lambda| = |(a_i - a) + \lambda|$, and we take $+ \lambda$;
4) if $(a_i - a) \leq 0$ and $(a - \mu_n) \leq 0$, then $|(a_i - a) + (a - \mu_n)| \geq | a - a_i| > |(a_i - a) + \lambda|$ as $a_i - a < - \varepsilon$, and we take $+ \lambda$.

In the expression (2.8), λ^2 converges to 0 because the sequence $\{\mu_n ;\ n = 1, 2, 3, \ldots \}$ converges to a when n tends to ∞. The sum $2\lambda\ (1/n) \sum_{i=1}^{n} \{(a_i - a);\ |a_i - a| \geq \varepsilon\}$ also converges to 0 because λ converges to 0 and $(1/n) \sum_{i=1}^{n} (a_i - a) < (1/n) \sum_{i=1}^{n} (|a_i| + |a|) \leq m + |a|$. At the same time, the sequence $\{\sigma_n;\ n = 1, 2, 3, \ldots\}$ also converges to 0. Thus, $\lim_{n\to\infty} (1/n) \sum_{i=1}^{n} \{(a_i - \mu_n)^2;\ |a_i - a| \geq \varepsilon\} = 0$. This implies that $\lim_{n\to\infty} (1/n)\sum_{i=1}^{n} \{|a_i - \mu_n|^2;\ |a_i - a| \geq \varepsilon\} = 0$. At the same time, $\lim_{n\to\infty} (1/n) \Sigma_{i=1}^{n} \{|a_i - \mu_n|^2;\ |a_i - a| \geq \varepsilon\} \geq \varepsilon^2 \cdot (\lim_{n\to\infty} (1/n)\ |\{i \leq n,\ i \in N;\ |a_i - a| \geq \varepsilon\}|)$ because $(1/n) \Sigma_{i=1}^{n} \{(u_i - \mu_n)^2 ;\ |u_i - u| \geq \varepsilon\} > (1/n) \Sigma_{i=1}^{n} \{((a_i - a) \perp \lambda)^2 ;\ |a_i - a| \geq \varepsilon\}$ (cf. formula (2.7)), λ tends to 0 when $n\to\infty$, and in the sum (2.7), all differences $(a_i -$

$a)^2$ are larger than or equal to ε^2 when $|a_i - a| \geq \varepsilon$. As ε is a fixed number, we have $\lim_{n\to\infty}(1/n)$ $|\{i \leq n, i \in N; | a_i - a| \geq \varepsilon\}| = 0$ for any $\varepsilon > 0$ as ε is an arbitrary positive number, i.e., a = *stat*-lim l.

Theorem is proved.

2.8.3. Statistical Fuzzy Convergence

As before, r denotes a non-negative real number and $l = \{a_i ; i = 1, 2, 3, \dots\}$ represents a sequence of real numbers.

Let us consider the set $L_{r,\varepsilon}(a) = \{i \in N; |a_i - a| \geq r + \varepsilon\}$ and a non-negative real number $r \geq 0$.

Definition 2.8.6. A sequence l *statistically r-converges* to a number a if $d(L_{r,\varepsilon}(a)) = 0$ for every $\varepsilon > 0$. The number (point a) is called a *statistical r-limit* of l and is denoted by a = *stat*-r-lim l. If a sequence l is not statistically r-convergent, it is called *statistically r-divergent*.

Definition 2.8.6 implies the following results.

Lemma 2.8.5. a) a = *stat*-r-lim $l \Leftrightarrow \forall\, \varepsilon > 0$, $\lim_{n\to\infty}(1/n)\, |\{i \in N; i \leq n ; |a_i - a| \geq r + \varepsilon\}| = 0$.

b) a = *stat*-r-lim $l \Leftrightarrow \forall\, \varepsilon > 0$, $\lim_{n\to\infty}(1/n)\, |\{i \in N; i \leq n ; |a_i - a| < r + \varepsilon\}| = 1$.

If a = lim l, then for any $r \geq 0$, we have a = r-lim l (cf. Section 2.2). In a similar way, using Definition 2.8.6, we obtain the following result since every finite subset of the natural numbers has density zero.

Lemma 2.8.6. If a = r-lim l, then a = *stat*-r-lim l

Remark 2.8.2. However, the converse of Lemma 2.8.6 is not true as the following example of a sequence that is statistically r-convergent, but is not r-convergent and also is not statistically convergent, shows.

Example 2.8.4. Let us consider the sequence $l = \{a_i ; i = 1, 2, 3, \dots \}$ whose terms are

$$a_i = \begin{cases} i \text{ when } i = n^2 \text{ for all } n = 1, 2, 3, \dots \\ (-1)^i \text{ otherwise.} \end{cases}$$

Then, it is easy to see that the sequence l is divergent in the ordinary sense. Even more, the sequence l has no r-limits for any r since it is unbounded above (see Section 2.2).

On the other hand, we see that the sequence x is not statistically convergent because it does not satisfy the Cauchy condition for statistical convergence (Fridy, 1985).

At the same time, 0 = *stat*-1-lim l since $d(K) = 0$ where $K = \{n^2$ for all $n = 1, 2, 3, \dots\}$.

Lemma 2.8.7. Statistical 0-convergence coincides with the concept of statistical convergence.

This result shows that statistical fuzzy convergence is a natural extension of statistical convergence.

As in the case of fuzzy convergence (cf. Section 2.2), we can prove the following results.

Lemma 2.8.8. If a = *stat*-lim l, then a = *stat*-r-lim l for any $r \geq 0$.

Lemma 2.8.9. If $a = stat\text{-}r\text{-}\lim l$, then $a = stat\text{-}q\text{-}\lim l$ for any $q > r$.

Lemma 2.8.10. If $a = stat\text{-}r\text{-}\lim l$ and $|b - a| = p$, then $b = stat\text{-}q\text{-}\lim l$ where $q = p + r$.

It is known that a subsequence of a fuzzy convergent sequence is fuzzy convergent (cf. Section 2.2). However, for statistical fuzzy convergence this is not true. Indeed, the sequence $h = \{i^2 ; i = 1, 2, 3,\ldots\}$ is a subsequence of the statistically fuzzy convergent sequence l from Example 2.8.4. At the same time, h is statistically fuzzy divergent.

However, if we consider dense subsequences of statistically fuzzy convergent sequences, it is possible to prove the following result.

Theorem 2.8.7. A sequence is statistically r-convergent if and only if any its statistically dense subsequence is statistically r-convergent.

Proof. Necessity. Let us take a statistically r-convergent sequence $l = \{a_i ; i = 1, 2, 3, \ldots \}$ and a statistically dense subsequence $h = \{b_k ; k = 1, 2, 3, \ldots\}$ of l. Let us also assume that h statistically r-diverges. Then we show that l is also statistically r-divergent.

Indeed, for any real number a, there is some $\varepsilon > 0$ such that $d(L_{r,\varepsilon}(h, a)) = \lim_{n\to\infty} (1/n) | H_{n,r,\varepsilon}(a)| = d > 0$ for some $d \in (0, 1)$, where $H_{n,r,\varepsilon}(a) = \{k \leq n; |b_k - a| > r + \varepsilon\}$. Let us put $k_n = | \{ a_i ; a_i \in l \setminus h \}|$. As h is a statistically dense subsequence of l, we have $\lim_{n\to\infty} (k_n /n) = 0$. Then

$$\begin{gathered} d(L_{r,\varepsilon}(l, a)) = \lim_{n\to\infty} (1/n) |\{i \in N; | a_i - a| \geq r + \varepsilon \}| \geq \\ \lim_{n\to\infty} (1/n) |\{i \in N; | a_i - a| \geq \varepsilon \text{ and } a_i \in h \}| = \\ \lim_{n\to\infty} (1/(n + k_n)) |H_{n,r,\varepsilon}(a)| = \\ \lim_{n\to\infty} (1/n)(1/(1+ (k_n/n))) | H_{n,r,\varepsilon}(a)| = \\ \lim_{n\to\infty} (1/n) | H_{n,r,\varepsilon}(a)| = d \end{gathered}$$

Thus, l is also statistically r-divergent.

Sufficiency follows from the fact that l is a statistically dense subsequence of itself.

Theorem is proved.

A statistically r-convergent sequence contains not only dense statistically r-convergent subsequences, but also dense r-convergent subsequences.

Theorem 2.8.8. $a = stat\text{-}r\text{-}\lim l$ if and only if there exists an increasing index sequence $K = \{k_n ; k_n \in N, n = 1, 2, 3, \ldots\}$ of the natural numbers such that $d(K) = 1$ and $a = r\text{-}\lim l_K$ where $l_K = \{a_i ; i \in K\}$.

Proof. Necessity. Suppose that $a = stat\text{-}r\text{-}\lim l$. Let us consider sets $L^{r,j}(a) = \{i \in N; |a_i - a| < r + (1/j) \}$ for all $j = 1, 2, 3, \ldots$ By the definition, we have

$$L^{r,j+1}(a) \subseteq L^{r,j}(a) \tag{2.9}$$

and as $a = stat\text{-}r\text{-}\lim l$, by Lemma 2.8.5, we have

$$d(L^{r,j}(a)) = 1 \tag{2.10}$$

for all $j = 1, 2, 3,\ldots$ Let us take some number i_1 from the set $L^{r,1}(a)$. Then, by (2.9) and (2.10), there is a number i_2 from the set $L^{r,2}(a)$ such that $i_1 < i_2$ and

$$(1/n) |\{i \in N; i \leq n ; |a_i - a| < r + 1/2\}| > 1/2 \text{ for all } n \geq i_2.$$

In a similar way, we can find a number i_3 from the set $L^{r,3}(a)$ such that $i_2 < i_3$ and

$$(1/n)\,|\{i \in N;\ i \leq n\ ;\ |a_i - a| < r + 1/3\}| > 2/3 \text{ for all } n \geq i_3.$$

We continue this process and construct an increasing sequence $\{i_j \in N, j = 1, 2, 3, \dots\}$ of the natural numbers such that each number i_j belongs to $L^{r,j}(a)$ and

$$(1/n)\,|\{i \in N;\ i \leq n\ ;\ |a_i - a| < r + 1/j\}| > (j - 1)/j \text{ for all } n \geq i_j \qquad (2.11)$$

Now we construct the increasing sequence of indices K as follows:

$$K = \{i \in N;\ 1 \leq i \leq i_1\} \cup (\textstyle\bigcup_{j\in N} \{\ i \in L^{r,j}(a);\ i_j \leq i \leq i_{j+1}\}) \qquad (2.12)$$

Then from (2.9), (2.11) and (2.12), we conclude that for all n from the interval $i_j \leq n \leq i_{j+1}$ and all $j = 1, 2, 3, \dots$, we have

$$(1/n)\{k \in K;\ k \leq n\} = (1/n)\,|\{i \in N;\ i \leq n\ ;\ |a_i - a| < r + 1/j\}| > (j - 1)/j \qquad (2.13)$$

In turn, (2.13) implies that $d(K) = 1$. Now let us denote $l_K = \{a_i\ ;\ i \in K\}$, take some $\varepsilon > 0$ and choose a number $j \in N$ such that $1/j < \varepsilon$. If $n \in K$ and $n \geq i_j$, then, by the definition of K, there exists a number $m \geq j$ such that $i_m \leq n \leq i_{m+1}$ and thus, $n \in L_{r,m}(a)$. Hence, we have

$$|a_n - a| < r + 1/j < r + \varepsilon$$

As this is true for all $n \in K$, we see that $a = r\text{-lim } l_K$

Thus, the proof of necessity is completed.

Sufficiency. Suppose that there exists an increasing index sequence $K = \{k_n\ ;\ k_n \in N;\ n = 1, 2, 3, \dots\}$ of the natural numbers such that $d(K) = 1$ and $a = r\text{-lim } l_K$ where $l_K = \{a_i\ ;\ i \in K\}$. Then there is a number n such that for each i from K such that $i \geq n$, the inequality $|a_i - a| < r + \varepsilon$ holds. Let us consider the set

$$L_{r,\varepsilon}(a) = \{i \in N;\ |a_i - a| \geq r + \varepsilon\}$$

Then we have

$$L_{r,\varepsilon}(a) \subseteq N \setminus \{k_i\ ;\ k_i \in N;\ i = n, n+1, n+2, \dots\}$$

Since $d(K) = 1$, by Lemma 2.8.1, we have $d(N \setminus \{k_i\ ;\ k_i \in N;\ i = n, n+1, n+2, \dots\}) = 0$, which yields $d(L_{r,\varepsilon}(a)) = 0$ for every $\varepsilon > 0$. Therefore, we conclude that $a = stat\text{-}r\text{-lim } l$.

Theorem is proved.

Corollary 2.8.4 (Šalat, 1980). $a = stat\text{-lim } l$ if and only if there exists an increasing index sequence $K = \{k_n\ ;\ k_n \in N;\ n = 1, 2, 3, \dots\}$ of the natural numbers such that $d(K) = 1$ and $a = \lim l_K$ where $l_K = \{a_i\ ;\ i \in K\}$.

Corollary 2.8.5. a = *stat-r*-lim l if and only if there exists a sequence $h = \{b_i\,;\, i = 1, 2, 3, \ldots\}$ such that $d(\{i;\, a_i = b_i\}) = 1$ and $a = r$-lim h.

Corollary 2.8.6. The following statements are equivalent:

(i) a = *stat-r*-lim l.

(ii) There is a set $K \subseteq \boldsymbol{N}$ such that $d(K) = 1$ and $a = r$-lim l_K where $l_K = \{a_i\,;\, i \in K\}$.

(iii) For every $\varepsilon > 0$, there exist a subset $K \subseteq \boldsymbol{N}$ and a number m $\in$ K such that $d(K) = 1$ and $|a_n - a| < r + \varepsilon$ for all $n \in K$ and $n \geq m$.

Let $l = \{a_i \in \boldsymbol{R};\, i = 1, 2, 3, \ldots\}$and $h = \{b_i \in \boldsymbol{R};\, i = 1, 2, 3, \ldots\}$. Then their sum $l + h$ is equal to the sequence $\{a_i + b_i\,;\, i = 1, 2, 3, \ldots\}$ and their difference $l - h$ is equal to the sequence $\{a_i - b_i\,;\, i = 1, 2, 3, \ldots\}$. Lemma 2.8.5 allows us to prove the following result.

Theorem 2.8.9. Let a = *stat-r*-lim l and b = *stat-q*-lim h. Then:

(a) $a + b$ = *stat*-$(r+q)$-lim$(l+h)$;

(b) $a - b$ = *stat*-$(r+q)$-lim$(l - h)$;

(c) ka = *stat*-$(|k|\cdot r)$-lim (kl) for any $k \in \boldsymbol{R}$ where $kl = \{ka_i\,;\, i = 1, 2, 3, \ldots\}$.

Theorem 2.8.9 implies Theorem 2.8.2 as a direct corollary.

Let us denote the set of all statistical r-limits of a sequence l by $L_{r\text{-}stat}(l)$, that is,

$$L_{r\text{-}stat}(l) = \{a \in \boldsymbol{R};\, a = stat\text{-}r\text{-}\lim l\}.$$

Then we have the following result.

Theorem 2.8.10. For any sequence l and number $r \geq 0$, $L_{r\text{-}stat}(l)$ is a convex subset of the real numbers.

Proof. Let $c, d \in L_{r\text{-}stat}(l)$, $c < d$ and $a \in [c, d]$. Then it is enough to prove that $a \in L_{r\text{-}stat}(l)$. Since $a \in [c, d]$, there is a number $\lambda \in [0, 1]$ such that $a = \lambda c - (1-\lambda)d$. As $c, d \in L_{r\text{-}stat}(l)$, then for every $\varepsilon > 0$, there exist index sets K_1 and K_2 with $d(K_1) = d(K_2) = 1$ and the numbers n_1 and n_2 such that $|a_i - c| < r + \varepsilon$ for all i from K_1 and $i \geq n_1$ and $|a_i - d| < r + \varepsilon$ for all i from K_2 and $i \geq n_2$. Let us put $K = K_1 \cap K_2$ and $n = \max\{n_1, n_2\}$. Then $d(K) = 1$ and for all $i \geq n$ with i from K, we have

$$\begin{aligned} |\, a_i - a \,| &= |\, a_i - \lambda c - (1-\lambda)d| \\ &= |(\lambda a_i - \lambda c) + ((1-\lambda)\, a_i - (1-\lambda)d)| \\ &\leq |\lambda a_i - \lambda c| + |(1-\lambda)\, a_i - (1-\lambda)d\,| \\ &\leq \lambda(r + \varepsilon) + (1-\lambda)(r + \varepsilon) = r + \varepsilon. \end{aligned}$$

So, by Theorem 2.8.6, we conclude a = *stat-r*-lim l, which implies $a \in L_{r\text{-}stat}(l)$.

Theorem is proved.

Lemmas 2.8.9 and 2.8.10 imply the following result.

Proposition 2.8.1. If $q > r$, then $L_{r\text{-}stat}(l) \subset L_{q\text{-}stat}(l)$

An important property in calculus is the Cauchy criterion of convergence, while an important property in neoclassical analysis is the extended Cauchy criterion of fuzzy convergence. Here we find an extended statistical Cauchy criterion for statistical fuzzy convergence.

Definition 2.8.7. A sequence l is called *statistically r-fundamental* if for any $\varepsilon > 0$ there is $n \in N$ such that $d(L_{n,r,\varepsilon}) = 0$ where $L_{n,r,\varepsilon} = \{i \in N;\ i \le n \text{ and } |a_i - a_n| \ge 2r + \varepsilon\}$.

Definition 2.8.8. A sequence l is called *statistically fuzzy fundamental* or a *statistically fuzzy Cauchy sequence* if it is statistically r-fundamental for some $r \ge 0$.

Lemma 2.8.11. If $r \le p$, then any statistically r-fundamental sequence is statistically p-fundamental.

Lemma 2.8.12. A sequence l is a statistically Cauchy sequence if and only if it is statistically 0-fundamental.

This result shows that the property to be a statistically fuzzy fundamental sequence is a natural extension of the property to be a statistically Cauchy sequence.

Lemma 2.8.13. Any r-fundamental sequence l is a statistically r-fundamental.

This result shows that the property to be a statistically fuzzy fundamental sequence is a natural extension of the property to be a fuzzy fundamental sequence.

Using the similar technique as in proof of Theorem 2.8.8, one can obtain the following result.

Theorem 2.8.11. A sequence l is statistically r-fundamental if and only if there exists an increasing index sequence $K = \{k_n;\ k_n \in N;\ n = 1, 2, 3, \dots\}$ of the natural numbers such that $d(K) = 1$ and the subsequence l_K is r-fundamental, that is, for every $\varepsilon > 0$, there is a number i such that $|\, a_{k_n} - a_{k_i} \,| < 2r + \varepsilon$ for all $n \ge i$.

Corollary 2.8.7. A sequence l is statistically fuzzy fundamental if and only if there exists a statistically dense subsequence u such that u is fuzzy fundamental.

Theorem 2.8.11 yields the following result.

Corollary 2.8.8 (Fridy, 1985). A sequence l is a statistically Cauchy sequence if and only if there exists an increasing index sequence $K = \{k_n;\ k_n \in N;\ n = 1, 2, 3, \dots\}$ of the natural numbers such that $d(K) = 1$ and the subsequence l_K is a Cauchy sequence.

Theorem 2.8.12 (the Extended Statistical Cauchy Criterion). A sequence l has a statistical r-limit if and only if it is statistically r-fundamental.

Proof. Necessity. Let $a = stat\text{-}r\text{-}\lim l$. Then by the definition, for any $\varepsilon > 0$, we have $d(L_{r,\varepsilon}(a)) = 0$, in other words, $\lim_{n\to\infty} (1/n)\, |\{i \in N;\ i \le n\ ;\ |a_i - a| \ge r + \varepsilon/2\}| = 0$. This implies that given $\varepsilon > 0$, we find $n \in N$ such that for any $i > n$, we have $|\, a_i - a_n| \le |\, a - a_i| + |\, a - a_n|$. As a result, $|\, a_i - a_n|$ cannot be larger than or equal to $2r + \varepsilon$ when $|\, a - a_n| \ge r + \varepsilon/2$ and $|\, a - a_i| \ge r + \varepsilon/2$. Consequently, $d(L_{n,\,2r,\varepsilon}) \le d(L_{r,\varepsilon/2}(a)) + d(L_{r,\varepsilon/2}(a)) = 0$, i.e., $d(L_{n,r,\varepsilon}) = 0$. Thus, l is a statistically r-fundamental sequence.

Sufficiency. Let assume that l is a statistically r-fundamental sequence. Then, by Theorem 2.8.11, we have an increasing index sequence $K = \{k_n;\ k_n \in N;\ n = 1, 2, 3, \dots\}$ of the natural numbers such that $d(K) = 1$ and the subsequence $l_K = \{a_i;\ i \in K\}$ is r-fundamental. Then, by Theorem 2.2.7, the sequence $l_K = \{a_i;\ i \in K\}$ is r-convergent. Let $a = r\text{-}\lim u$. Then, by Theorem 2.8.8, $a = stat\text{-}r\text{-}\lim u$.

The proof is completed.

Theorem 2.8.12 directly implies the following results.

Corollary 2.8.9 (the General Statistical Fuzzy Convergence Criterion). The sequence l statistically fuzzy converges if and only if it is statistically fuzzy fundamental.

Corollary 2.8.10 (the Statistical Cauchy Criterion) (Fridy, 1985). A sequence l statistically converges if and only if it is statistically fundamental, i.e., for any $\varepsilon > 0$ there is $n \in \boldsymbol{N}$ such that $d(L_{n,r,\varepsilon}) = 0$ where $L_{n,r,\varepsilon} = \{i \in N; \mid a_i - a_n \mid \geq \varepsilon\}$.

In turn, this result implies the classical Cauchy Criterion (cf. Section 2.1).

Let us consider a sequence l and a real nuber a.

Definition 2.8.9. The quantity

$$\inf\{r \,;\, a = stat\text{-}r\text{-}\lim l\}$$

is called the *statistical defect* $\delta(a = stat\text{-}\lim l)$ of statistical convergence of l to the number a.

Proposition 2.8.2. If $q = \inf\{r; a = stat\text{-}r\text{-}\lim l\}$, then $a = stat\text{-}q\text{-}\lim l$.

Definition 2.8.10. The quantity

$$\frac{1}{1 + \delta(a = stat\text{-}\lim l)}$$

is called the *statistical measure* $\mu(a = stat\text{-}\lim l)$ of statistical convergence of l to a number a.

Theorem 2.8.7 implies the following result.

Lemma 2.8.14. If h is a dense subsequence of a sequence l, then

$$\mu(a = stat\text{-}\lim h) \geq \mu(a = stat\text{-}\lim l)$$

and

$$\delta(a = stat\text{-}\lim h) \leq \delta(a = stat\text{-}\lim l)$$

The statistical measure of statistical convergence of l defines the fuzzy set $L_{stat\text{-}\lim}(l) = (\boldsymbol{R}, \mu(a = stat\text{-}\lim l), [0,1])$, which is called the *complete statistical fuzzy limit* of the sequence l.

Example 2.8.2. Let us consider the complete statistical fuzzy limit $L_{stat\text{-}\lim}(l)$ of the sequence l from Example 3.1. For this sequence and a real number a, we have

$$\mu(a = stat\text{-}\lim l) = 1/(2 + |a|)$$

Complete statistical fuzzy limits have the following property.

Theorem 2.8.13. The complete statistical fuzzy limit $L_{stat\text{-}\lim}(l)$ of a sequence l is a convex fuzzy set.

Proof. Let $c, d \in L_{stat\text{-}\lim}(l)$, $c < d$ and $a \in [c, d]$. Then it is enough to prove that $\mu(a = stat\text{-}\lim l) = \mu((\lambda c + (1 - \lambda)d) = stat\text{-}\lim l) \geq \min\{\mu(c = stat\text{-}\lim l), \mu(d = stat\text{-}\lim l)\}$. This is equivalent to the inequality $\delta(a = stat\text{-lim } l) = \delta((\lambda c + (1 - \lambda)d) = stat\text{-lim } l) \leq \max\{\delta(c = stat\text{-lim } l), \delta(d = stat\text{-lim } l)\}$.

Let us assume, for convenience, that $q = \delta(c = stat\text{-lim } l) \geq r = \delta(d = stat\text{-lim } l)\}$. Then by Lemma 2.8.8, $d = stat\text{-}q\text{-lim } l$. Then by Theorem 2.8.8, $d = stat\text{-}q\text{-lim } l$ as the set $L_{r\text{-}stat}(l)$ is convex. Thus, $\delta(a = stat\text{-lim } l) \leq q = \max\{\delta(c = stat\text{-lim } l), \delta(d = stat\text{-lim } l)\}$.

Theorem is proved.

Theorem 2.8.13 allows us to prove the following result.

Theorem 2.8.14. The complete statistical fuzzy limit $L_{stat\text{-lim}}(l)$ of a sequence l is a normal fuzzy set if and only if l statistically converges.

Relations between statistical convergence and convergence of statistical characteristics (such as mean and standard deviation) found above explain why this kind of convergence is called statistical. However, when data are obtained in experiments, they come from measurement and computation. As a result, we never have and never will be able to have absolutely precise convergence of statistical characteristics. It means that instead of ideal classical convergence, which exists only in pure mathematics, we have to deal with fuzzy convergence, which is closer to real life and gives more realistic models. That is why in this section, we consider relations between statistical fuzzy convergence and fuzzy convergence of statistical characteristics.

Let $l = \{a_i ; i = 1, 2, 3, \dots \}$ be a bounded sequence, i.e., there is a number m such that $|a_i| < m$ for all $i \in N$. This condition is usually true for all sequences generated by measurements or computations.

Theorem 2.8.15. If $a = stat\text{-}r\text{-lim } l$, then $a = r\text{-lim } \mu(l)$ where $\mu(l) = \{\mu_n = (1/n) \Sigma_{i=1}^{n} a_i ; n = 1, 2, 3, \dots \}$.

Proof. Since $a = stat\text{-}r\text{-lim } l$, for every $\varepsilon > 0$, we have

$$\lim_{n\to\infty} (1/n) \,|\{i \leq n, i \in N; |a_i - a| \geq r + \varepsilon\}| = 0 \qquad (2.14)$$

If $|a_i| < M$ for all $I \in N$, then there is a number k such that $|a_i - a| < k$ for all $i \in N$. Namely, $|a_i - a| \leq |a_i| + |a| \leq m + |a| = k$. Taking the set $L_{n,r,\varepsilon}(a) = \{i \in N; i \leq n \text{ and } |\, a_i - a \,| \geq r + \varepsilon\}$, denoting $|L_{n,r,\varepsilon}(a)|$ by u_n, and using the hypothesis $|a_i| < m$ for all $i \in N$, we have the following system of inequalities

$$\begin{aligned} |\mu_n - a| &= |(1/n) \Sigma_{i=1}^{n} a_i - a| \\ &\leq (1/n) \Sigma_{i=1}^{n} |a_i - a| \\ &\leq (1/n) (ku_n + (n - u_n)(r + \varepsilon)) \\ &\leq (1/n) (ku_n + n(r + \varepsilon)) \\ &= r + \varepsilon + (1/n) (ku_n). \end{aligned}$$

From the equality (2.14), we get, for sufficiently large n, that the inequality $|\mu_n - a| \leq r + 2\varepsilon$ holds. Thus, $a = r\text{-lim } \mu(l)$.

Theorem is proved.

Remark 2.8.3. Statistical r-convergence of a sequence does not imply r-convergence of this sequence even if all elements are bounded as the following example demonstrates.

Example 2.8.1. Let us consider the sequence $l = \{a_i ; i = 1, 2, 3, \dots\}$ whose terms are

$$a_i = \begin{cases} (-1)^i \cdot 1000 \text{ when } i = n^2 \text{ for all } n = 1, 2, 3, \dots \\ (-1)^i \text{ otherwise.} \end{cases}$$

By the definition, $0 = stat\text{-}1\text{-}\lim l$ since $d(K) = 0$ where $K = \{n^2$ for all $n = 1, 2, 3, \dots \}$. At the same time, this sequence does not have 1-limits.

Theorems 2.8.15, 2.8.12 and 2.2.7 imply the following result.

Corollary 2.8.11. If sequence l is statistically fuzzy fundamental, then the sequence of its partial means is fuzzy fundamental.

Finally, we get the following result.

Theorem 2.8.16. If $a = stat\text{-}r\text{-}\lim l$ and $|a_i| < m$ for all $i = 1, 2, 3, \dots$, then $0 = [2pr]^{½}\text{-}\lim \sigma(l)$ where $p = \max \{m^2 + |a|^2, m + |a|\}$.

Proof. We will first show that $\lim \sigma^2(l) = 0$. By the definition, $\sigma_n^2 = (1/n) \Sigma_{i=1}^n (a_i - \mu_n)^2 = (1/n) \Sigma_{i=1}^n (a_i)^2 - \mu_n^2$. Thus, $\lim \sigma^2(l) = \lim_{n\to\infty} (1/n) \Sigma_{i=1}^n (a_i)^2 - \lim_{n\to\infty} \mu_n^2$. Since $|a_i| < m$ for all $i \in N$, there is a number p such that $|a_i^2 - a^2| < p$ for all $i \in N$. Namely, $|a_i^2 - a^2| \leq |a_i|^2 + |a|^2 < m^2 + |a|^2 < \max \{m^2 + |a|^2, m + |a|\} = p$. Taking the set $L_{n,r,\varepsilon}(a) = \{i \in N; i \leq n \text{ and } |a_i - a| \geq r + \varepsilon\}$, denoting $|L_{n,r,\varepsilon}(a)|$ by u_n, and using the hypothesis $|a_i| < m$ for all $i \in N$, we have the following system of inequalities:

$$\begin{aligned} |\sigma^2_n| &= |(1/n) \Sigma_{i=1}^n (a_i)^2 - \mu_n^2| \\ &= |(1/n) \Sigma_{i=1}^n (a_i^2 - a^2) - (\mu_n^2 - a^2)| \\ &\leq (1/n) \Sigma_{i=1}^n |a_i^2 - a^2| + |\mu_n^2 - a^2| \\ &< (p/n) \Sigma_{i=1}^n |a_i - a| + |\mu_n - a|\,|\mu_n + a| \\ &< (p/n) (u_n + (n - u_n)(r + \varepsilon)) + |\mu_n - a|\,(|\mu_n| + |a|) \\ &\leq (p/n) (u_n + n(r + \varepsilon)) + |\mu_n - a|\,((1/n)\Sigma_{i=1}^n |a_i| + |a|) \\ &< p\,(u_n/n) + p\,(r + \varepsilon) + p\,|\mu_n - a|. \end{aligned}$$

By hypothesis and Theorem 2.8.15, we have $a = r\text{-}\lim \mu(l)$. At the same time, by (2.14), $\lim (u_n/n) = 0$. Then, for every $\varepsilon > 0$ and sufficiently large n, we may write that

$$|\sigma^2_n| < p\,\varepsilon + p\,(r + \varepsilon) + p\,(r + \varepsilon) = 2pr + 3p\varepsilon \tag{2.15}$$

Using the fact that $(x + y)^{½} \leq x^{½} + y^{½}$ for any $x, y > 0$, it follows from (2.15) that

$$|\sigma_n| \leq [2pr]^{½} + (3p\varepsilon)^{½},$$

which yields that $0 = [2pr]^{½}\text{-}\lim \sigma(l)$.

The proof is completed.

2.9. Fuzzy Convergence in Metric Spaces and Normed Linear Spaces

If we limit our vision to the real world,
we will forever be fighting on the minus side of things,
working only too make our photographs equal
to what we see out there, but no better.
Galen Rowell (1940-2002)

The majority of results obtained in previous sections are valid for metric spaces or in normed linear spaces. However, we demonstrate this only for some of the most important properties of sequences because our goal here is not to develop neoclassical analysis for metric and normed linear spaces in its entirety but to show how to do this. As any normed linear space is a metric space (cf. Appendix C; Proposition C.2), we primarily consider metric spaces, using linear spaces only when we study operations with sequences.

Let $\boldsymbol{E}$ be a metric space with a metric (distance) $\mathbf{d}$ and $l = \{a_i ; i = 1, 2, 3, \ldots\}$ be a sequence of elements from $\boldsymbol{E}$.

Definition 2.9.1. a) An element a from $\boldsymbol{E}$ is called the *limit of a sequence l* (it is denoted by $a = \lim_{i\to\infty} a_i$ or $a = \lim l$) if for any $\varepsilon \in \boldsymbol{R}^{++}$ the inequality $\mathbf{d}(a, a_i) < \varepsilon$ is valid for almost all a_i , i.e., there is such n that for any $i > n$, we have $\mathbf{d}(a, a_i) < \varepsilon$.

b) A sequence l that has a limit is called *convergent* and it is said that l *converges* to its limit a.

In Section 2.1, there are many examples of numerical sequences and their limits in such a metric space as $\boldsymbol{R}$ with the distance $\mathbf{d}(a, b) = |a - b|$.

Many properties of limits of numerical sequences remain true for limits of sequences in general metric spaces. In particular, such a limit is always unique. However, as limits of sequences in metric spaces have been studied for a long time, we do not go here into detail of the theory of sequences in metric spaces and start with fuzzy limits in metric spaces.

Let $r \in \boldsymbol{R}^{+}$.

Definition 2.9.2. a) An element a from $\boldsymbol{E}$ is called an *r-limit of a sequence l* (it is denoted by $a = r\text{-}\lim_{i\to\infty} a_i$ or $a = r\text{-}\lim l$) if for any $\varepsilon \in \boldsymbol{R}^{++}$ the inequality $\mathbf{d}(a, a_i) < r + \varepsilon$ is valid for almost all a_i , i.e., there is such n that for any $i > n$, we have $\mathbf{d}(a, a_i) < r + \varepsilon$.

b) A sequence l that has an r-limit is called *r-convergent* and it is said that *l r-converges* to its r-limit a.

Informally, a is an r-limit of a sequence l if for an arbitrarily small ε, the distance between a and all but a finite number of elements from l is smaller than $r + \varepsilon$. In other words, an element a is an r-limit of a sequence l if for any $\varepsilon \in \boldsymbol{R}^{++}$ almost all a_i belong to the neighborhood $O(a) = c \in \boldsymbol{E}; \mathbf{d}(a, c) < r + \varepsilon)$.

In Section 2.2, there are many examples of numerical sequences and their r-limits in such a metric space as $\boldsymbol{R}$ with the distance $\mathbf{d}(a, b) = |a - b|$.

Lemma 2.9.1. For any sequence l, we have $a = \lim l$ if and only if $a = 0\text{-}\lim l$.

It means that as in the case of numbers, the sequence l converges if and only if it 0-converges.

Definition 2.9.3. a) An element a from $\boldsymbol{E}$ is called a *fuzzy limit* of a sequence l if it is an r-limit of l for some $r \in \boldsymbol{R}^{+}$.

b) a sequence l *fuzzy converges* and is called *fuzzy convergent* if it has a fuzzy limit.

As in the case of limits of sequences in metric spaces, many properties of fuzzy limits of numerical sequences remain true for fuzzy limits of sequences in general metric spaces. That is why, we consider here only some of properties obtained for fuzzy limits of numerical sequences in the previous sections.

Lemma 2.9.2. If $a = r\text{-lim } l$, then $a = q\text{-lim } l$ for any $q > r$.

Definition 2.9.4. A sequence in a metric space is *bounded* if the distances between pairs of its elements are bounded, i.e., there is a number $k \in \boldsymbol{R}^{++}$ such that for any i, j, we have $\mathbf{d}(a_j, a_i) < k$.

Theorem 2.9.1. A sequence l in $\boldsymbol{E}$ fuzzy converges if and only if it is bounded.

Proof. Necessity. Let us take a fuzzy convergent sequence $l = \{a_i ; i = 1, 2, 3, \dots\}$. In this case, there is an element a from $\boldsymbol{E}$ such that $a = r\text{-lim}_{i\to\infty} a_i$. By Definition 2.9.2, for any $\varepsilon \in \boldsymbol{R}^{++}$, there is such n that for any $i > n = n(\varepsilon)$, we have $\mathbf{d}(a, a_i) < r + \varepsilon$. Let us fix some $\varepsilon < ½$ and take the corresponding number $n = n(½)$ and the number $q = \sup \{ \mathbf{d}(a, a_i); i = 1, 2, 3, \dots, n\}$. Then the distance between arbitrary elements a_i and a_j from l is less than p where $p = 2(q + r + 1)$. Really,

$$\mathbf{d}(a_j, a_i) \leq \mathbf{d}(a, a_i) + \mathbf{d}(a, a_j) < 2(q + r + 1)$$

as $\mathbf{d}(a, a_i) < q + 1$ if $i < n + 1$ and $\mathbf{d}(a, a_i) < r + 1$ for all $i > n$.

Thus, by Definition 2.9.4, the sequence l is bounded. This completes the proof of necessity.

Sufficiency. Let us consider a bounded sequence $l = \{a_i ; i = 1, 2, 3, \dots\}$. Then there is a number c such that $\mathbf{d}(a_j, a_i) < c < c + \varepsilon$ for all $i, j = 1, 2, 3, \dots$. In this case, the element a_1 is an r-limit of the sequence l where $r = c$ as $\mathbf{d}(a_1, a_i) < c$ for all $i = 1, 2, 3, \dots$ and any $\varepsilon > 0$. Consequently, the sequence l fuzzy converges.

Theorem is proved.

From Theorem 2.9.1, we obtain the following result.

Proposition 2.9.1. The following conditions are equivalent:

1) a sequence l is not fuzzy convergent;
2) the sequence l is not bounded;
3) some subsequence of l has no fuzzy limits;

Definition 2.9.2 implies the following result.

Proposition 2.9.2. If $a = q\text{-lim } l$, then $a = q\text{-lim } k$ for any subsequence k of l.

Lemma 2.9.1 and Definition 2.9.2 imply the following result.

Corollary 2.9.1. If $a = \lim l$, then $a = \lim k$ for any subsequence k of l.

Let $\boldsymbol{L}$ be a normed vector space. In it, distance $\mathbf{d}$ is defined by means of the norm as $\mathbf{d}(x, y) = \| x - y \|$ for any $x, y \in \boldsymbol{L}$. Properties of normed vector spaces allow us to prove the following results for sequences in the space $\boldsymbol{L}$.

Theorem 2.9.2. If $a = r\text{-lim } l$ and $b = q\text{-lim } h$, then:

a) $a + b = (r + q)\text{-lim}(l + h)$;
b) $a - b = (r + q)\text{-lim}(l - h)$;
c) $ka = |k|{\cdot}r\text{-lim}\,(kl)$ for any $k \in \boldsymbol{R}$ where $kl = \{k{\cdot}a_i\,;\, i = 1, 2, 3, \ldots\}$.

Proof. a) Let $a = r\text{-lim}\, l$ and $b = r\text{-lim}\, h$. Then by Definition 2.9.2, for any $\varepsilon \in \boldsymbol{R}^{++}$, there is such n that for any $i > n$, we have $\mathbf{d}(a, a_i) < r + \varepsilon/2$ and there is such m that for any $i > m$, we have $\mathbf{d}(b, b_i) < q + \varepsilon/2$. Taking $p = \max\{m, n\}$, we have $\mathbf{d}(a, a_i) < r + \varepsilon/2$ and $\mathbf{d}(b, b_i) < q + \varepsilon/2$ for any $i > p$. By the properties of the norm (cf. Appendix C), we have for any $i > p$

$$\mathbf{d}((a + b), (a_i + b_i)) = \|(a + b) - (a_i + b_i)\| =$$
$$\|(a - a_i) + (b - b_i)\| \leq \| a - a_i \| + \|b - b_i\| =$$
$$\mathbf{d}(a, a_i) + \mathbf{d}(b, b_i) < r + q + \varepsilon$$

By Definition 2.9.2, it means that $a + b = (r + q)\text{-lim}(l + h)$ as $l + h = \{ a_i + b_i\,;\, i = 1, 2, 3, \ldots\}$.

b) Let $a = r\text{-lim}\, l$ and $b = q\text{-lim}\, h$. Then by Definition 2.9.2, for any $\varepsilon \in \boldsymbol{R}^{++}$, there is such n that for any $i > n$, we have $\mathbf{d}(a, a_i) < r + \varepsilon/2$ and there is such m that for any $i > m$, we have $\mathbf{d}(b, b_i) < q + \varepsilon/2$. Taking $p = \max\{m, n\}$, we have $\mathbf{d}(a, a_i) < r + \varepsilon$ and $\mathbf{d}(b, b_i) < q + \varepsilon/2$ for any $i > p$. By the properties of the norm (cf. Appendix C), we have for any $i > p$

$$\mathbf{d}((a - b), (a_i - b_i)) = \|(a - b) - (a_i - b_i)\| =$$
$$\|(a - a_i) + (-(b - b_i))\| \leq \| a - a_i \| + \|(-(b - b_i))\| =$$
$$\| a - a_i\| + \|b - b_i\| = \mathbf{d}(a, a_i) + \mathbf{d}(b, b_i) < r + q + \varepsilon$$

By Definition 2.9.2, it means that $a - b = (r + q)\text{-lim}(l - h)$ as $l - h = \{ a_i - b_i;\, i = 1, 2, 3, \ldots\}$.

c) Let $a = r\text{-lim}\, l$ and $k \in \boldsymbol{R}$. Then by Definition 2.9.2, for any $\varepsilon \in \boldsymbol{R}^{++}$, there is such n that for any $i > n$, we have $\mathbf{d}(a, a_i) < r + \varepsilon/k$. By the properties of the norm (cf. Appendix C), we have for any $i > p$

$$\mathbf{d}(ka, ka_i) = \| ka - ka_i \| = |k|{\cdot}\| a - a_i \| = |k|{\cdot}\mathbf{d}(a, a_i) < |k|{\cdot}\, r + \varepsilon$$

By Definition 2.9.1, it means that $ka = |k|{\cdot}r\text{-lim}\,(kl)$ as $kl = \{k{\cdot}a_i\,;\, i = 1, 2, 3, \ldots\}$.

Theorem is proved.

Let us assume that $\boldsymbol{B}$ is a normed algebra and consider the scalar product of sequences in $\boldsymbol{B}$. If $l = \{a_i \in \boldsymbol{B};\, i = 1, 2, 3, \ldots\}$ and $h = \{b_i \in \boldsymbol{B};\, i = 1, 2, 3, \ldots\}$, then their *scalar product* $l \cdot h$ is equal to the sequence $\{ a_i \cdot b_i;\, i = 1, 2, 3, \ldots\}$.

Theorem 2.9.3. If $a = r\text{-lim}\, l$ and $b = q\text{-lim}\, h$, then $a{\cdot}b = (r{\cdot}q + \|b\|{\cdot}r + \|a\|{\cdot}q)\text{-lim}(l{\cdot}h)$.

Proof. Let us consider some $\varepsilon \in \boldsymbol{R}^{++}$. Then there is $\delta \in \boldsymbol{R}^{++}$ such that

$$(r + q + \|b\| + \|a\| + \delta)\delta < \varepsilon$$

By the definition of fuzzy limits, we have the following property:

$$\exists\, n\in\omega\ \forall i > n\ (\ \mathbf{d}(a, a_i) < r + \delta\ ) \text{ and } \exists\, m \in \omega\ \forall i > m\ (\ \mathbf{d}(b, b_i) < q + \delta\ ).$$

Let us take $p = \max\ \{m, n\}$. Then $\forall i > p\ (\ \mathbf{d}(a, a_i) < r + \delta\ )$ and $\forall i > p\ (\ \mathbf{d}(b, b_i) < q + \delta\ )$. This and properties of metric imply the following sequence of equalities and inequalities for any $i > p$:

$$\begin{gathered}
\mathbf{d}(ab\ , a_ib_i) \leq \mathbf{d}(a_ib_i\ , a_ib) + \mathbf{d}(a_ib, ab) \leq \|\ a_i\|\cdot \mathbf{d}(b_i\ , b) + \|\ b\ \|\ \cdot \mathbf{d}(a_i\ , a) < \\
\|\ a_i\|\cdot(\ q + \delta\ ) + \|\ b\ \|\cdot(\ r + \delta\ ) = \|\ a_i - a + a\ \|\cdot(\ q + \delta\ ) + \|\ b\ \|\cdot(\ r + \delta\ ) \leq \\
(\|\ a_i - a\| + \|\ a\|)\cdot(\ q + \delta\ ) + \|\ b\ \|\cdot(\ r + \delta\ ) = \\
\|\ a_i - a\|\cdot(\ q + \delta\ ) + \|\ a\|\cdot(\ q + \delta\ ) + \|\ b\ \|\cdot(\ r + \delta\ ) < \\
(r + \delta)\cdot(\ q + \delta\ ) + \|\ a\|\cdot(\ q + \delta\ ) + \|\ b\ \|\cdot(\ r + \delta\ ) = \\
r\cdot q + \delta\cdot q + \delta\cdot r + \delta^2 + \|\ a\|\cdot(\ q + \delta\ ) + \|\ b\ \|\cdot(\ r + \delta\ ) = \\
r\cdot q + \delta\cdot q + \delta\cdot r + \delta^2 + \|\ a\|\cdot q + \|\ a\|\cdot\delta + \|\ b\ \|\cdot r + \|\ b\ \|\cdot\delta = \\
r\cdot q + \|\ a\|\cdot q + \|b\ \|\cdot r + \delta\cdot q + \delta r + \delta^2 + \|\ a\|\cdot\delta + \|\ b\ \|\cdot\delta = \\
(\ r\cdot q + \|\ a\|\cdot q + \|b\ \|\cdot r\ ) + (\ q + r + \delta + \|\ a\| + \|\ b\ \|)\cdot\delta < \\
(\ r\cdot q + |\ a|\cdot q + |b\ |\cdot r\ ) + \varepsilon.
\end{gathered}$$

This means that $a\cdot b = (r\cdot q + \|b\|\cdot r + \|a\|\cdot q)\text{-}\lim(l\cdot h)$.

Theorem is proved.

As we know, the main criterion of number sequence convergence is the Cauchy criterion. The same is true for sequences in complete metric spaces. We remind that a metric space is called *complete* if any Cauchy (fundamental) sequence has a limit.

Definition 2.9.5. a) A sequence l in $\boldsymbol{E}$ is called *r-fundamental* if for any $\varepsilon \in \boldsymbol{R}^{++}$ there is $n \in \omega$ such that for any $i, j \geq n$, we have $\mathbf{d}(a_j, a_i) < 2r + \varepsilon$.

b) A sequence l in $\boldsymbol{E}$ is called *fuzzy fundamental* if it is r-fundamental for some $r \in \boldsymbol{R}^+$.

Lemma 2.9.3. If $r \leq p$, then any r-fundamental sequence is p-fundamental.

Lemma 2.9.4. A sequence l in $\boldsymbol{E}$ is fundamental (in the ordinary sense, i.e., it is a Cauchy sequence) if and only if it is 0-fundamental.

Lemma 2.9.5. A subsequence of an r-fundamental sequence is r-fundamental.

We prove the Extended Cauchy Criterion for Euclidean spaces $\boldsymbol{E}^n$. However, it is possible to prove it for arbitrary Banach spaces.

Theorem 2.9.4 (The Extended Cauchy Criterion). A sequence l in $\boldsymbol{E}$ is fuzzy convergent if and only if it is fuzzy fundamental.

Proof. Necessity. If a sequence l is fuzzy convergent, then for some element a, we have $a = r\text{-}\lim l$. Let us consider some number $\varepsilon > 0$. Then by the definition of an r-limit, we have

$$\exists\, n \in \omega\ \forall i > n\ (\mathbf{d}(a, a_i) \leq r + \varepsilon/2$$

Consequently, for any $i, j > n$, we obtain

$$\mathbf{d}(a_i\ , a_j) \leq \mathbf{d}(a, a_i) + \mathbf{d}(a, a_j) \leq 2r + 2(\varepsilon/2) < 2r + \varepsilon$$

Thus, l is an r-fundamental sequence.

Sufficiency. Let l be a fuzzy fundamental sequence, that is, an r-fundamental sequence for some r and assume that $k_m=1/m$. To prove that the sequence l is fuzzy convergent, we build a closed ball B_m with the radius $R_m < 2r + k_m$ for each number k_m in such a way that almost all elements from l belong to B_m.

We start with B_1. As l is an r-fundamental sequence, there is a number n, such that for all $i, j > n$, we have $\mathbf{d}(a_i , a_j) < 2r + 1$. We take a number j such that $j > n$ and determine in the space $\mathbf{E}$ the closed ball B_1 that has a_j as its center and radius $2r + 1$. As $\mathbf{d}(a_i , a_j) < 2r + 1$ when $i > n$, almost all elements from l belong to B_1.

Now let us assume that we have built such closed balls B_t for all $t = 1, 2, 3, \ldots , m - 1$ and build the ball B_m.

Properties of r-fundamental sequences imply that taking the number m, we can find a number n, which is dependent on m and thus, denoted by $n(m)$, such that for all $i, j > n(m)$, we have $\mathbf{d}(a_i , a_j) < 2r + (1/m)$. Then we can find a number $j = j(m)$ such that $j(m) > n(m)$ and the element a_j belongs to the closed ball B_{m-1}. Thus, for all elements a_i with $i > n(m)$, we have $\mathbf{d}(a_i , a_j) < 2r + (1/m)$. Consequently, all elements a_i with $i > n(m)$ belong to the closed ball B_m with the center a_j and radius p where $p = \min\{ 2r + (1/m), \mathbf{d}(a_j , S_{m-1}) \}$ and S_{m-1} is the boundary of B_{m-1}. We can continue this process and the radius R_m of the ball B_m is always less than $2r + (1/m)$ for all $m = 1, 2, 3, \ldots$.

As a result of this construction, we obtain a sequence of closed balls

$$B_1 \supseteq B_2 \supseteq B_3 \supseteq \ldots \supseteq B_{m-1} \supseteq B_m \supseteq \ldots .$$

Let $B = \bigcap_{m=1}^{\infty} B_m$. By properties of complete metric spaces (cf. Appendix C and (Kuratowski, 1966)), $B \neq \emptyset$. Thus, there is an element $a \in B$. Let us show that a is a fuzzy limit of l.

Indeed, taking $\varepsilon > 0$, we find a number m such that $1/m < \varepsilon$. By our construction, a belong to all balls B_t and thus, to the ball B_{2m}. In addition, almost all elements from l belong to B_{2m}. Then for $t > n(2m)$, we have

$$\mathbf{d}(a , a_t) \leq \mathbf{d}(a , a_{j(2m)}) + \mathbf{d}(a_{j(2m)} , a_t) \leq$$
$$2r + (1/2m) + 2r + (1/2m) = 4r + (1/m) \leq 4r + \varepsilon$$

As ε is an arbitrary positive number, $a = 4r\text{-}\lim l$.

Theorem is proved.

Remark 2.9.1. We have demonstrated that r-convergence of a sequence in a metric space implies r-fundamentality of this sequence. For real numbers, the converse is also true (cf. Theorem 2.2.7). However, this is not true for arbitrary (even complete) metric spaces as the following example demonstrates.

Example 2.9.1. Let us consider the sequence $l = \{ a_i \in \boldsymbol{E}_2; i = 1, 2, 3, \ldots\}$ such that $\boldsymbol{E}_2$ is a two dimensional Euclidean space, i.e., the space of complex numbers, $a_{3n} = (0, \sqrt{3})$ with $n = 1, 2, 3, \ldots$, $a_{3m+1} = (1, 0)$ and $a_{3m+2} = (-1, 0)$. Using definitions and properties of the two dimensional Euclidean geometry (geometry of the Euclidean plane), it is possible to show that the sequence l is 1-fundamental (the distance between any two elements is equal to 2), but this sequence does not have 1-limits. At the same time, the sequence l has the $(2/\sqrt{3})$-limit, i.e., it is fuzzy convergent.

Remark 2.9.2. Relations between r-convergence of a sequence in a normed linear space and r-fundamentality of this sequence are studied by Phu (2003). He shows that an r-fundamental sequence in a normed linear space L q-converges for all $q > ½ J(L)r$ where $J(L)$ is the Jung constant of the space L. The Jung constant of a normed linear space was introduced in (Jung, 1901).

Chapter 3

FUZZY CONTINUOUS FUNCTIONS

I was searching for what I believed in: continuity.
I found only discontinuity.
Frédéric Brenner (1959 -)

In relativity, movement is continuous,
causally determinate and well defined,
while in quantum mechanics it is discontinuous,
not causally determinate and not well defined.
David Bohm (1917-1992)

The concept of continuity is one of the most important in the whole calculus. In (Larson and Edwards, 2006), we can find the following explanation of the idea of continuity and its mathematical meaning:

> "In mathematics the term "continuous" has much the same meaning as it does in everyday use. To say that a function is continuous at $x = a$ means that there is no interruptions in the graph of f at a. The graph of f is unbroken at a, and there are no holes, jumps, or gaps. As simple as this concept may seem, precise definition eluded mathematicians for many years" (p. 61).

This explanation makes an emphasis on gapelessness as the essence of continuity with an appeal to graphs of functions. However, any graph always uses some material (physical) carrier, such as paper, board or screen of a calculator or computer. As a result, the graph inherits properties of these physical carriers. Such general properties are studied by physics, and physics teaches us that all such things are built (consist) of molecules and there are microscopic gaps between molecules. Being microscopic in the human scale, these gaps are very big in comparison with the molecule size. Thus, any graph inevitably has gaps and only imperfection of people's vision or idealization of the classical mathematics allows us not to see these gaps. Classical analysis simply ignores these gaps, while neoclassical analysis does not only take even very small gaps into consideration, but also teaches us when and how it is possible to disregard the impact of those gaps.

Neoclassical analysis studies functions that may be continuous only to some extent and are called fuzzy continuous. Continuous functions form a subclass of fuzzy continuous functions. This chapter gives an exposition of the theory of fuzzy continuous functions.

Presence of gaps changes situation with functions. A function with gaps is not continuous according to the classical calculus and topology. At the same time, due to their good properties, continuous functions are basic in the calculus and form a cornerstone of the contemporary topology. However, when gaps are very small, it becomes impossible to verify absence of gaps. Thus, it is natural to consider functions without big gaps (leaps or holes in their graphs) as fuzzy continuous.

Another situation when fuzzy continuity is useful is related to the correspondence between gaps in the function domain and gaps in the function range. When this relation is moderate, i.e., small gaps do not grow essentially, it is also reasonable to consider such functions as fuzzy continuous.

We start this chapter with an exposition of the theory of continuous functions, which form the base for all functions and operations studied in the calculus. Basic properties of continuous functions are presented in Section 3.1, comprising the material in this area given in many popular textbooks on calculus. The goal of this detailed exposition is to make exposition of this book self-contained, allowing the reader to learn calculus from the very beginning without references to other sources. The main results of Section 3.1 are Theorems 3.1.1 - 3.1.7.

To make a rigorous transition from continuous to discontinuous functions, we introduce several measures of continuity and discontinuity in Section 3.2. These measures show to what extent it is possible to assume that a given function is continuous or what is the likeness of a given function to continuous functions. Properties of measures of continuity and discontinuity are studied and relations between them are explicated. The main results of Section 3.2 are construction of measures and finding their properties.

Then, in Section 3.3, the concept of continuity is extended to the concept of fuzzy continuity (r-continuity) and 2-fuzzy continuity ((q, r)-continuity). Informally, r-continuous functions are such functions that do not have gaps larger than r. Fuzzy continuity of a function $f(x)$ may represent either absence of knowledge what is going with the function $f(x)$ when the values of its arguments come too close to some points or existence of actual but reasonably small gaps in the values of $f(x)$. The first situation reflects boundedness of measurement when in reality, we can measure only with some finite precision. The second situation reflects limits of computation when in reality, we can compute and operate with only some part of rational numbers, while mathematical models are dealing, as a rule with all real numbers (cf. Section 1.2).

The main results of Section 3.3 are different concepts of fuzzy continuity and relations between these concepts.

The study of fuzzy continuous functions in neoclassical analysis does not determine any restrictions on measures of continuity and discontinuity. These measures can be very small or very big. The goal of neoclassical analysis is to give tools for a researcher (physicist, biologist, economist or computer scientist) to determine herself or himself what measures of discontinuity are acceptable in her or his research because what is big and what is small is relative to the scale. For instance, absolute, or chronometric, time in geology is comprised of numerical ages in "millions of years" according to the numerically calibrated geologic time scale. This scale has been continuously refined since approximately the 1930s, although the amount of change with each revision has become smaller over the decades (Harland, et al, 1990). Experts think that because of continual refinement, none of the values depicted in the geological time scale should be considered definitive, even though some have not changed significantly in a long time and are very well constrained. The overall duration and relative length of these large geologic intervals is unlikely to change much, but the precise numbers may "wiggle" a bit as a result of new data (MacRae, 1996-1997). In this context, "wiggle" a bit can mean millions of years. For instance, with the Precambrian/Cambrian boundary modified according to the most recently-published radiometric dates on that interval, revising the boundary from 570 +/- 15 million years to 543 +/- 1 million years ago (Grotzinger, et al, 1995).

Properties of fuzzy continuous functions are studied in Section 3.4. In a broader context of fuzzy continuity, it is possible to essentially extend many classical results. For instance, the Intermediate Value Theorem, is extended from the class of continuous functions to the class of fuzzy continuous functions. The latter contains a lot of functions that are not continuous, including functions on discrete sets, which comprise all functions that are processed by computers.

In some cases, such a transition to a fuzzy context allows us to complete some basic classical results. For example, one of the basic results of the classical calculus, the first Weierstrass Theorem, states that any continuous function on a closed interval is bounded. The converse is not true. So, continuity is only a sufficient but not a necessary condition for boundedness of a function. However, fuzzy continuity makes attainable to obtain a complete criterion of boundedness of a function. Namely, as it is demonstrated in this section, *a function on a closed interval is bounded if and only if it is fuzzy continuous.*

The main results of Section 3.4 are Theorems 3.4.1 - 3.4.3, 3.4.5 - 3.4.9.

As fuzzy continuous functions better represent and model reality than continuous functions, they are used in diverse practical applications, especially in those that are related to computations. For instance, an interesting interplay between discrete structures in Artificial Intelligence and continuous processes and characteristics in medicine is analyzed by Steimann (2001). Let us consider some of his arguments.

Firstly, even when medical actions are inherently discrete (e.g., a surgery is either performed or not), their grounds usually involve some continuity and gradations.

Secondly, discretization can lead to the undue amplifications of differences. Would it not seem natural that patients with comparable symptoms be given comparable diagnoses?

Analogously and thirdly, discreteness may cause erratic behavior in the context of change. Would it not seem likely that a slight alteration over time in the vital parameters of a patient changes the diagnosis only slightly? Instead, however, discrete dynamic systems usually respond to the continuous change in the patient condition with an abrupt change in state. It is necessary to materialize an informal concept of continuity: a system is continuous (or more exactly, behaves continuously) if it maps close inputs to close outputs.

These arguments show that continuity in medicine is not the classical continuity with its absolute precision. Only fuzzy continuity gives an adequate model for continuous systems and processes in medicine.

In Section 3.5, fuzzy uniformly continuous functions are introduced and studied. Uniform continuity is a strengthening of continuity. While continuity is a local property, uniform continuity is a global property of a function. In a similar way, fuzzy uniform continuity is a strengthening of fuzzy continuity. While fuzzy continuity is a local property, fuzzy uniform continuity is a global property of a function. Fuzzy uniform continuity gives a better formalization of the informal idea existing in mathematical analysis of how continuity can be uniform. The main results of Section 3.5 are Theorems 3.4.1 - 3.4.3, 3.4.5 - 3.4.9.

In Section 3.6, we study properties of sequences of functions. One of the most important properties is convergence. Convergence of sequences of functions is used to define sums of series of functions, while series of functions are frequently used to represent functions, to build computational methods of function evaluation and to solve differential and integral equations. In the classical case, two types of convergence are considered: pointwise and uniform. Pointwise convergence, or simply convergence, of sequences of functions is the most natural extension of convergence of number sequences. However, (pointwise) convergence does not preserve many important properties of functions, for instance, continuity. In contrast to this, several properties of the functions $f_n(x)$, such as continuity, differentiability and Riemann integrability, are transferred to the limit $f(x)$ if the convergence is uniform.

However, it is, as a rule, impossible to test pointwise or uniform convergence for real life functions, i.e., for functions that come from measurement or computation. Moreover, such functions are usually defined on discrete sets and classical concept of convergence cannot be applied at all.

To remedy this situation, we introduce and study (pointwise) fuzzy convergence and uniform fuzzy convergence of sequences of functions. As in the case of classical convergence, (pointwise) fuzzy convergence is more natural, but uniform fuzzy convergence preserves more properties of converging functions. The main results of Section 3.6 are Theorems 3.4.1 - 3.4.3, 3.4.5 - 3.4.14.

In Section 3.7, fuzzy continuous mappings of metric and normed vector spaces are introduced and studied. The goal of this section is to show how it is possible to extend the theory of fuzzy continuous functions from real numbers to general metric spaces. It is necessary to remark that even more general theory is developed in scalable topology (Burgin, 2004a; 2005a; 2006). The main results of Section 3.7 are Theorems 3.4.1 - 3.4.3, 3.4.5 - 3.4.9.

3.1. Continuous Functions

> The self is not something ready-made,
> but something in continuous formation
> through choice of action.
> *John Dewey* (1859-1952)

We start with local continuity that defines behavior of a function at a point. Then we go to intermediate continuity that defines behavior of a function in a set. This brings us to global continuity that defines behavior of a function on the function domain.

Definition 3.1.1. a) A partial function $f: \mathbf{R} \to \mathbf{R}$ is called *continuous at a point* $a \in \mathbf{R}$ if $f(x)$ is defined at a and for any sequence $l = \{ a_i \in \mathbf{R}; i = 1, 2, 3, \ldots\}$ that converges to a, the point $f(a)$ is the limit of the sequence $\{ f(a_i) \in \mathbf{R}; i = 1, 2, 3, \ldots\}$ or formally, for any sequence $l = \{ a_i \in \mathbf{R}; i = 1, 2, 3, \ldots\}$, $\lim_{i\to\infty} a_i = a$ implies $\lim_{i\to\infty} f(a_i) = f(a)$. b) A partial function $f: \mathbf{R} \to \mathbf{R}$ that is not continuous at a is called *discontinuous at a*.

This is the sequential definition of continuity. It is also possible to define continuous functions without explicit use of limits as it is done by means of the (ε, δ)-construction.

Note that a function is continuous at a point $a \in \boldsymbol{R}$ only it is defined at this point.

Definition 3.1.2. A partial function f: $\boldsymbol{R} \to \boldsymbol{R}$ is called *continuous at a point* $a \in \boldsymbol{R}$ if $f(x)$ is defined at a and for any $\varepsilon > 0$, there is $\delta > 0$ such that the inequality $| a - x | < \delta$ implies the inequality $| f(x) - f(a) | < \varepsilon$, or in other words, for any x with $| a - x | < \delta$, we have $| f(x) - f(a) | < \varepsilon$. b) A partial function f: $\boldsymbol{R} \to \boldsymbol{R}$ that is not continuous at a is called *discontinuous at a*.

Example 3.1.1. Functions tan x and sec x are continuous at the point 0 according to both definitions, while functions cot x, $[x]$ and $x - [x]$ are discontinuous at the point 0 according to both definitions.

One more definition of local continuity is essentially topological (cf. (Kelly, 1957)).

Definition 3.1.3. A partial function f: $\boldsymbol{R} \to \boldsymbol{R}$ is called *continuous at a point* $a \in \boldsymbol{R}$ if $f(x)$ is defined at a and for any open interval (b, c) that contains $f(a)$, there is an open interval (u, v) such that it contains a and f maps (u, v) into (b, c). b) A partial function f: $\boldsymbol{R} \to \boldsymbol{R}$ that is not continuous at a is called *discontinuous at a*.

This definition implies the following result because any open interval contains a closed interval and vice versa.

Proposition 3.1.1. If function $f(x)$ is continuous at a point a, then for any closed interval $[b, c]$ that contains $f(a)$, there is a closed interval $[u, v]$ such that $[u, v]$ contains a and f maps $[u, v]$ into $[b, c]$.

Proposition 3.1.2. Definitions 3.3.1, 3.1.2 and 3.1.3 are equivalent.

Proof. a) Definition 3.1.1 ⇒ Definition 3.1.2. Let us assume that $f(x)$ is not continuous at the point a according to Definition 3.1.2. It means that there is $\varepsilon > 0$ such that for any $\delta > 0$ there is a point x with $| x - a| < \delta$ but $| f(x) - f(a) | \geq \varepsilon$. Let us take a sequence of such δ equal to ½ , 1/3 , ¼ , ... , $1/n$, ...

This gives us the sequence $l = \{x_i ; i = 1, 2, 3, \dots \}$ in which $| x_i - a| < 1/i$, but $| f(_i) - f(a) | \geq \varepsilon$. Consequently (cf., Section 2.1), $a = \lim l$, but the sequence $h = \{ f(x_i) ; i = 1, 2, 3, \dots \}$ does not converge to $f(a)$. This violates the condition from Definition 3.1.1. So, Definition 3.1.1 implies Definition 3.1.2 because we have demonstrated that it is impossible that for an arbitrary function $f(x)$, Definition 3.1.1 is true for $f(x)$ at the point a and at the same time, Definition 3.1.2 is not true for $f(x)$ at the point a.

b) Definition 3.1.2 ⇒ Definition 3.1.3. Let us assume that $f(x)$ is not continuous at the point a according to Definition 3.1.3. It means that there is an interval (b, c) that contains $f(a)$ such that in any interval (u, v) containing a there is, at least, one point d such that $f(d)$ does not belong to the interval (b, c).

Let us put $\varepsilon = \min \{ |b - f(a) | ; | c - f(a) | \}$. Then for any $\delta > 0$ there is a point e with $| e - a| < \delta$ but $| f(e) - f(a) | \geq \varepsilon$ because we can take the interval $(a - \delta, a + \delta)$ and as it is demonstrated, it has a point e the image of which $f(e)$ does not belong to the interval (b, c). This violates the condition from Definition 3.1.2. So, Definition 3.1.2 implies Definition 3.1.3 because we have demonstrated that it is impossible that for an arbitrary function $f(x)$, Definition 3.1.2 is true for $f(x)$ at the point a and at the same time, Definition 3.1.3 is not true for $f(x)$ at the point a.

c) Definition 3.1.3 ⇒ Definition 3.1.1. Let us assume that $f(x)$ is not continuous at the point a according to Definition 3.1.1. It means that there is a sequence $l = \{x_i ; i = 1, 2, 3, \dots \}$ such that $a = \lim l$, but the sequence $h = \{ f(x_i) ; i = 1, 2, 3, \dots \}$ does not converge to $f(a)$. By

Definition 2.1.1, this implies that outside some neighborhood Oa of the point a, there are infinitely many elements from h.

Then whatever small interval (u, v) containing a is taken, only a finite number of elements from l will outside (u, v). Thus, an infinite number of its elements will be outside the neighborhood Oa. It means that the condition from Definition 3.1.3 is not satisfied and $f(x)$ is not continuous at the point a according to Definition 3.1.3. So, Definition 3.1.3 implies Definition 3.1.1 because we have demonstrated that it is impossible that for an arbitrary function $f(x)$, Definition 3.1.3 is true for $f(x)$ at the point a and at the same time, Definition 3.1.1 is not true for $f(x)$ at the point a.

Thus, we have demonstrated that Definition 3.1.1 $\Rightarrow$ Definition 3.1.2, Definition 3.1.2 $\Rightarrow$ Definition 3.1.3, and Definition 3.1.3 $\Rightarrow$ Definition 3.1.1. As the logical implication $\Rightarrow$ is a transitive relation, all these definitions are equivalent.

Proposition is proved.

Proposition 3.1.2 shows that it is possible to take one of the conditions from Definitions 3.1.1 - 3.1.3 as a condition that defines continuity and to consider other conditions as derived properties of continuous functions or as criteria of continuity.

Remark 3.1.1. We define continuity by utilizing three different approaches: limits, neighborhoods and by the most popular in calculus (ε, δ)-definition. For real functions, all these definitions result in the same concept of continuity. However, in a more general situation of topological spaces, these three definitions bring us to different structures. Neighborhood continuity is directly extended to continuous mappings of topological spaces. Limits define continuity only for sequential topology, while the (ε, δ)-definition can be extended only to metric spaces but not to general topological spaces.

Let $X, Y \subseteq \mathbf{R}$. Usually the set of all continuous at a point a functions from X into Y is denoted by $C_a(X, Y)$.

From local continuity, we naturally come to global continuity.

Let X be a subset of $\mathbf{R}$.

Definition 3.1.4. A partial function $f: \mathbf{R} \to \mathbf{R}$ is called *continuous in* X if it is continuous at each point a from $X \cap \text{Dom} f$.

Example 3.1.2. Functions $[x]$ and $x - [x]$ are continuous neither in the interval $[0, 1)$ nor in any interval $[m, n)$ where m and n are integer numbers. At the same time, both functions are continuous in the interval $(0, 1)$.

Definition 3.1.5. A partial function $f: \mathbf{R} \to \mathbf{R}$ is called *continuous inside* X if for any element a from $X \cap \text{Dom} f$ and any $\varepsilon > 0$ there is $\delta > 0$ such that for any x from $X \cap \text{Dom} f$ the inequality $|a - x| < \delta$ implies the inequality $|f(x) - f(a)| < \varepsilon$.

Example 3.1.3. Functions $\tan x$ and $\sec x$ are continuous inside the interval $[-\pi/2, \pi/2]$, while functions $[x]$ and $x - [x]$ are continuous inside the interval $[0, 1)$. At the same time, functions $\tan x$ and $\sec x$ are not continuous inside the interval $[-\pi, \pi]$, while functions $[x]$ and $x - [x]$ are not continuous inside the interval $[-1, 1]$ or $[0, 1]$.

Remark 3.1.2. When a is a density point of X, functions continuous inside X at a coincide with the approximately continuous at a functions introduced by Denjoy (1915). Here density is defined relative to a subset Z of a measurable space X with a measure μ for points in X. Namely, the *density* $d(Z)$ of the function $f(x)$ at the point a is equal to

$$\lim_{\delta\to 0} (1/\delta)\, \mu(\{ Z\cap\Delta ; a \in \Delta \subseteq X, \mu(\Delta) = \delta \})$$

if this limit exists and coincides for all Δ from X containing a.

Definition 3.1.6. A partial function f: $\mathbf{R} \to \mathbf{R}$ is called *continuous inside* X if for any $\varepsilon > 0$ and for any a from X there is $\delta > 0$ such that for any x from X, the inequality $| a - x | < \delta$ implies the inequality $| f(x) - f(a) | < \varepsilon$.

Example 3.1.4. Functions tan x and sec x are continuous inside the interval $[-\pi, \pi]$, while they are discontinuous on the same interval.

Lemma 3.1.1. A continuous in X function $f(x)$ is continuous inside X.

Indeed, if the function $f(x)$ is continuous in X, $a \in X$, and a sequence $l = \{x_i ; x_i \in X, i = 1, 2, 3, \ldots \}$ is converging to a, then by Definitions 3.1.1 and 3.1.4, the sequence $h = \{ f(x_i) ; i = 1, 2, 3, \ldots \}$ converges to $f(a)$. Thus, by Definitions 3.1.1, 3.1.5 and 3.1.6, $f(x)$ is continuous inside X.

Remark 3.1.3. When the function $f(x)$ is defined only on X, then continuity in and inside X coincide and $f(x)$ is called (cf., for example, (Ross, 1996)) *continuous on* X.

Usually the set of all continuous functions from X into Y is denoted by $C(X, Y)$.

Theorem 3.1.1 (the First Weierstrass Theorem). A continuous on a closed interval function $f(x)$ is bounded.

Proof. Let us take a continuous function f: $[a, b] \to \mathbf{R}$ and assume that it is not bounded. A function can be unbounded either from above or from below or from both sides. For convenience, we consider the first case. Two others are treated in a similar way. It means that for any number n, there is a number $c_n \in [a, b]$ such that $f(c_n) > n$. Thus, we can choose a sequence $l = \{c_i ; i = 1, 2, 3, \ldots\}$ such that $f(c_i) > i$ for all $i = 1, 2, 3, \ldots$. Properties of a closed interval imply that the sequence l has a converging subsequence $h = \{a_i ; i = 1, 2, 3, \ldots\}$. Let $d = \lim h$. As the function $f(x)$ is continuous in $[a, b]$, $f(x)$ is continuous at the point d. It means (cf. Definition 3.1.3) that for any $\varepsilon > 0$ there is $\delta > 0$ such that the inequality $| d - x | < \delta$ implies the inequality $| f(x) - f(d) | < \varepsilon$. This inequality implies that with $| d - x | < \delta$, we have $f(x) < f(d) + \varepsilon$. As the sequence $h = \{a_i ; i = 1, 2, 3, \ldots\}$ converges to d, we have $| d - a_i | < \delta$ for almost all its elements. This implies that $f(a_i) < f(d) + \varepsilon$ for almost all $i = 1, 2, 3, \ldots$. At the same time, by the construction of the sequence h, the sequence $k = \{f(a_i); i = 1, 2, 3, \ldots\}$ is unbounded. This contradiction shows that the function $f(x)$ has to be bounded.

Theorem is proved.

Theorem 3.1.2 (the Local Weierstrass Theorem). A function $f(x)$ continuous at a point a is bounded at a.

Proof. Let us take a function $f(x)$ continuous at a point a and assume that it is not bounded at a. A function can be unbounded either from above or from below or from both sides. For convenience, we consider the first case. Two others are treated in a similar way. It means that in any neighborhood $(a - k, a + k)$ and any number n, there is a number $c_n \in (a - k, a + k)$ such that $f(c_n) > n$.

Let us take $k = 1/2, 1/3, 1/4, 1/5, \ldots$ Then we can choose a sequence $l = \{c_i ; i = 1, 2, 3, \ldots\}$ such that $f(c_i) > i$ and $c_i \in (a - 1/i, a + 1/i)$ for all $i = 1, 2, 3, \ldots$. By Definition 2.1.1, the sequence l converges to a. At the same time, by Definition 3.1.1, the function $f(x)$ is not continuous at the point a because the sequence $h = \{ f(c_i); i = 1, 2, 3, \ldots\}$ is diverging. This contradiction shows that the function $f(x)$ has to be bounded at a.

Theorem is proved.

Remark 3.1.4. When the function $f(x)$ is defined on an open interval, both Theorems 3.1.1 and 3.1.2 are incorrect in a general case. For instance, the function tan x is unbounded and does reach either maximum or minimum in the interval $(-\pi/2, \pi/2)$.

Theorem 3.1.3 (the Second Weierstrass Theorem). Any continuous in a closed interval function $f(x)$ reaches its global maximum and minimum.

Proof. Let $f(x)$ be a function continuous in $[a, b]$. By Theorem 3.1.1, any continuous in a closed interval function is bounded. Properties of real numbers imply that any bounded set of real numbers has supremum and infimum (cf. Appendix B, Proposition B.3). Consequently, for some numbers u and v, we have $u = \sup \{ f(x); x \in X\}$ and $v = \inf \{ f(x); x \in X\}$. If $u = f(c)$ for some point c from the interval $[a, b]$, then u is the global maximum of $f(x)$ in $[a, b]$.

Let us assume that $u \neq f(c)$ for any point c from the interval $[a, b]$. Then in this case, for any $\varepsilon \in \boldsymbol{R}^{++}$, there is an element c from $[a, b]$ such that $| u - f(c) | < \varepsilon$ (cf. Appendix B). In particular, there are elements c_i from $[a, b]$ such that $| u - f(c_i) | < 1/i$ for all $i = 1, 2, 3, \dots$ Properties of the closed interval $[a, b]$ imply that there is a converging subsequence $h = \{ d_i ; i = 1, 2, 3, \dots\}$ of the sequence $l = \{ c_i ; i = 1, 2, 3, \dots\}$ (cf. Section 2.1).

Let $e = \lim h$. Then $e \in [a, b]$ and by the definition of a continuous function $\lim_{i\to\infty} f(d_i) = f(e)$. However, $\lim_{i\to\infty} f(d_i) = u$. Thus, our assumption was incorrect and $u = f(e)$. Consequently, $f(x)$ reaches its global maximum in $[a, b]$, which is equal to u.

The proof that $f(x)$ reaches its infimum v and thus, its global minimum is similar. Consequently, Theorem 3.1.2 is proved.

Proposition 3.1.3. If there is a number q such that $|x - z| > q$ for any x and z from X, then any function $f: \boldsymbol{R} \to \boldsymbol{R}$ is continuous inside X.

Really, Definition 3.1.1 is formally true for any point a from $\boldsymbol{R}$ as there are no converging sequences in X but only trivial sequences, i.e., sequences of the form $\{a, a, a, \dots , a, \dots\}$.

Corollary 3.1.1. If there is a number q such that $|x - z| > q$ for any x and z from X, then any function $f: X \to \boldsymbol{R}$ is continuous on X.

Definition 3.1.7. A function $f: \boldsymbol{R} \to \boldsymbol{R}$ continuous in $\boldsymbol{R}$ is called *continuous*.

Example 3.1.5. The function x is continuous, while the function $[x]$ is not.

Let $f: [a, b] \to \boldsymbol{R}$ be a continuous function defined for all elements from the interval $[a, b]$.

Theorem 3.1.4 (Intermediate Value Theorem). If $f(a) = k$, $f(b) = h$, and $k < h$ ($k > h$), then for any number l from the interval $[k, h]$ (from the interval $[h, k]$) there is [at least one] point $c \in [a, b]$ such that $f(c) = l$.

We do not give here a proof of this important result of analysis as Theorem 3.1.4 is a direct corollary of Theorem 3.4.5.

Theorem 3.1.5. If the functions $f(x)$ and $g(x)$ are continuous at a point a, then:

a) the function $(f + g)(x)$ is continuous at the point a;
b) the function $(f - g)(x)$ is continuous at the point a;
c) the function $(f \cdot g)(x)$ is continuous at the point a;
d) the function $(k \cdot f)(x)$ is continuous at the point a.

Proof. We do not need to prove parts a), b) and d) because they are direct corollaries of the corresponding parts from Theorem 3.4.8 (cf., Section 3.4).

c) Let us take a sequence $l = \{a_i ; i = 1, 2, 3, \ldots\}$ such that $a = \lim l$. By Definition 3.1.1, the point $f(a)$ is the limit of the sequence $\{ f(a_i) \in \boldsymbol{R};\ i = 1, 2, 3, \ldots\}$ and the point $g(a)$ is the limit of the sequence $\{ g(a_i) \in \boldsymbol{R};\ i = 1, 2, 3, \ldots\}$. Then by Theorem 2.1.4, the point $(f \cdot g)(a) = f(a) \cdot g(a)$ is an limit of the sequence $\{ (f \cdot g)(a_i) = f(a_i) \cdot g(a_i);\ i = 1, 2, 3, \ldots\}$. Thus, by Definition 3.1.1, the function $(f \cdot g)(x)$ is continuous at the point a.

Corollary 3.1.2. The set of all continuous at a point a real functions is a linear space.

Theorem 3.1.6. If the functions $f(x)$ and $g(x)$ are continuous in (inside) X, then:

a) the function $(f + g)(x)$ is continuous in (inside) X;
b) the function $(f - g)(x)$ is continuous in (inside) X;
c) the function $(f \cdot g)(x)$ is continuous in (inside) X;
d) the function $(k \cdot f)(x)$ is continuous in (inside) X.

We do not give here proofs of these important results as parts a), b) and d) are direct corollaries of the corresponding parts from Corollary 3.4.7, while part c) can be easily deduced from part c) of Theorem 3.1.5.

Corollary 3.1.3. The set of all continuous in (inside) X real functions is a linear space.

Let $X, Y, Z \subseteq \boldsymbol{R}$ and $q, p, r \in \boldsymbol{R}^{++}$.

Theorem 3.1.7. If a mapping $f: X \to Y$ is continuous (at a point x from X) and a mapping $g: Y \to Z$ is continuous (at the point $f(x)$), then the mapping $gf: X \to Z$ is continuous (at the point x).

We do not give here a proof of this important result because Theorem 3.1.7 is a direct corollary of Theorem 3.4.9.

Definition 3.1.8. A function $f: \boldsymbol{R} \to \boldsymbol{R}$ is called *left* (*right*) *continuous* or *continuous from the left* (*right*) *at a point* $a \in \boldsymbol{R}$ if for any sequence $l = \{ a_i \in \boldsymbol{R};\ a_i < a\ (a_i > a),\ i = 1, 2, 3, \ldots\}$ that converges to a, the point $f(a)$ is a limit of the sequence $\{ f(a_i) \in \boldsymbol{R};\ i = 1, 2, 3, \ldots\}$ or formally, for any sequence $l = \{ a_i \in \boldsymbol{R};\ a_i < a\ (a_i > a),\ i = 1, 2, 3, \ldots\}$, $\lim_{i\to\infty} a_i = a$ implies $\lim_{i\to\infty} f(a_i) = f(a)$.

Example 3.1.6. The function $]x[$ is continuous from the left at the point 0 and is not continuous from the right at the point 0. The function $[x]$ is continuous from the right at the point 0 and is not continuous from the left at the point 0.

Remark 3.1.5. One-sided local continuity of functions allows one to define one-sided global continuity of functions in a natural way.

Theorem 3.1.8 (the One-sided Local Weierstrass Theorem). A function $f(x)$ continuous from the left (right) at a point a is bounded from the left (right) at a.

The proof is similar to the proof of Theorem 3.1.2.

Remark 3.1.6. Global Weierstrass theorems are not valid, in general, for left and right continuous functions when they are not continuous.

Example 3.1.7. Let us consider the function $f(x)$ that is equal to 0 when $x \leq 0$ and $x \geq \pi/2$ and equal to $\cot x$ when $0 < x < \pi/2$. This function is continuous from the left in $\boldsymbol{R}$ and in any interval $[a, b]$. However, $f(x)$ is not bounded in $[-1, 2]$.

Theorem 3.1.9. If the functions $f(x)$ and $g(x)$ are left (right) continuous at a point a, then:

a) the function $(f + g)(x)$ is left (right) continuous at the point a;
b) the function $(f - g)(x)$ is left (right) continuous at the point a;

c) the function $(f \cdot g)(x)$ is left (right) continuous at the point a;

d) the function $(k \cdot f)(x)$ is left (right) continuous at the point a.

Proof is similar to the proof of Theorem 3.4.8.

Corollary 3.1.4. The set of all left (right) continuous at a point a real functions is a linear space.

Theorem 3.1.10. The composition of two continuous (at a point) functions is continuous (at the same point).

We do not give here a proof of this result because it is a direct corollary of the corresponding result for fuzzy continuous functions (cf. Section 3.4).

3.2. MEASURES OF CONTINUITY AND DISCONTINUITY

I don't strive for perfect continuity,
which is good, because I'd never achieve it.
Lynn Abbey (1948-)

3.2.1. General Principles

Neoclassical analysis studies functions that are continuous only to some extent and are called fuzzy continuous. This idea is formalized by means of measures of continuity and discontinuity. Measures of discontinuity (also called, continuity defects) show to what extent a function is discontinuous. Measures of continuity show to what extent a function is continuous. Examples of such measures are considered in the next section 3.2.2.

Measures of continuity play the role of the membership function for the fuzzy set of fuzzy continuous functions. These measures allow us to define a fuzzy continuous function as a function the membership function (measure of continuity) of which is not equal to zero.

There are two types of measures of continuity and continuity defects: *local* and *global*. Local measures and defects characterize the function at a point, while global measures and defects characterize the function in some set or in its whole domain. Global measures and defects also can be divided into two classes: *selective* and *cumulative*. Selective measures and defects are constructed from local measures and defects by choosing a representative point in the domain and taking the value of a local measure (defect) at this point. For instance, we can take maximum or minimum of the considered local measure. Cumulative measures and defects are constructed by applying some integral operation (see Appendix A) to the values of local measures and defects. For instance, if $\alpha(x)$ is local measure, we can take the integral $\int\alpha(x)dx$ as the corresponding global measure.

Let $F(\boldsymbol{R}, \boldsymbol{R})$ be the class of all real functions with one variable.

Definition 3.2.1. A partial function α: $F(\boldsymbol{R}, \boldsymbol{R}) \times \boldsymbol{R} \to \boldsymbol{R}^+$ is called a local measure of discontinuity (or continuity defect) if two following conditions are satisfied:

(DM1) α is undefined for the pair (f, a) when the function $f(x)$ has an infinite gap at the point a.

(DM2) If the function $f(x)$ has a larger gap at the point a than the function $g(x)$ has at this point, then $\alpha(f, a) \geq \alpha(g, a)$.

Example 3.2.1. Functions tan x and sec x have infinite gaps, while functions $[x]$ and x - $[x]$ have only finite gaps. Thus, any continuity defect α is, for example, undefined for (tan, $\pi/2$) or (sec, - $\pi/2$).

Definition 3.2.2. A local measure of discontinuity α is called *adequate* if the following condition is satisfied:

(DM3) A function $f(x)$ is continuous at a point $a \in \boldsymbol{R}$ if and only if the measure $\alpha(f, a)$ at the point $a \in \boldsymbol{R}$ is equal to 0.

It means that an adequate local measure of discontinuity is equal to zero when there are no discontinuities at this point.

Lemma 3.2.1. If α and β are local measures of discontinuity, α is adequate and $\alpha \leq \beta \leq k\alpha$ or $k\alpha \leq \beta \leq \alpha$ for some positive number k, then β is adequate.

Indeed, under the initial conditions, $\beta(f, a)$ is equal to 0 if and only if $\alpha(f, a)$ is equal to 0. Thus, β satisfies the condition (DM3).

Definition 3.2.3. A local measure of discontinuity α is called *strict* if the following condition is satisfied:

(DMS2) If the function $f(x)$ has a larger gap at the point a than the function $g(x)$ has at this point, then $\alpha(f, a) > \alpha(g, a)$.

Definition 3.2.4. A partial function τ: $\mathrm{F}(\boldsymbol{R}, \boldsymbol{R}) \times \boldsymbol{R} \rightarrow [0, 1]$ is called a local measure of continuity if two following conditions are satisfied:

(CM1) $\tau(f, a) = 0$ when the function $f(x)$ has an infinite gap at the point a.

(CM2) If the function $f(x)$ has a larger gap at the point a than the function $g(x)$ has at this point, then $\tau(f, a) \leq \tau(g, a)$.

Definition 3.2.5. A local measure of continuity τ is called *adequate* if the following condition is satisfied:

(CM3) A function $f(x)$ is continuous at a point $a \in \boldsymbol{R}$ if and only if the measure $\tau(f, a)$ at the point $a \in \boldsymbol{R}$ is equal to 1.

It means that an adequate local fuzzy continuity is a natural extension of the conventional local continuity.

Lemma 3.2.2. If τ and σ are local measures of continuity, is τ is adequate and $\sigma^k \leq \tau \leq \sigma$, $\sigma \leq \tau \leq \sigma^k$, $\tau^k \leq \sigma \leq \tau$ or $\tau \leq \sigma \leq \tau^k$ for some natural number k, then σ is adequate.

Indeed, under the initial conditions, $\sigma(f, a)$ is equal to 1 if and only if $\tau(f, a)$ is equal to 1. Thus, σ satisfies the condition (CM3).

Definition 3.2.6. A local measure of continuity τ is called *strict* if the following condition is satisfied:

(CMS2) If the function $f(x)$ has a larger gap at the point a than the function $g(x)$ has at this point, then $\tau(f, a) < \tau(g, a)$.

If τ is a local measure of continuity, then it defines the set $C_{\tau,a}(X, Y)$ of τ-continuous at a point $a \in \boldsymbol{R}$ functions, namely, a function $f(x)$ is *τ-continuous* at a point $a \in \boldsymbol{R}$ if $\tau(f, a) > 0$.

Definition 3.2.7. A local measure of continuity τ is *weakly connected* to a local measure of discontinuity α if

$$\alpha(f, a) = 0 \Leftrightarrow \tau(f, a) = 1$$

and

$$\alpha(f, a) = \infty \Leftrightarrow \tau(f, a) = 0$$

An example of a natural weak connection between τ and α is given by the formula

$$\tau(f, a) = \frac{1}{1 + \alpha(f, a)} \tag{3.1}$$

It is possible to write relation (3.1) in a different form when $\tau(f, a)$ is normed, i.e., $0 \leq \tau(f, a) \leq 1$,

$$\alpha(f, a) = \frac{1 - \tau(f, a)}{\tau(f, a)} \tag{3.2}$$

If the connection between τ and α is given by the formula (3.1) or by the formula (3.2), then $f(x) \notin C_{\tau,a}(X, Y)$ if and only if $f(x)$ has an infinite gap at the point a.

Proposition 3.2.1. If a local measure of continuity τ is weakly connected to a local measure of discontinuity α, then τ is adequate if and only if α is adequate.

Indeed, under the initial conditions, $\alpha(f, a)$ is equal to 0 if and only if $\tau(f, a)$ is equal to 1. Thus, α satisfies the condition (DM3) if and only if τ satisfies the condition (CM3).

Let us consider properties of discontinuity measures for piece-wise continuous functions.

Definition 3.2.8. A local measure of discontinuity α is called *grounded* if for a function $f(x)$ that has a gap at a point $a \in \boldsymbol{R}$, the value $\alpha(f, a)$ is equal to the measure of this gap.

As continuous functions have zero gaps, we have the following result.

Proposition 3.2.2. Any grounded local measure of discontinuity α (local measure of continuity τ) is adequate.

Proposition 3.2.3. If α is a local measure of discontinuity (τ is a local measure of continuity) and $h(x)$ is an increasing function such that $h(0) = 0$ ($g(1) = 1$ and $g(0) = 0$), then $h(\alpha)$ is also a local measure of discontinuity ($g(\tau)$ is also a local measure of continuity).

Indeed, under the initial conditions, if $\alpha(f, a)$ is equal to 0, then $h(\alpha)(f, a)$ is equal to 0 and if $\tau(f, a)$ is equal to 1, then $g(\tau)(f, a)$ is equal to 1. Thus, $h(\alpha)$ satisfies the condition (DM3) if and only if $g(\tau)$ satisfies the condition (CM3).

Remark 3.2.1. Assuming that a defect of continuity α is normed, i.e., $0 \leq \alpha(f, a) \leq 1$ for all $f \in \mathrm{F}(\boldsymbol{R}, \boldsymbol{R})$ and $a \in \boldsymbol{R}$, we can determine a related to α measure of continuity τ by a formula different from (3.1), namely, $\tau(f, a) = 1 - \alpha(f, a)$. It satisfies only the first condition from Definition 3.2.7. Such measures are studied in (Burgin, 1999).

3.2.2. Local Measures of Continuity

Let us assume that a function $f(x)$ is defined at a point $a \in \boldsymbol{R}$.

Definition 3.2.9. The *local sequential continuity defect* (*local sequential measure of discontinuity*) $\delta(f, a)$ of the function $f(x)$ at the point $a \in \boldsymbol{R}$ is defined by the following formula

$$\delta(f, a) = \sup \{ \lim_{i\to\infty} | f(a) - f(a_i)| \text{ ; where } a_i \in \boldsymbol{R} \text{ and } \lim_{i\to\infty} a_i = a \} \quad (3.3)$$

Note that if there is, at least, one sequence $\{ a_i ; i = 1, 2, 3, \dots \}$ such that $\lim_{i\to\infty} a_i = a$ and the sequence $\{ f(a_i); i = 1, 2, 3, \dots \}$ is unbounded, then $\delta(f, a) = \infty$. When the only sequences that converge to a have almost all elements coinciding with a, then $\delta(f, a) = 0$.

The local continuity defect $\delta(f, a)$ was introduced in (Burgin and Šostak, 1992).

Definitions 3.2.9 and 2.5.1 imply the following results.

Lemma 3.2.3. $\delta(f, a) = \sup \{ \delta(f(a) = \lim_{i\to\infty} f(a_i)) ; a_i \in \boldsymbol{R} \text{ and } \lim_{i\to\infty} a_i = a \}$.

Proposition 3.2.4 (Reflection Symmetry). For any functions $f(x)$ and at any point $a \in \boldsymbol{R}$ where this function is defined, we have $\delta(f, a) = \delta(-f, a)$.

Proposition 3.2.5 (Subadditivity). For any functions $f(x)$ and $g(x)$ and at any point $a \in \boldsymbol{R}$ where both functions are defined, the following inequality holds:

$$\delta(f + g, a) \leq \delta(f, a) + \delta(g, a) \quad (3.4)$$

Proof. If one of the values $\delta(f, a)$ or $\delta(g, a)$ is infinite, then the inequality (3.4) is trivial. So, we can consider only the case when $\delta(f, a) < \infty$ and $\delta(g, a) < \infty$.

Let us take some sequence $\{ a_i ; i = 1, 2, 3, \dots \}$ such that $\lim_{i\to\infty} a_i = a$ and the sequence $\{ (f + g)(a_i); i = 1, 2, 3, \dots \}$ either diverges or has a finite limit. At the same time, inequalities $\delta(f, a) < \infty$ and $\delta(g, a) < \infty$ imply that both sequences $\{ f(a_i); i = 1, 2, 3, \dots \}$ and $\{ g(a_i); i = 1, 2, 3, \dots \}$ are bounded. Consequently, the sequence $\{ (f + g)(a_i) = f(a_i) + g(a_i); i = 1, 2, 3, \dots \}$ is also bounded and thus, there exist a finite limit $b = \lim_{i\to\infty} (f + g)(a_i)$.

By the Bolzano-Weierstrass theorem (Theorem 2.1.7), we can find a subsequence $\{ c_i ; i = 1, 2, 3, \dots \}$ of the sequence $\{ a_i ; i = 1, 2, 3, \dots \}$ such that all three sequences $\{ f(c_i); i = 1, 2, 3, \dots \}$, $\{ g(c_i); i = 1, 2, 3, \dots \}$ and $\{ (f + g)(c_i); i = 1, 2, 3, \dots \}$ are convergent. Note that $\lim_{i\to\infty} (f + g)(a_i) = \lim_{i\to\infty} (f + g)(c_i) = b$ by Proposition 2.1.9. Then by Definition 3.2.9, Theorem 2.1.4 and properties of the absolute value, we have

$$\begin{gathered}
\lim_{i\to\infty} | (f + g)(a) - (f + g)(a_i)| = | \lim_{i\to\infty} ((f + g)(a) - (f + g)(a_i))| = \\
| (f + g)(a) - \lim_{i\to\infty} (f + g)(a_i))| = | (f + g)(a) - \lim_{i\to\infty} (f + g)(c_i))| = \\
| f(a) + g(a) - \lim_{i\to\infty} (f(c_i) + g(c_i))| = \\
| (f(a) - \lim_{i\to\infty} (f(c_i)) + (g(a) - \lim_{i\to\infty} g(c_i))| \leq \\
| f(a) - \lim_{i\to\infty} (f(c_i)| + |g(a) - \lim_{i\to\infty} g(c_i)| \leq \delta(f, a) + \delta(g, a)
\end{gathered}$$

As $\{ a_i ; i = 1, 2, 3, \dots \}$ is an arbitrary sequence in $\boldsymbol{R}$ such that $\lim_{i\to\infty} a_i = a$ and the sequence $\{ (f + g)(a_i); i = 1, 2, 3, \dots \}$ has a finite limit, the equality (3.3), properties of supremums and the inequality

$$\lim_{i\to\infty} | (f + g)(a) - (f + g)(a_i)| \leq \delta(f, a) + \delta(g, a)$$

imply the inequality

$$\delta(f, a) \leq \delta(f, a) + \delta(g, a)$$

proposition is proved.

Remark 3.2.2. The local continuity defect $\delta(f, a)$ is not additive, i.e., it is impossible to change the inequality sign in (3.4) into by the inequality sign as the following example demonstrates.

Example 3.2.2. Let us consider functions

$$f(x) = \begin{cases} 1 & \text{for } x < 0, \\ 0 & \text{for } x \geq 0. \end{cases}$$

and

$$g(x) = \begin{cases} -1 & \text{for } x < 0, \\ 0 & \text{for } x \geq 0. \end{cases}$$

They have the following parameters: $\delta(f, 0) = \delta(g, 0) = 1$. At the same time, $(f + g)(x)$ is identically equal to 0. Thus, $\delta(f + g, 0) = 0 < \delta(f, 0) + \delta(g, 0) = 1 + 1 = 2$.

Corollary 3.2.1. For any functions $f(x)$ and $g(x)$ and at any point $a \in \boldsymbol{R}$ where both functions are defined, the following inequality holds:

$$\delta(f - g, a) \geq \delta(f, a) - \delta(g, a)$$

Indeed, by Proposition 3.2.5, we have

$$\delta(f, a) = \delta((f - g) + g, a) \leq \delta(f - g, a) + \delta(g, a)$$

This implies

$$\delta(f - g, a) \geq \delta(f, a) - \delta(g, a)$$

Corollary 3.2.2. For any functions $f(x)$ and $g(x)$ and at any point $a \in \boldsymbol{R}$ where both functions are defined, the following inequalities hold:

$$|\delta(f, a) - \delta(g, a)| \leq \delta((f + g, a) \leq \delta(f, a) + \delta(g, a)$$

and

$$|\delta(f, a) - \delta(g, a)| \leq \delta((f - g, a) \leq \delta(f, a) + \delta(g, a)$$

Proof. By Proposition 3.2.5, we have

$$\delta((f+g, a) \leq \delta(f, a) + \delta(g, a)$$

At the same time, by Corollary 3.2.1 and Proposition 3.2.4, we have

$$\delta((f+g, a) = \delta(f-(-g), a) \geq \delta(f, a) - \delta(-g, a) = \delta(f, a) - \delta(g, a)$$

and

$$\delta((f+g, a) = \delta((g+f, a) = \delta(g-(-f), a) \geq \delta(g, a) - \delta(-f, a) = \delta(g, a) - \delta(f, a)$$

This implies

$$\delta(f+g, a) \geq |\delta(f, a) - \delta(g, a)|$$

Consequently, we have

$$|\delta(f, a) - \delta(g, a)| \leq \delta((f+g, a) \leq \delta(f, a) + \delta(g, a)$$

The proof of the second part of Corollary 3.2.2 is similar as $f - g = f + (-g)$ and $\delta(-g, a) = \delta(\cdot g, a)$.

Corollary is proved.

Proposition 3.2.6. For any functions $f(x)$ and $g(x)$ at any point $a \in \boldsymbol{R}$ where both functions are defined, the inequalities

$$\delta(f{\cdot}g, a) \leq \delta(f, a) \, \| g(x) \| + \delta(g, a) \, |f(a)| \tag{3.5}$$

where $\| g(x) \| = \sup \{ |g(x)| \, ; x$ belongs to some neighborhood Oa of $a\}$, and

$$\delta(f{\cdot}g, a) \leq \delta(f, a) \, | g(a) | + \delta(g, a) \, |f(a)| + \delta(f, a)\delta(g, a) \tag{3.6}$$

are valid.

Proof. Note that we need to prove inequalities (3.5) and (3.6) only at points a where both defects $\delta(f, a)$ and $\delta(g, a)$ are finite.

Let us take a point $a \in \boldsymbol{R}$ where both defects $\delta(f, a)$ and $\delta(g, a)$ are finite. Then there is some neighborhood Oa of a where both functions $f(x)$ and $g(x)$ are bounded. We define $\|g(x)\| = \sup \{ |g(x)| \, ; x \in Oa \}$. Properties of absolute values allow us to derive the following chain of equalities and inequalities when all elements a_i belong to the neighborhood Oa of a

$$\begin{gathered}
|(f{\cdot}g)(a) - (f{\cdot}g)(a_i)| = | f(a){\cdot}g(a) - f(a_i){\cdot}g(a_i) | = \\
| f(a){\cdot}g(a) - f(a){\cdot}g(a_i) + f(a){\cdot}g(a_i) - f(a_i){\cdot}g(a_i) | \leq \\
| f(a){\cdot}g(a) - f(a){\cdot}g(a_i)| + | f(a){\cdot}g(a_i) - f(a_i){\cdot}g(a_i) | = \\
| (f(a) - f(a_i)){\cdot}g(a_i)| + | f(a){\cdot}(g(a) - g(a_i)) | = \\
| f(a) - f(a_i)|{\cdot}| g(a_i)| + | f(a)|{\cdot}| g(a) - g(a_i)| < \\
| f(a) - f(a_i)|{\cdot}\| g(x)\| + | f(a)|{\cdot}| g(a) - g(a_i)|
\end{gathered}$$

Taking some sequence $l = \{ a_i ; i = 1, 2, 3, \dots \}$ such that $\lim_{i\to\infty} a_i = a$ and the sequence $\{ (f+g)(a_i); i = 1, 2, 3, \dots \}$ converges, we can find a subsequence $h = \{ c_i ; i = 1, 2, 3, \dots \}$ of l such that both sequences $\{ f(c_i); i = 1, 2, 3, \dots \}$ and $\{ g(c_i); i = 1, 2, 3, \dots \}$ converge because both sequences $\{ f(a_i); i = 1, 2, 3, \dots \}$ and $\{ g(a_i); i = 1, 2, 3, \dots \}$ are bounded (cf. Section 2.1). Note that $\lim_{i\to\infty} c_i = a$. This means that we can define all three defects $\delta(f\cdot g, a)$, $\delta(f, a)$ and $\delta(g, a)$ utilizing only sequences $\{ a_i ; i = 1, 2, 3, \dots \}$ such that $\lim_{i\to\infty} a_i = a$ and sequences $\{ (f+g)(a_i); i = 1, 2, 3, \dots \}$, $\{ f(a_i); i = 1, 2, 3, \dots \}$ and $\{ g(a_i); i = 1, 2, 3, \dots \}$ converge.

Consequently, as supremum preserves non-strict inequalities, we have

$$\begin{gathered}
\delta(f\cdot g, a) = \sup \{ \lim_{i\to\infty} |(f\cdot g)(a) - (f\cdot g)(a_i)| \text{ ; where } a_i \in \boldsymbol{R} \text{ and } \lim_{i\to\infty} a_i = a \} \leq \\
\sup \{\lim_{i\to\infty} (|f(a) - f(a_i)|\cdot\| g(x)\| + |f(a)|\cdot| g(a) - g(a_i)|)\text{; where } a_i\in\boldsymbol{R} \text{ and } \lim_{i\to\infty} a_i = a\} \leq \\
\sup \{ \lim_{i\to\infty} (|f(a) - f(a_i)|\cdot\| g(x)\|)\text{; where } a_i \in \boldsymbol{R} \text{ and } \lim_{i\to\infty} a_i = a \} + \\
\sup \{ \lim_{i\to\infty} (|f(a)|\cdot| g(a) - g(a_i)|) \text{ ; where } a_i \in \boldsymbol{R} \text{ and } \lim_{i\to\infty} a_i = a \} = \\
\sup \{ \lim_{i\to\infty} |f(a) - f(a_i)|\cdot \lim_{i\to\infty} \| g(x)\| \text{ ; where } a_i \in \boldsymbol{R} \text{ and } \lim_{i\to\infty} a_i = a \} + \\
\sup \{ \lim_{i\to\infty} |f(a)|\cdot \lim_{i\to\infty} | g(a) - g(a_i)| \text{ ; where } a_i \in \boldsymbol{R} \text{ and } \lim_{i\to\infty} a_i = a \} = \\
\sup \{ \| g(x)\| \cdot\lim_{i\to\infty} |f(a) - f(a_i)|\text{; where } a_i \in \boldsymbol{R} \text{ and } \lim_{i\to\infty} a_i = a \} + \\
\sup \{ |f(a)|\cdot \lim_{i\to\infty} | g(a) - g(a_i)| \text{ ; where } a_i \in \boldsymbol{R} \text{ and } \lim_{i\to\infty} a_i = a \} = \\
\| g(x)\| \cdot \sup \{ \lim_{i\to\infty} |f(a) - f(a_i)|\text{; where } a_i \in \boldsymbol{R} \text{ and } \lim_{i\to\infty} a_i = a \} + \\
|f(a)|\cdot \sup \{ \lim_{i\to\infty} | g(a) - g(a_i)| \text{ ; where } a_i \in \boldsymbol{R} \text{ and } \lim_{i\to\infty} a_i = a \} = \\
\| g(x)\| \cdot\delta(f, a) + |f(a)|\cdot \delta(g, a)
\end{gathered}$$

Consequently, we have

$$\delta(f\cdot g, a) \leq \delta(f, a) \| g(x) \| + \delta(g, a) |f(a)|$$

i.e., the inequality (3.5) is proved.

By definition and properties of absolute values, we have

$$\begin{gathered}
\| g(x) \| = \sup \{ |g(x)| \text{ ; } x \in \mathrm{O}a \} = \sup \{ | g(a) - g(a) + g(x)| \text{ ; } x \in \mathrm{O}a \} \leq \sup \{ (| g(a)| + | \\
g(a) - g(x)|)\text{; } x \in \mathrm{O}a \} = | g(a)| + \sup \{| g(a) - g(x)|)\text{; } x \in \mathrm{O}a \}
\end{gathered}$$

As $\delta(g, a)$ is finite, $\sup \{| g(a) - g(x)|); x \in \mathrm{O}a \}$ converges to $\delta(g, a)$ when the diameter of the neighborhood $\mathrm{O}a$ tends to 0.

As the inequality (3.5) is true for any the neighborhood $\mathrm{O}a$ of the point a and $\sup \{| g(a) - g(x)|); x \in \mathrm{O}a \}$ converges to $\delta(g, a)$, we have

$$\begin{gathered}
\delta(f\cdot g, a) \leq \delta(f, a) (| g(a)| + \delta(g, a)) + \delta(g, a) |f(a)| = \\
\delta(f, a) | g(a) | + \delta(g, a) |f(a)| + \delta(f, a)\delta(g, a)
\end{gathered}$$

This concludes the proof of Proposition as the inequality (3.6) is also proved.

It is interesting to note that the inequalities (3.5) and (3.6) are similar to the product rule for differentiation (see Section 4.1).

It is possible to define similar local sequential continuity defect even when a function $f(x)$ is not defined at a point $a \in \boldsymbol{R}$.

Definition 3.2.10. The *local sequential continuity defect* (*local sequential measure of discontinuity*) $\delta_1(f, a)$ of the function $f(x)$ at the point $a \in \boldsymbol{R}$ is defined by the following formula

$$\delta_1(f, a) = \sup \{ \lim_{i,j\to\infty} | f(a_j) - f(a_i)| \; ; \text{ where } a_i \, , \, a_i \in \boldsymbol{R} \text{ and } \lim_{i\to\infty} a_i = \lim_{j\to\infty} a_j = a \}$$

Proposition 3.2.7. For any functions $f(x)$ at any point $a \in \boldsymbol{R}$ where this function is defined, $\delta_1(f, a) \leq 2\delta(f, a)$.

Remark 3.2.3. In some cases, $\delta_1(f, a) = 2\delta(f, a)$, while in other cases, $\delta_1(f, a) < 2\delta(f, a)$.

Indeed, if $f(x)$ is a continuous function, then $\delta_1(f, a) = 2\delta(f, a) = 0$ at any point $a \in \boldsymbol{R}$. If $f(x) = 0$ for all negative x and $f(x) = 1$ for 0 and all positive x, then $\delta_1(f, 0) = \delta(f, 0) = 1 < 2\delta(f, 0)$. If $f(x) = -1$ for all negative x, $f(0) = 0$, and $f(x) = 1$ for all positive x, then $\delta_1(f, 0) = 2$, $\delta(f, 0) = 1$, and thus, $\delta_1(f, 0) = 2\delta(f, 0)$.

Many properties of the sequential continuity defect $\delta_1(f, a)$ are similar to properties of the sequential continuity defect $\delta(f, a)$.

It is necessary to remark that the sequential continuity defect $\delta_1(f, a)$ is related to the concept of the oscillation $o(f, a)$ of a function $f(x)$ at a point $a \in \boldsymbol{R}$ (cf., for example, (Saks, 1964)).

Definition 3.2.11. The *local sequential continuity measure* $\lambda(f, a)$ of a function $f(x)$ at a point $a \in \boldsymbol{R}$ is defined by the following formula

$$\lambda(f, a) = \frac{1}{1 + \delta(f, a)} \tag{3.7}$$

Let us assume that a function $g(x)$ is defined at a point $a \in \boldsymbol{R}$.

Definition 3.2.12. The *sequential distance* $d_a(f, g)$ between functions $f(x)$ and $g(x)$ at a point $a \in \boldsymbol{R}$ is defined by the following formula

$$d_a(f, g) = \sup \{ \lim_{i\to\infty} | g(a) - f(a_i)| \; ; \text{ where } a_i \in \boldsymbol{R} \text{ and } \lim_{i\to\infty} a_i = a \} \tag{3.8}$$

Note that the sequential distance $d_a(f, g)$ is not a metric (see Appendix C) as it is not symmetric, i.e., in a general case, $d_a(f, g) \neq d_a(g, f)$ as the following example demonstrates.

Example 3.2.3. Let us consider functions $f(x) = x/|x|$ when $x \neq 0$, $f(x) = 0$, and $g(x) = 0$ for all $x \in \boldsymbol{R}$. Then $d_0(f, g) = 1$, $d_0(g, f) = 0$ and $d_0(f, f) = 1$. Here we can see that the first axiom M1 for metric spaces is also invalid.

Lemma 3.2.4. $d_a(f, g) \geq | g(a) - f(a)|$.

Indeed, $| g(a) - f(a)| = \lim_{i\to\infty} | g(a) - f(a_i)|$ when $a_i = a$ for all $i = 1, 2, 3, \ldots$

It is possible that almost all elements from all sequences $\{ a_i \; ; \; i = 1, 2, 3, \ldots \}$ that converge to a coincide with a, then $d_a(f, g) = | g(a) - f(a)|$.

Lemma 3.2.5. $d_a(f, f) = \delta(f, a)$.

To see this, it is necessary to compare formulas (3.3) and (3.8).

Let $C_a(\boldsymbol{R}, \boldsymbol{R})$ be the class of all real functions continuous at a.

Definition 3.2.13. The *local metric continuity defect* (*local metric measure of discontinuity*) $\gamma(f, a)$ of a function $f(x)$ at a point $a \in \boldsymbol{R}$ is defined by the following formula

$\gamma(f, a) = \inf \{ d_a(f, g); g(x) \in C_a(\boldsymbol{R}, \boldsymbol{R})$, i.e., $g(x)$ is continuous at $a\}$

Lemma 3.2.6. For any real function $f(x)$ and point $a \in \boldsymbol{R}$, there is a continuous at a function $g(x)$ such that $\gamma(f, a) = d_a(f, g)$.

Proof. There is a number $d \in \boldsymbol{R}$ such that $\gamma(f, a) = \inf \{ \lim_{i\to\infty} | d - f(a_i)| ;$ where $a_i \in \boldsymbol{R}$ and $\lim_{i\to\infty} a_i = a \}$. Really, by the definition if infimum, there is a sequence $\{ g_n(x) ; n = 1, 2, 3, \ldots\}$ of continuous functions $g_n(x)$ such that $\gamma(f, a) = \lim_{n\to\infty} d_a(f, g_n)$. Then the sequence $\{ g_n(a) ; n = 1, 2, 3, \ldots\}$ converges because this sequence cannot go to infinity as the set of all values $d_a(f, g_n)$ is bounded and if u and v are two different partial limits of $\{ g_n(a) ; n = 1, 2, 3, \ldots\}$, then the sequence $\{ d_a(f, g_n) ; n = 1, 2, 3, \ldots\}$ also has two different partial limits and does not converge contrary to our choice of functions $g_n(x)$.

We can take a continuous function $g(x)$ for which $g(a) = d$. Then $d_a(f, g) = \lim_{n\to\infty} d_a(f, g_n) = \gamma(f, a)$.

Lemma is proved.

To derive properties of continuity measures from properties of continuity defects, we use the following simple lemma.

Lemma 3.2.7. If $a \le b$ and $a \le 2c$, then $1/(1+a) \ge 1/(1+ b)$ and $2/(1+a) \ge 1/(1+c)$.

Indeed, if $b \ge a$, then $1+ b \ge 1+ a$. Thus, $1/(1+a) \ge 1/(1+ b)$.

If $2c \ge a$, then $2 + 2c \ge 1+ a$. Thus, $1/(1+a) \ge 1/(2 + 2c)$ and $2/(1+a) \ge 1/(1+c)$.

Proposition 3.2.8. For any function $f(x)$ and any point a, we have $\gamma(f, a) \ge \inf \{ |f(a) - g(a)| ; g(x) \in C_a(\boldsymbol{R}, \boldsymbol{R})$, i.e., $g(x)$ is continuous at $a\} = 0$.

Proposition 3.2.9. For any function $f(x)$ and any point a, we have $\gamma(f, a) \le \delta(f, a) \le 2\gamma(f, a)$

Proof. 1) There is a continuous function $g(a)$ such that $g(a) = f(a)$. Then $d_a(f, g) = \delta(f, a) \ge \inf \{ d_a(f, g); g(x) \in C_a(\boldsymbol{R}, \boldsymbol{R})$, i.e., $g(x)$ is continuous at $a\} = \gamma(f, a)$.

2) Let us take the continuous function $g(x)$ for which $\gamma(f, a) = d_a(f, g)$ (cf. Lemma 3.2.6). Then we have

$$\begin{gathered}
\delta(f, a) = \sup \{ \lim_{i\to\infty} | f(a) - f(a_i)| ; a_i \in \boldsymbol{R} \text{ and } \lim_{i\to\infty} a_i = a \} = \\
\sup \{ \lim_{i\to\infty} | f(a) - g(a) + g(a) - f(a_i)| ; a_i \in \boldsymbol{R} \text{ and } \lim_{i\to\infty} a_i = a \} \le \\
\sup \{ \lim_{i\to\infty} | f(a) - g(a)| + | g(a) - f(a_i)| ; a_i \in \boldsymbol{R} \text{ and } \lim_{i\to\infty} a_i = a \} \le \\
\sup \{ \lim_{i\to\infty} | f(a) - g(a)| \} + \sup \{ | g(a) - f(a_i)| ; a_i \in \boldsymbol{R} \text{ and } \lim_{i\to\infty} a_i = a \} = \\
| f(a) - g(a)| + d_a(f, g) = | f(a) - g(a)| + \gamma(f, a) \le \\
\gamma(f, a) + \gamma(f, a) = 2\gamma(f, a)
\end{gathered}$$

because by Lemma 3.2.4, $| f(a) - g(a)| \le d_a(f, g) = \gamma(f, a)$.

Proposition is proved.

Definition 3.2.14. The *local metric continuity measure* $\sigma(f, a)$ of a function $f(x)$ at a point $a \in \boldsymbol{R}$ is defined by the following formula

$$\sigma(f, a) = \frac{1}{1 + \gamma(f, a)} \tag{3.9}$$

The local continuity defect $\gamma(f, a)$ and continuity measure $\sigma(f, a)$ were introduced in (Burgin, 1993b).

Proposition 3.2.10. For any function $f(x)$ and any point a, we have $\lambda(f, a) \leq \sigma(f, a) \leq 2\lambda(f, a)$.

Indeed, as by Proposition 3.2.9, $\gamma(f, a) \leq \delta(f, a) \leq 2\gamma(f, a)$ for any function $f(x)$ and any point a. Thus, Lemma 3.2.7 implies $\lambda(f, a) \leq \sigma(f, a) \leq 2\lambda(f, a)$ for any function $f(x)$ and any point a.

Definition 3.2.15. The *local Kuratowski continuity defect* (*local Kuratowski measure of discontinuity*) $\omega(f, a)$ of a function $f(x)$ at a point $a \in \boldsymbol{R}$ is defined by the following formula

$$\omega(f, a) = \inf \{ \text{Diam}(f(\text{O}_a)); \text{O}_a \text{ is an open neighborhood of } a\}$$

where $\text{Diam}(f(\text{O}_a))$ is the diameter of the set $f(\text{O}_a)$ (cf. Appendix C).

The measure $\omega(f, a)$ is considered in (Kuratowski, 1966) and called the *oscillation* of $f(x)$ at a point a. However, its properties were not studied.

Oscillation is a construction often used in analysis and denoted by $o(f, a)$ or $\omega f(a)$ (cf., for example, (Saks, 1964; Choquet, 1969; Kolmogorov and Fomin, 1999)).

To define oscillation $o(f, a)$, two values $M(f, a)$ and $m(f, a)$ are used where

$$M(f, a) = \sup \{ \lim_{i\to\infty} f(a_i) \text{ ; where } a_i \in \boldsymbol{R} \text{ and } \lim_{i\to\infty} a_i = a \}$$

and

$$m(f, a) = \inf \{ \lim_{i\to\infty} f(a_i) \text{ ; where } a_i \in \boldsymbol{R} \text{ and } \lim_{i\to\infty} a_i = a \}$$

Then oscillation of a function $f(x)$ is defined by the following formula:

$$o(f, a) = M(f, a) - m(f, a)$$

Proposition 3.2.11. For any function $f(x)$ and any point a, we have $\delta(f, a) \leq \omega(f, a) \leq 2\delta(f, a)$.

Proof. 1) $\text{Diam}(f(\text{O}a)) \geq | f(a) - f(a_i)|$ for all $a_i \in \text{O}a$. Thus, $\omega(f, a) = \inf \{ \text{Diam}(f(\text{O}a));$ $\text{O}a$ is an open neighborhood of $a\} \geq \lim |f(a) - f(a_i)|$. In turn, this implies

$$\omega(f, a) \geq \sup \{ \lim_{i\to\infty} | f(a) - f(a_i)| \text{ ; } a_i \in \boldsymbol{R} \text{ and } \lim_{i\to\infty} a_i = a \} = \delta(f, a)$$

2) If we apply limit to both sides of the inequality

$$| f(b) - f(c)| \leq | f(b) - f(a)| + | f(a) - f(c)| \tag{3.10}$$

that is true for arbitrary $b, c \in Oa$, then we see that the inequality $\omega(f, a) \leq 2\delta(f, a)$ follows from the inequality (3.9).

It is also possible to show that the measure $\omega(f, a)$ coincides with the measure $\delta_1(f, a)$ and oscillation $o(f, a)$.

Definition 3.2.16. The *local Kuratowski continuity measure* $\kappa(f, a)$ of a function $f(x)$ at a point $a \in \mathbf{R}$ is defined by the following formula

$$\kappa(f, a) = \frac{1}{1 + \omega(f, a)}$$

Proposition 3.2.11 and Lemma 3.2.7 give us the following result.

Proposition 3.2.12. For any function $f(x)$ and any point a, we have $\kappa(f, a) \leq \lambda(f, a) \leq 2\kappa(f, a)$.

Proposition 3.2.13. All three local continuity defects δ, γ and ω do not coincide.

It is demonstrated by the following example.

Example 3.2.3. Let us consider the following functions:

$$f(x) = \begin{cases} \sin(1/x) & \text{for } x \neq 0, \\ 0 & \text{for } x = 0. \end{cases}$$

and

$$g(x) = \begin{cases} \sin(1/x) & \text{for } x \neq 0, \\ 1 & \text{for } x = 0. \end{cases}$$

They have the following parameters: $\omega(f, 0) = \omega(g, 0) = 2$, $\delta(f, 0) = \gamma(f, 0) = 1$, $\delta(g, 0) = 2$, and $\gamma(g, 0) = 1$. Thus, $\omega(f, 0) \neq \gamma(f, 0)$, $\omega(f, 0) \neq \delta(f, 0)$, and $\gamma(g, 0) \neq \delta(g, 0)$.

Let $a_i \in \mathbf{R}$ and $b_i = 1/(1 + a_i)$ for $i = 1, 2, 3, \dots$.

Lemma 3.2.7. $a_i = a_j \Leftrightarrow b_i = b_j$.

Example 3.3.2, Lemma 3.2.5, and Definitions 3.2.4, 3.2.7 and 3.2.9 give us the following result.

Proposition 3.2.14. All three local continuity measures λ, κ and σ do not coincide.

Remark 3.2.4. When the range of considered functions is a subset of a set Y with a finite diameter Diam(Y), e.g., $Y = [a, b]$, it is possible to build normed counterparts of the local continuity defects δ, γ and ω by the formulas

$$\delta_H(f, a) = \delta(f, a)/\text{Diam}(Y),$$
$$\gamma_H(f, a) = \gamma(f, a)/\text{Diam}(Y),$$
$$\omega_H(f, a) = \omega(f, a)/\text{Diam}(Y),$$

It is also possible to measures of continuity weakly connected to these defects:

$$\lambda_H(f, a) = 1 - \delta_H(f, a).$$
$$\kappa_H(f, a) = 1 - \gamma_H(f, a).$$
$$\sigma_H(f, a) = 1 - \omega_H(f, a).$$

Such measures are studied in (Burgin, 1999).

Remark 3.2.5. It is also possible to introduce one-sided (right and left) local and global measures of continuity and discontinuity. Such measures are studied in (Burgin and Glushchenko, 1998b).

3.2.3. Global Measures

Local measures of continuity and discontinuity induce global measures.

If $\alpha(f, a)$ is local measure of discontinuity and $X \subseteq \boldsymbol{R}$, then the measure $\alpha(f, X)$ of continuity of $f(x)$ on X is equal to

$$\alpha(f, X) = \sup \{ \alpha(f, a); a \in X\}$$

If $\rho(f, a)$ is local measure of continuity and $X \subseteq \boldsymbol{R}$, then the measure $\rho(f, X)$ of continuity of $f(x)$ on X is equal to

$$\rho(f, X) = \inf \{ \rho(f, a); a \in X\}$$

In such a way, we obtain the following measure of continuity and discontinuity:

- the *sequential continuity defect (sequential measure of discontinuity)*

$$\delta(f, X) = \sup \{ \delta(f, a); a \in X\}$$

- the *metric continuity defect (metric measure of discontinuity)*

$$\gamma(f, X) = \sup \{ \gamma(f, a); a \in X\}$$

- the *Kuratowski continuity defect (Kuratowski measure of discontinuity)*

$$\omega(f, X) = \sup \{ \omega(f, a); a \in X\}$$

- the *sequential continuity measure*

$$\lambda(f, X) = \inf \{ \lambda(f, a); a \in X\}$$

- the *metric continuity measure*

$$\sigma(f, X) = \inf \{ \sigma(f, a); a \in X\}$$

- the *Kuratowski continuity measure*

$$\kappa(f, X) = \inf \{ \kappa(f, a); a \in X\}$$

Besides, Cromme and Diener (1991) defined two global measures of discontinuity and applied them to a study of fuzzy fixed-point properties of functions. Namely, if M is a subset of a normed linear space, $B_r(x)$ is a ball with radius r and center x and $f: M \rightarrow M$ is a mapping, then:

$$\delta(f) = \sup \lim_{x \in M} \sup \lim_{r \to 0} \sup_{y \in Br(x)} \|f(x) - f(y)\|$$

and

$$\delta'(f) = \sup \lim_{x \in M} \sup \lim_{r \to 0} \sup_{y \in Br(x) \setminus \{x\}} \|f(x) - f(y)\|$$

Defined here global measures preserve many relations between local measures due to the fact that sup and inf are monotone integral operations.

Proposition 3.2.15. For any function $f(x)$ and any set X, we have $\gamma(f, X) \leq \delta(f, X) \leq 2\gamma(f, X)$.

Proposition 3.2.16. For any function $f(x)$ and any set X, we have $\delta(f, X) \leq \omega(f, X) \leq 2\delta(f, X)$.

Proposition 3.2.17. For any function $f(x)$ and any set X, we have $\lambda(f, a) \leq \sigma(f, a) \leq 2\lambda(f, a)$.

Proposition 3.2.18. For any function $f(x)$ and any set X, we have $\kappa(f, a) \leq \lambda(f, a) \leq 2\kappa(f, a)$

Definition 3.2.17. A global measure of continuity ψ is called *adequate* (*almost adequate*) if the following condition is satisfied:

(CM3) A function $f(x)$ is continuous (almost everywhere) in X if and only if its measure $\psi(f, X)$ is equal to 1.

It means that an adequate fuzzy continuity is a natural extension of the conventional continuity.

The sequential distance $d_a(f, g)$ allows us to define one more global continuity defect $\gamma_0(f, a)$. It is related to $\gamma(f, X)$ but is different from it.

Definition 3.2.18. The sequential distance $d_X(f, g)$ of $f(x)$ to $f(x)$ on X is defined as

$$d_X(f, g) = \sup\{ d_a(f, g); a \in X \}$$

Definition 3.2.19. The *global metric continuity defect* (*global metric measure of discontinuity*) $\gamma_0(f, a)$ of a function $f(x)$ at a point $a \in \boldsymbol{R}$ is defined by the following formula

$\gamma_0(f, X) = \inf \{ d_X(f, g); g(x) \in C(X, \boldsymbol{R})$, i.e., $g(x)$ is continuous in $X\}$

Proposition 3.2.19. If X is the domain of a function $f(x)$, then $\gamma(f, X) \leq \gamma_0(f, X)$.

Remark 3.2.5. It is possible to define all considered here measures of continuity and discontinuity for functions (mappings) in metric spaces. To do this, it is possible to take the distance in the metric space instead of the absolute value of the difference, which is the most common metric in $\boldsymbol{R}$. Functions in general metric spaces are considered in Section 3.7.

As supremum is a subadditive operation on sets of real numbers, Proposition 3.2.5 implies the following result.

Proposition 3.2.20 (Global Subadditivity). For any functions $f(x)$ and $g(x)$ and any set X from $\boldsymbol{R}$ where both functions are bounded, the following inequality holds:

$$\delta(f + g, X) \le \delta(f, X) + \delta(g, X)$$

Corollary 3.2.3. For any functions $f(x)$ and $g(x)$ at any point $a \in \boldsymbol{R}$ where both functions are defined, the following inequality holds:

$$\delta(f - g, X) \ge \delta(f, X) - \delta(g, X)$$

Corollary 3.2.4. For any functions $f(x)$ and $g(x)$ at any point $a \in \boldsymbol{R}$ where both functions are defined, the following inequalities hold:

$$|\delta(f, X) - \delta(g, X)| \le \delta((f + g, X) \le \delta(f, X) + \delta(g, X)$$

and

$$|\delta(f, X) - \delta(g, X)| \le \delta((f - g, X) \le \delta(f, X) + \delta(g, X)$$

Proposition 3.2.6 implies the following result.

Proposition 3.2.21. For any functions $f(x)$ and $g(x)$ and any set X from $\boldsymbol{R}$ where both functions are bounded, the inequalities

$$\delta(f \cdot g, X) \le \delta(f, X) \, \| g(x) \| + \delta(g, X) \, \| f(x) \| \qquad (3.11)$$

is valid where $\| f(x) \| = \sup \{ \, | f(x) | \, ; x \in X\}$ and $\| g(x) \| = \sup \{ \, |g(x)| \, ; x \in X\}$.

3.3. Fuzzy Continuous Functions

And the continuity of our science
has not been affected by all these turbulent happenings,
as the older theories have always been included
as limiting cases in the new ones.
Max Born (1882-1970)

Similar to the situation with continuity (cf. Section 3.1), there are three approaches to the definition of fuzzy continuous functions:

- using fuzzy limits, which can be either fuzzy limits of sequences (cf., Section 2.2) or fuzzy limits of functions (cf., Section 2.3);
- deriving and applying discontinuity measures (cf., Section 3.2);
- by the most popular in calculus (ε, δ)-construction.

These three approaches are represented in Definitions 3.3.1 and 3.3.2 (the first approach), Definition 3.3.3 (the second approach), and Definition 3.3.4 (the third approach).

Definition 3.3.1. A partial function f: $\boldsymbol{R} \to \boldsymbol{R}$ is called *r-continuous at a point* $a \in \boldsymbol{R}$ if $f(x)$ is defined at a and for any sequence $l = \{ a_i \in \boldsymbol{R}; i = 1, 2, 3, \dots\}$ that converges to a, the point $f(a)$ is an r-limit of the sequence $\{ f(a_i) \in \boldsymbol{R}; i = 1, 2, 3, \dots\}$.

Example 3.3.1. Let us consider the following function:

$$f(x) = \begin{cases} x^2 & \text{if } x \in [0, 1/2]; \\ x & \text{otherwise.} \end{cases}$$

This function is (1/4)-continuous at the point ½ and 0-continuous at all other points.

It is possible to define r-continuity, using fuzzy limits of functions (Section 2.3).

Definition 3.3.2. A partial function f: $\boldsymbol{R} \to \boldsymbol{R}$ is called *r-continuous at a point* $a \in \boldsymbol{R}$ if $f(x)$ is defined at a and $f(a) = r\text{-}\lim_{x\to a} f(x)$.

It is also possible to give an (ε, δ)-definition of r-continuity similar to the traditional (ε, δ)-definition of continuity.

Definition 3.3.3. A partial function f: $\boldsymbol{R} \to \boldsymbol{R}$ is called *r-continuous at a point* $a \in \boldsymbol{R}$ if $f(x)$ is defined at a and for any $\varepsilon > 0$, there is $\delta > 0$ such that the inequality $| a - x | < \delta$ implies the inequality $| f(x) - f(a) | < r + \varepsilon$, or in other words, for any x with $| a - x | < \delta$, we have $| f(x) - f(a) | < r + \varepsilon$.

One more definition of local fuzzy continuity is essentially topological (cf. (Burgin, 2004a)).

Definition 3.3.4. A partial function f: $\boldsymbol{R} \to \boldsymbol{R}$ is called *r-continuous at a point* $a \in \boldsymbol{R}$ if $f(x)$ is defined at a and for any open interval (b, c) that contains the interval $[a - r, a + r]$, there is an open interval (u, v) such that (u, v) contains a and $f(x)$ maps (u, v) into (b, c).

This implies the following result because any open interval contains a closed interval and vice versa.

Corollary 3.3.1. If function $f(x)$ is continuous at a point a, then for any closed interval $[b, c]$ that properly contains the interval $[a - r, a + r]$, there is a closed interval $[u, v]$ such that $[u, v]$ contains a and f maps $[u, v]$ into $[b, c]$.

Taking the function $f(x)$ from Example 3.3.1, we see that it is (1/4)-continuous at the point ½ and 0-continuous at all other points with respect to any of the given definitions. This is a general case as the following result shows.

Proposition 3.3.1. Definitions 3.3.1, 3.3.2, 3.3.3 and 3.3.4 are equivalent, i.e., if a function $f(x)$ is r-continuous at a point according to one of these definitions, then $f(x)$ is r-continuous at the same point according to another of these definitions.

Proof. a) Equivalence of Definitions 3.3.1 and 3.3.2 follows from the definition of an r-limit of a function (cf., Section 2.3).

b) Definition 3.3.1 $\to$ Definition 3.3.3. Let us assume that $f(x)$ is not r-continuous at the point a according to Definition 3.3.2. It means that there is $\varepsilon > 0$ such that for any $\delta > 0$ there is a point x with $| x - a| < \delta$ but $| f(x) - f(a) | \geq r + \varepsilon$. Let us take a sequence of such δ equal to ½ , 1/3 , ¼ , … , $1/n$, …

This gives us a numerical sequence $l = \{x_i ; i = 1, 2, 3, \dots \}$ in which $| x_i - a| < 1/i$, but $|f(x_i) - f(a) | \geq r + \varepsilon$. Consequently (cf., Sections 2.1 and 2.2), $a = \lim l$, but the sequence $h = \{ f(x_i) ; i = 1, 2, 3, \dots \}$ does not r-converge to $f(a)$. This violates the condition from Definition 3.3.1. So, Definition 3.3.1 implies Definition 3.3.3 because we have demonstrated that it is impossible that for an arbitrary function $f(x)$, Definition 3.3.1 is true for $f(x)$ at the point a and at the same time, Definition 3.3.3 is not true for $f(x)$ at the point a.

c) Definition 3.3.3 → Definition 3.3.4. Let us assume that $f(x)$ is not r-continuous at the point a according to Definition 3.3.4. It means that there is an interval (b, c) that contains the interval $[a - r, a + r]$ such that in any interval (u, v) containing a there is, at least, one point d such that $f(d)$ does not belong to the interval (b, c).

Let us put $\varepsilon = \min \{ |b - f(a) - r| ; | c - f(a) + r | \}$. Then for any $\delta > 0$ there is a point e with $| e - a| < \delta$ but $| f(e) - f(a) | \geq r + \varepsilon$ because we can take the interval $(a - \delta, a + \delta)$ and as it is demonstrated, it has a point e the image of which $f(e)$ does not belong to the interval (b, c). This violates the condition from Definition 3.3.3. So, Definition 3.3.3 implies Definition 3.3.4 because we have demonstrated that it is impossible that for an arbitrary function $f(x)$, Definition 3.3.3 is true for $f(x)$ at the point a and at the same time, Definition 3.3.4 is not true for $f(x)$ at the point a.

d) Definition 3.3.4 → Definition 3.3.1. Let us assume that $f(x)$ is not r-continuous at the point a according to Definition 3.3.1. It means that there is a sequence $l = \{x_i ; i = 1, 2, 3, \dots \}$ such that $a = \lim l$, but the sequence $h = \{ f(x_i) ; i = 1, 2, 3, \dots \}$ does not r-converge to $f(a)$. By Definition 2.2.1, this implies that outside some neighborhood Oa of the point a that contains the interval $[a - r, a + r]$, there are infinitely many elements from h.

Then whatever small interval (u, v) containing a is taken, only a finite number of elements from l will outside (u, v). Thus, an infinite number of elements from (u, v) will be outside the neighborhood Oa. It means that the condition from Definition 3.3.4 is not satisfied and $f(x)$ is not r-continuous at the point a according to Definition 3.3.4. So, Definition 3.3.4 implies Definition 3.3.1 because we have demonstrated that it is impossible that for an arbitrary function $f(x)$, Definition 3.3.4 is true for $f(x)$ at the point a and at the same time, Definition 3.3.1 is not true for $f(x)$ at the point a.

Thus, we have demonstrated that Definitions 3.3.1 and 3.3.2 are equivalent, Definition 3.3.1 → Definition 3.3.3, Definition 3.3.3 → Definition 3.3.4, and Definition 3.3.4 → Definition 3.3.1. As the logical implication → is a transitive relation, all these definitions are equivalent.

Proposition is proved.

Proposition 3.3.1 shows that it is possible to take one of the conditions from Definitions 3.3.1 - 3.3.4 as a condition that defines r-continuity and to consider other conditions as derived properties of r-continuous functions or as criteria of r-continuity.

Remark 3.3.1. We define fuzzy continuity by utilizing three different approaches: fuzzy limits, neighborhoods and by the (ε, δ)-definition. For real functions, all these definitions result in the same concept of fuzzy continuity. However, in a more general situation of topological spaces, these three definitions bring us to different structures. Neighborhood fuzzy continuity is directly extended to fuzzy continuous mappings of scalable topological spaces (Burgin, 2004a; 2005a; 2006). Fuzzy limits define fuzzy continuity only for sequential scalable topology, while the (ε, δ)-definition can be extended only to metric spaces (Burgin, 1995) but not to general topological spaces.

Remark 3.3.2. It is possible to define fuzzy continuity even when a function $f(x)$ is not defined at a point a.

Lemma 3.3.1. A function $f(x)$ is continuous at a point $a \in \boldsymbol{R}$ if and only if it is 0-continuous at the point a.

Indeed, if $f(x)$ is 0-continuous at the point a, then for any $\varepsilon > 0$ there is $\delta > 0$ such that the inequality $| a - x | < \delta$ implies the inequality $| f(x) - f(a) | < 0 + \varepsilon = \varepsilon$. This means that $f(x)$ is continuous at the point a. The inverse is also true. Thus, the condition of 0-continuity coincides with the condition of continuity.

This result shows that the concept of an r-continuity is a natural extension of the concept of conventional continuity.

Remark 3.3.3. Some can argue that only such fuzzy continuity is interesting (and meaningful) where the parameter r is small. However, understanding what is small is relative. For instance, for an individual, 1000 years is a very big period of time. At the same time, like any good scientific measurement, every dated boundary in the geological time scale has an uncertainty associated with it. This uncertainty is expressed as "± X millions of years" (cf. (Harland *et al.*, 1990)). It means that the case of the geological time scale the "small" parameter r is measured in millions of years.

Definition 3.3.1 implies the following result.

Lemma 3.3.2. If $q > r$, then any r-continuous at a point a function is also q-continuous at a.

Indeed, if the inequality $| a - x | < \delta$ implies the inequality $| f(x) - f(a) | < r + \varepsilon$, then the inequality $| a - x | < \delta$ implies the inequality $| f(x) - f(a) | < q + \varepsilon$ because $q > r$. Thus, by Definition 3.3.3, if $f(x)$ is r-continuous at a point a function, then it is also q-continuous at a.

The measure $\delta\ (f, a)$ of discontinuity of a function $f(x)$ at a point considered in Section 3.2 gives an exact measure for r-continuity of $f(x)$.

Proposition 3.3.2. If $\delta\ (f, a) = p$, then the function $f(x)$ is p-continuous at the point a and is not r-continuous at the point a for any $r < p$.

Proof. 1. If a sequence $\{a_i\ ;\ i = 1, 2, 3, \dots \}$ converges to a, then by the definition of supremum (cf. Appendix B), for all partial limits c_j of the sequence $l = \{ f(a_i);\ i = 1, 2, 3, \dots \}$, we have $| f(a) - c_j | < p = \delta(f, a)$. Thus (cf. Section 2.2), $f(a) = r\text{-}\lim_{i\to\infty} f(a_i)$. By Definition 3.3.1, the function $f(x)$ is p–continuous.

2. Now let us assume that $f(x)$ is a p–continuous function. If $r < p$, then $p - r = k > 0$. As $\delta(f, a) = \sup \{ \lim_{i\to\infty} | f(a) - f(a_i)|\ ;$ where $a_i \in \boldsymbol{R}$ and $\lim_{i\to\infty} a_i = a \}$, there is a sequence $\{a_i\ ;\ i = 1, 2, 3, \dots \}$ such that $a = \lim_{i\to\infty} a_i$, the limit $\lim_{i\to\infty} f(a_i)$ exists and by the definition of supremum (cf. Appendix B), if $\lim_{i\to\infty} f(a_i) = c$, then $| f(a) - c| > p - k/3$. Consequently, $| f(a) - c| > r + k/3$. This contradicts to Definition 3.3.1 of an r–limit. Such a contradiction shows that the function $f(x)$ cannot be r–continuous.

Proposition is proved.

It is also possible to directly introduce fuzzy continuity based on some measure of discontinuity or of continuity. Taking some local measure of continuity τ or local measure of discontinuity ρ (cf., Section 3.2), we can define relative fuzzy continuous functions.

Definition 3.3.5. A function $f\colon \boldsymbol{R} \to \boldsymbol{R}$ is called $[\rho, r]$-*continuous* ($[\tau, r]$-*continuous*) at a point $a \in \boldsymbol{R}$ if $\rho(f, a) \leq r$ (correspondingly, $\tau(f, a) \geq 1/(1+ r)$) and is called ρ-*continuous* (τ-continuous) at a if $\rho(f, a) \leq \infty$ ($\tau(f, a) > 0$).

When we take r-continuous functions for all r, we come to the class of fuzzy continuous functions.

Definition 3.3.6. a) A partial function $f\colon \boldsymbol{R} \to \boldsymbol{R}$ is called *fuzzy continuous* at a point $a \in \boldsymbol{R}$ if $f(x)$ is r-continuous at a for some $r \in \boldsymbol{R}^+$.

b) A partial function f: $\boldsymbol{R} \rightarrow \boldsymbol{R}$ is called *fuzzy* ρ-*continuous* (τ-*continuous*) at a point $a \in \boldsymbol{R}$ if $f(x)$ is $[\rho, r]$-continuous ($[\tau, r]$-continuous) at a for some $r \in \boldsymbol{R}^+$.

Inequalities that exist between different measures of discontinuity considered in Section 3.2 (cf. Propositions 3.2.8 – 3.2.10) show that all these measure determine one and the same set of fuzzy continuous functions. Thus, we use only one term "a fuzzy continuous function".

Let us take the measure δ (f, x) of discontinuity of a function at a point considered in Section 3.2. Then Proposition 3.3.2 implies the following result.

Proposition 3.3.3. a) A function $f(x)$ is $[\delta, r]$-continuous at a point a if and only if $f(x)$ it is r-continuous at the point a.

b) A function $f(x)$ is δ-continuous at a point a if and only if $f(x)$ it is fuzzy continuous at the point a.

From local r-continuity, we naturally come to global r-continuity.

Let X be a subset of $\boldsymbol{R}$.

Definition 3.3.7. A partial function f: $\boldsymbol{R} \rightarrow \boldsymbol{R}$ is called *r-continuous in* (*inside*) X if $f(x)$ (the restriction of $f(x)$ on X) is r-continuous at each point a from $X \cap \text{Dom} f$.

Example 3.3.2. The function from example 3.3.1 is (1/4)-continuous on $\boldsymbol{R}$.

Example 3.3.3. To define the Riemann integral for a continuous function $f(x)$, step functions are utilized (cf., Chapter 7). If the integral of $f(x)$ exists, then any such step function is fuzzy continuous.

Lemma 3.3.1 implies the following result.

Corollary 3.3.2. A function $f(x)$ is continuous at a point $a \in \boldsymbol{R}$ if and only if it is 0-continuous at the point a.

Lemma 3.3.2 implies the following result.

Corollary 3.3.3. If $q > r$, then any r-continuous (in X) function is also q-continuous (in X).

Definition 3.3.8. A partial function f: $\boldsymbol{R} \rightarrow \boldsymbol{R}$ is called *locally fuzzy continuous in* (*inside*) X if at each point a from $X \cap \text{Dom} f$, there is a number $r \in \boldsymbol{R}^+$ such that $f(x)$ (the restriction of $f(x)$ on X) is r-continuous at a for some $r \in \boldsymbol{R}^+$.

From local fuzzy continuity, we naturally come to global fuzzy continuity.

Definition 3.3.9. A partial function f: $\boldsymbol{R} \rightarrow \boldsymbol{R}$ is called *globally fuzzy continuous in* (*inside*) X if there is a number $r \in \boldsymbol{R}^+$ such that at each point a from $X \cap \text{Dom} f$, the function $f(x)$ (the restriction of $f(x)$ on X) is r-continuous at a.

Remark 3.3.4. If X contains more than one point, then for each $r \in \boldsymbol{R}^+$ every r-continuous on X function is fuzzy continuous on X, but not every fuzzy continuous on X function is r-continuous on X for some $r \in \boldsymbol{R}^+$. It means that in contrast to continuous functions where functions continuous at each point are globally continuous, it is possible that a fuzzy continuous at each point function is not globally fuzzy continuous.

Remark 3.3.5. Fuzzy continuous functions are also related to roughly continuous functions studied in (Pawlak, 1995).

Example 3.3.4. Let us consider the function $f(x) = x^n$ when $x \in [n, n + 1)$, $n \in \boldsymbol{Z}$ and the function $g(x) = [x]^n$. These functions are fuzzy continuous at each point of $\boldsymbol{R}$, but they are not fuzzy continuous on $\boldsymbol{R}$.

Remark 3.3.6. It is necessary to make a distinction between fuzzy continuous function defined and studied in this section and fuzzy functions that are continuous with respect to some fuzzy topology or convergence. There are several constructions that define fuzzy topology in a set and continuous mappings of fuzzy topological spaces (cf. (Chang, 1968; Goguen, 1973; Lowen, 1976; Šostak, 1985; 1989)). In some cases, fuzzy topology in a set X is a family of fuzzy subsets, while in others, a fuzzy topology in a set X is a mapping of the set of all subsets of X into the unit interval [0, 1]. Consequently, there are even more different kinds of continuous fuzzy functions. For instance, Ekici (2004) analyzes nine types of continuity for fuzzy functions.

However, there are concepts closely related to fuzzy continuity in the sense of this Chapter. For instance, Klee (1961) introduced r-continuous (or ε–continuous or nearly continuous) mappings and studied fixed points of such mappings.

If we want to understand continuity of functions on discrete sets, we need more general construction than fuzzy continuity. Such a construction is extended fuzzy continuity introduced in (Burgin, 1999a). There are three ways to define extended fuzzy continuous functions: the sequential approach presented in Definitions 3.3.8, the most popular in calculus (ε, δ)-construction presented in Definitions 3.3.9, and the topological definition based on the concept of a scalable topological space (Burgin, 2004a; 2006). Here we consider only two first definitions because scalable topological spaces are highly abstract mathematical objects that go far beyond the scope of this book.

At first, we give the sequential definition.

Definition 3.3.10. a) A partial function f: $\boldsymbol{R} \to \boldsymbol{R}$ is called (q, r)-*continuous at a point* $a \in \boldsymbol{R}$ if for any sequence $l = \{ a_i \in \boldsymbol{R}; i = 1, 2, 3, \ldots\}$, for which a is an q-limit, the point $f(a)$ is an r-limit of the sequence $\{f(a_i) \in \boldsymbol{R}; i = 1, 2, 3, \ldots\}$.

b) A partial function f: $\boldsymbol{R} \to \boldsymbol{R}$ is called 2-*fuzzy continuous at a point* $a \in \boldsymbol{R}$ if it is (q, r)-continuous at a for any some q an r.

It is possible to define (q, r)-continuous functions without explicit use of limits as it is done by means of the (ε, δ)-construction.

Definition 3.3.11. a) A function f: $\boldsymbol{R} \to \boldsymbol{R}$ is called (q, r)-*continuous at a point* $a \in \boldsymbol{R}$ if for any $\varepsilon > 0$ there is $\delta > 0$ such that the inequality $| a - x | < q + \delta$ implies the inequality $| f(x) - f(a) | < r + \varepsilon$, or in other words, for any x with $| a - x | < q + \delta$, we have $| f(x) - f(a) | < r + \varepsilon$.

b) A partial function f: $\boldsymbol{R} \to \boldsymbol{R}$ is called 2-*fuzzy continuous at a point* $a \in \boldsymbol{R}$ if it is (q, r)-continuous at a for any some q an r.

In a similar way as Proposition 3.3.1, we can prove the following statement.

Proposition 3.3.3. Definitions 3.3.10 and 3.3.11 are equivalent.

Lemma 3.3.3. a) A function $f(x)$ is r-continuous at a point $a \in \boldsymbol{R}$ if and only if it is $(0, r)$-continuous at the point a.

b) A function $f(x)$ is continuous at a point $a \in \boldsymbol{R}$ if and only if it is (0,0)-continuous at the point a.

These results show that the concept of (q, r)-continuity is a natural extension of the concepts of conventional continuity and r-continuity.

Definition 3.3.11 implies the following result.

Lemma 3.3.4. If $t \geq r$, and $p \leq q$, then any (q, r)-continuous at a function $f(x)$ is also (p, t)-continuous at a.

Indeed, if that the inequality $| a - x | < q + \delta$ implies the inequality $| f(x) - f(a) | < r + \varepsilon$, then the inequality $| a - x | < p + \delta$ implies the inequality $| a - x | < q + \delta$ because $p \leq q$. The inequality $| a - x | < q + \delta$ implies the inequality $| f(x) - f(a) | < r + \varepsilon$ as $f(x)$ is (q, r)-continuous at a, which, in turn, implies the inequality $| f(x) - f(a) | < t + \varepsilon$ because $t \geq r$. Thus, by Definition 3.3.11, if $f(x)$ is (q, r)-continuous at a point a function, then it is also (p, t)-continuous at a.

Corollary 3.3.4. If $q > l$ ($r < p$), then any (q, r)-continuous at a point $a \in \boldsymbol{R}$ function is also (l, r)-continuous (also (q, p)-continuous) at the point a.

Corollary 3.3.5. If $q > l$ and $r < p$, then any (q, r)-continuous function is also (l, p)-continuous.

Corollary 3.3.6. If a function $f(x)$ is (q, r)-continuous at a point $a \in \boldsymbol{R}$, then $f(x)$ is r-continuous at the point a.

Note that if $q < p$, then it is possible that a (q, r)-continuous at a function is not (p, r)-continuous at a. For instance, the function $f(x) = x$ is $(0,0)$-continuous at the point 0, but for any $p > 0$, it is not $(p,0)$-continuous at 0.

Lemma 3.3.5. Any function $f: X \rightarrow \boldsymbol{R}$ is (q, r)-continuous at a point $a \in X$, if $\mathbf{d}(a, X \backslash \{a\}) = u$ and $q < u$.

Let us assume that $\varepsilon \in \boldsymbol{R}^{++}$.

Definition 3.3.12 (Klee, 1961). A function $f: \boldsymbol{R} \rightarrow \boldsymbol{R}$ is called ε-*continuous* at a point $a \in \boldsymbol{R}$ if there is $\delta > 0$ such that the inequality $| a - x | \leq \delta$ implies the inequality $| f(x) - f(a) | \leq \varepsilon$, or in other words, for any x with $| a - x | \leq \delta$, we have $| f(x) - f(a) | \leq \varepsilon$.

There is a transparent relation between ε-continuous functions in the sense of Klee and 2-fuzzy continuous functions defined in this book.

Proposition 3.3.4. A function $f: \boldsymbol{R} \rightarrow \boldsymbol{R}$ is ε-continuous at a point $a \in \boldsymbol{R}$ if and only if there is a number $q \in \boldsymbol{R}$ such that the function $f(x)$ is (q, r)-continuous at a point a for any $r > \varepsilon$.

Proof. 1. Let us assume that $f: \boldsymbol{R} \rightarrow \boldsymbol{R}$ is an ε-continuous at a point $a \in \boldsymbol{R}$ function and $\varepsilon < r$. Then by Definition 3.3.12, there is $\delta > 0$ such that the inequality $| a - x | \leq \delta$ implies the inequality $| f(x) - f(a) | \leq \varepsilon$. Let us take some $q < \delta$ and show that the function $f(x)$ is (q, r)-continuous at the point a. To show this, we need to take an arbitrary $\eta > 0$ and to find a number $\nu > 0$ such that the inequality $| a - x | < q + \nu$ implies the inequality $| f(x) - f(a) | < r + \eta$, Indeed, taking any $\nu > 0$ such that $q + \nu < \delta$, we have that the inequality $| a - x | < q + \nu$ implies the inequality $| f(x) - f(a) | \leq \varepsilon$. Then $| a - x | < q + \nu$ implies $|f(x) - f(a) | < r + \eta$ as $\varepsilon < r + \eta$. Thus by Definition 3.3.11, the function f is (q, r)-continuous at the point a.

2. Let us assume that $f: \boldsymbol{R} \rightarrow \boldsymbol{R}$ is a (q, r)-continuous at a point a function for some $q > 0$ and any $r > \varepsilon$. Then by Definition 3.3.11, for any $\nu > 0$ there is $\delta > 0$ such that the inequality $| a - x | < q + \delta$ implies the inequality $| f(x) - f(a) | < r + \nu$. We show that the function $f(x)$ is ε-continuous at the point a as the inequality $| a - x | \leq q$ implies the inequality $| f(x) - f(a) | \leq \varepsilon$. Assume that this is not true. Then there is a number x such that $| a - x | \leq q$ and $| f(x) - f(a) | > \varepsilon$. In this case, we can take some $r > \varepsilon$ and $\eta > 0$ such that $| f(x) - f(a) | > r + \eta$. This contradicts our assumption that the function $f(x)$ is (q, r)-continuous at the point a as $q < q + \delta$ for any $\delta > 0$.

Proposition is proved.

Definition 3.3.13 (Klee, 1961). A function $f: \boldsymbol{R} \rightarrow \boldsymbol{R}$ is called *nearly continuous* at a point $a \in \boldsymbol{R}$ if $f(x)$ is ε-continuous at a for some $\varepsilon \in \boldsymbol{R}^{+}$.

Proposition 3.3.4 implies the following result.

Corollary 3.3.7. A function $f: \boldsymbol{R} \to \boldsymbol{R}$ is fuzzy 2-continuous at a point $a \in \boldsymbol{R}$ if and only if $f(x)$ is nearly continuous at this point.

From fuzzy continuity at a point, we naturally come to fuzzy continuity in a set. We introduce here several concepts of fuzzy continuity for further study. Different properties determine corresponding classes of functions that more or less preserve some properties of continuous functions. We do not study here properties of these constructions not because they are not important but due to the limitations on the space in the book. At first, we consider local fuzzy continuity in a set.

Definition 3.3.14. A partial function $f: \boldsymbol{R} \to \boldsymbol{R}$ is called *locally fuzzy r-continuous in* (*inside*) X if for each point a from X, there is a number $q \in \boldsymbol{R}^+$ such that the function $f(x)$ (the restriction of the function $f(x)$ on X) is (q, r)-continuous at a for some $r \in \boldsymbol{R}^+$.

Lemma 3.3.6. Any locally fuzzy r-continuous in X (inside X) function $f(x)$ is r-continuous in X (inside X).

Corollary 3.3.5. Any locally fuzzy r-continuous in X (inside X) function $f(x)$ is fuzzy continuous in X (inside X).

Lemma 3.3.7. If $r < p$, then any locally fuzzy r-continuous in X (inside X) function is also locally fuzzy p-continuous in X (inside X).

Definition 3.3.15. A partial function $f: \boldsymbol{R} \to \boldsymbol{R}$ is called *locally 2-fuzzy q-continuous in* (*inside*) X if for each point a from $X \cap \text{Dom} f$, there is a number $r \in \boldsymbol{R}^+$ such that the function $f(x)$ (the restriction of the function $f(x)$ on X) is (q, r)-continuous at a for some $q \in \boldsymbol{R}^+$.

Lemma 3.3.8. If $q > l$, then any locally 2-fuzzy q-continuous in X (inside X) function is also locally 2-fuzzy l-continuous in X (inside X).

Definition 3.3.16. A partial function $f: \boldsymbol{R} \to \boldsymbol{R}$ is called *locally 2-fuzzy continuous* in X (inside X) if for each point a from $X \cap \text{Dom} f$, there are numbers $r, q \in \boldsymbol{R}^+$ such that the function $f(x)$ (the restriction of the function $f(x)$ on X) is (q, r)-continuous at a for some $r, q \in \boldsymbol{R}^+$.

Lemma 3.3.9. Any locally 2-fuzzy continuous in X (inside X) function $f(x)$ is fuzzy continuous in X (inside X).

Definition 3.3.17. A function $f: \boldsymbol{R} \to \boldsymbol{R}$ is called (q, r)*-continuous in* (*inside*) set $X \subseteq \boldsymbol{R}$ if $f(x)$ (the restriction of $f(x)$ on X) is (q, r)-continuous at each point a from $X \cap \text{Dom} f$.

Remark 3.3.8. More general (with respect to spaces) definitions of fuzzy continuity are given and studied for general metric spaces in Section 3.7.

Lemma 3.3.10. If $X = \{ ku; k = 0, \pm 1, \pm 2, \pm 3, \dots \}$ or $X = \{ ku; k \in \boldsymbol{Z}$ and $m < k < n \}$ and $q < u$, then any function $f: X \to \boldsymbol{R}$ is (q, r)-continuous inside X.

Corollary 3.3.8. Any function $f: X \to \boldsymbol{R}$ defined only on a discrete set $X = \{ ku; k = 0, \pm 1, \pm 2, \pm 3, \dots \}$ or $X = \{ ku; k \in \boldsymbol{Z}$ and $m < k < n \}$ is (q, r)-continuous if $q < u$.

Definition 3.3.18. A function $f: \boldsymbol{R} \to \boldsymbol{R}$ is called *2-fuzzy continuous in* (*inside*) X if for some $r, q \in \boldsymbol{R}^+$ and each point a from X, the function $f(x)$ (the restriction of the function $f(x)$ on X) is (q, r)-continuous at a.

Lemma 3.3.11. Any 2-fuzzy continuous in X (inside X) function $f(x)$ is fuzzy continuous in X (inside X).

From local ε-continuity, we naturally come to global ε-continuity.

Definition 3.3.19 (Klee, 1961; Klee and Yandl, 1974)**.** a) A function $f: \boldsymbol{R} \to \boldsymbol{R}$ is called ε-*continuous* in a set $X \subseteq \boldsymbol{R}$ if there is $\delta > 0$ such that for any points x and y from X, the inequality $|x - y| \leq \delta$ implies the inequality $|f(x) - f(y)| \leq \varepsilon$, or in other words, for any x with $|x - y| \leq \delta$, we have $|f(x) - f(y)| \leq \varepsilon$.

b) A function f: $\boldsymbol{R} \to \boldsymbol{R}$ is called *globally* ε*-continuous* in a set $X \subseteq \boldsymbol{R}$ if it is ε-continuous at each point a from X.

c) A function f: $\boldsymbol{R} \to \boldsymbol{R}$ is called *nearly continuous* in a set X if $f(x)$ is ε-continuous in X for some $\varepsilon \in \boldsymbol{R}^+$.

d) A function f: $\boldsymbol{R} \to \boldsymbol{R}$ is called *globally nearly continuous* in a set X if $f(x)$ is globally ε-continuous in X for some $\varepsilon \in \boldsymbol{R}^+$.

Proposition 3.3.4 implies the following result.

Corollary 3.3.9. a) A function f: $\boldsymbol{R} \to \boldsymbol{R}$ is globally ε-continuous in a set X if and only if for each point a in X, there is a number $q \in \boldsymbol{R}^+$ such that the function $f(x)$ is (q, r)-continuous at a for any $r > \varepsilon$.

b) A function f: $\boldsymbol{R} \to \boldsymbol{R}$ is 2-fuzzy continuous in a set X if and only if $f(x)$ is globally nearly continuous in a set X.

c) A function f: $\boldsymbol{R} \to \boldsymbol{R}$ is ε-continuous in a set X if and only if there is a number $q \in \boldsymbol{R}^+$ such that the function $f(x)$ is uniformly (q, r)-continuous in X for any $r > \varepsilon$.

d) A function f: $\boldsymbol{R} \to \boldsymbol{R}$ is uniformly 2-fuzzy continuous in a set X if and only if $f(x)$ is globally nearly continuous in a set X.

Definition 3.3.20. a) A function f: $\boldsymbol{R} \to \boldsymbol{R}$ is called *left* (*right*) *r-continuous* or *r-continuous from the left* (*right*) *at a point* $a \in \boldsymbol{R}$ if for any sequence $l = \{ a_i \in \boldsymbol{R};\ a_i < a\ (a_i > a),\ i = 1, 2, 3, \ldots \}$ that converges to a, the point $f(a)$ is a r-limit of the sequence $\{ f(a_i) \in \boldsymbol{R};\ i = 1, 2, 3, \ldots \}$ or formally, for any sequence $l = \{ a_i \in \boldsymbol{R};\ a_i < a\ (a_i > a),\ i = 1, 2, 3, \ldots \}$, $\lim_{i\to\infty} a_i = a$ implies $r\text{-}\lim_{i\to\infty} f(a_i) = f(a)$.

b) A function f: $\boldsymbol{R} \to \boldsymbol{R}$ is called *left* (*right*) *(q, r)-continuous* or *(q, r)-continuous from the left* (*right*) *at a point* $a \in \boldsymbol{R}$ if for any sequence $l = \{ a_i \in \boldsymbol{R};\ a_i < a\ (a_i > a),\ i = 1, 2, 3, \ldots \}$ that q-converges to a, the point $f(a)$ is a r-limit of the sequence $\{ f(a_i) \in \boldsymbol{R};\ i = 1, 2, 3, \ldots \}$ or formally, for any sequence $l = \{ a_i \in \boldsymbol{R};\ a_i < a\ (a_i > a),\ i = 1, 2, 3, \ldots \}$, $q\text{-}\lim_{i\to\infty} a_i = a$ implies $r\text{-}\lim_{i\to\infty} f(a_i) = f(a)$.

Remark 3.3.9. One-sided local fuzzy and 2-fuzzy continuity of functions allows one to define one-sided global fuzzy and 2-fuzzy continuity of functions in a natural way.

Proposition 3.3.5. a) A function $f(x)$ is r-continuous from the left (right) at a point $a \in \boldsymbol{R}$ if and only if it is $(0, r)$-continuous from the left (right) at the point a.

b) A function $f(x)$ is continuous from the left (right) at a point $a \in \boldsymbol{R}$ if and only if it is $(0,0)$-continuous from the left (right) at the point a.

These results show that the concept of left (right) (q, r)-continuity is a natural extension of the concept of conventional left (right) continuity.

Definition 3.3.20 implies the following result.

Proposition 3.3.6. If $t > r$, and $p < q$, then any (q, r)-continuous from the left (right) at a function $f(x)$ is also (p, t)-continuous from the left (right) at a.

Corollary 3.3.10. If $q > l$ ($r < p$), then any (q, r)-continuous from the left (right) at a point $a \in \boldsymbol{R}$ function is also (l, r)-continuous (also (q, p)-continuous) from the left (right) at the point a.

Corollary 3.3.11. If $q > l$ and $r < p$, then any (q, r)-continuous from the left (right) function is also (l, p)-continuous from the left (right).

Corollary 3.3.12. If a function $f(x)$ is (q, r)-continuous from the left (right) at a point $a \in \mathbf{R}$, then $f(x)$ is r-continuous from the left (right) at the point a.

3.4. Properties of Fuzzy Continuous Functions

The movement of humanity, arising as it does
from innumerable arbitrary human wills, is continuous.
Lev Tolstoy (1828 -1910)

The Local Weierstrass Theorem (Theorem 3.1.2) gives only a sufficient condition of boundedness of a function at a point. Examples show that this is not a necessary condition for boundedness. Coming to the realm of fuzzy continuous functions, we are able to make the Weierstrass' result complete. The following theorem attains completion of the Local Weierstrass Theorem, giving a criterion of function boundedness at a point.

Theorem 3.4.1. A function $f(x)$ is fuzzy continuous at a point a if and only if it is defined and bounded at this point.

Proof. Necessity. Let us take a function $f(x)$ and assume that it is r-continuous at a point a, but not bounded at this point. A function can be unbounded either from above or from below or from both sides. For convenience, we consider the first case. Two others are treated in a similar way.

Let us consider the sequence of closed intervals $\{[a - 1/n, a + 1/n] ; n = 1, 2, 3, \dots \}$. As $f(x)$ is not bounded at the point a, for any number n, there is a number $c_n \in [a - 1/n, a + 1/n]$ such that $f(c_n) > n$. Thus, we can choose a sequence $l = \{c_i ; i = 1, 2, 3, \dots\}$ such that $f(c_i) > i$ for all $i = 1, 2, 3, \dots$. The lengths of the intervals $\{[a - 1/n, a + 1/n]$ converge to zero. Consequently, the sequence l converges to a. The function $f(x)$ is r-continuous at the point a. It means (cf. Definition 3.3.3) that for any $\varepsilon > 0$ there is $\delta > 0$ such that the inequality $| a - x | < \delta$ implies the inequality $| f(x) - f(a) | < r + \varepsilon$. This inequality implies that $f(x) < f(a) + 2r$ in some small neighborhood of a. However, this not true as the sequence of $f(c_i)$ tends to infinity. This contradiction implies that $f(x)$ has to be bounded at the point a.

Sufficiency. Let us consider a bounded at a point a function $f(x)$. It means that $f(x)$ is a bounded both above and below at the point a. The first condition means that there are a number M and an interval $[b, c]$ such that a belongs to this interval and $f(x) < M$ for all x from the interval $[b, c]$. The second condition means that there are a number m and an interval $[u, v]$ such that a belongs to this interval and $f(x) > m$ for all x from the interval $[u, v]$. Consequently, in the interval $[p, q]$ where $p = \max \{b, u\}$ and $q = \min \{c, v\}$, the equalities $m < f(x) < M$ hold for all $x \in [p, q]$.

Let us take $r = M - m$. In this case, taking any point x inside the interval $[p, q]$, we have $| f(x) - f(a) | < r$. This allows us to conclude that for any $\varepsilon > 0$, there is $\delta > 0$ (namely, $\delta = \min \{|p - a|, |q - a|\}$) such that the inequality $| a - x | < \delta$ implies the inequality $| f(x) - f(a) | < r + \varepsilon$. By Definition 3.3.1, $f(x)$ is r-continuous at the point a.

Theorem is proved.

As any continuous at a point function is fuzzy continuous at this point (Lemma 3.3.1), Theorem 3.4.1 implies the following classical result.

Corollary 3.4.1. A continuous at a point function is bounded at this point.

The first Weierstrass Theorem (Theorem 3.1.1) gives a sufficient condition of boundedness of a function in a closed interval. The following result completes the first Weierstrass Theorem, giving a criterion of boundedness of a function in a closed interval.

Theorem 3.4.2 (the First Fuzzy Weierstrass Theorem). A function f: $[a, b] \to \boldsymbol{R}$ is fuzzy continuous in $[a, b]$ if and only if it is bounded.

Proof. Necessity. Let us take a fuzzy continuous function f: $[a, b] \to \boldsymbol{R}$ and assume that it is not bounded. A function can be unbounded either from above or from below or from both sides. For convenience, we consider the first case. Two others are treated in a similar way. It means that for any number n, there is a number $c_n \in [a, b]$ such that $f(c_n) > n$. Thus, we can choose a sequence $l = \{c_i ; i = 1, 2, 3, \ldots\}$ such that $f(c_i) > i$ for all $i = 1, 2, 3, \ldots$. Properties of a closed interval imply that the sequence l has a converging subsequence $h = \{a_i ; i = 1, 2, 3, \ldots\}$. Let $d = \lim h$. As the function $f(x)$ is fuzzy continuous in $[a, b]$, $f(x)$ is r-continuous at the point d. It means (cf. Definition 3.3.3) that for any $\varepsilon > 0$ there is $\delta > 0$ such that the inequality $| d - x | < \delta$ implies the inequality $| f(x) - f(d) | < r + \varepsilon$. This inequality implies that with $| d - x | < \delta$, we have $f(x) < f(d) + r + \varepsilon$. As the sequence $h = \{a_i ; i = 1, 2, 3, \ldots\}$ converges to d, we have $| d - a_i | < \delta$ for almost all its elements. This implies that $f(a_i) < f(d) + r + \varepsilon$ for almost all $i = 1, 2, 3, \ldots$. At the same time, by the construction of the sequence h, the sequence $k = \{f(a_i); i = 1, 2, 3, \ldots\}$ is unbounded. This contradiction completes the proof of necessity, i.e., it shows that the function $f(x)$ has to be bounded.

Sufficiency. Let us consider a bounded function f: $[a, b] \to \boldsymbol{R}$. Then there are such numbers c and d such that $c < f(x) < d$ for all $x \in [a, b]$. Let us take $p = d - c$. In this case, taking any point u inside the interval $[a, b]$, we have $| f(x) - f(u) | < p$ for all $x \in [a, b]$. This allows us to conclude that for any $\varepsilon > 0$ there is $\delta > 0$ such that the inequality $| u - x | < \delta$ implies the inequality $| f(x) - f(u) | < p + \varepsilon$, or in other words, for any x with $| u - x | < \delta$, we have $| f(x) - f(u) | < p + \varepsilon$. As u is an arbitrary point in $[a, b]$, by Definition 3.3.3, $f(x)$ is p-continuous in $[a, b]$.

Theorem is proved.

Remark 3.4.1. For functions defined in $\boldsymbol{R}$ or in an open interval, Theorem 3.4.2 is not in general true as even a continuous function can be unbounded. For instance, we can take such functions as x^2 in $\boldsymbol{R}$ or $\tan x$ in the interval $(-\pi/2, \pi/2)$.

Such a classical result as the first Weierstrass Theorem (Theorem 3.1.1) is a direct corollary of Theorem 3.4.2 because any continuous function is fuzzy continuous (cf. Lemma 3.3.1).

Let $X \subseteq \boldsymbol{R}$.

Theorem 3.4.3. Any bounded function f: $X \to \boldsymbol{R}$ is fuzzy continuous in X.

Proof is similar to the proof of sufficiency in Theorem 3.4.2.

However, not every fuzzy continuous in X function is bounded when X is not a compact space. For instance, the function $f(x) = [x]$ is fuzzy continuous on $\boldsymbol{R}$, i.e., is 1- continuous, but it is not bounded. In particular, it means that fuzzy continuous functions do not coincide with bounded functions.

Let f: $[a, b] \to \boldsymbol{R}$ be an r-continuous function defined for all elements from the interval $[a, b]$.

Theorem 3.4.4. If $f(a) < 0$ and $f(b) > 0$, then there is [at least one] point $c \in [a, b]$ such that $| f(c)| < r$.

Proof. We consider several possibilities. First, there is some point $c \in [a, b]$, for which $f(c) = 0$. In this case, the statement of the theorem is true.

Second option is that there is no such a point, but the function $f(x)$ is larger (or less) than zero only on a finite set of points x_0 , x_1 , … , x_n from $[a, b]$. Then there is a sequence $l = \{ z_i; i = 1, 2, 3, \dots \}$ of points from $[a, b]$ such that $\lim_{i\to\infty} z_i = c = x_0$ and the value $f(z_i) < 0$ for all $i = 1, 2, 3, \dots$.

As $f(x)$ is an r-continuous function, by Theorem 3.4.2, it is bounded on $[a, b]$. As a closed interval is a compact space, it is possible to choose such a subsequence of the sequence $l = \{ z_i; i = 1, 2, 3, \dots \}$ that the values of $f(x)$ on the elements of this subsequence have a limit and $\lim_{i\to\infty} x_i = c$. Thus, it is possible to assume that the sequence $f(l) = \{ f(x_i) ; i = 1, 2, 3, \dots \}$ has a limit.

Let $\lim_{i\to\infty} f(z_i) = u$. Then by the definition of fuzzy continuity, $|u - f(c)| < r$. All numbers $f(z_i)$ are less than zero. Consequently, $u \leq 0$, while $f(c) > 0$. As a result, we have $u \leq 0 < f(c)$. This implies $|f(c)| = |0 - f(c)| \leq |u - f(c)| < r$. So, in this case, the theorem is also true.

The case when the function $f(x)$ is less than zero only on a finite set of points x_0 , x_1 , … , x_n from $[a, b]$ is considered in a similar way with the same result of validity the theorem.

The last option is when there are infinitely many points in which $f(x)$ is larger than zero and there are infinitely many points in which f is less than zero. Then we can take the subset $A = \{ x \in [a, b] ; f(x) > 0 \}$ and $B = \{ x \in [a, b] ; f(x) < 0 \}$. Then $[a, b] = A \cup B$. Any interval is a connected set. Consequently, there is such a point x_0 that belongs to the closures both of A and of B. By the definition of a closure in $\mathbf{R}$, there are a sequence $l = \{ z_i; i = 1, 2, 3, \dots \}$ of points from B and a sequence $h = \{ x_i; i = 1, 2, 3, \dots \}$ of points from A such that $\lim_{i\to\infty} z_i = c$ and $\lim_{i\to\infty} x_i = c$.

As $f(x)$ is a bounded function on a closed interval $[a, b]$, it is possible to assume that both sequences $f(l) = \{ f(z_i) ; i = 1, 2, 3, \dots \}$ and $f(h) = \{ f(x_i) ; i = 1, 2, 3, \dots \}$ have limits (cf. Section 2.1).

Let $\lim_{i\to\infty} f(z_i) = u$ and $\lim_{i\to\infty} f(x_i) = v$. Then by the definition of fuzzy continuity and properties of the absolute value, we have $|u - f(c)| < r$ and $|v - f(c)| < r$. As $f(c) \neq 0$ (cf. case 1), then either $f(c) > 0$ or $f(c) < 0$.

Let us consider at first the first case. All numbers $f(z_i)$ are less than zero. Consequently, $u \leq 0$, while $f(c) > 0$. As a result, we have $u \leq 0 < f(c)$. This implies the following inequalities $|f(c)| = |0 - f(c)| \leq | u - f(c)| < r$. So in this case, the theorem is true.

In the second case, $f(c) < 0$ and all numbers $f(x_i)$ are larger than zero. Consequently, $v \geq 0$, while $f(c) < 0$. As a result, we have $f(c) < 0 \leq v$. This implies the following inequalities $|f(c)| = |0 - f(c)| \leq | v - f(c)| < r$. So in this case, the theorem is also true.

Theorem is proved because we have considered all possible cases.

Let $f: [a, b] \to \mathbf{R}$ be an r-continuous function defined for all elements from the interval $[a, b]$.

Theorem 3.4.5 (The Fuzzy Intermediate Value Theorem). If $f(a) = k$, $f(b) = h$, and $k < h$ ($k > h$), then for any number l from the interval $[k, h]$ (from the interval $[h, k]$) there is [at least one] point $c \in [a, b]$ such that $|l - f(c)| < r$.

Proof. We prove the theorem for the case when $k < h$ because the case when $k > h$ is proved in a similar way. Let us take a number l from the interval $[k, h]$. If $l = k$, then $f(a) = l$ and we have the necessary result. If $l = h$, then $f(b) = l$ and we have the necessary result.

If $k < l < h$, we consider the function $g(x) = f(x) - l$. For this function, we have $g(a) < 0$ and $g(b) > 0$. Besides, by Definitions 3.3.1 and 3.3.7, g: $[a, b] \to \boldsymbol{R}$ is also an r-continuous function. By Theorem 3.4.4, there is such a point $c \in [a, b]$ that $|0 - g(c)| < r$. The natural metric on $\boldsymbol{R}$ is invariant with respect to translations, i.e., $|a - b| = |(a + c) - (b + c)|$. Consequently, $|l - f(c)| = |l - g(c) + l| = |0 - g(c)| < r$.

Theorem is proved.

In the case when $r = 0$, such a classical result as the Intermediate Value Theorem (Theorem 3.1.4) is a direct corollary of Theorem 3.4.5.

By Lemmas 3.3.3 and 3.3.4, any (q, r)-continuous function is r-continuous. Thus, Theorem 3.4.5 implies the following result.

Corollary 3.4.2. If a function f: $X \to Y$ is (q, r)-continuous inside the interval $[a, b]$, $f(a) = k$, $f(b) = h$, and $k < h$ ($k > h$), then for any number l from the interval $[k, h]$ (from the interval $[h, k]$) there is [at least one] point $c \in [a, b]$ such that $|l - f(c)| < r$.

Functions on discrete sets are very important for numerical computations (cf. Section 1.2). However, we cannot directly apply Theorems 3.4.4 and 3.4.5 to functions on discrete sets because functions treated in these theorems have to be defined for all points of the interval $[a, b]$. To deal with functions on discrete sets, we need to use the concept of (q, r)-continuity, which is finer than the concept of r-continuity and better suits computations with discrete sets.

Let us look what properties of functions on discrete sets can be deduced from Theorem 3.4.5.

Let us consider two discrete sets $X = \{ ku; k = 0, \pm1, \pm2, \pm3, \dots \}$ (or $X = \{ ku; k \in \boldsymbol{Z}$ and $m < k < n \}$) and $Y = \{ kv; k = 0, \pm1, \pm2, \pm3, \dots \}$.

Theorem 3.4.6 (The Discrete Intermediate Value Theorem). For any (q, r)-continuous function f: $X \to Y$, any $a = k_1u$, $b = k_2v$ with $k_1 < k_2$ $(m < k_1 < k_2 < n)$ and any element l that belongs both to the set Y and to the interval $[f(a), f(b)]$ (or to the interval $[f(b), f(a)]$ when $f(a) > f(b)$), there is an element c from X such that $a < c < b$ and $l = f(c)$ if $q \geq u$ and $v \geq r$.

Proof. By Lemma 3.3.4, the function $f(x)$ is r-continuous. Let us extend this function to a function $g(x)$ defined on the whole space $\boldsymbol{R}$ when $X = \{ ku; k = 0, \pm1, \pm2, \pm3, \dots \}$ and to a function defined in the interval $[(m+1)u, nu]$ when $X = \{ ku; m < k < n \}$). The function $g(x)$ is equal to $f(ku)$ when $ku \leq x < (k + 1)u$ and $k = 0, \pm1, \pm2, \pm3, \dots$ (or $k \in \boldsymbol{Z}$ and $m < k < n$), provides such an extension. By Definition 3.4.10, $g(x)$ is (q, r)-continuous and thus, by Lemma 3.3.4, the function $g(x)$ is r-continuous.

Then by Theorem 3.4.5, there is an element c from the interval $[a, b]$ such that $a < c < b$ and $|l - g(c)| < r$. By the definition of the function $g(x)$, we have

$$g(c) = g(ku) = f(ku)$$

for some $k \in \boldsymbol{Z}$ and ku that belongs to the interval $[a, b]$.

Thus we have

$$|l - f(ku)| < r \tag{3.12}$$

As l and $f(ku)$ both belong to the set Y and the distance between points in Y is larger than or equal to r (because $v \geq r$), the inequality (3.12) is possible only if $l = f(ku)$. Thus, we can take $c = ku$ and have $a < c < b$ and $l = f(c)$.

Theorem is proved.

Corollary 3.4.3. If a function $f: X \to Y$ is (u, v)-continuous inside the interval $[a, b]$ with $a = k_1 u$, $b = k_2 v$ and $k_1 < k_2$ ($m < k_1 < k_2 < n$), then for any element l that belongs both to the set Y and to the interval $[f(a), f(b)]$ (or to the interval $[f(b), f(a)]$ when $f(a) > f(b)$), there is an element c from X such that $a < c < b$ and $l = f(c)$.

This implies the Intermediate Value Theorem that was obtained for the case $u = 1$ and $v = 1$ in digital topology (cf. (Rosenfeld, 1979; 1986; Hamlet, 2002)).

Remark 3.4.2. One of the principal results of the classical analysis (and topology) states that the composition of continuous functions is continuous. For fuzzy continuous functions this result is not true.

Example 3.4.1. Let us consider the function $g(x) = n$ when $x \in [n, n + 1)$, $n \in \boldsymbol{Z}$ and the function $f(x) = x^2$. The function $g(x)$ is 1-continuous and the function $f(x)$ is continuous and consequently, 1-continuous and 0-continuous on $\boldsymbol{R}$. However, their composition $f(g(x))$ is not fuzzy continuous on $\boldsymbol{R}$. It means that even a composition of a fuzzy continuous function and a continuous function can be not fuzzy continuous.

However, in a closed interval, the property of the composition of continuous functions remains true for fuzzy continuous functions.

Theorem 3.4.7. The composition of fuzzy continuous functions $f: [a, b] \to \boldsymbol{R}$ and $g: [a, b] \to \boldsymbol{R}$ is fuzzy continuous in $[a, b]$.

Proof. As both functions $f(x)$ and $g(x)$ are fuzzy continuous in $[a, b]$, then by Theorem 3.4.2, they are bounded in $[a, b]$. The composition of bounded functions is a bounded function. Thus, by Theorem 3.4.2, the function $f(g(x))$ is fuzzy continuous in $[a, b]$.

Theorem is proved.

Remark 3.4.3. Topological spaces with continuous mappings as morphisms build a category (Herrlich and Strecker, 1973). In particular, taking the space $\boldsymbol{R}$ with continuous mappings as its morphisms, we obtain a category. However, Example 3.4.1 shows that we cannot build a category with fuzzy continuous mappings as morphisms.

Example 3.4.2. Let us consider the function $g(x) = 2n$ when $x \in [n, n + 1)$, $n \in \boldsymbol{Z}$. This function is 2-continuous in $\boldsymbol{R}$, but its composition $g(g(x))$ with itself is not 2-continuous.

However, composition of functions preserves local fuzzy continuity.

Theorem 3.4.8. If a function $g(x)$ is continuous at a point a and a function $f(x)$ is fuzzy continuous at a point $g(a)$, then the composition $f(g(x))$ is fuzzy continuous at a point a.

Proof. Let us assume that a function $g(x)$ is continuous at a point a and a function $f(x)$ is fuzzy continuous at a point $g(a)$. Then by Theorem 3.4.1, there is an interval $[b, c]$ such that $g(a)$ belongs to this interval and $f(x)$ is bounded in the interval $[b, c]$. At the same time, as $g(x)$ is continuous at a point a, there is an interval $[u, v]$ such that a belongs to this interval and g maps $[u, v]$ into $[b, c]$ (cf. Corollary 3.1.1). Thus, the composition $f(g(x))$ is bounded in the interval $[u, v]$. Then, by theorem 3.4.1, the composition $f(g(x))$ is fuzzy continuous at a point a.

Theorem is proved.

Theorem 3.4.2 and the proof of Theorem 3.4.8 show that taking all finite closed intervals in the space $\boldsymbol{R}$ and their total fuzzy continuous mappings as morphisms, we obtain a category.

(q, r)-continuity of $f(x)$ implies a definite uniformity of $f(x)$ in the interval with the length q and center in a.

Lemma 3.4.6. If $f(x)$ is (q, r)-continuous at a point a function, then for any points x_1 and x_2 from the interval $[a - q, a + q]$, we have $| f(x_1) - f(x_2) | < 2r$, i.e., the variance $\omega(f)$ of $f(x)$ in the interval $[a - q, a + q]$ is not larger than $2r$.

Let f: $\boldsymbol{R} \to \boldsymbol{R}$ and g: $\boldsymbol{R} \to \boldsymbol{R}$ be two functions and assume that k is a real number.

Theorem 3.4.9. If the function $f(x)$ is (q, r)-continuous at a point a and the function $g(x)$ is (p, h)-continuous at the point a, then:

a) the function $(f + g)(x)$ is $(u, r + h)$-continuous at the point a where $u = \min\{q, p\}$;
b) the function $(f - g)(x)$ is $(u, r + h)$-continuous at the point a where $u = \min\{q, p\}$;
c) the function kf is $(q, |k|\cdot r)$-continuous at the point a.

Proof. a) Let us take a sequence $l = \{a_i\,;\, i = 1, 2, 3, \ldots\}$ such that $a = q$-lim l and $a = p$-lim l. Then $a = u$-lim l where $u = \min\{q, p\}$. By Definition 3.3.8, the point $f(a)$ is an r-limit of the sequence $\{ f(a_i) \in \boldsymbol{R};\, i = 1, 2, 3, \ldots\}$ and the point $g(a)$ is an h-limit of the sequence $\{ g(a_i) \in \boldsymbol{R};\, i = 1, 2, 3, \ldots\}$. Then by Theorem 2.2.5, the point $f(a) + g(a)$ is an $(r+h)$-limit of the sequence $\{ f(a_i) + g(a_i);\, i = 1, 2, 3, \ldots\}$. Thus, by Definition 3.3.8, the function$(f + g)(x)$ is $(u, r + h)$-continuous at the point a.

Proofs of parts b) and c) are similar and are based on Theorem 2.2.5.

Remark 3.4.4. When the conditions of Theorem 3.4.9 are satisfied, the function $(f - g)(x)$ is not necessarily $(r - q)$-continuous at a. However, in some cases, it is so. Besides, $(f + g)(x)$ and $(f - g)(x)$ can be continuous at some point where neither of them is continuous.

Example 3.4.3. Let us consider the following functions

$$f(x) = \begin{cases} 0 & \text{if } x \text{ is a rational number;} \\ 1 & \text{if } x \text{ is an irrational number} \end{cases}$$

and

$$g(x) = \begin{cases} 0 & \text{if } x \text{ is a rational number;} \\ -1 & \text{if } x \text{ is an irrational number.} \end{cases}$$

Both functions $f(x)$ and $g(x)$ are 1-continuous but not continuous at the point 0. At the same time, $(f + g)(x)$ is continuous at 0 and $(f - f)(x)$ is (1-1)-continuous, that is, 0-continuous at 0.

Theorem 3.4.9 and Lemma 3.4.3 imply the following result because any (0,0)-continuous function is continuous.

Corollary 3.4.4. (any course of the calculus, cf., for example, (Ribenboim, 1964; Fihtengoltz, 1955) and Theorem 3.1.5). If the functions $f(x)$ and $g(x)$ are continuous at a point a, then:

a) the function $(f + g)(x)$ is continuous at the point a;
b) the function $(f - g)(x)$ is continuous at the point a;

c) the function $(kf)(x)$ is continuous at the point a.

Theorem 3.4.9 and Lemma 3.4.4 imply the following result.

Corollary 3.4.5. If the function $f(x)$ is r-continuous at a point a and the function $g(x)$ is h-continuous at a point a, then:

a) the function $(f + g)(x)$ is $(r + h)$-continuous at a point a;
b) the function $(f - g)(x)$ is $(r + h)$-continuous at a point a;
c) the function $(kf)(x)$ is $|k|\cdot r$-continuous at a point a.

Theorem 3.4.9 and Definition 3.3.17 imply the following result.

Corollary 3.4.6. If the function $f(x)$ is (q, r)-continuous in (inside) X and the function $g(x)$ is (p, h)-continuous in (inside) X, then:

a) the function $(f + g)(x)$ is $(u, r + h)$-continuous in (inside) X where $u = \min \{q, p\}$;
b) the function $(f - g)(x)$ is $(u, r + h)$-continuous in (inside) X where $u = \min \{q, p\}$;
c) the function $(kf)(x)$ is $(q, |k|\cdot r)$-continuous in (inside) X.

Corollary 3.4.6 and Lemma 3.4.4 imply the following result.

Corollary 3.4.7. If a function $f(x)$ is r-continuous in (inside) X and a function $g(x)$ is h-continuous in (inside) X, then:

a) the function $(f + g)(x)$ is $(r + h)$-continuous in (inside) X where $u = \min \{q, p\}$;
b) the function $(f - g)(x)$ is $(r + h)$-continuous in (inside) X where $u = \min \{q, p\}$;
c) the function $(kf)(x)$ is $|k|\cdot r$-continuous in (inside) X.

Corollary 3.4.8. If functions $f(x)$ and $g(x)$ are fuzzy (2-fuzzy) continuous at a point a, then functions $(f + g)(x)$, $(f - g)(x)$, and $(kf)(x)$ are fuzzy (2-fuzzy) continuous at the point a.

Corollary 3.4.9. If functions $f(x)$ and $g(x)$ are fuzzy (2-fuzzy) continuous in (inside) X, then functions $(f + g)(x)$, $(f - g)(x)$, and $(kf)(x)$ are fuzzy (2-fuzzy) continuous in (inside) X.

Corollary 3.4.10. The set of all fuzzy (2-fuzzy) continuous at a point a real functions is a linear space.

Corollary 3.4.11. The set of all fuzzy (2-fuzzy) continuous (inside) X real functions is a linear space.

Theorem 3.4.10. If the function $f(x)$ is (q, r)-continuous from the left (right) at a point a and the function $g(x)$ is (p, h)-continuous from the left (right) at the point a, then:

a) the function $(f + g)(x)$ is $(u, r + h)$-continuous from the left (right) at the point a where $u = \min \{q, p\}$;
b) the function $(f - g)(x)$ is $(u, r + h)$-continuous from the left (right) at the point a where $u = \min \{q, p\}$;
c) the function $(kf)(x)$ is $(q, |k|\cdot r)$-continuous from the left (right) at the point a.

Proof is similar to the proof of Theorem 3.4.9.

Corollary 3.4.12. If the functions $f(x)$ and $g(x)$ are continuous from the left (right) at a point a, then:

a) the function $(f+g)(x)$ is continuous from the left (right) at the point a;
b) the function $(f-g)(x)$ is continuous from the left (right) at the point a;
c) the function $(kf)(x)$ is continuous from the left (right) at the point a.

Theorem 3.4.10 implies the following result.

Corollary 3.4.13. If the function $f(x)$ is r-continuous from the left (right) at a point a and the function $g(x)$ is h-continuous from the left (right)at a point a, then:

a) the function $(f+g)(x)$ is $(r+h)$-continuous from the left (right) at a point a;
b) the function $(f-g)(x)$ is $(r+h)$-continuous from the left (right) at a point a;
c) the function $(kf)(x)$ is $|k|\cdot r$-continuous from the left (right) at a point a.

Remark 3.4.4. One of the principal results of the classical analysis (and topology) states that the product of continuous functions is continuous (cf. Theorem 3.1.5.c). For fuzzy continuous functions, this result is not true in a general case.

Example 3.4.4. Let us consider the function $g(x) = n$ when $x \in [n, n+1)$, $n \in \mathbf{Z}$ and the function $f(x) = x$. The function $g(x)$ is 1-continuous and the function $f(x)$ is continuous and consequently, 1-continuous and 0-continuous on $\mathbf{R}$. However, their product $(f{\cdot}g)(x)$ is not fuzzy continuous on $\mathbf{R}$. It means that the result of multiplication of a fuzzy continuous function and a continuous function can be not fuzzy continuous.

However, in a closed interval, the property of the product of continuous functions remains true for fuzzy continuous functions.

Theorem 3.4.11. The product of fuzzy continuous functions $f: [a, b] \to \mathbf{R}$ and $g: [a, b] \to \mathbf{R}$ is fuzzy continuous in $[a, b]$.

Proof. As both functions $f(x)$ and $g(x)$ are fuzzy continuous in $[a, b]$, then by Theorem 3.4.2, they are bounded in $[a, b]$. The product of bounded functions is a bounded function. Thus, by Theorem 3.4.2, the function $(f \cdot g)(x)$ is fuzzy continuous in $[a, b]$.

Theorem is proved.

Taking $r = 0$ in Theorem 3.4.11, we have the classical result that the product of two continuous at a point functions is continuous at the same point.

For 2-fuzzy continuous functions, statements from Remarks 3.4.2 and 3.4.3 are not true. Namely, we have the following result.

Let $X, Y, Z \subseteq \mathbf{R}$ and $q, p, r \in \mathbf{R}^{++}$.

Theorem 3.4.12. If a mapping $f: X \to Y$ is (q, r)-continuous (at a point x from X) and a mapping $g: Y \to Z$ is (r, p)-continuous (at the point $f(x)$), then the mapping $gf: X \to Z$ is (q, p)-continuous (at the point x).

Proof. If a mapping $f: X \to Y$ is (q, r)-continuous at a point c from X, then for any $\varepsilon > 0$ there is $\delta > 0$ such that the inequality $|a - c| < q + \delta$ implies the inequality $|f(c) - f(a)| < r + \varepsilon$. If a mapping $g: Y \to Z$ is (r, p)-continuous at the point $f(c)$ from Y, then for any $\alpha > 0$ there is $\gamma > 0$ such that the inequality $|d - b| < r + \gamma$ implies the inequality $|g(b) - f(d)| < p + \alpha$. Then for any $\alpha > 0$ there is $\delta > 0$ such that the inequality $|a - c| < q + \delta$ implies the inequality $| f(c) - f(a) | < r + \gamma$ and this, in turn, implies the inequality $| gf(c) - gf(a) | < p + \alpha$. Thus, the mapping $gf: X \rightarrow Z$ is (q, p)-continuous at the point c.

Consequently, when $f: X \rightarrow Y$ is (q, r)-continuous at all points c from X and a mapping $g: Y \rightarrow Z$ is (r, p)-continuous at all points $f(c)$, the mapping $gf: X \rightarrow Z$ is (q, p)-continuous at all points c from X. It means that the mapping $gf: X \rightarrow Z$ is (q, p)-continuous.

Theorem is proved.

Taking $r = q = 0$ in Theorem 3.4.12, we have the classical result (Theorem 3.1.10) that the composition of two continuous at a point functions is continuous at the same point.

Theorem 3.4.12 is a special case of more general results obtained in scalable topology (Burgin, 2004a; 2005a; 2006).

Proposition 3.4.2. Any $(q, 0)$-continuous at a point a function $f(x)$ is constant in some neighborhood of a with the diameter larger than or equal to q.

Indeed, by Definition 3.3.9, $(q, 0)$-continuity at the point a of the function $f(x)$ implies that for any number $\varepsilon > 0$ there is number $\delta > 0$ such that the inequality $| a - x | < q + \delta$ implies the inequality $| f(x) - f(a) | < \varepsilon$. As the number ε is arbitrarily small, the difference $f(x) - f(a)$ is equal to 0 when $| a - x | < q$. Thus, the function $f(x)$ is constant in the q-neighborhood of a.

Let (a, d) be a finite or infinite interval in $\boldsymbol{R}$.

Corollary 3.4.14. If a function $f(x)$ is $(q, 0)$-continuous in an interval (a, d), then $f(x)$ is constant in the interval (a, d).

Indeed, if a function $f(x)$ is $(q, 0)$-continuous in (a, d), then $f(x)$ is $(q, 0)$-continuous at each point a of X. Taking two points b and c from (a, d), we can cover the interval $[c, b]$ by a finite number of q-neighborhoods of points from this interval. In each of these intervals, $f(x)$ is constant. Thus, it is constant in the interval $[c, b]$. As we take arbitrary points b and c from (a, d), the function $f(x)$ is constant in the interval (a, d).

These results show that parameters q and r play complimentary roles for (q, r)-continuous mappings. The parameter r extends the scope of admissible (that is, continuous to some extent) functions, the parameter q restricts the scope of admissible functions. It is demonstrated here that, in general there are much more r-continuous functions that continuous. However, the class of (q, r)-continuous functions can be much less than the class of continuous ones.

As we can see, (q, r)-continuity only decreases the scope of continuous functions in comparison with r-continuity. A natural question arises: Why do we need the concept of (q, r)-continuity? One of the reasons to introduce this concept is necessity to study and utilize functions in discrete spaces (sets). Examples of such spaces are: all integer numbers, all zeros of the function $\sin x$, and all point of the form $1/n$ where $n = 1, 2, 3, \ldots$ As all points in a discrete space are isolated, there are no converging sequences there besides stabilizing sequences, i.e., such sequences in which almost all elements are equal to the same number. In turn, this implies that any function in a discrete space is r-continuous for any r and even continuous. It means that the concepts of continuity and r-continuity degenerate for functions on discrete sets.

At the same time, computable functions are discrete. They are defined and take values in discrete sets. Practical situations show that there is some kind of continuity of discrete functions. Thus, theoretical and practical problems demand a relevant mathematical definition of continuity for discrete functions (cf., for example, (Rosenfeld, 1979; 1986; Boxer, 1994; Hamlet, 2002)). Such a definition is provided by the structure of (q, r)-continuity.

3.5. Uniformly Continuous and Fuzzy Uniformly Continuous Functions

Strength and growth come only through
continuous effort and struggle.
Napoleon Hill (1883-1970)

In some cases, we need stronger conditions than continuity and fuzzy continuity to be able to get some important properties of functions

In (ε, δ)-definitions of continuity and fuzzy continuity, we see that δ depends on (is a function of) the chosen number ε, but in a general case, δ also depends on the point. For instance, let us take the function $f(x) = x^2$ and ε = 0.1. This is a continuous function and at the point 0, if we take x such that $| a - x | = | x | < \delta = 0.3$, then we have $|f(x) - f(a)| = |f(x)| = |x^2| < \varepsilon = 0.1$. At the same time, at the point 100, if we take $x = 100.1$, then we have $|100 - x| = 0.1 < \delta = 0.3$ but $|f(x) - f(a)| = | 100.1^2 - 100^2 | = 10020.01 - 10000 = 20.01 > \varepsilon = 0.1$. To get the difference $| f(x) - f(a) |$ smaller than ε = 0.1, we need much smaller δ than 0.3. For instance, if we take $|100 - x| < \delta = 0.0001$, then we have $|f(x) - f(a)| < \varepsilon = 0.1$.

Conditions that δ depends only on the chosen number ε (but not on the point) bring us to uniformly continuous and uniformly fuzzy continuous functions.

Let $X \subseteq \boldsymbol{R}$.

Definition 3.5.1. A partial function $f: \boldsymbol{R} \to \boldsymbol{R}$ is called *uniformly continuous in* X if for any $\varepsilon > 0$, there is $\delta > 0$ such that for any a from Dom f and c from $X \cap$ Dom f the inequality $| a - c | < \delta$ implies the inequality $|f(c) - f(a)| < \varepsilon$, or in other words, for any a from $\boldsymbol{R}$ and c from X with $| a - c | < \delta$, we have $|f(c) - f(a)| < \varepsilon$.

Example 3.5.1. The function $f(x) = x$ is uniformly continuous in $\boldsymbol{R}$. The function $f(x) = x^2$ is uniformly continuous in any interval $[d, b]$, is continuous in $\boldsymbol{R}$, but it is not uniformly continuous in $\boldsymbol{R}$.

Definition 3.5.2. A partial function $f: \boldsymbol{R} \to \boldsymbol{R}$ is called *uniformly continuous inside* X if for any $\varepsilon > 0$, there is $\delta > 0$ such that for any a and c from $X \cap$ Dom f the inequality $| a - c | < \delta$ implies the inequality $|f(c) - f(a)| < \varepsilon$, or in other words, for any a and x from X with $| a - c | < \delta$, we have $|f(c) - f(a)| < \varepsilon$.

Example 3.5.2. The function $f(c) = \tan x$ is continuous inside the interval $[0, \pi/2]$ but it is neither continuous in this interval nor uniformly continuous inside this interval.

Remark 3.5.1. When the function $f(x)$ is defined only on X, then uniform continuity in and inside X coincide and $f(x)$ is called (cf., for example, (Ross, 1996)) *uniformly continuous on* X.

Definition 3.5.3. A function $f: \boldsymbol{R} \to \boldsymbol{R}$ uniformly continuous in $\boldsymbol{R}$ is called *uniformly continuous*.

Definitions imply the following results.

Lemma 3.5.1. A uniformly continuous in a set X function $f(x)$ is uniformly continuous inside X.

Lemma 3.5.2. A uniformly continuous in a set X function $f(x)$ is continuous in X.

For closed intervals, the inverse statement is also true.

Theorem 3.5.1 (the Heine–Cantor Theorem). A continuous function f: $[a, b] \to \boldsymbol{R}$ is uniformly continuous in $[a, b]$.

We do not give a proof of this important result, as it is a direct corollary of Theorem 3.5.2.

Remark 3.5.2. In some sources, Theorem 3.5.1 is called the Cantor Theorem.

Remark 3.5.3. If we take intuition that exists in mathematical analysis, then we see that a function $f(x)$ is called *uniformly continuous* if, roughly speaking, small changes in the input x effect small changes in the output $f(x)$ (an idea of continuity), and furthermore the *size* of the changes in $f(x)$ depends only on the size of the changes in x but not on x itself (an idea of uniformity). However, if we do not demand arbitrarily small changes (and who can measure such changes?), then we come to the concept of uniform fuzzy continuity instead of the classical uniform continuity

Uniform fuzzy continuity is a natural extension of the classical uniform continuity.

Definition 3.5.4. a) A function f: $\boldsymbol{R} \to \boldsymbol{R}$ is called *uniformly r-continuous in X* if for any $\varepsilon > 0$, there is $\delta > 0$ such that for any a from $\boldsymbol{R}$ and c from X with $| a - c| < \delta$, we have $| f(c) - f(a) | < r + \varepsilon$.

b) A function f: $\boldsymbol{R} \to \boldsymbol{R}$ is called *uniformly fuzzy continuous in X* if it is uniformly r-continuous in X for some $r > 0$.

c) A function f: $\boldsymbol{R} \to \boldsymbol{R}$ is called *uniformly* (q, r)*-continuous in X* if for any $\varepsilon > 0$, there is $\delta > 0$ such that for any a from $\boldsymbol{R}$ and c from X the inequality $| a - c | < q + \delta$ implies the inequality $| f(c) - f(a) | < r + \varepsilon$.

d) A function f: $\boldsymbol{R} \to \boldsymbol{R}$ is called *uniformly 2-fuzzy continuous in X* if it is uniformly (q, r)-*continuous* in X for some $r \geq 0$ and $q \geq 0$.

Example 3.5.3. The function $f(x) = kx$ with $k > 0$ is uniformly r-continuous for all $r \in \boldsymbol{R}^+$ on the whole $\boldsymbol{R}$.

Example 3.5.4. The function $f(x) = kx^2$ with $k > 0$ is uniformly r-continuous for all r in any interval $[d, b]$, but it is not uniformly r-continuous for any r on the whole $\boldsymbol{R}$. Indeed, in the case $r = 0$, given $\varepsilon > 0$, we need to find $\delta > 0$ such that the inequality $|a - c| < \delta$ implies the inequality $| f(c) - f(a)| < \varepsilon$. This means that $k| c^2 - a^2| < \varepsilon$. Using properties of real numbers, we obtain the following sequence of equalities and inequalities

$$|kc^2 - ka^2| = k\cdot| c^2 - a^2| = | k(c - a)(c + a)| =$$
$$k | c - a| \cdot | c + a| \leq k\cdot| c - a| (| c| + |a|) \leq k| c - a| (2u)$$

where $a, c \in [d, b]$ and $u = \max \{|b|, |d|\}$.

Then taking $\delta = \varepsilon/[k(2u + 1)]$, we see that this δ satisfies the necessary condition, i.e.,

$$k\cdot| c^2 - a^2| \leq k\cdot| c - a| \cdot (2u) < \delta\cdot(2ku) = (2ku) \cdot (\varepsilon/[k(2u + 1)]) < \varepsilon$$

However, when the absolute values of numbers c and a are unbounded, we cannot find such a number δ.

Lemma 3.5.3. The following conditions are equivalent for a function $f(x)$:

1. $f(x)$ is uniformly continuous in X.
2. $f(x)$ is uniformly 0-continuous in X.

3. $f(x)$ is uniformly (0, 0)-continuous in X.

Indeed, if $f(x)$ is uniformly 0-continuous in X, then for any $\varepsilon > 0$ and any points a from $\boldsymbol{R}$ and b from X, there is $\delta > 0$ such that the inequality $| b - a | < \delta$ implies the inequality $| f(b) - f(a) | < 0 + \varepsilon = \varepsilon$. This means that $f(x)$ is uniformly continuous in X. The converse is also true. Thus, the condition of uniform 0-continuity coincides with the condition of uniform continuity.

If $f(x)$ is uniformly (0, 0)-continuous in X, then for any $\varepsilon > 0$ and any points a from $\boldsymbol{R}$ and b from X, there is $\delta > 0$ such that the inequality $| b - a | < 0 + \delta$ implies the inequality $| f(b) - f(a) | < 0 + \varepsilon = \varepsilon$. This means that $f(x)$ is uniformly continuous in X. The inverse is also true. Thus, the condition of uniform (0, 0)-continuity coincides with the condition of uniform continuity.

Lemma 3.5.4. If $q > p$ and $t > r$, then any uniformly r-continuous (uniformly (q, r)-continuous) in X function is also uniformly t-continuous (uniformly (p, t)-continuous) in X.

Definition 3.5.5. a) A function $f: \boldsymbol{R} \to \boldsymbol{R}$ is called *uniformly r-continuous inside X* if for any $\varepsilon > 0$ there is $\delta > 0$ such that for any a and c from X with $| a - c | < \delta$, we have $| f(c) - f(a) | < r + \varepsilon$.

b) A function $f: \boldsymbol{R} \to \boldsymbol{R}$ is called *uniformly fuzzy continuous inside X* if it is uniformly r-continuous in X for some $r > 0$.

c) A function $f: \boldsymbol{R} \to \boldsymbol{R}$ is called *uniformly (q, r)-continuous inside X* if for any $\varepsilon > 0$ there is $\delta > 0$ such that for any a and c from X the inequality $| a - c| < q + \delta$ implies the inequality $| f(c) - f(a) | < r + \varepsilon$.

d) A function $f: \boldsymbol{R} \to \boldsymbol{R}$ is called *uniformly 2-fuzzy continuous inside X* if it is uniformly *(q, r)-continuous* in X for some $r \geq 0$ and $q \geq 0$.

Remark 3.5.4. When the function $f(x)$ is defined only on X, then uniform continuity in and inside X coincide and $f(x)$ is called *uniformly r-continuous on X*.

Lemma 3.5.5. If a function $f(x)$ is uniformly r-continuous in X, then it is uniformly r-continuous inside X.

Remark 3.5.5. The converse of Lemma 3.5.5 is not true as the following example demonstrates.

Example 3.5.5. Let us consider the function

$$f(x) = \begin{cases} 0 & \text{when } x \leq 0; \\ \cot x & \text{when } x > 0 \end{cases}$$

Then $f(x)$ is uniformly 0-continuous and uniformly (0, 0)–continuous inside [-1, 0], but it is neither uniformly r-continuous nor uniformly (q, r)–continuous in [-1, 0] for any $r \geq 0$ and $q \geq 0$.

Definitions imply the following results.

Lemma 3.5.6. A uniformly (q, r)-continuous in an interval function $f(x)$ is (q, r)-continuous in this interval.

Corollary 3.5.1. A 2-fuzzy uniformly continuous in an interval function $f(x)$ is 2-fuzzy continuous in this interval.

The same is true for fuzzy continuity.

Lemma 3.5.7. A uniformly r-continuous in an interval function $f(x)$ is r-continuous in this interval.

Corollary 3.5.2. A fuzzy uniformly continuous in an interval function $f(x)$ is fuzzy continuous in this interval.

For closed intervals, the inverse statement is also true.

Theorem 3.5.2 (the Fuzzy Heine–Cantor Theorem). An r-continuous function f: $[a, b] \to \boldsymbol{R}$ is uniformly $2r$-continuous on $[a, b]$.

Proof. Let $f(x)$ be an r-continuous function on $[a, b]$, but it is not uniformly $2r$-continuous on $[a, b]$. Then there exists $\varepsilon > 0$ such that the implication $| u - v | < \delta \Rightarrow | f(u) - f(v) | < 2r + \varepsilon$ fails for each $\delta > 0$. That is, for each $\delta > 0$, there are elements u and v from $[a, b]$ such that $| u - v | < \delta$ and yet $| f(u) - f(v) | \geq 2r + \varepsilon$. Then for each $n = 1, 2, 3, \ldots$, there are elements u_n and v_n from $[a, b]$ such that $| u_n - v_n | < 1/n$ and yet $| f(u_n) - f(v_n) | \geq 2r + \varepsilon$. By Theorem 2.1.7, it possible to assume that $l = \{ u_n ; n = 1, 2, 3, \ldots\}$ is a converging sequence as its elements belong to a finite interval. If $u = \lim l$, then $u = \lim h$ where $h = \{ v_n ; n = 1, 2, 3, \ldots\}$ because $| u_n - v_n |$ tends to 0.

As $f(x)$ is an r-continuous function on $[a, b]$, we have that $f(u) = r\text{-}\lim_{n\to\infty} f(u_n) = r\text{-}\lim_{n\to\infty} f(v_n)$. Then

$$| f(u_n) - f(v_n) | = | f(u_n) - f(u) + f(u) - f(v_n) | \leq$$
$$| f(u_n) - f(u) | + | f(u) - f(v_n) | < r + \varepsilon/3 + r + \varepsilon/3$$

when the number n is larger than some m.

Thus, $| f(u_n) - f(v_n) | < 2r + 2\varepsilon/3$. We come to a contradiction with our assumption.

Theorem is proved.

Corollary 3.5.3. A fuzzy continuous function f: $[a, b] \to \boldsymbol{R}$ is uniformly fuzzy continuous on $[a, b]$.

Another corollary from Theorem 3.5.2 is such a classical result as the Heine–Cantor theorem that states that a continuous function f: $[a, b] \to \boldsymbol{R}$ is uniformly continuous on $[a, b]$.

Let us consider two functions f: $\boldsymbol{R} \to \boldsymbol{R}$ and g: $\boldsymbol{R} \to \boldsymbol{R}$ and a real number k.

Theorem 3.5.3. If the function $f(x)$ is uniformly (q, r)-continuous in (inside) X and the function $g(x)$ is uniformly (p, h)-continuous in (inside) X, then:

a) the function $(f + g)(x)$ is uniformly $(u, r + h)$-continuous in (inside) X where $u = \min \{q, p\}$;
b) the function $(f - g)(x)$ is uniformly $(u, r + h)$-continuous in (inside) X where $u = \min \{q, p\}$;
c) the function $(k \cdot f)(x)$ is uniformly $(q, |k|\cdot r)$-continuous in (inside) X.

Proof. a) Let us assume that $f(x)$ is uniformly (q, r)-continuous inside X and the function $g(x)$ is uniformly (p, h)-continuous inside X, $q \leq p$, and take some $\varepsilon > 0$. Then there is $\delta > 0$ such that the condition $| a - c | < q + \delta$ for points $a, c \in X$ implies the inequality

$$| f(a) - f(c) | < r + \varepsilon/2$$

As the inequality $| a - c | < q + \delta$ implies the inequality $| a - c | < p + \delta$, we also have

$$g(a) - g(c) | < h + \varepsilon/2$$

Combining both inequalities, we obtain

$$|(f + g)(a) - (f + g)(c)| = |(f(a) + g(a)) - (f(c) + g(c))| =$$
$$|f(a) + g(a) - f(c) - g(c)| = |f(a) - f(c) + g(a) - g(c)| \leq$$
$$|f(a) - f(c)| + |g(a) - g(c)| < r + \varepsilon/2 + h + \varepsilon/2 = r + h + \varepsilon$$

As a and c are arbitrary points from X subject to the condition $|a - c| < q + \delta$, Definition 3.5.4 shows that the function $(f + g)(x)$ is uniformly $(q, r + h)$-continuous inside X. The case when $q > p$ is treated in the same way. Thus, we proved uniform $(u, r + h)$-continuity of the function $(f + g)(x)$ inside X with $u = \min \{q, p\}$.

By a similar reasoning, only taking an arbitrary point from the domains of $f(x)$ and $g(x)$ as c, we show that the function $(f + g)(x)$ is uniformly $(u, r + h)$-continuous in X with $u = \min \{q, p\}$ when $f(x)$ is uniformly (q, r)-continuous in X and the function $g(x)$ is uniformly (p, h)-continuous in X.

This concludes the proof of the statement (a) from the theorem.

b) Let us assume that $f(x)$ is uniformly (q, r)-continuous inside X and the function $g(x)$ is uniformly (p, h)-continuous inside X, $q \leq p$, and take some $\varepsilon > 0$. Then there is $\delta > 0$ such that the condition $|a - c| < q + \delta$ for points $a, c \in X$ implies the inequality

$$|f(a) - f(c)| < r + \varepsilon/2$$

As the inequality $|a - c| < q + \delta$ implies the inequality $|a - c| < p + \delta$, we also have

$$|g(a) - g(c)| < h + \varepsilon/2$$

Combining both inequalities, we obtain

$$|(f - g)(a) - (f - g)(c)| = |(f(a) - g(a)) - (f(c) - g(c))| =$$
$$|f(a) - g(a) - f(c) + g(c)| = |(f(a) - f(c)) - (g(a) - g(c))| \leq$$
$$|f(a) - f(c)| + |g(c) - g(a)| < r + \varepsilon/2 + h + \varepsilon/2 = r + h + \varepsilon$$

As a and c are arbitrary points from X subject to the condition $|a - c| < q + \delta$, Definition 3.5.4 shows that the function $(f - g)(x)$ is uniformly $(q, r + h)$-continuous inside X. The case when $q > p$ is treated in the same way. Thus, we proved uniform $(u, r + h)$-continuity of the function $(f - g)(x)$ inside X with $u = \min \{q, p\}$.

By a similar reasoning, only taking an arbitrary point from the domains of $f(x)$ and $g(x)$ as c, we show that the function $(f - g)(x)$ is uniformly $(q, r + h)$-continuous in X with $u = \min \{q, p\}$ when $f(x)$ is uniformly (q, r)-continuous in X and the function $g(x)$ is uniformly (p, h)-continuous in X.

This concludes the proof of the statement (b) from the theorem.

c) Let us assume that $f(x)$ is uniformly (q, r)-continuous inside X, k is a real number, and take some $\varepsilon > 0$. Then there is $\delta > 0$ such that the condition $|a - c| < q + \delta$ for points $a, c \in X$ implies the inequality

$$| f(a) - f(c) | < r + \varepsilon/| k|$$

The we have

$$|(kf)(a) - (kf)(c) | = | kf(a) + kf(c) | = | k|\cdot| f(a) + f(c) | \leq$$
$$| k|\cdot(r + \varepsilon/| k|) = r + \varepsilon$$

As a and c are arbitrary points from X subject to the condition $| a - c | < q + \delta$, Definition 3.5.4 shows that the function $(kf)(x)$ is uniformly $(q, |k|r)$-continuous inside X.

By a similar reasoning, only taking an arbitrary point from the domain of $f(x)$ as c, we show that the function $(kf)(x)$ is uniformly $(q, |k|r)$-continuous in X when $f(x)$ is uniformly (q, r)-continuous in X.

This concludes the proof of the theorem.

Theorem 3.4.1, Lemma 3.5.8 and Definition 3.5.5 imply the following result.

Corollary 3.5.4. If a function $f(x)$ is uniformly r-continuous in (inside) X and a function $g(x)$ is uniformly h-continuous in (inside) X, then:

a) the function $(f + g)(x)$ is uniformly $(r + h)$-continuous in (inside) X;
b) the function $(f - g)(x)$ is uniformly $(r + h)$-continuous in (inside) X;
c) the function $(k \cdot f)(x)$ is uniformly $|k|\cdot r$-continuous in (inside) X.

Corollary 3.5.5. If functions $f(x)$ and $g(x)$ are uniformly fuzzy (2-fuzzy) continuous in (inside) X, then functions $(f + g)(x)$, $(f - g)(x)$, and $(k\cdot f)(x)$ are uniformly fuzzy (2-fuzzy) continuous in (inside) X.

Corollary 3.5.6. The set of all uniformly fuzzy (2-fuzzy) continuous (inside) X real functions is a linear space.

Remark 3.5.6. One of the important results of the classical analysis states that the product of uniformly continuous functions is uniformly continuous. For uniformly fuzzy continuous functions, this result is not true.

Example 3.5.6. Let us consider the function $g(x) = n$ when $x \in [n, n + 1)$, $n \in \boldsymbol{Z}$ and the function $f(x) = x$. The function $g(x)$ is uniformly 1-continuous and the function $f(x)$ is uniformly continuous and consequently, uniformly 1-continuous and uniformly 0-continuous on $\boldsymbol{R}$. However, their product $(f\cdot g)(x)$ is not fuzzy continuous on $\boldsymbol{R}$, and thus, not uniformly fuzzy continuous on $\boldsymbol{R}$. It means that the result of multiplication of a uniformly fuzzy continuous function and a continuous function can be not uniformly fuzzy continuous.

However, in a closed interval, the property of the product of uniformly continuous functions remains true for uniformly fuzzy continuous functions.

Theorem 3.5.4. The product of uniformly fuzzy continuous functions f: $[a, b] \to \boldsymbol{R}$ and g: $[a, b] \to \boldsymbol{R}$ is uniformly fuzzy continuous in $[a, b]$.

Proof. By Corollary 3.5.2, both functions $f(x)$ and $g(x)$ are fuzzy continuous in $[a, b]$. By Theorem 3.4.2, both functions $f(x)$ and $g(x)$ are bounded in $[a, b]$. The product of bounded functions is a bounded function. Thus, by Theorem 3.4.2, the function $(f \cdot g)(x)$ is fuzzy continuous in $[a, b]$. Then by Theorem 3.5.3, the function $(f \cdot g)(x)$ is uniformly fuzzy continuous in $[a, b]$.

Theorem is proved.

3.6. Fuzzy Convergence and Uniform Fuzzy Convergence of Functions

Life is the continuous adjustment
of internal relations to external relations.
Herbert Spencer (1820-1903)

Let X and Y be subsets of $\boldsymbol{R}$ and $f: X \to Y, f_n : X \to Y$ be real functions for all $n = 1, 2, 3, \ldots$.

Definition 3.6.1. A sequence of functions $\{ f_n(x) ; n = 1, 2, 3, \ldots\}$ *converges* to a function $f(x)$ *at a point* $a \in X$ if $f(a)$ is an limit of the sequence $\{ f_n(a); n = 1, 2, 3, \ldots\}$.

Example 3.6.1. The sequence of functions $\{ \cos nx ; n = 1, 2, 3, \ldots\}$ converges to the function x^2 at the point 0.

Let $Z \subseteq X$.

Definition 3.6.2. A sequence of functions $\{ f_n(x) ; n = 1, 2, 3, \ldots\}$ *converges* to a function $f(x)$ in Z if this sequence converges to $f(x)$ at all points from Z.

It is denoted by $f(x) = \lim^Z{}_{n\to\infty} f_n(x)$.

This is *pointwise convergence* of function sequences.

Example 3.6.2. The sequence of functions $\{ \sin x/n ; n = 1, 2, 3, \ldots\}$ converges to the function $f(x) \equiv 0$.

Corollary 2.3.8 implies the following result.

Theorem 3.6.1. If a sequence of functions $\{ f_n(x) ; n = 1, 2, 3, \ldots\}$ converges to a function $f(x)$ in X and a sequence of functions $\{ g_n(x) ; n = 1, 2, 3, \ldots\}$ converges to a function $g(x)$ in X, i.e., $f(x) = \lim^X{}_{n\to\infty} f_n(x)$ and $g(x) = \lim^X{}_{n\to\infty} g_n(x)$, then:

a) the sequence of functions $\{ (f_n + g_n)(x) ; n = 1, 2, 3, \ldots\}$ converges to a function $(f + g)(x)$ in X, i.e., $(f + g)(x) = \lim^X{}_{n\to\infty} (f_n + g_n)(x)$;
b) the sequence of functions $\{ (f_n - g_n)(x) ; n = 1, 2, 3, \ldots\}$ converges to a function $(f - g)(x)$ in X, i.e., $(f - g)(x) = \lim^X{}_{n\to\infty} (f_n - g_n)(x)$;
c) the sequence of functions $\{ kf_n(x) ; n = 1, 2, 3, \ldots\}$ converges to a function $kf(x)$ in X, i.e., $kf(x) = \lim^X{}_{n\to\infty} kf_n(x)$.

It is natural to ask a question whether a convergent sequence of continuous functions always converges to a continuous function. In a general case, the answer is negative as the following example demonstrates.

Example 3.6.3. Let us consider the sequence of functions $\{ f_n(x) = (1 - |x|)^n ; n = 1, 2, 3, \ldots; x \in (-1, 1) \}$ and the function $f(x) = 0$ when $x \neq 0$ and $f(0) = 1$. Then this sequence converges to the function $f(x)$, which is not continuous at the point 0 and thus, not continuous in (-1, 1).

As we know from history of mathematics, Cauchy in 1821 published a theorem that the pointwise limit of a sequence of continuous functions is always continuous. Fourier and Abel found counter examples to this theorem. Dirichlet then analyzed Cauchy's proof and found that to make this theorem true, it is possible to replace pointwise convergence by uniform convergence. This result is proved in Theorem 3.6.4.

Definition 3.6.3. A sequence of functions $\{ f_n(x) ; n = 1, 2, 3, \ldots\}$ *uniformly converges* to a function $f(x)$ if for any $\varepsilon > 0$, there is $m \in \omega$ such that for any point $a \in X$, the inequality $| f_n(a) - f(a) | < \varepsilon$ is true whenever $n > m$.

As uniform convergence is a stronger condition than convergence, we have the following result.

Lemma 3.6.1. If a sequence $\{ f_n(x) ; n = 1, 2, 3, \ldots\}$ uniformly converges to a function $f(x)$ (at a point a), then the sequence $\{f_n(x) ; n = 1, 2, 3, \ldots\}$ converges to $f(x)$ (at a point a).

However, not every converging sequence of functions converges uniformly.

Example 3.6.1. Let us consider the sequence of functions $\{ f_n(x); n = 1, 2, 3, \ldots\}$ where $f_n(x) = \sin x/n$ when $n\pi/6 \leq x \leq n\pi$ and equal to 0 at all other points from $\boldsymbol{R}$. For any point $a \in \boldsymbol{R}$, there is m such that $f_n(a) = 0$ for all $n > m$. It means that this sequence converges to the function $f(x) \equiv 0$. However, this convergence is not uniform because for any $n \in \omega$, there is always a point $a \in \boldsymbol{R}$ such that $f_n(a) = 1$, namely, $a = n\pi/2$ because $\sin \pi/2 = 1$.

Uniformly converging sequences of functions have different good properties.

Theorem 3.6.2. If a sequence of functions $\{ f_n(x) ; n = 1, 2, 3, \ldots\}$ uniformly converges to a function $f(x)$ in X and a sequence of functions $\{ g_n(x) ; n = 1, 2, 3, \ldots\}$ uniformly converges to a function $g(x)$ in X, then:

a) the sequence of functions $\{ (f_n + g_n)(x) ; n = 1, 2, 3, \ldots\}$ uniformly converges to a function $(f + g)(x)$ in X;
b) the sequence of functions $\{ (f_n - g_n)(x) ; n = 1, 2, 3, \ldots\}$ uniformly converges to a function $(f - g)(x)$ in X;
c) the sequence of functions $\{ kf_n(x) ; n = 1, 2, 3, \ldots\}$ uniformly converges to a function $kf(x)$ in X.

We do not give proofs of these statements because they are direct corollaries of the corresponding results for uniform fuzzy convergence (cf. Theorem 3.6.18).

Theorem 3.6.3. If a sequence of continuous at a point $a \in X$ functions $\{ f_n(x); n = 1, 2, 3, \ldots\}$ uniformly converges in some neighborhood of the point a to a function $f(x)$, then $f(x)$ is continuous at the point a.

Proof. Let us assume that the sequence of functions $f_n(x)$ uniformly converges to the function $f(x)$ in some interval $[u, v]$ that contains a and consider some $\varepsilon > 0$. Taking, if necessary, a smaller interval, we can assume that a is the midpoint of the interval $[u, v]$ and its length is 2α. Then by Definition 3.6.3, there is such a number m that for any number $n > m$ and any point d from $[u, v]$, we have $|f(d) - f_n(d)| < \varepsilon/3$. Let us take such a number n, e.g., $n = m + 1$. As all functions $f_n(x)$ are continuous at the point a, there is $\delta > 0$ such that if $| c - a| < \delta$, then $|f_n(c) - f_n(a)| < \varepsilon/3$.

Let us consider the difference $| f(c) - f(a) |$. If $| c - a| < \lambda = \min \{\alpha, \delta\}$, then c belongs to the interval $[u, v]$ and from the properties of absolute values, we have

$$| f(c) - f(a) | = | f(c) - f_n(c) + f_n(c) - f_n(a) + f_n(a) - f(a) | \leq$$
$$| f(c) - f_n(c)| + |f_n(c) - f_n(a)| + |f_n(a) - f(a) | < \varepsilon/3 + \varepsilon/3 + \varepsilon/3 = \varepsilon$$

because $| f(c) - f_n(c)| < \varepsilon/3$, $|f_n(a) - f(a) | < \varepsilon/3$, and $|f_n(c) - f_n(a)| < \varepsilon/3$.

It means that the function $f(x)$ is continuous at the point a.

Theorem is proved.

Analyzing this proof, we see that if a sequence of functions $f_n(x)$ with $n = 1, 2, 3, \ldots$ uniformly converges to the function $f(x)$ in some interval $[u, v]$, then $f(x)$ is continuous at any point d from (u, v) where all functions $f_n(x)$ are continuous. This gives us the following result.

Theorem 3.6.4. If a sequence of continuous inside an interval $[a, b]$ (in $\mathbf{R}$) functions $\{f_n(x)\ ;\ n = 1, 2, 3, \ldots\}$ uniformly converges to a function $f(x)$, then $f(x)$ is continuous inside the interval $[a, b]$ (in $\mathbf{R}$).

In some cases, convergence coincides with uniform convergence.

Definition 3.6.4. a) A sequence of functions $\{ f_n(x)\ ;\ n = 1, 2, 3, \ldots\}$ is *equicontinuous at a point* $a \in \mathbf{R}$ if for any $\varepsilon > 0$, there is $\delta > 0$ such that the inequality $|a - x| < \delta$ implies the inequality $|f_n(x) - f_n(a)| < \varepsilon$ for any $n = 1, 2, 3, \ldots$ and any $x \in X$.

b) A sequence of functions $\{ f_n(x)\ ;\ n = 1, 2, 3, \ldots\}$ is *equicontinuous* in (inside) X if the sequence of $\{ f_n(x)\ ;\ n = 1, 2, 3, \ldots\}$ is equicontinuous at all points $a \in X$.

c) A sequence of functions $\{ f_n(x)\ ;\ n = 1, 2, 3, \ldots\}$ is *uniformly equicontinuous* in (inside) a set X if for any $\varepsilon > 0$, there is $\delta > 0$ such that the inequality $|a - x| < \delta$ implies the inequality $|f_n(x) - f_n(a)| < \varepsilon$ for any $n = 1, 2, 3, \ldots$ and any points $a, x \in X$.

Note that in an equicontinuous at a point $a \in \mathbf{R}$ (in or inside a set X) sequence of functions all functions are continuous at the same point (in or inside the same set X).

Lemma 3.6.2. A sequence of continuous at a point $a \in \mathbf{R}$ functions $\{ f_n(x)\ ;\ n = 1, 2, 3, \ldots\}$ is equicontinuous at the point a if and only if for any $\varepsilon > 0$, there is $\delta > 0$ such that the inequality $|a - x| < \delta$ implies the inequality $|f_n(x) - f_n(a)| < \varepsilon$ for almost all $n = 1, 2, 3, \ldots$.

Theorem 3.6.5. If an equicontinuous at a point a sequence of functions $\{ f_n(x)\ ;\ n = 1, 2, 3, \ldots\}$ converges at a to a continuous at a function $f(x)$, then functions $f_n(x)$ uniformly converge to $f(x)$ at the point a.

Proof. Let us assume that an equicontinuous at a point a sequence of functions $f_n(x)$ converges at the point a to a continuous at the point a function $f(x)$. If this convergence is not a uniform convergence at the point a, then the sequence $\{ f_n(x);\ n = 1, 2, 3, \ldots \}$ has the following property: for any interval $[u, v]$ that contains a, there is $\varepsilon > 0$ such that for any $n = 1, 2, 3, \ldots$, there is a point d_n such that $|f(d_n) - f_n(d_n)| > \varepsilon$.

Let us take the sequence of intervals $[a - 1/m , a + 1/m]$ where $m = 1, 2, 3, \ldots$. Properties of the sequence $\{ f_n(x);\ n = 1, 2, 3, \ldots \}$ allow us to find a the sequence of points $\{ d_m;\ m = 1, 2, 3, \ldots \}$ such that each point d_m belongs to the interval $[a - 1/m , a + 1/m]$ and $|f(d_n) - f_n(d_n)| > \varepsilon$.

As each point d_m belongs to $[a - 1/m , a + 1/m]$, we have $a = \lim_{m\to\infty} d_m$. Then there is $\nu > 0$ such that $|d_m - a| < \nu$ implies $|f(d_m) - f(a)| < \varepsilon/3$ because the function $f(x)$ is continuous at the point a. In addition, there is $\kappa > 0$ such that $|d_m - a| < \kappa$ implies $|f_n(d_m) - f_n(a)| < \varepsilon/3$ for all $n = 1, 2, 3, \ldots$ because the sequence of functions $f_n(x)$ is equicontinuous at a.

Then when $1/m < \delta = \min \{\nu, \kappa\}$, we have

$$|f(d_n) - f(a)| < \varepsilon/3$$

and

$$|f_n(d_n) - f_n(a)| < \varepsilon/3 \text{ for all } n > m.$$

The sequence $\{ f_n(x); n = 1, 2, 3, \ldots \}$ converges at the point a to the function $f(x)$. Consequently, there is $k \in \omega$ such that $|f(a) - f_n(a)| < \varepsilon/3$ for all $n > k$.

Let us take $n > \max \{k, m\}$. Then we have

$$|f(d_n) - f_n(d_n)| = |f(d_n) - f(a) + f(a) - f_n(a) + f_n(a) - f_n(d_n)| \leq$$
$$|f(d_n) - f(a)| + |f(a) - f_n(a)| + |f_n(a) - f_n(d_n)| <$$
$$\varepsilon/3 + \varepsilon/3 + \varepsilon/3 = \varepsilon.$$

This contradicts to the possibility that the sequence $\{ f_n(x); n = 1, 2, 3, \ldots \}$ does not uniformly converges at the point a to the function $f(x)$. By contradiction, we have that functions $f_n(x)$ uniformly converge to $f(x)$ at the point a.

Theorem is proved.

From local convergence, we come to global convergence.

Theorem 3.6.6. If an equicontinuous inside an interval $[a, b]$ sequence of functions $\{f_n(x) ; n = 1, 2, 3, \ldots\}$ converges inside $[a, b]$ to a continuous inside $[a, b]$ function $f(x)$, then functions $f_n(x)$ uniformly converge to $f(x)$ inside the interval $[a, b]$.

We do not give a proof of this important result of the classical analysis, as it is a direct corollary of Theorem 3.6.11.

Definition 3.6.5. A sequence $\{ f_n(x) ; n = 1, 2, 3, \ldots\}$ is *monotone* if $f_n(x) \geq f_{n-1}(x)$ for all x and all $n = 1, 2, 3, \ldots$ or $f_n(x) \leq f_{n-1}(x)$ for all $n = 1, 2, 3, \ldots$

Theorem 3.6.7 (the Dini Theorem). If a monotone sequence of continuous functions $\{f_n(x) ; n = 1, 2, 3, \ldots\}$ converges to a continuous function $f(x)$ inside a closed interval $[a, b]$, then functions $f_n(x)$ uniformly converge to $f(x)$ inside the interval $[a, b]$.

Proof. Let us consider a monotone sequence of continuous functions $\{ f_n(x) ; n = 1, 2, 3, \ldots\}$ converges to a continuous function $f(x)$ inside a closed interval $[a, b]$.

Let us assume that convergence of functions $f_n(x)$ is not uniform and $f_n(x) \leq f_{n+1}(x)$ for all $n = 1, 2, 3, \ldots$. The case when $f_n(x) \geq f_{n+1}(x)$ for all $n = 1, 2, 3, \ldots$ is considered in a similar way. Then the assumed conditions imply that $f_n(x) \leq f(x)$ for all $n = 1, 2, 3, \ldots$ as $f(x) = \lim_{n\to\infty} f_n(x)$.

Let us define $g_n(x) = f(x) - f_n(x)$ for all $n = 1, 2, 3, \ldots$. Then by Theorem 3.1.5, all functions $g_n(x)$ are continuous in the interval $[a, b]$. In addition, $g_n(x) \geq 0$ for all x from the interval $[a, b]$ because $f_n(x) \leq f(x)$ for all $n = 1, 2, 3, \ldots$. By Theorem 3.6.1, we have

$$\lim^{[a, b]}{}_{n\to\infty} g_n(x) = \lim^{[a, b]}{}_{n\to\infty} (f(x) - f_n(x)) =$$
$$\lim^{[a, b]}{}_{n\to\infty} f(x) - \lim^{[a, b]}{}_{n\to\infty} f_n(x) = f(x) - f(x) = 0$$

i.e., functions $g_n(x)$ converge to the function identically equal to 0 in the interval $[a, b]$.

In addition, $g_n(x) \geq g_{n+1}(x)$ for all $n = 1, 2, 3, \ldots$ as $f_n(x) \leq f_{n+1}(x) \leq f(x)$ for all $n = 1, 2, 3, \ldots$.

If convergence of $f_n(x)$ is not uniform, then convergence of $g_n(x)$ also is not uniform. In turn, this implies that there is $\varepsilon > 0$ such that for any number $n \in \boldsymbol{N}$, there is a number $m > n$ and a point c_m from the interval $[a, b]$ such that

$$g_m(c_m) > \varepsilon \tag{3.13}$$

This gives us an infinite sequence of functions $g_m(x)$ with $m \in N$. Note that in this sequence, indices (numbers m) do not necessarily constitute the whole set N of natural numbers. Besides, in this sequence $\{ g_m(x) \}$, we have $g_m(x) \geq g_k(x)$ for all points x from the interval $[a, b]$ when $m < k$ because $g_n(x) \geq g_{n+1}(x)$ for all $n = 1, 2, 3, \ldots$ and the order relation $\geq$ is transitive. Consequently, $g_m(c_k) > \varepsilon$ for all points c_k when $m < k$.

More over, we can change the numbering of the chosen points c_m so that in the new numbering, we have the sequence $c_1, c_2, c_3, \ldots, c_i, \ldots$ and the order of points c_i is the same in the old and new orderings, i.e., with old and new indices. Besides, if i is the new index of the point c_k, then $i \leq k$ and consequently, $g_i(c_i) \geq g_k(c_i) > \varepsilon$ for all $i = 1, 2, 3, \ldots$

As all points c_i belong to the interval $[a, b]$, the sequence $l = \{ c_1, c_2, c_3, \ldots, c_i, \ldots \}$ contains a converging subsequence h (cf. Section 2.1). Taking this sequence h and re-enumerating its members, we obtain a converging sequence $h = \{d_i \in \mathbf{R}; i = 1, 2, 3, \ldots\}$. By the same token as for the $l = \{ c_1, c_2, c_3, \ldots, c_i, \ldots \}$, we have

$$g_i(d_i) > \varepsilon \text{ for all } i = 1, 2, 3, \ldots$$

Let $a = \lim_{i\to\infty} d_i$. As the function sequence $F = \{ g_n(x); n = 1, 2, 3, \ldots\}$ converges to the function identically equal to 0 in the interval $[a, b]$, there is a number m such that $g_n(a) < \varepsilon/3$ for all $n > m$.

Let us fix some $n > m$ and consider the function $g_n(x)$. By the initial conditions, it is a continuous function. So, for the number $\varepsilon/3 > 0$, there is $\delta > 0$ such that the inequality $|a- c| < \delta$ implies the inequality $|g_n(a) - g_n(c)| < \varepsilon/3$.

As the sequence $h = \{d_i \in \mathbf{R}; i = 1, 2, 3, \ldots\}$ converges to a, there is a number k such that $|a- d_i| < \delta$ for all $i > k$. Taking $p = \max \{n, k\}$ and $i > p$, we have

$$g_n(d_i) = g_n(d_i) - g_n(a) + g_n(a) = | g_n(d_i) - g_n(a) + g_n(a) | \leq$$
$$| g_n(d_i) - g_n(a)| + |g_n(a)| < \varepsilon/3 + \varepsilon/3 = 2\varepsilon/3 < \varepsilon$$

This contradicts to our choice of the points d_i because for these points we have $g_i(d_i) > \varepsilon$. Thus, we come to a contradiction to the assumption that convergence of $f_n(x)$ is not uniform. By the principle of excluded middle, the sequence $\{ f_n(x); n = 1, 2, 3, \ldots\}$ converges uniformly in the interval $[a, b]$.

Theorem is proved.

Now we come to fuzzy convergence of function sequences.

Let X and Y be subsets of $\mathbf{R}$ and $f: X \to Y$, $f_n : X \to Y$ be real functions for all $n = 1, 2, 3, \ldots$.

Definition 3.6.6. A sequence of functions $\{ f_n(x) ; n = 1, 2, 3, \ldots\}$ *r-converges* to a function $f(x)$ *at a point* $a \in X$ if $f(a)$ is an r-limit of the sequence $\{ f_n(a); n = 1, 2, 3, \ldots\}$.

Example 3.6.5. The sequence of functions $\{ f_n(x) = (1 - |x|)^n ; n = 1, 2, 3, \ldots; x \in [-1, 1] \}$ does not converge at the point 0 to the function $f(x)$ identically equal to 0. However, this sequence 1-converges at the point 0 to the function $f(x)$.

Let $Z \subset X$.

Definition 3.6.7. A sequence of functions $\{ f_n(x) ; n = 1, 2, 3, \ldots\}$ *r-converges in* Z to a function $f(x)$ if this sequence r-converges to $f(x)$ at all points from Z.

It is denoted by $f(x) = r\text{-}\lim^{Z}_{n\to\infty} f_n(x)$.

Example 3.6.6. The sequence of functions $\{ f_n(x) = (1 - |x|)^n ; n = 1, 2, 3, \ldots; x \in [-1, 1] \}$ does not converge in the interval [-1, 1] to the function $f(x)$ identically equal to 0. However, this sequence 1-converges in [-1, 1] to the function $f(x)$.

Example 3.6.7. The sequence of functions $\{ f_n(x); n = 1, 2, 3, \ldots \}$ where $f_n(x) = \sin x/n$ when $n\pi/6 \leq x \leq n\pi$ and equal to 0 at all other points from $\boldsymbol{R}$ does not converge in $\boldsymbol{R}$ to the function $f(x)$ identically equal to 0. However, this sequence 1-converges in $\boldsymbol{R}$ to the function $f(x)$.

Definition 3.6.8. A sequence of functions $\{ f_n(x) ; n = 1, 2, 3, \ldots \}$ *uniformly r-converges in Z* to a function $f(x)$ if for any $\varepsilon > 0$, there is $m \in \omega$ such that for any point $a \in Z$, the inequality $| f_n(a) - f(a) | < r + \varepsilon$ is true whenever $n > m$.

Example 3.6.8. Both sequences of functions from Examples 3.6.6 and 3.6.7 do not uniformly converge, but they uniformly 1-converge.

Lemma 3.6.3. A sequence $\{ f_n(x) ; n = 1, 2, 3, \ldots \}$ [uniformly] 0-converges in Z (at a point a) to a function $f(x)$ if and only if this sequence [uniformly] converges in Z (at the point a) to $f(x)$.

Proof directly follows from definitions.

Lemma 3.6.4. If a sequence $\{ f_n(x) ; n = 1, 2, 3, \ldots \}$ [uniformly] r-converges in Z (at a point a) to a function $f(x)$, then the sequence $\{ f_n(x) ; n = 1, 2, 3, \ldots \}$ [uniformly] q-converges in Z (at the point a) to $f(x)$ for any $q > r$.

Indeed, if $f(a)$ is an r-limit of the sequence $\{ f_n(a); n = 1, 2, 3, \ldots \}$ and $q > r$, then by Lemma 2.2.4, $f(a)$ is a q-limit of the sequence $\{ f_n(a); n = 1, 2, 3, \ldots \}$. In the same way, if a sequence of functions $\{ f_n(x) ; n = 1, 2, 3, \ldots \}$ uniformly r-converges in Z to a function $f(x)$, then for any $\varepsilon > 0$, there is $m \in \omega$ such that for any point $a \in Z$, the inequality $| f_n(a) - f(a) | < r + \varepsilon$ is true whenever $n > m$. As $q > r$, it implies that for any $\varepsilon > 0$, there is $m \in \omega$ such that for any point $a \in Z$, the inequality $| f_n(a) - f(a) | < q + \varepsilon$ is true whenever $n > m$. It means that the sequence $\{ f_n(x) ; n = 1, 2, 3, \ldots \}$ uniformly q-converges in Z (at the point a) to $f(x)$.

Proposition 3.6.1. If a sequence $\{ f_n(x) ; n = 1, 2, 3, \ldots \}$ r-converges in Z to a function $f(x)$ and $| f_n(x) - f(x)| < q$ for all points $x \in Z$ and all $n = 1, 2, 3, \ldots$, then the sequence $\{ f_n(x) ; n = 1, 2, 3, \ldots \}$ uniformly q-converges in Z to $f(x)$.

Proof directly follows from definitions.

Lemma 3.6.5. If a sequence $\{ f_n(x) ; n = 1, 2, 3, \ldots \}$ uniformly r-converges in Z (at a point a) to a function $f(x)$, then the sequence $\{ f_n(x) ; n = 1, 2, 3, \ldots \}$ r-converges in Z (at the point a) to $f(x)$.

Theorem 2.3.7 implies the following result.

Theorem 3.6.8. If a sequence of functions $\{ f_n(x) ; n = 1, 2, 3, \ldots \}$ r-converges to a function $f(x)$ in X and a sequence of functions $\{ g_n(x) ; n = 1, 2, 3, \ldots \}$ q-converges to a function $g(x)$ in X, i.e., $f(x) = r\text{-}\lim^{X}_{n\to\infty} f_n(x)$ and $g(x) = q\text{-}\lim^{X}_{n\to\infty} g_n(x)$, then:

a) the sequence of functions $\{ (f_n + g_n)(x) ; n = 1, 2, 3, \ldots \}$ $(r + q)$-converges to a function $(f + g)(x)$ in X, i.e., $(f + g)(x) = (r + q)\text{-}\lim^{X}_{n\to\infty} (f_n + g_n)(x)$;
b) the sequence of functions $\{ (f_n - g_n)(x) ; n = 1, 2, 3, \ldots \}$ $(r + q)$-converges to a function $(f - g)(x)$ in X, i.e., $(f - g)(x) = (r + q)\text{-}\lim^{X}_{n\to\infty} (f_n - g_n)(x)$;
c) the sequence of functions $\{ kf_n(x) ; n = 1, 2, 3, \ldots \}$ $|k|\cdot r$-converges to a function $kf(x)$ in X, i.e., $kf(x) = |k|\cdot r\text{-}\lim^{X}_{n\to\infty} kf_n(x)$.

Proof. a) Let us assume that a sequence of functions $\{ f_i(x) ; i = 1, 2, 3, \ldots\}$ r-converges to a function $f(x)$ in X and a sequence of functions $\{ g_i(x) ; i = 1, 2, 3, \ldots\}$ q-converges to a function $g(x)$ in X, i.e., $f(x) = r\text{-}\lim^X_{i\to\infty} f_n(x)$ and $g(x) = q\text{-}\lim^X_{i\to\infty} g_n(x)$.

Then by Theorem 2.3.7, the sequence of functions $\{ (f_n + g_n)(x) ; n = 1, 2, 3, \ldots\}$ $(r + q)$-converges to a function $(f + g)(x)$ in X, i.e., $(f + g)(x) = (r + q)\text{-}\lim^X_{n\to\infty} (f_n + g_n)(x)$, the sequence of functions $\{ (f_n - g_n)(x) ; n = 1, 2, 3, \ldots\}$ $(r + q)$-converges to a function $(f - g)(x)$ in X, i.e., $(f - g)(x) = (r + q)\text{-}\lim^X_{n\to\infty} (f_n - g_n)(x)$, and the sequence of functions $\{ kf_n(x) ; n = 1, 2, 3, \ldots\}$ $|k|\cdot r$-converges to a function $kf(x)$ in X, i.e., $kf(x) = |k|\cdot r\text{-}\lim^X_{n\to\infty} kf_n(x)$.

Theorem is proved.

Theorem 3.6.9. If a sequence of r-continuous at a point $a \in X$ functions $\{ f_n(x) ; n = 1, 2, 3, \ldots\}$ uniformly q-converges to a function $f(x)$, then $f(x)$ is $(r + 2q)$-continuous at the point a.

Proof. Let us assume that the sequence of functions $f_n(x)$ uniformly q-converges to the function $f(x)$ in some interval $[u, v]$ that contains a and consider some $\varepsilon > 0$. Taking, if necessary, a smaller interval, we can assume that a is the midpoint of the interval $[u, v]$ and its length is 2α. Then by Definition 3.6.8, there is such a number m that for any number $n > m$ and any point d from $[u, v]$, we have $| f(d) - f_n(d)| < q + \varepsilon/3$. Let us take such a number n. As all functions $f_n(x)$ are r-continuous at the point a, there is $\delta > 0$ such that if $| c - a| < \delta$, then $|f_n(c) - f_n(a)| < r + \varepsilon/3$.

Let us consider the difference $| f(c) - f(a) |$. If $\lambda = \min \{\alpha, \delta\}$ and $| c - a| < \lambda$, then c belongs to the interval $[u, v]$ and from the properties of absolute values, we have

$$| f(c) - f(a) | = | f(c) - f_n(c) + f_n(c) - f_n(a) + f_n(a) - f(a) | \leq$$
$$| f(c) - f_n(c)| + | f_n(c) - f_n(a)| + | f_n(a) - f(a) | <$$
$$q + \varepsilon/3 + r + \varepsilon/3 + q + \varepsilon/3 = r + 2q + \varepsilon$$

because $| f(c) - f_n(c)| < q + \varepsilon/3$, $|f_n(a) - f(a) | < q + \varepsilon/3$, and $|f_n(c) - f_n(a)| < r + \varepsilon/3$.

It means that the function $f(x)$ is $(r + 2q)$-continuous at the point a.

Theorem is proved.

Corollary 3.6.1. If a sequence of r-continuous at a point $a \in X$ functions $\{ f_n(x) ; n = 1, 2, 3, \ldots\}$ uniformly r-converges to a function $f(x)$, then $f(x)$ is $3r$-continuous at the point a.

Analyzing the proof of Theorem 3.6.4, we see that if a sequence of functions $f_n(x)$ with $n = 1, 2, 3, \ldots$ uniformly q-converges to the function $f(x)$ in some interval $[u, v]$, then $f(x)$ is $(r + 2q)$-continuous at any point d from (u, v) where all functions $f_n(x)$ are r-continuous. This gives us the following result.

Theorem 3.6.10. If a sequence of r-continuous inside an interval $[a, b]$ (in $\boldsymbol{R}$) functions $\{f_n(x) ; n = 1, 2, 3, \ldots\}$ uniformly q-converges to a function $f(x)$, then $f(x)$ is $(r + 2q)$-continuous inside the interval $[a, b]$ (in $\boldsymbol{R}$).

Corollary 3.6.2. If a sequence of r-continuous inside an interval $[a, b]$ (in $\boldsymbol{R}$) functions $\{f_n(x) ; n = 1, 2, 3, \ldots\}$ uniformly r-converges to a function $f(x)$, then $f(x)$ is $3r$-continuous inside the interval $[a, b]$ (in $\boldsymbol{R}$).

If we have a sequence of continuous or fuzzy continuous functions, their continuity can be coordinated or non-coordinated. In the first case, this sequence has better properties. The next definition gives a formalization of continuity coordination in a sequence.

Definition 3.6.9. a) A sequence of functions $\{ f_n(x) ; n = 1, 2, 3, \ldots\}$ is (q, r)-*equicontinuous at a point* $a \in \mathbf{R}$ if for any $\varepsilon > 0$, there is $\delta > 0$ such that the inequality $| a - x | < q + \delta$ implies the inequality $| f_n(x) - f_n(a) | < r + \varepsilon$ for any $n = 1, 2, 3, \ldots$ and any $x \in X$.

b) A sequence of functions $\{ f_n(x) ; n = 1, 2, 3, \ldots\}$ is *r-equicontinuous at a point* $a \in \mathbf{R}$ if for any $\varepsilon > 0$, there is $\delta > 0$ such that the inequality $| a - x | < \delta$ implies the inequality $| f_n(x) - f_n(a) | < r + \varepsilon$ for any $n = 1, 2, 3, \ldots$ and any $x \in X$.

c) A sequence of functions $\{ f_n(x) ; n = 1, 2, 3, \ldots\}$ is (q, r)-*equicontinuous* in (inside) X if the sequence of $\{ f_n(x) ; n = 1, 2, 3, \ldots\}$ is (q, r)-equicontinuous at all points $a \in X$;

d) A sequence of functions $\{ f_n(x) ; n = 1, 2, 3, \ldots\}$ is *r-equicontinuous* in (inside) X if the sequence of $\{ f_n(x) ; n = 1, 2, 3, \ldots\}$ is r-equicontinuous at all points $a \in X$.

Lemma 3.6.6. A sequence of (q, r)-continuous (r-continuous) at a point $a \in \mathbf{R}$ functions $\{ f_n(x) ; n = 1, 2, 3, \ldots\}$ is (q, r)-equicontinuous (r-equicontinuous) at a point $a \in \mathbf{R}$ if and only if for any $\varepsilon > 0$, there is $\delta > 0$ such that the inequality $| a - x | < q + \delta$ (the inequality $| a - x | < \delta$) implies the inequality $| f_n(x) - f_n(a) | < r + \varepsilon$ for almost all $n = 1, 2, 3, \ldots$.

Theorem 3.6.11. If an r-equicontinuous at a point a sequence of functions $\{ f_n(x) ; n = 1, 2, 3, \ldots\}$ q-converges at a to an r-continuous at a function $f(x)$, then functions $f_n(x)$ uniformly $(2r + q)$-converge to $f(x)$ at the point a.

Proof. Let us assume that an r-equicontinuous at a point a sequence of functions $f_n(x)$ q-converges at the point a to an r-continuous at the point a function $f(x)$. If this convergence is not a uniform $(2r + q)$-convergence at the point a, then the sequence $\{ f_n(x); n = 1, 2, 3, \ldots \}$ has the following property: for any interval $[u, v]$ that contains a, there is $\varepsilon > 0$ such that for any $n = 1, 2, 3, \ldots$, there is a point d_n such that $| f(d_n) - f_n(d_n)| > 2r + q + \varepsilon$.

Let us take the sequence of intervals $[a - 1/m , a + 1/m]$ where $m = 1, 2, 3, \ldots$. Properties of the sequence $\{ f_n(x); n = 1, 2, 3, \ldots \}$ allow us to find a the sequence of points $\{ d_m; m = 1, 2, 3, \ldots \}$ such that each point d_m belongs to $[a - 1/m , a + 1/m]$ and $| f(d_n) - f_n(d_n)| > 2r + q + \varepsilon$.

As each point d_m belongs to $[a - 1/m , a + 1/m]$, we have $a = \lim_{m\to\infty} d_m$. Then there is $\nu > 0$ such that $| d_m - a| < \nu$ implies $| f(d_m) - f(a)| < r + \varepsilon/3$ because the function $f(x)$ is r-continuous at the point a. In addition, there is $\kappa > 0$ such that $| d_m - a| < \kappa$ implies $| f_n(d_m) - f_n(a)| < r + \varepsilon/3$ for all $n = 1, 2, 3, \ldots$ because the sequence of functions $f_n(x)$ is r-equicontinuous at a.

Then when $1/m < \delta = \min \{\nu, \kappa\}$, we have

$$| f(d_n) - f(a)| < r + \varepsilon/3$$

and

$$| f_n(d_n) - f_n(a)| < r + \varepsilon/3 \text{ for all } n > m.$$

The sequence $\{ f_n(x); n = 1, 2, 3, \ldots \}$ q-converges at the point a to the function $f(x)$. Consequently, there is $k \in \omega$ such that $| f(a) - f_n(a)| < q + \varepsilon/3$ for all $n > k$.

Let us take $n > \max \{k, m\}$. Then we have

$$| f(d_n) - f_n(d_n)| = | f(d_n) - f(a) + f(a) - f_n(a) + f_n(a) - f_n(d_n)| \leq$$
$$| f(d_n) - f(a)| + | f(a) - f_n(a)| + | f_n(a) - f_n(d_n)| <$$

$$r + \varepsilon/3 + q + \varepsilon/3 + r + \varepsilon/3 = 2r + q + \varepsilon.$$

This contradicts to the possibility that the sequence $\{ f_n(x);\ n = 1, 2, 3, \ldots \}$ does not uniformly $(2r + q)$-converges at the point a to the function $f(x)$. By contradiction, we have that functions $f_n(x)$ uniformly $(2r + q)$-converge to $f(x)$ at the point a.

Theorem is proved.

Corollary 3.6.3. If an r-equicontinuous at a point a sequence of functions $\{ f_n(x) ;\ n = 1, 2, 3, \ldots\}$ r-converges at a to an r-continuous at a function $f(x)$, then functions $f_n(x)$ uniformly $3r$-converge to $f(x)$ at the point a.

Theorem 3.6.12. If an r-equicontinuous inside an interval $[a, b]$ sequence of functions $\{f_n(x) ;\ n = 1, 2, 3, \ldots\}$ q-converges inside $[a, b]$ to an r-continuous inside $[a, b]$ function $f(x)$, then functions $f_n(x)$ uniformly $(2r + q)$-converge to $f(x)$ inside the interval $[a, b]$.

Proof. Let us assume that an r-equicontinuous inside an interval $[a, b]$ sequence of functions $f_n(x)$ q-converges inside $[a, b]$ to an r-continuous inside $[a, b]$ function $f(x)$. If this convergence is not a uniform $(2r + q)$-convergence inside $[a, b]$, then the sequence $\{ f_n(x);\ n = 1, 2, 3, \ldots \}$ has the following property: there is $\varepsilon > 0$ such that for any $n = 1, 2, 3, \ldots$, there is a point a_n from the interval $[a, b]$ such that $|f(d_n) - f_n(d_n)| > 2r + q + \varepsilon$.

The set $\{ a_n;\ n = 1, 2, 3, \ldots \}$ is bounded. Consequently (cf. Section 2.1), it contains a converging subsequence of points $\{ d_m;\ m = 1, 2, 3, \ldots \}$. Let us consider the limit $a = \lim_{m\to\infty} d_m$. Then there is $\nu > 0$ such that $| d_m - a| < \nu$ implies $| f(d_m) - f(a)| < r + \varepsilon/3$ because the function $f(x)$ is r-continuous inside the interval $[a, b]$ and thus, at the point a. In addition, there is $\kappa > 0$ such that $| d_m - a| < \kappa$ implies $| f_n(d_m) - f_n(a)| < r + \varepsilon/3$ for all $n = 1, 2, 3, \ldots$ because the sequence of functions $f_n(x)$ is r-equicontinuous inside the interval $[a, b]$ and thus, r-equicontinuous at a.

Then when $1/m < \delta = \min \{\nu, \kappa\}$, we have $| f(d_n) - f(a)| < r + \varepsilon/3$ and $| f_n(d_n) - f_n(a)| < r + \varepsilon/3$ for all $n > m$.

The sequence $\{ f_n(x);\ n = 1, 2, 3, \ldots \}$ q-converges inside the interval $[a, b]$ and thus, at the point a, to the function $f(x)$. Consequently, there is $k \in \omega$ such that $| f(a) - f_n(a)| < q + \varepsilon/3$ for all $n > k$.

Let us take $n > \max \{k, m\}$. Then we have

$$\begin{gathered} | f(d_n) - f_n(d_n)| = | f(d_n) - f(a) + f(a) - f_n(a) + f_n(a) - f_n(d_n)| \leq \\ | f(d_n) - f(a)| + | f(a) - f_n(a)| + | f_n(a) - f_n(d_n)| < \\ r + \varepsilon/3 + q + \varepsilon/3 + r + \varepsilon/3 = 2r + q + \varepsilon. \end{gathered}$$

This contradicts to the possibility that the sequence $\{ f_n(x);\ n = 1, 2, 3, \ldots \}$ does not uniformly $(2r + q)$-converges inside the interval $[a, b]$ to the function $f(x)$. By contradiction, we have that functions $f_n(x)$ uniformly $(2r + q)$-converge to $f(x)$ inside the interval $[a, b]$.

Theorem is proved.

Corollary 3.6.4. If an r-equicontinuous inside an interval $[a, b]$ sequence of functions $\{ f_n(x) ;\ n = 1, 2, 3, \ldots\}$ r-converges inside $[a, b]$ to an r-continuous inside $[a, b]$ function $f(x)$, then functions $f_n(x)$ uniformly $3r$-converge to $f(x)$ inside the interval $[a, b]$.

Corollary 3.6.5. If a sequence of fuzzy uniformly continuous functions $\{ f_n(x) ;\ n = 1, 2, 3, \ldots\}$ converges to a fuzzy continuous inside X function $f(x)$, then functions $f(x)$ fuzzy uniformly converge to $f(x)$ inside X.

We already know that if a monotone sequence of real numbers $l = \{ a_n ; n = 1, 2, 3, \ldots\}$ q-converges for some $q \geq 0$, then by Proposition 2.2.3, the sequence l converges in the classical sense. As convergence of functions is induces by convergence of numbers, we have a similar result for sequences of functions.

Lemma 3.6.7. If a monotone sequence of functions $\{ f_n(x) ; n = 1, 2, 3, \ldots\}$ q-converges, then this sequence converges in the classical sense.

This allows us to prove a fuzzy counterpart of the Dini Theorem.

Theorem 3.6.13 (The Fuzzy Dini Theorem). If a monotone sequence of r-continuous functions $\{ f_n(x) ; n = 1, 2, 3, \ldots\}$ q-converges to a p-continuous function $f(x)$ inside a closed interval $[a, b]$, then functions $f_n(x)$ uniformly $(r + p)$-converge to $f(x)$ inside the interval $[a, b]$.

Proof. Let us consider a monotone sequence of continuous functions $\{ f_n(x) ; n = 1, 2, 3, \ldots\}$ q-converges to a continuous function $f(x)$ inside a closed interval $[a, b]$. As the q-convergence is monotone, by Lemma 3.6.7, the sequence $\{ f_n(x); n = 1, 2, 3, \ldots \}$ converges in the interval $[a, b]$ to $f(x)$ in the classical sense and this convergence is monotone.

Let us assume that the sequence $\{ f_n(x); n = 1, 2, 3, \ldots \}$ does not uniformly $(r + p)$-converge to $f(x)$ and $f_n(x) \leq f_{n+1}(x)$ for all $n = 1, 2, 3, \ldots$. The case $f_n(x) \geq f_{n+1}(x)$ is considered in a similar way. Then the assumed conditions imply that $f_n(x) \leq f(x)$ for all $n = 1, 2, 3, \ldots$ in the interval $[a, b]$.

Let us define $g_n(x) = f(x) - f_n(x)$ for all $n = 1, 2, 3, \ldots$. Then by Theorem 3.4.9 and Corollary 3.4.5, all functions $g_n(x)$ are $(r + p)$-continuous in the interval $[a, b]$. In addition, $g_n(x) \geq 0$ for all x from the interval $[a, b]$ because $f_n(x) \leq f(x)$ for all $n = 1, 2, 3, \ldots$. By Theorem 3.6.1, we have

$$\lim^{[a, b]}{}_{n\to\infty} g_n(x) = \lim^{[a, b]}{}_{n\to\infty} (f(x) - f_n(x)) =$$
$$\lim^{[a, b]}{}_{n\to\infty} f(x) - \lim^{[a, b]}{}_{n\to\infty} f_n(x) = f(x) - f(x) = 0$$

i.e., functions $g_n(x)$ converge to the function identically equal to 0 in the interval $[a, b]$.

In addition, $g_n(x) \geq g_{n+1}(x)$ for all $n = 1, 2, 3, \ldots$ as $f_n(x) \leq f_{n+1}(x) \leq f(x)$ for all $n = 1, 2, 3, \ldots$.

If convergence of $f_n(x)$ is not a uniform $(r + p)$-convergence, then convergence of $g_n(x)$ is not a uniform $(r + p)$-convergence as $|f(x) - f_n(x)| = | g_n(x) - 0|$ for all x from the interval $[a, b]$. In turn, this implies that there is $\varepsilon > 0$ such that for any number $n \in N$, there is a number $m > n$ and a point c_m from the interval $[a, b]$ such that

$$g_m(c_m) > r + p + \varepsilon \qquad (3.14)$$

This gives us an infinite sequence of functions $g_m(x)$ with $m \in N$. Note that in this sequence, indices (numbers m) do not necessarily constitute the whole set N of natural numbers. Besides, in this sequence $\{ g_m(x) \}$, we have $g_m(x) \geq g_k(x)$ for all points x from the interval $[a, b]$ when $m < k$ because $g_n(x) \geq g_{n+1}(x)$ for all $n = 1, 2, 3, \ldots$ and the order relation $\geq$ is transitive. Consequently, $g_m(c_k) > r + p + \varepsilon$ for all points c_k when $m < k$.

Moreover, we can change the numbering of the chosen points c_m so that in the new numbering, we have the sequence $c_1 , c_2 , c_3 , \ldots , c_i , \ldots$ and the order of points c_i is the same in the old and new orderings, i.e., with old and new indices. Besides, if i is the new index of the point c_k , then $i \leq k$ and consequently, $g_i(c_i) \geq g_k(c_i) > r + p + \varepsilon$ for all $i = 1, 2, 3, \ldots$

As all points c_i belong to the interval $[a, b]$, the sequence $l = \{ c_1, c_2, c_3, \dots, c_i, \dots \}$ contains a converging subsequence h (cf. Section 2.1). Taking this sequence h and re-enumerating its members, we obtain a converging sequence $h = \{d_i \in \mathbf{R};\ i = 1, 2, 3, \dots\}$. By the same token as for the $l = \{ c_1, c_2, c_3, \dots, c_i, \dots \}$, we have

$$g_i(d_i) > r + p + \varepsilon \text{ for all } i = 1, 2, 3, \dots$$

Let $a = \lim_{i\to\infty} d_i$. As the function sequence $F = \{ g_n(x);\ n = 1, 2, 3, \dots\}$ converges to the function identically equal to 0 in the interval $[a, b]$, there is a number m such that $g_n(a) < \varepsilon/3$ for all $n > m$.

Let us fix some $n > m$ and consider the function $g_n(x)$. As we have demonstrated, it is an $(r + p)$-continuous function. So, for the number $\varepsilon/3 > 0$, there is $\delta > 0$ such that the inequality $|a- c| < \delta$ implies the inequality $|g_n(a) - g_n(c)| < r + p + \varepsilon/3$.

As the sequence $h = \{d_i \in \mathbf{R};\ i = 1, 2, 3, \dots\}$ converges to a, there is a number k such that $|a- d_i| < \delta$ for all $i > k$. Taking $m = \max \{n, k\}$ and $i > m$, we have

$$g_n(d_i) = g_n(d_i) - g_n(a) + g_n(a) = |\, g_n(d_i) - g_n(a) + g_n(a)\, | \leq$$
$$|\, g_n(d_i) - g_n(a)| + |g_n(a)| < r + p + \varepsilon/3 + \varepsilon/3 = r + p + 2\varepsilon/3 < r + p + \varepsilon$$

This contradicts to our choice of the points d_i because for these points we have $g_i(d_i) > r + p + \varepsilon$. Thus, we come to a contradiction to the assumption that $(r + p)$-convergence of $f_n(x)$ is not uniform. By the principle of excluded middle, the sequence $\{ f_n(x);\ n = 1, 2, 3, \dots\}$ uniformly $(r + p)$-converges in the interval $[a, b]$.

Theorem is proved.

Remark 3.6.1. Although, the sequence $\{ f_n(x);\ n = 1, 2, 3, \dots \}$ converges in the interval $[a, b]$ to $f(x)$ in the classical sense and this convergence is monotone, in general, it is not true that the sequence $\{ f_n(x);\ n = 1, 2, 3, \dots \}$ uniformly converges to $f(x)$ as the following example demonstrates.

Example 3.6.9. The monotone (increasing) sequence of continuous functions $\{ f_n(x) = (1 - |x|)^n;\ n = 1, 2, 3, \dots;\ x \in [0, 1] \}$ converges in the interval $[0, 1]$ to the function $f(x)$ that is identically equal to 1 for all x from the interval $[0, 1)$ and $f(1) = 0$. However, this sequence does not uniformly converges to $f(x)$. As it is easy to show, this sequence uniformly 1-converges in $[0, 1]$ to the function $f(x)$. Note that $f(x)$ is 1-continuous, what well correlates with the result of the Fuzzy Dini Theorem.

Theorem 3.6.14. If a sequence of functions $\{ f_n(x);\ n = 1, 2, 3, \dots\}$ uniformly r-converges to a function $f(x)$ in X and a sequence of functions $\{ g_n(x);\ n = 1, 2, 3, \dots\}$ uniformly q-converges to a function $g(x)$ in X, then:

a) the sequence of functions $\{ (f_n + g_n)(x);\ n = 1, 2, 3, \dots\}$ uniformly $(r + q)$-converges to a function $(f + g)(x)$ in X, i.e., $(f + g)(x) = (r + q)\text{-}\lim_{n\to\infty} (f_n + g_n)(x)$;
b) the sequence of functions $\{ (f_n - g_n)(x);\ n = 1, 2, 3, \dots\}$ uniformly $(r + q)$-converges to a function $(f - g)(x)$ in X, i.e., $(f - g)(x) = (r + q)\text{-}\lim_{n\to\infty} (f_n - g_n)(x)$;
c) the sequence of functions $\{ kf_n(x);\ n = 1, 2, 3, \dots\}$ uniformly $|k|\cdot r$-converges to a function $kf(x)$ in X, i.e., $kf(x) = |k|\cdot r\text{-}\lim_{n\to\infty} kf_n(x)$.

Proof. a) Let us assume that a sequence of functions $\{ f_i(x) ; i = 1, 2, 3, \ldots\}$ uniformly r-converges to a function $f(x)$ in X and a sequence of functions $\{ g_i(x) ; i = 1, 2, 3, \ldots\}$ uniformly q-converges to a function $g(x)$ in X. By Definition 3.6.8, it means that for any $\varepsilon > 0$, there is $m \in \omega$ such that for any point $a \in X$, the inequality $| f_i(a) - f(a) | < r + \varepsilon/2$ is true whenever $i > m$ and there is $n \in \omega$ such that for any point $a \in X$, the inequality $| g_i(a) - g(a) | < q + \varepsilon/2$ is true whenever $i > m$.

Then for all $i > \max \{n, m\}$ and any point $a \in X$, we have

$$| (f_i + g_i)(a) - (f + g)(a) | = | f_i(a) - f(a) + g_i(a) - g(a) | \leq$$
$$| f_i(a) - f(a)| + | g_i(a) - g(a) | < r + \varepsilon/2 + q + \varepsilon/2 < r + q + \varepsilon$$

By Definition 3.6.8, it means that the sequence of functions $\{ (f_n + g_n)(x) ; n = 1, 2, 3, \ldots\}$ uniformly $(r + q)$-converges to a function $(f + g)(x)$ in X, i.e., $(f + g)(x) = (r + q)\text{-}\lim_{n\to\infty} (f_n + g_n)(x)$. Part a) from Theorem 3.6.14 is proved.

Parts b) and c) are proved in a similar way.

Let us assume that $X \subseteq \text{Dom } f_i(x)$ for all functions $f_i(x)$ with $i = 1, 2, 3, \ldots$.

Definition 3.6.10. a) A sequence of functions $l = \{ f_i(x) ; i = 1, 2, 3, \ldots\}$ is called *uniformly fundamental* (also called a *uniformly Cauchy sequence*) in X if for any $\varepsilon \in \mathbf{R}^{++}$, there is $n \in \mathbf{N}$ such that for any $x \in X$ and any $i, j \geq n$, we have $| f_j(x) - f_i(x) | < \varepsilon$.

b) A sequence of functions $l = \{ f_i(x) ; i = 1, 2, 3, \ldots\}$ is called *uniformly r-fundamental* in X if for any $\varepsilon \in \mathbf{R}^{++}$, there is $n \in \mathbf{N}$ such that for any $x \in X$ and any $i, j \geq n$, we have $| f_j(x) - f_i(x) | < 2r + \varepsilon$.

c) A sequence $l = \{ f_i(x) ; i = 1, 2, 3, \ldots\}$ is called *uniformly fuzzy fundamental* if it is uniformly r-fundamental for some $r \in \mathbf{R}^{+}$.

Lemma 3.6.8. If $r \leq p$, then any uniformly r-fundamental sequence is uniformly p-fundamental.

Lemma 3.6.9. A sequence $l = \{ f_i(x) ; i = 1, 2, 3, \ldots\}$ is uniformly fundamental (a uniformly Cauchy sequence) if and only if it is uniformly 0-fundamental.

Lemma 3.6.10. A subsequence of a uniformly r-fundamental sequence is uniformly r-fundamental.

Theorem 3.6.15 (The Extended Cauchy Criterion for sequences of functions). A sequence $l = \{ f_i(x) ; i = 1, 2, 3, \ldots\}$ uniformly r-converges in X if and only if it is uniformly r-fundamental in X.

Proof. Necessity. Let sequence of functions $l = \{ f_i(x) ; i = 1, 2, 3, \ldots\}$ uniformly r-converges in X, $f(x) = r\text{-}\lim l$ and $\varepsilon \in \mathbf{R}^{++}$. Then by Definition 3.6.10 there is a number $n \in \omega$ such that for any $x \in X$ and any $i > n$, we have $| f(x) - f_i(x) | \leq r + \varepsilon/2$. Consequently, for any $x \in X$ and any $i, j > n$, we have $| f_i(x), - f_j(x) | \leq | f(x) - f_i(x) | + | f(x) - f_j(x) | \leq 2r + 2(\varepsilon/2) < 2r + \varepsilon$. Thus, l is a uniformly r-fundamental sequence.

Sufficiency. Let $l = \{ f_i(x) ; i = 1, 2, 3, \ldots\}$ be a uniformly r-fundamental sequence of functions. Then by the Extended Cauchy Criterion for sequences (Theorem 2.2.7), each sequence of numbers $\{ f_i(a) ; i = 1, 2, 3, \ldots\}$ r-converges at each point a from X. Consequently, the sequence of functions $l = \{ f_i(x) ; i = 1, 2, 3, \ldots\}$ r-converges in X, and we need only to prove that this convergence is uniform.

Let us fix some number m and take $k_m = 1/m$. Then there is a number n, which is dependent on m and thus, denoted by $n(m)$, such that for all $i, j > n(m)$ and any $x \in X$, we have

$| f_i(x) - f_j(x) | < 2r + (1/m)$. That is, all numbers $f_i(x)$ with $i > n(m)$ and $x \in X$ belong to a closed interval I_m. Really, let us put $T_m(x) = \{ f_i(x) ; i > n(m) \}$, $b(x) = \sup T_m(x)$, and $c(x) = \inf T_m(x)$ for all $x \in X$. Then all values $f_i(x)$ belong to the interval $I_m(x) = [c(x), b(x)]$ for $i > n(m)$. Note that the number $n(m)$ does not depend on x.

Let us estimate the length of this interval $I_m(x)$ for an arbitrary. Suppose that $| b(x) - c(x) | > 2r + (1/m)$. It means that for some $h \in \boldsymbol{R}^+$, we have $| b - c | > 2r + (1/m) + h$. At the same time, there are elements $f_p(x)$ and $f_q(x)$, for which $p, q > n(m)$, $| b(x) - f_p(x) | \leq (1/3)\cdot h$, and $| c(x) - f_q(x) | \leq (1/3)\cdot h$ because $b(x)$ is the supremum and $c(x)$ is the infimum of all elements from the set $T_m(x)$. Consequently, for these $f_p (x)$ and $f_q (x)$, we have

$$| b(x) - c(x) | \leq | b(x) - f_p(x) | + | f_p(x) - f_q(x) | + |f_q (x) - c(x) | <$$
$$(1/3)\cdot h + (2r + (1/m)) +(1/3)\cdot h =$$
$$2r + (1/m) +(2/3)h < 2r + (1/m) + h.$$

This contradicts our supposition that $| b(x) - c(x) | > 2r + (1/m) + h$. Thus, the length of $I_m(x) = [c(x), b(x)]$ is not larger than $2r + (1/m)$.

We can choose these intervals $I_m(x)$ so that the inclusion $I_{m+1}(x) \subseteq I_m(x)$ is valid for all $m = 1, 2, 3, \ldots$. In such a way, we obtain a sequence of nested closed intervals $\{ I_m(x); m = 1, 2, 3, \ldots \}$. The space $\boldsymbol{R}$ is a complete metric space (cf. Appendices B and C). Consequently, the intersection $I(x) = \bigcap_{m=1}^{\infty} I_m(x)$ is non-empty. That is, the interval $I(x)$ consists of one point d or $I(x)$ is a closed interval having the length not larger than $2r$ as the length of $I_m(x)$ is not larger than $2r + (1/m)$. When $I(x) = \{d(x) \}$, the sequence $l(x)$ converges, and thus, by Lemma 2.2.11, $l(x)$ has an r-limit $d(x)$ and $| d(x) - f_i(x) | < r + 1/m$ when $i > n(m)$.

When I(x) is a non-trivial interval, we consider its midpoint $e(x)$. Then for any number $k \in \boldsymbol{R}^{++}$, there is some $I_m(x) \supseteq I(x)$ such that $k > (1/m)$, $| e(x) - e_m(x) | < (1/3) k$ for the center $e_m(x)$ of $I_m(x)$ and almost all $f_i(x)$ belong to $I_m(x)$. Besides, $| e_m(x) - f_i(x) | < r + (1/2m)$ when $i > n(m)$. Consequently, as $k > (1/m)$, we have

$$| e(x) - f_i(x) | \leq | e_m(x) - f_i(x) | + | e(x) - e_m(x) | <$$
$$(r + (1/2m)) + (1/3)k < r + (1/2)k + (1/3)k < r + k$$

when $i > n(m)$, i.e., the point $e(x)$ is an r-limit of l and $| e(x) - f_i(x) | < r + k$.

Thus, in both cases, the sequence of functions l uniformly r-converges in X as the number $n(m)$ does not depend on x.

Theorem is proved.

From Theorem 3.6.15, we obtain the following result.

Corollary 3.6.6 (The General Fuzzy Convergence Criterion for sequences of functions). A sequence of functions $l = \{ f_i(x) ; i = 1, 2, 3, \ldots\}$ uniformly fuzzy converges if and only if it is uniformly fuzzy fundamental (a uniformly fuzzy Cauchy sequence).

Theorem 3.6.15 and Lemmas 3.6.3 and 3.6.9 imply such classical result as the Cauchy Criterion for sequences of functions.

Corollary 3.6.7 (The Cauchy Criterion for sequences of functions). A sequence $l = \{f_i(x) ; i = 1, 2, 3, \ldots\}$ uniformly converges if and only if it is a uniformly Cauchy sequence (a uniformly fundamental sequence).

Remark 3.6.2. Summation of function series is traditionally defined by convergence of the sequence of its partial sums (cf. Section 2.7). Thus, it is possible to apply all results obtained here for convergence, uniform convergence, fuzzy convergence, and uniform fuzzy convergence of sequences of functions to convergence (summation), uniform convergence, fuzzy convergence (fuzzy summation), and uniform fuzzy convergence of function series. This is very important for many applications because function series are used in physics, engineering and other areas for approximation of functions and computing these approximations.

3.7. Fuzzy Continuous Mappings of Metric Spaces and Normed Linear Spaces

Continuity is one of the things
I like about New England.
Tracy Kidder (1945-)

As it is explained in Appendix C (cf. Proposition C.2), any normed linear space is a metric space. That is why, we primarily consider metric space. We use linear spaces only when arithmetical operations with functions (mappings) are involved.

Let $\boldsymbol{E}$ and $\boldsymbol{L}$ be a metric spaces with a metrics (distances), which we denote by the same letter $\mathbf{d}$.

Definition 3.7.1. A mapping f: $\boldsymbol{E} \to \boldsymbol{L}$ is called *continuous at a point* $a \in \boldsymbol{E}$ if for any $\varepsilon > 0$, there is $\delta > 0$ such that the inequality $\mathbf{d}(a, x) < \delta$ implies the inequality $\mathbf{d}(f(x), f(a)) < \varepsilon$.

In Section 3.1, there are many examples of continuous functions in such a metric space as $\boldsymbol{R}$ with the distance $\mathbf{d}(a, b) = |a - b|$.

Definition 3.7.2. A mapping f: $\boldsymbol{E} \to \boldsymbol{L}$ is called *r-continuous at a point* $a \in \boldsymbol{E}$ if for any $\varepsilon > 0$, there is $\delta > 0$ such that the inequality $\mathbf{d}(a, x) < \delta$ implies the inequality $\mathbf{d}(f(x), f(a)) < r + \varepsilon$, or in other words, for any x with $\mathbf{d}(a, x) < \delta$, we have $\mathbf{d}(f(x), f(a)) < r + \varepsilon$.

In Sections 3.3 and 3.4, there are many examples of r-continuous functions in such a metric space as $\boldsymbol{R}$ with the distance $\mathbf{d}(a, b) = | a - b|$.

Lemma 3.7.1. A mapping f is continuous at a point $a \in \boldsymbol{E}$ if and only if it is 0-continuous at the point a.

Indeed, on the one hand, if f is 0-continuous at the point a, then for any $\varepsilon > 0$ there is $\delta > 0$ such that the inequality $\mathbf{d}(a, x) < \delta$ implies the inequality $\mathbf{d}(f(x), f(a)) < 0 + \varepsilon = \varepsilon$. This means that f is continuous at the point a. On the other hand, any continuous at a point a function is, by definition, 0-continuous at the point a. Thus, the condition of 0-continuity coincides with the condition of continuity.

This result shows that the concept of an r-continuity in metric spaces is a natural extension of the concept of conventional continuity in metric spaces.

Definition 3.7.2 implies the following result.

Lemma 3.7.2. If $q > r$, then any r-continuous at a mapping f is also q-continuous at a.

When we take r-continuous functions for all r, we come to the class of fuzzy continuous functions.

Definition 3.7.3. A mapping $f: \mathbf{E} \to \mathbf{L}$ is called *fuzzy continuous at* a point $a \in \mathbf{E}$ if f is r-continuous at a for some $r \in \mathbf{R}^+$.

As in the case of real functions, we extend r-continuity to (q, r)-continuity.

Definition 3.7.4. A mapping $f: \mathbf{E} \to \mathbf{L}$ is called (q, r)-continuous at a point $a \in \mathbf{E}$ if for any sequence $l = \{ a_i \in \mathbf{E}; i = 1, 2, 3, \dots\}$, for which a is an q-limit, the point $f(a)$ is an r-limit of the sequence $\{ f(a_i) \in \mathbf{L}; i = 1, 2, 3, \dots\}$.

It is possible to define (q, r)-continuous functions without explicit use of limits as it is done by means of the (ε, δ)-construction.

Definition 3.7.5. A mapping $f: \mathbf{E} \to \mathbf{L}$ is called (q, r)-continuous at a point $a \in \mathbf{E}$ if for any $\varepsilon > 0$, there is $\delta > 0$ such that the inequality $\mathbf{d}(a, x) < q + \delta$ implies the inequality $\mathbf{d}(f(x), f(a)) < r + \varepsilon$, or in other words, for any x with $\mathbf{d}(a, x) < q + \delta$, we have $\mathbf{d}(f(x), f(a)) < r + \varepsilon$.

In a similar way as Proposition 3.3.1, we can prove the following statement.

Proposition 3.7.2. Definitions 3.7.4 and 3.7.5 are equivalent.

Remark 3.7.1. By the same technique, it is possible to define concepts of r-continuity, fuzzy continuity, and (q, r)-continuity for mapping (functions) from an arbitrary topological space into a metric space (cf., for example, (Burgin, 1995)). Much more general constructions allow one to extend concepts of r-continuity, fuzzy continuity, and (q, r)-continuity to general topological spaces. It is done in scalable topology (Burgin, 1999b; 2004a; 2005a; 2006).

Lemma 3.7.3. a) A mapping f is r-continuous at a point $a \in \mathbf{R}$ if and only if it is $(0, r)$-continuous at the point a.

b) A mapping f is continuous at a point $a \in \mathbf{R}$ if and only if it is $(0,0)$-continuous at the point a.

These results show that the concept of an (q, r)-continuity is a natural extension of the concepts of r-continuity and conventional continuity in metric spaces.

Definition 3.7.4 implies the following result.

Lemma 3.7.4. If $t \geq r$, and $p \leq q$, then any (q, r)-continuous at a mapping f is also (p, t)-continuous at a.

Indeed, if that the inequality $\mathbf{d}(a, x) < q + \delta$ implies the inequality $\mathbf{d}(f(x), f(a)) < r + \varepsilon$, then the inequality $\mathbf{d}(a, x) < p + \delta$ implies the inequality $\mathbf{d}(a, x) < q + \delta$ because $p \leq q$. The inequality $\mathbf{d}(a, x) < q + \delta$ implies the inequality $\mathbf{d}(f(x), f(a)) < r + \varepsilon$, as $f(x)$ is (q, r)-continuous at a, which, in turn, implies the inequality $\mathbf{d}(f(x), f(a)) < t + \varepsilon$ because $t \geq r$. Thus, by Definition 3.7.4, if $f(x)$ is (q, r)-continuous at a point a function, then it is also (p, t)-continuous at a.

Corollary 3.7.1. If $q > l$ ($r < p$), then any (q, r)-continuous at a point $a \in \mathbf{R}$ mapping f is also (l, r)-continuous (also (q, p)-continuous) at the point a.

Corollary 3.7.2. If $q > l$ and $r < p$, then any (q, r)-continuous mapping f is also (l, p)-continuous.

Corollary 3.7.3. If a mapping f is (q, r)-continuous at a point $a \in \mathbf{R}$, then f is r-continuous at the point a.

It is also possible to introduce fuzzy continuity based on some measure of discontinuity or of continuity.

Definition 3.7.6. A mapping $f: \mathbf{E} \to \mathbf{L}$ is called *bounded* (in $X \subseteq \mathbf{E}$) if there is a number c such that for any points $a, b \in \mathbf{E}$ (any points $a, b \in X$), we have $\mathbf{d}(f(a), f(b)) < c$.

Lemma 3.7.5. A mapping $f: \boldsymbol{E} \to \boldsymbol{L}$ is bounded in X if and only if for any point $a \in X$, there is a number c_a such that for any point $b \in X$, we have $\mathbf{d}(f(a), f(b)) < c_a$.

Proof. Necessity. If a mapping $f: \boldsymbol{E} \to \boldsymbol{L}$ is bounded in X, then given point a, we can take the number c as c_a and get the inequality $\mathbf{d}(f(a), f(b)) < c_a$ for all $b \in X$.

Sufficiency. Let us consider a mapping $f: \boldsymbol{E} \to \boldsymbol{L}$ that satisfies the condition from the lemma. Then taking two points $d, b \in X$ and using properties of the metric (cf. Appendix C), we have

$$\mathbf{d}(f(d), f(b)) \le \mathbf{d}(f(d), f(a)) + \mathbf{d}(f(a), f(b)) < c_a + c_a = c$$

for all points $d, b \in X$ with $c = c_a + c_a$.

Lemma is proved.

Definition 3.7.7. A mapping $f: \boldsymbol{E} \to \boldsymbol{L}$ is called *bounded at a point* $a \in \boldsymbol{E}$ if there is a number c such that $\mathbf{d}(f(a), f(b)) < c$ for any points b from some open ball $\mathrm{B}_r(a) = \{x; \mathbf{d}(x, a) < r, r \in \boldsymbol{R}^+\}$.

Similar to Lemma 3.7.5, we can prove the following result.

Lemma 3.7.6. A mapping $f: \boldsymbol{E} \to \boldsymbol{L}$ is bounded in X if and only if f is bounded in some closed ball $\mathrm{B}_r(a)$.

Properties of fuzzy continuous mappings of metric spaces are very similar to properties of fuzzy continuous real functions obtained in Section 3.4.

Theorem 3.7.1. Any bounded at a point a mapping $f: \boldsymbol{E} \to \boldsymbol{L}$ is fuzzy continuous at this point.

Proof. Let us consider a bounded at a point a mapping $f: \boldsymbol{E} \to \boldsymbol{L}$. Its boundedness means that there are a number c and a ball $\mathrm{B}_q(a)$ such that $\mathbf{d}(f(x), f(a)) < c$ for all x from the ball $\mathrm{B}_q(a)$.

Let us take $r = c$. In this case, taking any point x inside the ball $\mathrm{B}_q(a)$, we have $\mathbf{d}(f(x), f(a)) < r$. This allows us to conclude that for any $\varepsilon > 0$ there is $\delta > 0$ such that the inequality $\mathbf{d}(a, x) < \delta$ implies the inequality $\mathbf{d}(f(x), f(a)) < r + \varepsilon$. By Definition 3.7.2, the mapping f is r-continuous at the point a.

Theorem is proved.

Definition 3.7.8. a) A mapping $f: \boldsymbol{E} \to \boldsymbol{L}$ is called uniformly *r-continuous at a point* $a \in \boldsymbol{E}$ if there is a ball $\mathrm{B}_q(a)$ such that for any $\varepsilon > 0$, there is $\delta > 0$ such that the inequality $\mathbf{d}(a, x) < \delta$ implies the inequality $\mathbf{d}(f(x), f(a)) < r + \varepsilon$ for any x from $\mathrm{B}_q(a)$.

b) A mapping $f: \boldsymbol{E} \to \boldsymbol{L}$ is called uniformly *fuzzy continuous at a point a* if it is uniformly *r-continuous at the point a* for some $r \ge 0$.

Corollary 3.7.3. Any bounded at a point a mapping $f: \boldsymbol{E} \to \boldsymbol{L}$ is uniformly fuzzy continuous at this point.

Let $\boldsymbol{E}$ be a complete metric space (cf. Appendix C).

Theorem 3.7.2. A mapping $f: \boldsymbol{E} \to \boldsymbol{L}$ is fuzzy continuous at a point a if and only if it is bounded at this point.

Proof. Necessity. Let us take a mapping $f: \boldsymbol{E} \to \boldsymbol{L}$, assume that it is r-continuous at a point a, but not bounded at this point, and consider the sequence of closed balls $\{\mathrm{B}_{1/n}(a)\ ; n = 1, 2, 3, \ldots \}$. As f is not bounded at the point a, then for any number n, there is a number $c_n \in \mathrm{B}_{1/n}(a)$ such that $\mathbf{d}(f(a), f(c_n)) > n$. Thus, we can choose a sequence $l = \{c_i \in \boldsymbol{E};\ i = 1, 2, 3,$

...} such that $\mathbf{d}(f(a), f(c_i)) > i$ for all $i = 1, 2, 3, \ldots$. The diameters of the balls $B_{1/n}$ converge to zero. Consequently, the sequence l converges to a (cf. Section 2.9). The mapping f is r-continuous at the point a. It means (cf. Definition 3) that for any $\varepsilon > 0$ there is $\delta > 0$ such that the inequality $\mathbf{d}(a, x) < \delta$ implies the inequality $\mathbf{d}(f(x), f(a)) < r + \varepsilon$. However, this not true for points $c_i \in \boldsymbol{E}$ ($i = 1, 2, 3, \ldots$) as the sequence of $\mathbf{d}(f(a), f(c_i))$ tends to infinity while the points c_i converge to a. By the principle of excluded middle, this contradiction implies that f has to be bounded at the point a.

Sufficiency follows from Theorem 3.7.1.

Theorem is proved.

Theorem 3.7.2 gives us the following classical result.

Corollary 3.7.4. A continuous at a point function is bounded at this point.

From local r-continuity, we naturally come to global r-continuity.

Let X be a subset of $\boldsymbol{E}$.

Definition 3.7.9. a) A mapping $f: \boldsymbol{E} \to \boldsymbol{L}$ is called *r-continuous* in X if it is r-continuous at each point a from X.

b) A mapping $f: \boldsymbol{E} \to \boldsymbol{L}$ is called (q, r)-continuous in X if it is (q, r)-continuous at each point a from X.

Lemma 3.7.2 implies the following result.

Corollary 3.7.5. If $q > r$, then any r-continuous (in X) mapping f is also q-continuous (in X).

Definition 3.7.10. a) A mapping $f: \boldsymbol{E} \to \boldsymbol{L}$ is called *locally fuzzy continuous* in X if at each point a from X, there is a number $r \in \boldsymbol{R}^+$ such that $f(x)$ is r-continuous at a for some $r \in \boldsymbol{R}^+$.

b) A mapping $f: \boldsymbol{E} \to \boldsymbol{L}$ is called *locally 2-fuzzy continuous* in X if at each point a from X, there are numbers $r, q \in \boldsymbol{R}^+$ such that $f(x)$ is (q, r)-continuous at a for some $r \in \boldsymbol{R}^+$.

From local fuzzy continuity, we naturally come to global fuzzy continuity.

Definition 3.7.11. a) A mapping $f: \boldsymbol{E} \to \boldsymbol{L}$ is called *globally fuzzy continuous* in X if there is a number $r \in \boldsymbol{R}^+$ such that at each point a from X, $f(x)$ is r-continuous at a.

b) A mapping $f: \boldsymbol{E} \to \boldsymbol{L}$ is called *locally 2-fuzzy continuous* in X if at each point a from X, there are numbers $r, q \in \boldsymbol{R}^+$ such that $f(x)$ is (q, r)-continuous at a for some $r \in \boldsymbol{R}^+$.

Remark 3.7.1. If X contains more than one point, then for each $r \in \boldsymbol{R}^+$ every r-continuous on X function is fuzzy continuous on X, but not every fuzzy continuous on X function is r-continuous on X for some $r \in \boldsymbol{R}^+$. It means that in contrast to continuous functions where functions continuous at each point are globally continuous, it is possible that a fuzzy continuous at each point function is not globally continuous.

Example 3.7.1. Let us consider the two dimensional normed vector space $\boldsymbol{C}$, two mappings $f, g: \boldsymbol{C} \to \boldsymbol{L}$ defined by formulas $f(x) = \|x\|^n$ when $x \in [n, n + 1)$, $n \in \boldsymbol{Z}$ and $g(x) = [\ \|x\|\ ]^n$. These mappings are fuzzy continuous at each point of $\boldsymbol{C}$, but they are not fuzzy continuous on $\boldsymbol{C}$.

Definition 3.7.12. a) A mapping $f: \boldsymbol{E} \to \boldsymbol{L}$ is called *uniformly r-continuous* in X if for any $\varepsilon > 0$, there is $\delta > 0$ such that for any x and y from X, the inequality $\mathbf{d}(x, y) < \delta$ implies the inequality $\mathbf{d}(f(x), f(y)) < r + \varepsilon$.

b) A mapping f: $\boldsymbol{E} \to \boldsymbol{L}$ is called *uniformly fuzzy continuous* in X if there is a number $r \in \boldsymbol{R}^+$ such that $f(x)$ is uniformly r-continuous in X.

Definition 3.7.13. a) A mapping f: $\boldsymbol{E} \to \boldsymbol{L}$ is called *uniformly* (q, r)*-continuous* in X if for any $\varepsilon > 0$, there is $\delta > 0$ such that for any x and y from X, the inequality $\mathbf{d}(x, y) < q + \delta$ implies the inequality $\mathbf{d}(f(x), f(y)) < r + \varepsilon$.

b) A mapping f: $\boldsymbol{E} \to \boldsymbol{L}$ is called *uniformly* 2*-fuzzy continuous* in X if at each point a from X, there are numbers r, $q \in \boldsymbol{R}^+$ such that $f(x)$ is uniformly (q, r)-continuous in X.

Let $X \subseteq \boldsymbol{E}$.

Theorem 3.7.3. Any bounded in a set X mapping f: $\boldsymbol{E} \to \boldsymbol{L}$ is fuzzy continuous in X.

Proof. Let us consider a bounded in a set X mapping f: $\boldsymbol{E} \to \boldsymbol{L}$. Its boundedness means that there are a number c and a point a in $\boldsymbol{E}$ such that $\mathbf{d}(f(x), f(a)) < c$ for all x from X.

Let us take $r = c$. In this case, taking any point x from X, we have $\mathbf{d}(f(x), f(a)) < r$. This allows us to conclude that for any $\varepsilon > 0$ there is $\delta > 0$ such that the inequality $\mathbf{d}(a, x) < \delta$ implies the inequality $\mathbf{d}(f(x), f(a)) < r + \varepsilon$. By Definition 3.7.2, $f(x)$ is r-continuous at the point a.

Theorem is proved.

Corollary 3.7.6. Any bounded in a set X mapping f: $\boldsymbol{E} \to \boldsymbol{L}$ is uniformly fuzzy continuous in this set.

Let $\boldsymbol{L}$ be a metric space and $\boldsymbol{B}$ be a compact space in a complete metric space $\boldsymbol{E}$ (cf. Appendix C).

Theorem 3.7.4 (The First Fuzzy Weierstrass Theorem for metric spaces). A mapping f: $\boldsymbol{B} \to \boldsymbol{L}$ is fuzzy continuous in $\boldsymbol{B}$ if and only if it is bounded.

Proof. Necessity. Let us take a fuzzy continuous mapping f: $\boldsymbol{B} \to \boldsymbol{L}$ and assume that it is not bounded. As f is not bounded, then there is a point $a \in \boldsymbol{B}$ such that for any number n, there is a point $c_n \in \boldsymbol{B}$ such that $\mathbf{d}(f(a), f(c_n)) > n$. Thus, we can choose a sequence $l = \{c_i \in \boldsymbol{E}; i = 1, 2, 3, \ldots\}$ such that $\mathbf{d}(f(a), f(c_i)) > i$ for all $i = 1, 2, 3, \ldots$.

Properties of compact sets imply (cf. Appendix C and (Kelly, 1957)) that the sequence l has a converging subsequence $h = \{a_i ; i = 1, 2, 3, \ldots\}$. Let $d = \lim h$. Then properties of metric imply the following inequality

$$\mathbf{d}(f(a), f(a_i)) \leq \mathbf{d}(f(d), f(a_i)) + \mathbf{d}(f(a), f(d))$$

Thus, we have

$$\mathbf{d}(f(d), f(a_i)) \geq \mathbf{d}(f(a), f(a_i)) - \mathbf{d}(f(a), f(d))$$

As the sequence $\{ \mathbf{d}(f(a), f(c_i)); i = 1, 2, 3, \ldots\}$ tends to infinity, the sequence $\{ \mathbf{d}(f(a), f(c_i)); i = 1, 2, 3, \ldots\}$ also tends to infinity.

The mapping f: $\boldsymbol{B} \to \boldsymbol{L}$ is fuzzy continuous in $\boldsymbol{B}$. Thus, f is fuzzy continuous at the point d. At the same time, f is not bounded at the point d because for any number $i \in \boldsymbol{N}$ and in any neighborhood (ball $\mathrm{B}_q(d)$) of d there is a point c_i such that $\mathbf{d}(f(a), f(c_i)) > i$. This contradicts to Theorem 3.7.2 and by the principle of excluded middle, concludes the proof of necessity.

Sufficiency follows from Theorem 3.7.3.

Theorem is proved.

Such a classical result as the first Weierstrass Theorem for metric spaces is a direct corollary of Theorem 3.7.4.

Remark 3.7.4. For functions defined not in a compact set, Theorem 3.7.4 is not in general true as even a continuous function can be unbounded. For instance, we can take such functions as x^2 on the whole real line or tan x in the interval $(-\pi/2\ ,\ \pi/2)$.

Let $\boldsymbol{E}$ and $\boldsymbol{L}$ be normed vector spaces with distances denoted by the letter **d** and defined by means of the norm as $\mathbf{d}(x, y) = \| x - y \|$. Properties of normed vector spaces allow us to prove the following results.

Let us assume that $f: \boldsymbol{E} \to \boldsymbol{L}$ and $g: \boldsymbol{E} \to \boldsymbol{L}$ are two mappings of normed vector spaces $\boldsymbol{E}$ and $\boldsymbol{L}$ and k is a real number.

Theorem 3.7.5. If the mapping f is (q, r)-continuous at a point a and the mapping g is (p, h)-continuous at the point a, then:

a) the mapping $f + g$ is $(u, r + h)$-continuous at the point a where $u = \min \{q, p\}$;
b) the mapping $f - g$ is $(u, r + h)$-continuous at the point a where $u = \min \{q, p\}$;
c) the mapping kf is $(q, |k|\cdot r)$-continuous at the point a.

Proof. a) Let us take a sequence $l = \{a_i\ ;\ i = 1, 2, 3, \dots\}$ such that $a = q\text{-lim } l$ and $a = p\text{-lim } l$. Then $a = u\text{-lim } l$ where $u = \min \{q, p\}$. By Definition 3.7.11, the point $f(a)$ is an r-limit of the sequence $\{ f(a_i) \in \boldsymbol{L};\ i = 1, 2, 3, \dots\}$ and the point $g(a)$ is an h-limit of the sequence $\{ g(a_i) \in \boldsymbol{L};\ i = 1, 2, 3, \dots\}$. Then by Theorem 2.9.2, the point $f(a) + g(a)$ is an $(r+h)$-limit of the sequence $\{ f(a_i) + g(a_i);\ i = 1, 2, 3, \dots\}$. Thus, by Definition 3.7.11, the mapping $f + g$ is $(u, r + h)$-continuous at the point a.

Proofs of parts b) and c) are similar and based on Theorem 2.9.2.

Theorem 3.7.5 and Lemma 3.7.3 imply the following classical result.

Corollary 3.7.7. If the mappings f and g are continuous at a point a, then:

a) the mapping $f + g$ is continuous at the point a;
b) the mapping $f - g$ is continuous at the point a;
c) the mapping kf is continuous at the point a.

Theorem 3.7.5 and Lemma 3.7.4 imply the following result.

Corollary 3.7.8. If the mapping f is r-continuous at a point a and the mapping g is h-continuous at a point a, then:

a) the mapping $f + g$ is $(r + h)$-continuous at a point a;
b) the mapping $f - g$ is $(r + h)$-continuous at a point a;
c) the mapping kf is $|k|\cdot r$-continuous at a point a.

There are concepts closely related to fuzzy continuity. For instance, Klee (1961) introduced r-continuous (or *ε–continuous* or *nearly continuous*) mappings and studied fixed points of such mappings.

Let ε be a fixed positive real number.

Definition 3.7.14 (Klee, 1961; Klee and Yandl, 1974). A mapping $f: \boldsymbol{E} \to \boldsymbol{L}$ is called *ε-continuous* at a point $a \in \boldsymbol{E}$ if there is $\delta > 0$ such that the inequality $\mathbf{d}(a, x) \le \delta$ implies the inequality $\mathbf{d}(f(x), f(a)) \leq \varepsilon$, or in other words, for any x with $\mathbf{d}(a, x) \leq \delta$, we have $\mathbf{d}(f(x), f(a)) \leq \varepsilon$.

Definition 3.7.15 (Klee, 1961; Klee and Yandl, 1974). A mapping $f: \boldsymbol{E} \to \boldsymbol{L}$ is called *nearly continuous* at a point $a \in \boldsymbol{E}$ if f is ε-continuous at a for some $\varepsilon \in \boldsymbol{R}^{+}$.

There is a transparent relation between ε-continuous mappings in the sense of Klee and (q, r)-fuzzy continuous functions defined here.

Proposition 3.7.3. A mapping $f: \boldsymbol{E} \to \boldsymbol{L}$ is ε-continuous at a point $a \in \boldsymbol{E}$ if and only if there is a number $q \in \boldsymbol{R}$ such that the function $f(x)$ is (q, r)-continuous at a point a for any $r > \varepsilon$.

Proof. 1. Let us assume that $f: \boldsymbol{E} \to \boldsymbol{L}$ is an ε-continuous at a point $a \in \boldsymbol{E}$ function and $\varepsilon < r$. Then by Definition 3.7.13, there is $\delta > 0$ such that the inequality $\mathbf{d}(a, x) \leq \delta$ implies the inequality $\mathbf{d}(f(x), f(a)) \leq \varepsilon$. Let us take some $q < \delta$ and show that the mapping f is (q, r)-continuous at the point a. To show this, we need to take an arbitrary $\eta > 0$ and to find a number $\nu > 0$ such that the inequality $\mathbf{d}(a, x) < q + \nu$ implies the inequality $\mathbf{d}(f(x), f(a)) < r + \eta$, Indeed, taking any $\nu > 0$ such that $q + \nu < \delta$, we have that the inequality $\mathbf{d}(a, x) < q + \nu$ implies the inequality $\mathbf{d}(f(x), f(a)) \leq \varepsilon$. Then $\mathbf{d}(a, x) < q + \nu$ implies $\mathbf{d}(f(x), f(a)) < r + \eta$ as $\varepsilon < r + \eta$. Thus by Definition 3.7.12, the mapping f is (q, r)-continuous at the point a.

2. Let us assume that $f: \boldsymbol{E} \to \boldsymbol{L}$ is a (q, r)-continuous at a point $a \in \boldsymbol{E}$ mapping for some $q > 0$ and any $r > \varepsilon$. Then by Definition 3.7.12, for any $\nu > 0$ there is $\delta > 0$ such that the inequality $\mathbf{d}(a, x) < q + \delta$ implies the inequality $\mathbf{d}(f(x), f(a)) < r + \nu$. We show that the mapping f is ε-continuous at the point a as the inequality $\mathbf{d}(a, x) \leq q$ implies the inequality $\mathbf{d}(f(x), f(a)) \leq \varepsilon$. Assume that this is not true. Then there is a number x such that $\mathbf{d}(a, x) \leq q$ and $\mathbf{d}(f(x), f(a)) > \varepsilon$. In this case, we can take some $r > \varepsilon$ and $\eta > 0$ such that $\mathbf{d}(f(x), f(a)) > r + \eta$. This contradicts our assumption that the mapping f is (q, r)-continuous at the point a as $q < q + \delta$ for any $\delta > 0$.

Proposition is proved.

Proposition 3.7.3 implies the following result.

Corollary 3.7.9. A mapping $f: \boldsymbol{E} \to \boldsymbol{L}$ is fuzzy continuous at a point $a \in \boldsymbol{E}$ if and only if f is nearly continuous at this point.

Global fuzzy continuity and nearly continuity are related in a similar way.

Let ε be a fixed positive real number and X be a set in a metric space $\boldsymbol{E}$.

Definition 3.7.16 (Klee, 1961; Klee and Yandl, 1974). a) A mapping $f: \boldsymbol{E} \to \boldsymbol{L}$ is called ε-*continuous* in a set X if there is $\delta > 0$ such that for any x and y from X, the inequality $\mathbf{d}(x, y) \leq \delta$ implies the inequality $\mathbf{d}(f(x), f(y)) \leq \varepsilon$, or in other words, for any x and y from X with $\mathbf{d}(x, y) \leq \delta$, we have $\mathbf{d}(f(x), f(y)) \leq \varepsilon$.

b) A mapping $f: \boldsymbol{E} \to \boldsymbol{L}$ is called *globally* ε-*continuous* in a set X if it is ε-continuous at each point a from X.

Definition 3.7.17 (Klee, 1961; Klee and Yandl, 1974). a) A mapping $f: \boldsymbol{E} \to \boldsymbol{L}$ is called *nearly continuous* in a set X if $f(x)$ is ε-continuous in X for some $\varepsilon \in \boldsymbol{R}^{+}$.

b) A function $f: \boldsymbol{R} \to \boldsymbol{R}$ is called *globally nearly continuous* in a set X if $f(x)$ is globally ε-continuous in X for some $\varepsilon \in \boldsymbol{R}^{+}$.

Proposition 3.3.4 implies the following result.

Proposition 3.7.4. A mapping f: $\boldsymbol{E} \to \boldsymbol{L}$ is globally ε-continuous in a set X if and only if for each point a in X, there is a number $q \in \boldsymbol{R}^+$ such that the function $f(x)$ is (q, r)-continuous at a for any $r > \varepsilon$.

Corollary 3.7.10. A mapping f: $\boldsymbol{E} \to \boldsymbol{L}$ is 2-fuzzy continuous in a set X if and only if $f(x)$ is globally nearly continuous in a set X.

Proposition 3.7.5. A mapping f: $\boldsymbol{E} \to \boldsymbol{L}$ is ε-continuous in a set X if and only if there is a number $q \in \boldsymbol{R}^+$ such that the function $f(x)$ is uniformly (q, r)-continuous in X for any $r > \varepsilon$.

Corollary 3.7.11. A mapping f: $\boldsymbol{E} \to \boldsymbol{L}$ is uniformly 2-fuzzy continuous in a set X if and only if $f(x)$ is globally nearly continuous in a set X.

Remark 3.7.5. Taking an interval $I = [a, b]$ of real numbers, a mapping f: $\boldsymbol{E} \to \boldsymbol{L}$, a mapping h: $I \to \boldsymbol{E}$, and a point $c = h(d)$ for some point d from the interval $[a, b]$, we can naturally define continuity, r-continuity, and fuzzy continuity of f at a point c along the line $h(I)$ and study properties of these concepts. When $\boldsymbol{E}$ is an n dimensional Euclidean space $\boldsymbol{R}^n$, then continuity, r-continuity, and fuzzy continuity of a mapping f: $\boldsymbol{R}^n \to \boldsymbol{L}$ at some point c as a function of one chosen variable give examples of continuity, r-continuity, and fuzzy continuity, correspondingly, of f at c along a line.

Chapter 4

FUZZY DIFFERENTIATION

America has believed that in differentiation,
not in uniformity, lies the path of progress.
It acted on this belief;
it has advanced human happiness,
and it has prospered.
Louis D. Brandeis (1856-1941)

Differential calculus is one of two central parts of the calculus and analysis. The main concept of differential calculus is the concept of a derivative. The main goal of this chapter is to extend the classical construction of derivatives and to make it more flexible and more relevant to real life conditions where data are obtained from measurement and computation.

Conventional derivative is defined in mathematics by a limit process. The basic idea of fuzzy differentiation is to use fuzzy limits instead of conventional limits. What does it give? On the one hand, it essentially extends the scope of differentiable functions. As a result, fuzzy derivatives allow one to investigate behavior of functions where classical derivatives do not exist or cannot be calculated (cf., for example, Chapter 5). As a result, approximations are often used as estimates of classical derivatives instead of exact values. For instance, method based on continuous wavelet transform with Haar wavelet function is suggested in (Shao, et al, 2000) for approximate derivative calculation of analytical signals. In addition, fuzzy derivatives allow one to better solve optimization problems (cf. Chapter 6). On the other hand, fuzzy derivatives are natural extensions of classical derivatives and as such, preserve many important properties of classical derivatives. For instance, fuzzy differentiability implies classical continuity of a function (cf. Section 4.5).

In this Chapter, a mathematical technique for working with differential models with uncertainty that emerges in computation and measurement is developed. It is based on the concept of a fuzzy limit studied in Chapter 2. To take into account the intrinsic uncertainty of a model, it is suggested to use fuzzy derivatives instead of conventional derivatives of functions in such models. This makes the concept of a derivative appropriate for management of imprecise, vague, uncertain, and incomplete information. Two kinds of fuzzy derivatives are introduced: weak and strong ones. Strong fuzzy derivatives are similar to ordinary derivatives of real functions, being their fuzzy extensions. Strong fuzzy derivatives are studied in Section 4.2. At the same time, weak fuzzy derivatives generate a new concept of a weak derivative even in the classical case of exact limits. Weak fuzzy derivatives are studied in Section 4.3. In addition, conditional and extended fuzzy derivatives are introduced and studied in Sections 4.4 and 4.5. Conditional fuzzy derivatives unify different kinds of strong and weak fuzzy derivatives and, at the same time, allow one to take into account computational and measurement procedures that are used for getting values of derivatives. Extended fuzzy derivatives may take infinite values and are useful for finding fuzzy minima and maxima (cf., Chapter 6). Many properties of fuzzy derivatives remain true for extended fuzzy derivatives.

When a function is differentiable, there is a unique instant rate at which the dependent variable changes relative to the independent variable. However, the exact value of this rate is often non-computable. At the same time, there are many functions that do not have derivatives at some or even at all points of their domain. Thus, in many cases, we either do not have a unique instant rate or cannot precisely evaluate this rate. It means that the classical derivative does not work in these cases and we need fuzzy derivatives. As in the classical case, a fuzzy derivative of a function represents an approximation to the rate at which the dependent variable changes relative to the independent variable. Strong fuzzy derivatives represent approximations of all possible instant rates, while a weak fuzzy derivative reflects an approximation of a particular instant rate of the variable change.

Rates of change are highly important in science. For instance, velocity is the rate of position change, and acceleration is the rate of velocity change. In some cases, the exact rate does not exist. In other cases, it exists but it is impossible to measure such exact rate. For example, if we take the rate of the particle position change, an intrinsic impossibility to measure this rate with full precision is one of the consequences of the Principle of Uncertainty introduced by Heisenberg. All measurement instruments can give only approximates values of continuous properties, such as lemgth, velocity, mass, force, acceleration, etc. Besides, there are cases when exact rate exists, it is feasible to measure it, but it is impossible to calculate the value of this exact rate with absolute precision. All these and many other situations imply usefulness and necessity to study fuzzy derivatives.

Later, in Chapter 6, we apply fuzzy derivatives to problems of optimization, with the emphasis on the mathematical context of these problems, and in Chapter 5, fuzzy derivatives are used to study monotonicity of functions. This allows us not only to extend but also to complete some classical results.

For instance, one of the basic theorems of calculus states (cf., for example, (Fihtengoltz, 1955; Marsden and Weinstein, 1981; Ross, 1996; Stewart, 2003)):

> If a real function f is differentiable in an open interval and $f'(x) > 0$ ($f'(x) < 0$) for all x from this interval, then f is strictly increasing (decreasing) on this interval.

This theorem gives only sufficient conditions for strict monotonicity and only for differentiable functions. Weak derivatives allow us to deduce a complete criterion for strict monotonicity of arbitrary functions.

Fuzzy derivatives in a form of derivative approximations or blurred derivatives emerge in different applications of the calculus. For example, in (Lindeberg, 1993), discrete derivative approximations are used as a basis for low-level feature extraction. Starting from a set of natural requirements on the first processing stages of a visual system, Lindeberg gives an axiomatic derivation of a multi-scale representation of derivative approximations from a discrete signal. This representation has an algebraic structure similar to that possessed by the derivatives of the traditional scale-space representation in the continuous domain.

It is necessary to remark that differentiation has been introduced and differential calculus developed for different kinds of fuzziness. Here we mention only some of the approaches. The attention of Zadeh (1978), Goetshel and Voxman (1986), and Puri and Ralescu (1983) is focused on functions that are not necessarily fuzzy but "carry" the possible fuzziness of their arguments (see also (Zimmermann, 2001)). The uncertainty of knowledge about the precise location of the argument induces an uncertainty about the value of the derivative of a function at this point. To achieve this, functions with fuzzy numbers as their domain and/or range are considered. Differentiation of conventional fuzzy functions is considered in (Kaleva, 1987; Buckley and Yunxia, 1991; Kalina, 1997). In (Kalina, 1998; 1999), three basic types of vagueness (on the y-axis, on the x-axis, and on both) are considered. It implies three constructions for fuzzy derivatives, which are investigated.

Fuzzy derivatives studied here are related to the nearness derivative of a function introduced by Kalina (1998) and developed by Janiš (1999), while weak fuzzy derivatives are related to the notion of weakly continuous (Collingwood and Lohwater, 1966) and weakly symmetrically continuous (Ciesielski and Larson, 1993-94; Ciesielski, 1995-96) functions.

However, there is a difference between classical approach to differentiation, derivatives of fuzzy functions and those constructions that are introduced here. Namely, computation of a classical derivative assumes that the result does not depend on the choice of points and procedures and may be obtained with an infinite precision because there is only one (if any) value for the classical derivative. Similar assumptions are made for derivatives of fuzzy functions. At the same time, a fuzzy derivative essentially depends on initial data and computational procedures, reflecting a finite precision of computation. Taking points closer and closer to some point a, we approach the value of the classical derivative of $f(x)$ at the point a if the classical derivative exists. All these approximations are fuzzy derivatives. When the classical derivative of $f(x)$ at a does not exist, we can often build and utilize a new generalized concept of the derivative of $f(x)$ at a point a, namely, the fuzzy derivative, which is applicable to a much larger universe of functions.

4.1. Classical Differentiation

Progress is measured by the degree of
differentiation within a society.
Herbert Read (1893-1968)

Let X, $Y \subseteq \boldsymbol{R}$, $f: X \rightarrow Y$ be a function, $a \in X$, $b \in \boldsymbol{R}$, and assume that X contains some open interval $(a - k, a + k)$.

Definition 4.1.1. A number b is called the *derivative* of $f(x)$ at a point $a \in X$ if $b = \lim_{i\to\infty} (f(a) - f(x_i))/(a - x_i)$ for all sequences $\{x_i ; x_i \in (a - k, a + k); x_i \neq a ; i = 1, 2, 3, \dots \}$ converging to a.

The derivative of $f(x)$ at a point a is denoted by $b = f'(a)$ or $b = d/_{dx} f(a)$.

Lemma 4.1.1. The derivative of $f(x)$ at a point a (if it exists) is unique.

Example 4.1.1. Let $f(x) = x^3$. Then the ratio $(f(0) - f(x_i)) / (0 - x_i) = x^3/x = x^2$ converges to 0 when x^2 converges to 0. Thus, the derivative of the function x^3 at 0 is equal to 0.

Definition 4.1.2. A function $f(x)$ is called *differentiable at a point* $a \in X$, if $f(x)$ has a derivative at a.

Example 4.1.2. The function $f(x) = x$ is differentiable at 0 and its derivative is equal to 1. However, the function $f(x) = |x|$ is not differentiable at 0 because the limit described in Definition 4.1.1 does not exist.

Theorem 4.1.1. If a function $f(x)$ is differentiable at a point $a \in X$, then $f(x)$ is continuous at a.

Proof. Let $f(x)$ be a differentiable at a point a from X function. It means that for any sequence $\{x_n; x_n \in X\}$, if $a = \lim_{n\to\infty} x_n$, then there is a number b such that $b = \lim l$ where $l = \{ (f(a) - f(x_n))/(a - x_n) ; n = 1, 2, 3, \dots \}$. By the definition of a limit (cf. Chapter 2), for any $\varepsilon > 0$, there is some natural number $m \in \omega$ such that for all $n > m$ the following inequality is valid: $| b - (f(a) - f(x_n))/(a - x_n)) | \leq \varepsilon$. It implies the inequality $| (f(a) - f(x_n))/(a - x_n) | \leq \varepsilon + |b|$. Consequently, we have

$$| f(a) - f(x_n) | \leq (\varepsilon + |b|)\cdot| a - x_n| \tag{4.1}$$

Thus, convergence of the sequence $\{x_n ; n = 1, 2, 3, \dots \}$ to the point a implies that the sequence $\{ f(x_n); n = 1, 2, 3, \dots \}$ converges to the number $f(a)$ because when the right part of the inequality (4.1) converges to 0, the left part of this inequality also converges to 0. As the sequence $\{x_n ; n = 1, 2, 3, \dots \}$ is an arbitrary sequence that converges to a, it means that the function $f(x)$ is continuous at a (cf. Section 3.1).

Theorem is proved.

Definition 4.1.3. A function $f(x)$ is called *differentiable* if it is differentiable at any point of its domain.

As a consequence of Theorem 4.1.1, we obtain the following classical result (cf., for example, (Ross, 1996; Stewart, 2003; or Fihtengoltz, 1955)).

Theorem 4.1.2. Any differentiable function $f(x)$ is continuous.

Theorem 4.1.3. If the derivative $f'(x)$ of a function $f(x)$ exists and is equal to zero in an interval $[a, b]$, then $f(x)$ is the constant function in the interval $[a, b]$.

Proof. Let us assume that the derivative $f'(x)$ of $f(x)$ exists and is equal to zero in an interval $[a, b]$. In this case, by the Rolle Theorem (Theorem 6.1.2), all points in the interval $[a, b]$ are points of maximum or minimum of the function $f(x)$. This is possible only when the value of $f(x)$ at any two points in the interval $[a, b]$ coincide. Thus, $f(x)$ is the constant function in the interval $[a, b]$.

Let us consider two differentiable functions $f(x)$ and $g(x)$.

Theorem 4.1.4. a) **(Local additivity of differentiation)** If $b = f'(a)$ and $c = g'(a)$, then $b \pm c = (f \pm g)'(a)$.

b) **(Local uniformity of differentiation)** If $b = f'(a)$ and $t \in \boldsymbol{R}$, then $t\cdot b = t\cdot f'(a)$.

We do not give proofs of these theorems because they are direct corollaries of the corresponding results for fuzzy derivatives (cf. Proposition 4.2.3 and Theorem 4.2.1).

Combined together, additivity and uniformity of differentiation give linearity of differentiation. In other words, differentiation at a given point is a linear functional.

Let $f(x)$ and $g(x)$ be differentiable at a point a functions.

Theorem 4.1.5. a) **(The Local Product Rule)** $(f \cdot g)'(a) = f(a)g'(a) + f'(a)g(a)$.

b) **(The Local Quotient Rule)** $(f / h)'(a) = (f'(a)h(a) - f(a)h'(a))/h^2(a)$ at a point a where $h(a) \neq 0$.

Proof. a) Let $b = f'(a)$, $c = g'(a)$, and $l = \{ x_i ; i = 1, 2, 3, \ldots \}$ be a sequence with $a = \lim_{i\to\infty} x_i$. Then by Definition 4.1.1,

$$(f \cdot g)'(a) = \lim_{i\to\infty} (f(x_i)\, g(x_i) - f(a)g(a))/(x_i - a))$$

Let us transform the expression under the limit sign.

$$(f(x_i) \cdot g(x_i) - f(a) \cdot g(a))/(x_i - a)) =$$
$$(f(x_i)\, g(x_i) - f(a)g(x_i) + f(a)\, g(x_i) - f(a)g(a))/(x_i - a)) =$$
$$(f(x_i)\, g(x_i) - f(a)g(x_i))/(x_i - a)) + (f(a)\, g(x_i) - f(a)g(a))/(x_i - a)) =$$
$$g(x_i) \cdot ((f(x_i) - f(a))/(x_i - a)) + f(a) \cdot ((g(x_i) - g(a))/(x_i - a))$$

Applying properties of operations with sequences, such as distributivity of the limit with respect to addition and scalar multiplication of sequences (see Section 2.1), we obtain the following sequence of equalities:

$$\lim_{i\to\infty} (f(x_i)\, g(x_i) - f(a)g(a))/(x_i - a)) =$$
$$\lim_{i\to\infty} g(x_i) \cdot ((f(x_i) - f(a))/(x_i - a)) + \lim_{i\to\infty} f(a) \cdot ((g(x_i) - g(a))/(x_i - a)) =$$
$$\lim_{i\to\infty} g(x_i) \cdot \lim_{i\to\infty} ((f(x_i) - f(a))/(x_i - a)) + f(a) \cdot \lim_{i\to\infty} ((g(x_i) - g(a))/(x_i - a))$$

Any differentiable at a point a function is continuous at this point (Theorem 4.1.1). Thus, we have

$$\lim_{i\to\infty} g(x_i) = g(a)$$

By Definition 4.1.1,

$$\lim_{i\to\infty} ((f(x_i) - f(a))/(x_i - a)) = f'(a)$$

and

$$\lim_{i\to\infty} ((g(x_i) - g(a))/(x_i - a)) = g'(a)$$

Thus, we have

$$(f \cdot g)'(a) = \lim_{i\to\infty} (f(x_i)\, g(x_i) - f(a)g(a))/(x_i - a)) = f(a)g'(a) + f'(a)g(a)$$

Part a) of the theorem is proved as $l = \{ x_i ; i = 1, 2, 3, \ldots \}$ is an arbitrary sequence that converges to a.

Part b) of the theorem becomes a consequence of the part a) if we put $g(x) - (g(x))^{-1}$ in the product rule and use such properties of differentiation as the Chain Rule and the equality $(x^{-1})' = - x^{-2}$ (cf., for example, (Stewart, 2003)).

Definition 4.1.4. A functions $g(x)$ is called the *derivative* of $f(x)$ if at each point $a \in$ Dom f if $g(a) = f'(a)$.

Derivative of $f(x)$ is also denoted by $f'(x)$.

Theorem 4.1.4 directly implies the following result.

Theorem 4.1.6. For any differentiable functions $f(x)$ and $g(x)$, we have:

a) **(Global additivity of differentiation)** $(f \pm g)'(x) = f'(x) \pm g'(x)$,.

b) **(Global uniformity of differentiation)** $(tf)'(x) = t{\cdot}f'(x)$.

Combined together, global additivity and uniformity of differentiation give global linearity of differentiation. In other words, differentiation is a linear operator.

Theorem 4.1.5 directly implies the following result.

Theorem 4.1.7 a) **(the Global Product Rule).** For any differentiable functions $f(x)$ and $g(x)$, we have $(f{\cdot}g)'(x) = f(x)g'(x) + f'(x)g(x)$.

b) **(the Global Quotient Rule)** $(f / g)'(x) = (f'(x)g(x) - f(x)g'(x))/g^2(x)$ at all points x where $g(x) \neq 0$.

At some points where the derivative of $f(x)$ does not exist, it is possible to define right or left derivatives.

Let X, $Y \subseteq \boldsymbol{R}$, $f: X \to Y$ be a function, $a \in X$, $b \in \boldsymbol{R}$., and X contains some interval $[a, a + k)$ (interval $(a - k, a]$).

Definition 4.1.5. a) A number b is called the *right* (*left*) *derivative* of $f(x)$ at a point $a \in X$ if $b = \lim_{i\to\infty} (f(a) - f(x_i))/(a - x_i)$ for all sequences $\{x_i ; x_i \in (a, a + k); i = 1, 2, 3, \dots\}$ (correspondingly, all sequences $\{x_i ; x_i \in (a - k, a); i = 1, 2, 3, \dots\}$) converging to a.

A function $f(x)$ is called *differentiable from the right* (*left*) *at a point* $a \in X$, if $f(x)$ has a right (left) derivative at a.

Example 4.1.3. The function $f(x) = [x]$ is differentiable from the right at all points from $\boldsymbol{R}$. However, it is not differentiable from the left at all points 0, 1, 2, 3, … .

Lemma 4.1.2. The right (left) derivative of $f(x)$ at a point a is unique (if it exists).

Let us consider two differentiable functions $f(x)$ and $g(x)$.

Theorem 4.1.8. a) **(Local additivity of one-sided differentiation)** If b is the right (left) derivative of $f(x)$ at a point a and c is the right (left) derivative of $g(x)$ at the point a, then $b \pm c$ is the right (left) derivative of the function $(f \pm g)'(x)$ at the point a.

b) **(Local uniformity of differentiation)** If b is the right (left) derivative of $f(x)$ at a point a and $t \in \boldsymbol{R}$, then $t{\cdot}b$ is the right (left) derivative of $t{\cdot}f'(x)$ at the point a.

We do not give proofs of these theorems because they are direct corollaries of the corresponding results for fuzzy derivatives (cf. Proposition 4.2.8 and Theorem 4.2.1).

Theorem 4.1.9. If a function $f(x)$ is differentiable from the right (left) at a point $a \in X$, then $f(x)$ is continuous from the right (left) at a.

Proof is similar to the proof of Theorem 4.1.1.

4.2. STRONG FUZZY DERIVATIVES

Well building hath three conditions.
Commodity, firmness, and delight.
Henry Wotton (1568-1639)

As it is traditionally done in courses of calculus, we begin with local fuzzy derivatives.

In contrast to the conventional, or crisp, derivative of real functions, which is defined in a unique way, there are many kinds of fuzzy derivatives: weak, strong centered, left, right, two-sided, conditional, and extended fuzzy derivatives. At first, we define strong fuzzy derivatives, which are closer to conventional derivatives of functions and inherit many their properties.

Let $X, Y \subseteq \boldsymbol{R}$, $f: X \to Y$ be a function, $a \in X$, $b \in \boldsymbol{R}$, and assume that X contains some open interval $(a - k, a + k)$ and $r \in \boldsymbol{R}^+$.

Definition 4.2.1. a) A number b is called a *strong centered r-derivative* of the function $f(x)$ at a point $a \in X$ if $b = r\text{-lim}\ \{(f(a) - f(x_i)) / (a - x_i)\ ;\ i = 1, 2, 3, \ldots\}$ for all sequences $\{x_i\ ;\ x_i \in (a - k, a + k); x_i \neq a\ ;\ i = 1, 2, 3, \ldots\}$ converging to a.

b) A number b is called a *strong left r-derivative* of the function $f(x)$ at a point $a \in X$ if $b = r\text{-lim}\ \{(f(a) - f(x_i)) / (a - x_i)\ ;\ i = 1, 2, 3, \ldots\}$ for all sequences $\{x_i\ ;\ x_i \in (a - k, a + k); x_i < a;\ i = 1, 2, 3, \ldots\}$ converging to a.

c) A number b is called a *strong right r-derivative* of the function $f(x)$ at a point $a \in X$ if $b = r\text{-lim}\ \{(f(a) - f(x_i)) / (a - x_i)\ ;\ i = 1, 2, 3, \ldots\}$ for all sequences $\{x_i\ ;\ x_i \in (a - k, a + k); x_i > a;\ i = 1, 2, 3, \ldots\}$ converging to a.

d) A number b is called a *strong two-sided r-derivative* of the function $f(x)$ at a point $a \in X$ if $b = r\text{-lim}\ \{(f(z_i) - f(x_i)) / (z_i - x_i)\ ;\ i = 1, 2, 3, \ldots\}$ for all sequences $\{x_i\ ;\ x_i \in (a - k, a + k);\ x_i < a;\ i = 1, 2, 3, \ldots\}$ and $\{z_i\ ;\ z_i \in (a - k, a + k); z_i > a;\ i = 1, 2, 3, \ldots\}$ converging to a.

A strong centered (left, right, two-sided) r-derivative of a function $f(x)$ at a point $a \in X$ is denoted by $b = \text{st}_r{}^{\text{ct}}\text{d}/_{\text{d}x}\, f(a)$ ($b = \text{st}_r{}^{\text{l}}\text{d}/_{\text{d}x}\, f(a)$, $b = \text{st}_r{}^{\text{r}}\text{d}/_{\text{d}x}\, f(a)$, and $b = \text{st}_r{}^{\text{t}}\text{d}/_{\text{d}x}\, f(a)$, correspondingly).

Remark 4.2.1. Strong centered and two-sided r-derivatives are fuzzy counterparts of classical derivatives, while strong right r-derivatives are fuzzy counterparts of right derivatives and strong left r-derivatives are fuzzy counterparts of left derivatives.

Remark 4.2.2. In what follows, $\text{st}_r{}^{z}\text{d}/_{\text{d}x}\, f(a)$ denotes one of the four defined types of strong r-derivatives of $f(x)$ at a point a. Here $\text{z} \in \{\text{ct, l, r, t}\}$.

Remark 4.2.3. In the denotation $b = \text{st}_r{}^{\text{ct}}\text{d}/_{\text{d}x}\, f(a)$ or $b = \text{st}_r{}^{\text{r}}\text{d}/_{\text{d}x}\, f(a)$ the symbol "=" means "is", that is, the number b is strong centered or b is strong right r-derivative of a function $f(x)$ at a point a. However, it is incorrect to write $\text{st}_r{}^{\text{ct}}\text{d}/_{\text{d}x}\, f(a) = b$ because, in a general case, an r-derivative of a function at a point is not unique.

Remark 4.2.4. Even in mathematics, there are many functions that fail to have a classical derivative at some point, but have a strong fuzzy derivative at this point.

Example 4.2.1. Let us take the function $f(x) = x$ if x is zero or a positive number, and $f(x) = -x$ if x is a negative number, that is, $f(x) = |x|$, the absolute value of x, and choose a to be 0. It follows that, if $x_i > 0$, then the quotient $(f(a) - f(x_i)) / (a - x_i)$ is equal to 1, as calculations show; whereas, if $x_i < 0$, then the quotient $(f(a) - f(x_i)) / (a - x_i)$ is equal to -1. Thus, $f(x)$ does not have a derivative at $a = 0$ because, coming arbitrarily close to 0, the quotient of differences assumes the values 1 and –1. In other words, this quotient of differences does not approach one unique number as x_i approaches zero

However, according to Definition 4.2.1, the number 0 is a strong two-sided 1-derivative of $f(x)$ at 0, the number 1 is also a strong right, while the number -1 is a strong left 0-derivatives of $f(x)$ at 0.

Example 4.2.2. Piecewise linear transformations on the interval have been widely studied in the theory of dynamical systems (Collet and Eckmann, 1980; Marcuard and Visinescu, 1992) and under different names as well: broken linear transformations (Gervois and Mehta, 1977) or weak unimodal maps (Misiurewicz, 1989). An example of piecewise linear transformations is given by a skew tent map $f_{a,b}(x)$ that is equal to $b + ((1 - b)/a)x$ when $0 \leq x < a$ and equal to $(1 - x)/(1 - a)$ when $a \leq x \leq 1$. Piecewise linear transformations do not have conventional derivatives at some points but they have strong fuzzy derivatives at all points. It provides for application of differential methods to these mappings as well as to dynamics generated by them.

Remark 4.2.5. In contrast to the classical derivative, it is possible that different numbers are strong centered (or left, right, two-sided) r-derivatives of a given function $f(x)$ at a point a.

Remark 4.2.6. An alternative approach to fuzzy differentiation of real functions is suggested by Kalina (1998; 1999; 1999a) and Janiš (1999). Their construction for differentiation is based on the concepts of fuzzy continuity from (Burgin and Šostak, 1992; 1994) and nearness on the set $\boldsymbol{R}$ of real numbers, which was introduced by Kalina (1997).

Let us consider the case when the space X has an isolated point a. Then there are no sequences converging in X to the point a but the sequences almost all elements of which are equal to a. However, we do not consider such sequences in the definition of derivatives and fuzzy derivatives because the denominator in the expression $(f(a) - f(x_i)) / (a - x_i)$ becomes equal to 0. Thus, the condition from the Definition 4.2.1 is satisfied in trivial way, and we have the following result.

Lemma 4.2.1. Any number $b \in \boldsymbol{R}$ is a strong centered (left, right, two-sided) r-derivative of $f(x)$ at an isolated point $a \in X$ for any $r \in \boldsymbol{R}^+$.

However, it is also possible to say that there is no strong centered (left, right, two-sided) r-derivative of $f(x)$ at the point $a \in X$ for any number $b \in \boldsymbol{R}$ and any $r \in \boldsymbol{R}^+$. This is a constructive understanding of a fuzzy derivative.

Taking limits of sets of sequences (cf. Section 2.6), it is possible to define strong derivatives in a different (but equivalent to the initial definition) way. For arbitrary points of X, Definition 4.2.1 implies the following results.

Lemma 4.2.2. For any point a from $\boldsymbol{R}$ and any real function $f(x)$, we have:

a. $b = \mathrm{st}_r^{ct} d/_{dx} f(a)$ if and only if $b = r\text{-lim } E_{ct}$ where $E_{ct} = \{\{ (f(a) - f(x_i)) /(a - x_i); i = 1, 2, 3, \dots \}; \{ x_i ; i = 1, 2, 3, \dots \}$ is a sequence converging to a $\}$.

b. $b = \mathrm{st}_r^{l} d/_{dx} f(a)$ if and only if $b = r\text{-lim } E_l$ where $E_l = \{\{ (f(a) - f(x_i)) /(a - x_i); i = 1, 2, 3, \dots\}; \{ x_i ; i = 1, 2, 3, \dots \}$ is a sequence converging to a and all $x_i < a\}$.

c. $b = st_r{}^{r}d/_{dx}f(a)$ if and only if $b = r$-lim E_r where $E_r = \{\{(f(a) - f(x_i)) / (a - x_i); i = 1, 2, 3, \ldots\}; \{x_i; i = 1, 2, 3, \ldots\}$ is a sequence converging to a and all $x_i > a\}$.

d. $b = st_r{}^{t}d/_{dx}f(a)$ if and only if $b = r$-lim E_t where $E_t = \{\{(f(a) - f(x_i)) / (a - x_i); i = 1, 2, 3, \ldots\}; \{x_i; i = 1, 2, 3, \ldots\}$ is a sequence converging to a and $z_i < a < x_i$ for all $i = 1, 2, 3, \ldots\}$.

Indeed, if $b = r$-lim $\{(f(a) - f(x_i)) / (a - x_i); i = 1, 2, 3, \ldots\}$ (and all $x_i < a$; and all $x_i > a$; $b = r$-lim $\{(f(z_i) - f(x_i)) / (z_i - x_i); i = 1, 2, 3, \ldots\}$, with $z_i < a < x_i$ for all $i = 1, 2, 3, \ldots$ and the sequences $\{x_i ; i = 1, 2, 3, \ldots\}$ and $\{z_i ; i = 1, 2, 3, \ldots\}$ converging to a, correspondingly) for all sequences $\{x_i ; x_i \neq a ; i = 1, 2, 3, \ldots\}$ converging to a, then by Definition 2.5.1, $b = r$-lim E_{ct} ($b = r$-lim E_l , $b = r$-lim E_r or $b = r$-lim E_t, correspondingly).

At the same time, if $b = r$-lim E_{ct} ($b = r$-lim E_l , $b = r$-lim E_r or $b = r$-lim E_t , correspondingly), then by Definition 4.2.1, $b = st_r{}^{ct}d/_{dx}f(a)$ ($b = st_r{}^{l}d/_{dx}f(a)$, $b = st_r{}^{r}d/_{dx}f(a)$ or $b = st_r{}^{t}d/_{dx}f(a)$, correspondingly).

Lemma 4.2.3. Any strong centered r-derivative of $f(x)$ at a point $a \in X$ is both a strong left and strong right r-derivative of $f(x)$ at the same point for any $r \in \boldsymbol{R}^+$.

Indeed, conditions for both left and right r-derivatives are included in the conditions for the centered r-derivative.

Lemma 2.2.4 implies the following result.

Lemma 4.2.4. If $d = st_r{}^{z}d/_{dx}f(a)$, then $d = st_q{}^{z}d/_{dx}f(a)$ for any $z \in \{ct, l, r, t\}$ and any $q > r$.

Lemma 4.2.5. If a number b is both a strong left and strong right r-derivative of $f(x)$ at a point $a \in X$, then b is a strong centered r-derivative of $f(x)$ at the same point for any $r \in \boldsymbol{R}^+$.

Proof. Let us consider a sequence $\{x_i \in \boldsymbol{R}; i = 1, 2, 3, \ldots\}$ converging to $a \in X$ and let b be both strong left and strong right r-derivatives of $f(x)$ at a. Then the sequence $\{x_i \in \boldsymbol{R}; i = 1, 2, 3, \ldots\}$ consists of two subsequences $\{v_i \in \boldsymbol{R}; i = 1, 2, 3, \ldots\}$ and $\{z_i \in \boldsymbol{R}; i = 1, 2, 3, \ldots\}$ such that $v_i < a$ and $z_i > a$ for all $i = 1, 2, 3, \ldots$. Each of them is either finite or converges to a. When one of these subsequences is finite, then the definition of a left or right fuzzy derivative implies that $b = r$-lim $\{(f(a) - f(x_i)) / (a - x_i) ; i = 1, 2, 3, \ldots, a = \lim_{i\to\infty} x_i\}$.

To prove the statement of the lemma, we have also to consider the case when both subsequences $\{v_i \in \boldsymbol{R}; i = 1, 2, 3, \ldots\}$ and $\{z_i \in \boldsymbol{R}; i = 1, 2, 3, \ldots\}$ are infinite. In this case, by the definition of strong r-derivatives, we have $b = r$-lim $\{(f(a) - f(v_i)) / (a - v_i)) ; i = 1, 2, 3, \ldots\}$ and $b = r$-lim $\{(f(a) - f(z_i)) / (a - z_i); i = 1, 2, 3, \ldots\}$. Then by Theorem 2.2.3 and Lemma 2.2.5, we have $b = r$-lim $\{(f(a) - f(x_i)) / (a - x_i)) ; i = 1, 2, 3, \ldots\}$.

As the sequence $\{x_i \in \boldsymbol{R}; i = 1, 2, 3, \ldots\}$ that converges to a is chosen arbitrarily, Lemma is proved.

Definition 4.2.2. A number b is called a *full r-derivative* of a function $f(x)$ at a point $a \in X$ if b is at the same time a strong centered, left, right, and two-sided r-derivative of $f(x)$ at the point a.

Proposition 4.2.1. If a number b is a strong centered r-derivative of a function $f(x)$ at a point $a \in X$, then b is a strong two-sided r-derivative of $f(x)$ at the point a.

Proof. Let us consider an arbitrary sequence

$$\{ (f(z_i) - f(x_i)) /(z_i - x_i); z_i > a > x_i , i = 1, 2, 3, \dots , \text{ and } a = \lim_{i\to\infty} x_i = \lim_{i\to\infty} z_i \}.$$

Geometrical considerations demonstrate that either

$$(f(a) - f(x_i)) /(a - x_i) \leq (f(z_i) - f(x_i)) /(z_i - x_i) \leq (f(a) - f(z_i)) /(a - z_i) \quad (4.2)$$

or

$$(f(a) - f(x_i)) /(a - x_i) \geq (f(z_i) - f(x_i)) /(z_i - x_i) \geq (f(a) - f(z_i)) /(a - z_i) \quad (4.3)$$

Indeed, let us assume that

$$(f(a) - f(x_i)) /(a - x_i) \leq (f(z_i) - f(x_i)) /(z_i - x_i) \quad (4.4)$$

Then we have

$$(f(a) - f(x_i))(z_i - x_i) \leq (f(z_i) - f(x_i))(a - x_i)$$

as $z_i > a > x_i$. Performing multiplication, we have

$$(f(a))z_i - (f(x_i))z_i - (f(a))x_i + (f(x_i))x_i \leq$$
$$(f(z_i))\, a - (f(x_i))a - (f(z_i))\, x_i + (f(x_i))x_i$$

cancelling $(f(x_i))x_i$ from both parts of this inequality, we obtain

$$(f(a))z_i - (f(x_i))z_i - (f(a))x_i \leq (f(z_i))\, a - (f(x_i))a - (f(z_i))\, x_i$$

Using properties of inequalities and real numbers, we derive the following sequence of inequalities:

$$(f(a))z_i - (f(x_i))z_i - (f(a))x_i \leq (f(z_i))\, a - (f(x_i))a - (f(z_i))\, x_i$$
$$(f(a))z_i - (f(z_i))z_i - (f(a))x_i + (f(z_i))\, x_i \leq (f(z_i))\, a - (f(x_i))a - (f(z_i))z_i + (f(x_i))z_i$$
$$(f(a) - f(z_i))(z_i - x_i) \leq (f(z_i) - f(x_i))(a - z_i)$$
$$(f(z_i) - f(x_i)) /(z_i - x_i) \leq (f(a) - f(z_i)) /(a - z_i)$$

as $(a - z_i) < 0$. It means that if the right part of the inequality (4.2) is true, then the left part of this inequality is also true, or that the inequality (4.4) implies the inequality (4.2) and the inequality (4.5) implies the inequality (4.3).

By a similar argument, the case when

$$(f(a) - f(x_i)) /(a - x_i) \geq (f(z_i) - f(x_i)) /(z_i - x_i) \quad (4.5)$$

implies the inequality

$$(f(z_i) - f(x_i)) /(z_i - x_i) \geq (f(a) - f(z_i)) /(a - z_i)$$

As all real numbers are linearly ordered, always either (4.4) or (4.5) is true. In addition, (4.4) implies (4.2), while (4.5) implies (4.3).

If the number b is a strong centered r-derivative of $f(x)$ at a, then b is an r-limit of both sequences $\{ (f(a) - f(z_i)) /(a - z_i); i = 1, 2, 3, \dots \}$ and $\{ (f(a) - f(x_i)) /(a - x_i); i = 1, 2, 3, \dots \}$ by inequalities (4.2) and (4.3). In addition, properties of r-limits (cf. the Fuzzy Squeeze Theorem 2.2.9) imply that b is an r-limit of the sequence $\{ (f(z_i) - f(x_i)) /(z_i - x_i); i = 1, 2, 3, \dots \}$. As $\{ (f(z_i) - f(x_i)) /(z_i - x_i); z_i > a > x_i, i = 1, 2, 3, \dots \}$ is an arbitrary sequence of this form, the point b is (by the definition) a strong two-sided r-derivative of $f(x)$ at the point $a \in X$.

Proposition 4.2.1 is proved.

Corollary 4.2.1. If a number b is a strong centered r-derivative of a function $f(x)$ at a point $a \in X$, then b is a strong full r-derivative of $f(x)$ at the point a.

Indeed, by Proposition 4.2.1, if a number b is a strong centered r-derivative of a function $f(x)$ at a point $a \in X$, then b is a strong two-sided r-derivative of $f(x)$ at the point $a \in X$. In addition, by Lemma 4.2.3, the number b is both a strong left and strong right r-derivative of $f(x)$ at the point a. Thus, by Definition 4.2.2, the number b is a strong full r-derivative of $f(x)$ at the point a.

Lemma 2.2.5 and Corollary 4.2.1 imply the following result.

Corollary 4.2.2. If a number b is a strong centered r-derivative of a function $f(x)$ at the point $a \in X$, then $2b$ is a strong full $(|b| + 2r)$-derivative of $f(x)$ at the point a.

Let $f(x)$ be a function, which is continuous at a point $a \in X$.

Proposition 4.2.2. If a strong two-sided r-derivative of $f(x)$ at a point $a \in X$ exists (and is equal to b), then both one-sided strong r-derivatives of $f(x)$ at a point a exist (and coincide with b).

Proof. Let us consider a sequence $\{x_i \in \boldsymbol{R}; i = 1, 2, 3, \dots \}$ that converges to $a \in X$ and in which all $x_i < a$. Let us consider a sequence $\{x_i \in \boldsymbol{R}; i = 1, 2, 3, \dots \}$, which converges to $a \in X$ and in which all $x_i < a$. As $f(x)$ is a continuous function at a and $a = \lim_{i\to\infty} x_i$, it is possible to correspond a number z_i to each number x_i such that $a < z_i$, $| a - z_i | < \delta_n (| a - x_i |)$ and $| f(a) - f(z_i)| < \varepsilon_n (| a - x_i |)$ when i is larger than some n. It is possible to do this in such a way that both sequences $\{ \delta_n; n = 1, 2, 3, \dots \}$ and $\{\varepsilon_n ; n = 1, 2, 3, \dots \}$ converge to zero. For example, we can take $\delta_n = 1/n$ and find corresponding numbers ε_n for all $n = 1, 2, 3, \dots$.

Let us take some $k \in \boldsymbol{R}^{++}$. Then as $a - x_i < z_i - x_i$, we have

$$
\begin{gathered}
| b - ((f(a) - f(x_i)) / (a - x_i)) | = \\
| b - ((f(z_i) - f(x_i)) /(a - x_i)) + ((f(z_i) - f(a)) /(a - x_i)) | \le | \\
b - ((f(z_i) - f(x_i)) /(a - x_i))| + | ((f(z_i) - f(a)) /(a - x_i)) | = \\
| b - ((f(z_i) - f(x_i)) /(z_i - x_i)) + ((f(z_i) - f(x_i)) /(z_i - x_i)) - \\
((f(z_i) - f(x_i)) /(a - x_i))| + |((f(z_i) - f(a)) /(a - x_i)) | \le \\
| b - ((f(z_i) - f(x_i)) /(z_i - x_i))| + |((f(z_i) - f(x_i)) /(z_i - x_i)) - \\
((f(z_i) - f(x_i)) /(a - x_i))| + | ((f(z_i) - f(a)) /(a - x_i)) | < \\
r + (1/3)k + |((f(z_i) - f(x_i)) /(z_i - x_i)) - ((f(z_i) - f(x_i)) /(a - x_i))| + \\
| ((f(z_i) - f(a)) /(a - x_i)) |
\end{gathered}
$$

as the number b is a strong two-sided r-derivative of f at the point $a \in X$.

At the same time, by the choice of the sequence $\{z_i \in \boldsymbol{R}; i = 1, 2, 3, \dots \}$, we have

$$|((f(z_i) - f(a)) /(a - x_i))| \leq |((f(z_i) - f(a))| / |(a - x_i))| < (\varepsilon_n (| a - x_i |))/(| a - x_i |) = \varepsilon_n$$

and

$$|((f(z_i) - f(x_i)) /(z_i - x_i)) - ((f(z_i) - f(x_i)) /(a - x_i))| \leq |(f(z_i) - f(x_i)| \, | ((1/(z_i - x_i)) - (1 /(a - x_i)) | = |(f(z_i) - f(x_i)| \, |(a - x_i - z_i + x_i)/((z_i - x_i)(a - x_i)) | = |(f(z_i) - f(x_i)| \, |(a - z_i)/((z_i - x_i)(a - x_i)) | \leq (|(f(z_i) - f(x_i)| / | z_i - x_i |)\cdot(|(a - z_i)/(a - x_i)| < \delta_n(|(f(z_i) - f(x_i)| / | z_i - x_i |).$$

There is some number k such that $(|(f(z_i) - f(x_i)| / | z_i - x_i |) < k$ because the number b is a strong two-sided r-derivative of $f(x)$ at the point $a \in X$.

Consequently, $| b - ((f(a) - f(x_i)) / (a - x_i)) | < r + (1/3)k + \delta_n + \varepsilon_n < r + k$ for some $n = 1, 2, 3, \ldots$ as both sequences $\{ \delta_n ; n = 1, 2, 3, \ldots\}$ and $\{\varepsilon_n ; n = 1, 2, 3, \ldots\}$ converge to zero. As a result, number $b \in \boldsymbol{R}$ is a strong left r-derivative of $f(x)$ at the point a.

In a similar way, we prove that b is a strong right r-derivative of $f(x)$ at the point a.

Proposition is proved.

Remark 4.2.7. Continuity of $f(x)$ is essential for the validity of Proposition 4.2.2. It is demonstrated by the following example.

Example 4.2.2. Let $f(x) = x$ for all $x > 0$, $f(x) = -x$ for all $x < 0$, and $f(0) = 1$. Then $f(x)$ has a strong two-sided 3-derivative at 0 but it has no strong one-sided r-derivatives at 0 for any $r \in \boldsymbol{R}^+$.

Corollary 4.2.3. If the strong two-sided r-derivative of $f(x)$ at a point $a \in X$ exists (and is equal to b), then a strong centered r-derivative of $f(x)$ at the point $a \in X$ exists (and coincides with b).

Lemma 2.2.5 and Corollary 4.2.3 imply the following result.

Corollary 4.2.4. If $f(x)$ is continuous at a point $a \in X$ and b is a strong centered r-derivative of $f(x)$ at the point $a \in X$, then $2b$ is a strong centered $(|b| + 2r)$-derivative of $f(x)$ at the point $a \in X$.

Proposition 4.2.3. a) If a strong centered 0-derivative $st_0^{ct}d/_{dx}\, f(a)$ of $f(x)$ at a point $a \in X$ exists, then it is unique and equal to the classical derivative $f'(a)$ of $f(x)$ at a.

b) If the classical derivative $f'(a)$ of $f(x)$ at a exists, then it is equal to the strong centered 0-derivative $st_0^{ct}d/_{dx} f(a)$ of $f(x)$ at a.

Proof follows from the definition of a strong centered 0-derivative and uniqueness of the classical derivative $f'(a)$.

Corollary 4.2.5. For any $b \in \boldsymbol{R}$, we have $b = st_0^{ct}d/_{dx}\, f(a)$ if and only if $b = f'(a)$,

This result demonstrates that the concept of a fuzzy derivative is a natural extension of the concept of the conventional derivative.

Lemma 2.2.8 implies the following result.

Proposition 4.2.4. If b is a strong r-derivative of any type of $f(x)$ at a and $| b - e | < k$, then e is a strong $(r+k)$-derivative of the same type of $f(x)$ at a.

Corollary 4.2.6. If $b = f\,'(a)$ and $|\, b - e \,| < k$, then e is a strong k-derivative of $f(x)$ at a.

Definition 4.2.3. Any strong r-derivative of $f(x)$ at a point $a \in X$ is called a *fuzzy strong derivative* of $f(x)$ at a of the same type (i.e., centered, right, left, or two-sided).

It is denoted by $b = \mathrm{stF}^{z}\mathrm{d}/_{\mathrm{d}x}f(a)$.

From Proposition 4.2.3, we have the following result.

Corollary 4.2.7. If $b = f\,'(a)$, then $b = \mathrm{stF}^{ct}\mathrm{d}/_{\mathrm{d}x}f(a)$.

In other words, classical derivative is a special case of strong fuzzy derivative.

Theorem 2.2.5 makes it possible to prove the following result demonstrating local linearity and additivity of strong fuzzy derivatives.

Let us assume that $r, q \in \boldsymbol{R}^{+}$, while $f: X \rightarrow \boldsymbol{R}$ and $g: X \rightarrow \boldsymbol{R}$ are arbitrary real functions.

Theorem 4.2.1. *a*) **(Local additivity of fuzzy differentiation)** If b is a strong centered (left, right, two-sided) r-derivative of a function $f(x)$ at a and c is a strong centered (correspondingly, left, right, two-sided) q-derivative of a function $g(x)$ at a, then $b \pm c$ is a strong centered (correspondingly, left, right, two-sided) $(r + q)$-derivative of the function $(f \pm g)(x)$ at the point a.

b) **(Local uniformity of fuzzy differentiation)** If b is a strong centered (left, right, two-sided) r-derivative of $f(x)$ at a point a and $t \in \boldsymbol{R}$, then $t \cdot b$ is a strong centered (correspondingly, left, right, two-sided) $|t| \cdot r$-derivative of the function $t \cdot f(x)$ at the point a.

We do not give proofs of these statements because they are direct corollaries of the corresponding results for conditional fuzzy derivatives (cf. Theorem 4.4.1).

As a consequence, Theorem 4.2.1 gives a well-known result of the classical analysis such as Theorem 4.1.4.

Corollary 4.2.8. For any function $f(x)$ and any $z \in \{\mathrm{ct, l, r, t}\}$, if $b = \mathrm{st}_{r}{}^{z}\mathrm{d}/_{\mathrm{d}x}\, f(a)$, then $- b = \mathrm{st}_{r}{}^{z}\mathrm{d}/_{\mathrm{d}x}\,(-f(a))$.

Remark 4.2.8. Theorem 4.2.1 shows that fuzzy differentiation is a linear multivalued operator on the linear space of real functions.

Remark 4.2.9. When the conditions of Theorem 4.2.1 are satisfied, the point $b - c$ is not necessarily a strong $(r - q)$-derivative of the function $(f - g)(x)$ at a. However, in some cases, it might be such a derivative, for example, when $r = q = 0$.

Let $f(x)$ and $g(x)$ be fuzzy differentiable at a point a functions.

Theorem 4.2.2 (The Fuzzy Local Product Rule). If $u = \mathrm{st}_{r}{}^{ct}d/_{\mathrm{d}x}\, f(a)$ and $v = \mathrm{st}_{q}{}^{ct}d/_{\mathrm{d}x}\, g(a)$, then $f(a)v + ug(a) = \mathrm{st}_{g(a)\cdot r+q}{}^{ct}d/_{\mathrm{d}x}\, (f \cdot g)'(a)$.

Proof. Let us assume that $u = \mathrm{st}_{r}{}^{ct}d/_{\mathrm{d}x}\, f(a)$ and $v = \mathrm{st}_{q}{}^{ct}d/_{\mathrm{d}x}\, g(a)$, $l = \{\, x_i\,;\, i = 1, 2, 3, \ldots \,\}$ is a with $a = \lim_{i\to\infty} x_i$. Then by the definition of a strong derivative, $u = r\text{-}\lim_{i\to\infty} ((f(x_i) - f(a))/(x_i - a)\,)$ and $v = q\text{-}\lim_{i\to\infty} ((g(x_i) - g(a))/(x_i - a)\,)$ for some sequence $h = \{\, x_i\,;\, i = 1, 2, 3, \ldots \,\}$ that converges to a and with $x_i \neq a$.

By Theorem 2.2.5, we have

$$ug(a) = r\text{-}\lim_{i\to\infty} [[(f(x_i) - f(a))/(x_i - a)\,)]\, g(a)]$$

and

$$f(a)v = q\text{-}\lim_{i\to\infty} [f(a) [(g(x_i) - g(a))/(x_i - a))]]$$

As the function $g(x)$ is r-differentiable at the point a (cf. Section 4.6), this function is continuous at the point a. Consequently, $g(a) = \lim_{i\to\infty} g(x_i)$.

By Corollary 2.2.11, we have if

$$ug(a) = r\text{-}\lim_{i\to\infty} [[(f(x_i) - f(a))/(x_i - a))] g(a)] \text{ and } g(a) = \lim_{i\to\infty} g(x_i),$$

then

$$ug(a) = (g(a)\cdot r)\text{-}\lim_{i\to\infty} [[(f(x_i) - f(a))/(x_i - a))] g(x_i)].$$

Thus, by Theorem 2.2.5, we have

$$f(a)v + ug(a) =$$
$$(g(a)\cdot r + q)\text{-}\lim_{i\to\infty} [[f(a) [(g(x_i) - g(a))/(x_i - a))]] + [[(f(x_i) - f(a))/(x_i - a))] g(x_i)]]$$

Let us transform the expression under the limit sign.

$$g(x_i) \cdot ((f(x_i) - f(a))/(x_i - a)) + f(a) \cdot ((g(x_i) - g(a))/(x_i - a) =$$
$$(f(x_i) g(x_i) - f(a)g(x_i))/(x_i - a)) + (f(a) g(x_i) - f(a)g(a))/(x_i - a) =$$
$$(f(x_i) g(x_i) - f(a)g(x_i) + f(a) g(x_i) - f(a)g(a))/(x_i - a) =$$
$$(f(x_i) \cdot g(x_i) - f(a) \cdot g(a))/(x_i - a)$$

This gives us

$$f(a)v + ug(a) =$$
$$(g(a)\cdot r + q)\text{-}\lim_{i\to\infty}(f(x_i) \cdot g(x_i) - f(a) \cdot g(a))/(x_i - a)) \tag{4.6}$$

The right part of the formula (4.6) is used in the definition of a fuzzy derivative. As $l = \{ x_i ; i = 1, 2, 3, \dots \}$ is an arbitrary sequence that converges to a and with $x_i \neq a$, we see that the function $f(x) \cdot g(x)$ is $(r + q)$-differentiable and $f(a)v + ug(a) = \mathrm{st}_{g(a)\cdot r+q}{}^{\mathrm{ct}}d/\mathrm{d}x\ (f \cdot g)'(a)$.

Theorem is proved.

Lemma 4.2.2 and Proposition 2.6.3 imply the following result.

Proposition 4.2.5. For any function $f(x)$ and any $z \in \{\mathrm{ct, l, r, t}\}$, if $r_0 = \inf \{ r ; f(x)$ has a strong r-derivative of the type z at a point $a\}$, then there is a number d_0 such that $d_0 = \mathrm{st}_r{}^z\mathrm{d}/\mathrm{d}x\ f(a)$.

Remark 4.2.10. By Proposition 2.2.3, the derivative $d_0 = \mathrm{st}r_o{}^z\mathrm{d}/\mathrm{d}x f(a)$ is unique for any function $f(x)$.

Corollary 4.2.9. For any function $f(x)$ and any $z \in \{\mathrm{ct, l, r, t}\}$, the set $\mathrm{FDer}^z f(x) = \{ r ; f(x)$ has a strong r-derivative of the type z at a point $a\}$ is empty or $\mathrm{FDer}^z f(x)$ is a closed ray.

Definition 4.2.4. The number

$$r_0 = \inf\{r;\ d = \text{st}_r{}^z\text{d}/_{\text{d}x} f(a)\}$$

(if it exists) is called the *defect* $\delta(\ d = \text{stF}^z\text{d}/_{\text{d}x} f(a)\ )$ of d being a strong derivative of the type z of the function $f(x)$ at the point a. When such a number r_0 does not exist, we put $\delta(\ d = \text{stF}^z\text{d}/_{\text{d}x} f(a)) = \infty$.

Proposition 2.6.4 and Definition 4.2.4 imply the following result.

Proposition 4.2.6. For any z ∈ {ct, l, r, t}, $\delta(d = \text{stF}^z\text{d}/_{\text{d}x} f(a))$ is a number if and only if the set of all weak fuzzy derivatives of the function $f(x)$ at the point a is bounded.

Theorem 4.2.1 and properties of infimum imply the following result.

Proposition 4.2.7. For any z ∈ {ct, l, r, t}, we have the following inequalities:

$$\delta((d + c) = \text{stF}^z\text{d}/_{\text{d}x}\ (f + g)(a)) \leq \delta(d = \text{stF}^z\text{d}/_{\text{d}x}\ f(a)) + \delta(c = \text{stF}^z\text{d}/_{\text{d}x}\ g(a)) \tag{4.7}$$

$$\delta((d - c) = \text{stF}^z\text{d}/_{\text{d}x}\ (f - g)(a)) \geq |\delta(d = \text{stF}^z\text{d}/_{\text{d}x} f(a)) - \delta(c = \text{stF}^z\text{d}/_{\text{d}x}\ g(a))| \tag{4.8}$$

$$\delta(d = \text{stF}^z\text{d}/_{\text{d}x}\ t\cdot f(a)) = |t|\cdot \delta(d = \text{stF}^z\text{d}/_{\text{d}x}\ g(a)) \tag{4.9}$$

Corollary 4.2.10. For any function $f(x)$ and any z ∈ {ct, l, r, t}, we have:

$$\delta(d = \text{stF}^z\text{d}/_{\text{d}x}\ \text{-}f(a)) = \delta(d = \text{stF}^z\text{d}/_{\text{d}x}\ f(a))$$

Remark 4.2.11. It possible that (4.7) and (4.8) become strict inequalities as the following example demonstrates.

Example 4.2.3. Let us consider the following functions:

$$f(x) = \begin{cases} 0 & \text{for } x < 0, \\ x & \text{for } x \geq 0. \end{cases}$$

and

$$g(x) = \begin{cases} 0 & \text{for } x < 0, \\ -x & \text{for } x \geq 0. \end{cases}$$

Then $(f + g)(x)$ is identically equal to 0 and consequently, $\delta(0 = \text{stF}^z\text{d}/_{\text{d}x}\ (f + g)(0)) = 0$. At the same time, $\delta(0 = \text{stF}^z\text{d}/_{\text{d}x} f(0)) = 1$ and $\delta(0 = \text{stF}^z\text{d}/_{\text{d}x}\ g(0)) = 1$. Thus,

$$\delta(0 = \text{stF}^z\text{d}/_{\text{d}x}\ (f + g)(0)) = 0 <$$

$$\delta(0 = \mathrm{stF}^z \mathrm{d}/_{\mathrm{d}x} f(0)) + \delta(0 = \mathrm{stF}^z \mathrm{d}/_{\mathrm{d}x} (f + g)(0)) = 1 + 1 = 2.$$

Local fuzzy derivatives allow us to define global fuzzy derivatives.

Definition 4.2.5. A function $g(x)$ is called a *strong centered* (*left*, *right*, *two-sided*) *r-derivative* of the function $f(x)$ on a set $D \subseteq X$ if for any $d \in D$, we have $g(d) = \mathrm{st}_r{}^{\mathrm{ct}}\, \mathrm{d}/_{\mathrm{d}x} f(d)$ ($g(d) = \mathrm{st}_r{}^{\mathrm{l}}\, \mathrm{d}/_{\mathrm{d}x} f(d)$, $g(d) = \mathrm{st}_r{}^{\mathrm{r}}\, \mathrm{d}/_{\mathrm{d}x} f(d)$, and $g(d) = \mathrm{st}_r{}^{\mathrm{t}}\, \mathrm{d}/_{\mathrm{d}x} f(d)$, correspondingly).

It is denoted by $g(x) = \mathrm{st}_r{}^{\mathrm{ct}}\mathrm{d}/_{\mathrm{d}x} f(x)$, $g(x) = \mathrm{st}_r{}^{\mathrm{l}}\mathrm{d}/_{\mathrm{d}x} f(x)$, $g(x) = \mathrm{st}_r{}^{\mathrm{r}}\mathrm{d}/_{\mathrm{d}x} f(x)$, and $g(x) = \mathrm{st}_r{}^{\mathrm{t}}\mathrm{d}/_{\mathrm{d}x} f(x)$, correspondingly.

Corollary 4.2.9. a) (**Global additivity of fuzzy differentiation**)) If h: $X \rightarrow \boldsymbol{R}$ and u: $X \rightarrow \boldsymbol{R}$ are real functions, for which we have $h(x) = \mathrm{st}_r{}^z\mathrm{d}/_{\mathrm{d}x} f(x)$ and $u(x) = \mathrm{st}_q{}^z\mathrm{d}/_{\mathrm{d}x} g(x)$, then $(h \pm u)(x) = \mathrm{st}_{r+q}{}^z\mathrm{d}/_{\mathrm{d}x} (f \pm g)(x)$.

b) (**Global uniformity of fuzzy differentiation**) If h: $X \rightarrow \boldsymbol{R}$, $h(x) = \mathrm{st}_r{}^z\mathrm{d}/_{\mathrm{d}x} f(x)$, and $q \in \boldsymbol{R}$, then $qh(x) = \mathrm{st}_{|q|\,r}{}^z\mathrm{d}/_{\mathrm{d}x}\, qf(x)$.

Let $f(x)$ and $g(x)$ be fuzzy differentiable at a point a functions. Then Theorem 4.2.2 implies the following result.

Theorem 4.2.3 (the Fuzzy Global Product Rule). If $u(x) = \mathrm{st}_r{}^{\mathrm{ct}} d/_{\mathrm{d}x} f(x)$ and $v(x) = \mathrm{st}_q{}^{\mathrm{ct}} d/_{\mathrm{d}x} g(x)$, then $f(x)v(x) + u(x)g(x) = \mathrm{st}_{g(a)\cdot r+q}{}^{\mathrm{ct}} d/_{\mathrm{d}x} (f \cdot g)(x)$.

4.3. Weak Fuzzy Derivatives

> The pleasure we derive from the representation of the present is due,
> not only to the beauty it can be clothed in,
> but also to its essential quality of being the present.
> *Charles Baudelaire* (1821-1867)

In addition to strong fuzzy derivatives, we define weak fuzzy derivatives,

Let $X, Y \subseteq \boldsymbol{R}$, f: $X \rightarrow Y$ be a function, $a \in X$, $b \in \boldsymbol{R}$, and $r \in \boldsymbol{R}^+$.

Definition 4.3.1. a) A number b is called a *weak centered r-derivative* of the function $f(x)$ at a point $a \in X$ for a sequence $\{x_i\, ;\, x_i \in \mathrm{Dom}\, f;\, x_i \neq a\, ;\, i = 1, 2, 3, \dots\}$ converging to a if $b = r\text{-lim}\, \{(f(a) - f(x_i))\, /(a - x_i)\, ;\, i = 1, 2, 3, \dots\}$.

b) A number b is called a *weak left r-derivative* of the function $f(x)$ at a point $a \in X$ for a sequences $\{x_i\, ;\, x_i \in \mathrm{Dom}\, f;\, x_i < a;\, i = 1, 2, 3, \dots\}$ converging to a if $b = r\text{-lim}\, \{(f(a) - f(x_i))\, /(a - x_i)\, ;\, i = 1, 2, 3, \dots\}$.

c) A number b is called a *weak right r-derivative* of the function $f(x)$ at a point $a \in X$ for a sequences $\{x_i\, ;\, x_i \in \mathrm{Dom}\, f;\, x_i > a;\, i = 1, 2, 3, \dots\}$ converging to a if $b = r\text{-lim}\, \{(f(a) - f(x_i))\, /(a - x_i)\, ;\, i = 1, 2, 3, \dots\}$.

d) A number b is called a *weak two-sided r-derivative* of the function $f(x)$ at a point $a \in X$ for sequences $\{x_i ; x_i \in \text{Dom} f ; x_i < a; i = 1, 2, 3, \ldots\}$ and $\{z_i ; z_i \in \text{Dom} f ; z_i > a; i = 1, 2, 3, \ldots\}$ converging to a if $b = r\text{-lim}\ \{(f(z_i) - f(x_i)) / (z_i - x_i) ; i = 1, 2, 3, \ldots\}$.

Weak r-derivatives are denoted by $b = \mathrm{w}_r^{\mathrm{ct}}\, \mathrm{d}/_{\mathrm{d}x}\, f(a)$, $b = \mathrm{w}_r^{\mathrm{l}}\, \mathrm{d}/_{\mathrm{d}x} f(a)$, $b = \mathrm{w}_r^{\mathrm{r}}\, \mathrm{d}/_{\mathrm{d}x}\, f(a)$, and $b = \mathrm{w}_r^{\mathrm{t}}\, \mathrm{d}/_{\mathrm{d}x} f(a)$, correspondingly.

When r is not specified, we call weak centered (left, right, two-sided) r-derivatives of $f(x)$ at a point $a \in X$ *fuzzy weak centered* (*left*, *right*, *two-sided*, correspondingly) *derivatives* of $f(x)$ at a point $a \in X$.

Example 4.3.1. Let us take the membership function $\mathrm{m_Q}(x)$ of the set of rational numbers, i.e., $\mathrm{m_Q}(x)$ is equal to 1 when x is a rational number and $\mathrm{m_Q}(x)$ is equal to 0 when x is an irrational number. This function is not even continuous, consequently it does not have derivatives neither in the classical sense (cf., for example, (Goldstein *et al*, 1987)) nor as a generalized function (Shwartz, 1950-51). However, at any point x from $\boldsymbol{R}$, $\mathrm{m_Q}(x)$ has a weak derivative, which is equal to 0.

Remark 4.3.1. It is useful to know that in modern functional analysis, there are other kinds of concepts that bear names of strong and weak derivatives. First, when differentiation is developed in functional normed (or topological linear) spaces, it is possible to consider two kinds of norms (topologies): *strong* and *weak norm* (*topology*) (cf., for example, (Kolmogorov and Fomin, 1999)). A derivative of a function is, as a rule, defined by means of some limit process. Consequently, the result depends on the utilized topology. Thus, when the strong topology is used, we have the concept of the strong derivative or, as it is even more frequently called, the Frechet derivative. When the weak topology is utilized, we come to the concept of the weak derivative or, as it is frequently called, the Gataux derivative.

Second, Sobolev spaces imply a more generalized concept of a derivative, which gave birth to the construction of distributions or generalized functions. These derivatives are defined through integration and are called weak derivatives (cf., for example, Evans, 1998). They are used extensively in the theory of differential equations.

Third, in some texts on mathematical physics and functional analysis (cf., for example, (Reed and Simon, 1972)), the derivative of a function in the sense of distributions is also called the weak derivative.

To separate all these kinds of weak and strong derivatives, we may call derivatives of operators in normed (topological linear) spaces *weak and strong by norm* (*by topology*) *derivatives*. Weak derivatives in Sobolev spaces and in distributions, we may call *weak derivatives by integration*, while weak and strong derivatives in neoclassical analysis, we may call *weak and strong derivatives by sequences*.

However, in this Chapter, we have only derivatives in the latter sense. So, we will not use additional adjectives and call them simply weak and strong derivatives.

It is useful to understand that, although mathematicians tend to eliminate similar naming of different structures and give an exact meaning to each mathematical concept, many mathematical terms also have several meanings. For example a field in algebra, which can have characteristic zero or prime (cf., for example, (Van der Varden, 1971)), and a field in analysis, which can be scalar, vector or tensor (cf., for example, (Stewart, 2003)), are essentially different structures. Distributions in probability theory and distributions in functional analysis (Shwartz, 1950-51) give another example of such situation.

Remark 4.3.2. In what follows, $w_r{}^z d/_{dx} f(a)$ denotes one of the four defined types of weak r-derivatives of $f(x)$ at a point a. Here $z \in \{ct, l, r, t\}$.

Lemma 4.3.1. $b = st_r{}^z d/_{dx} f(a)$ if and only if b is a unique weak r-derivative $b = w_r{}^z d/_{dx} f(a)$ of the function $f(x)$ at a point $a \in X$ for any two sequences $\{x_i ; i = 1, 2, 3, \ldots \}$ and $\{z_i ; i = 1, 2, 3, \ldots \}$ converging to a where $z \in \{ct, l, r, t\}$.

It means that strong fuzzy derivatives are special cases of weak fuzzy derivatives.

Lemma 4.3.2. Any weak centered r-derivative of $f(x)$ at a point $a \in X$ is either a weak left or weak right or both a weak left and weak right r-derivative of $f(x)$ at the same point for any $r \in \boldsymbol{R}^+$.

Lemma 4.3.3. If a number b is both a weak left and weak right r-derivative of $f(x)$ at a point $a \in X$, then b is a weak centered r-derivative of $f(x)$ at the same point a for any $r \in \boldsymbol{R}^+$.

Proof. Let us consider a sequence $\{x_i \in \boldsymbol{R}; i = 1, 2, 3, \ldots \}$ converging to a and let b be both strong (weak) left and weak right r-derivatives of $f(x)$ at a. Then the sequence $\{x_i \in \boldsymbol{R}; i = 1, 2, 3, \ldots \}$ consists of two subsequences $\{v_i \in \boldsymbol{R}; i = 1, 2, 3, \ldots \}$ and $\{z_i \in \boldsymbol{R}; i = 1, 2, 3, \ldots \}$ such that $v_i < a$ and $z_i > a$ for all $i = 1, 2, 3, \ldots$. Each of them is either finite or converges to a. When one of these subsequences is finite, then the definition of a left or right fuzzy derivative implies that $b = r\text{-lim} \{ (f(a) - f(x_i)) / (a - x_i) ; i = 1, 2, 3, \ldots , a = \lim_{i\to\infty} x_i \}$.

To prove the statement of the lemma, we have to consider the case when both subsequences $\{v_i \in \boldsymbol{R}; i = 1, 2, 3, \ldots \}$ and $\{z_i \in \boldsymbol{R}; i = 1, 2, 3, \ldots \}$ are infinite. By the definition of weak r-derivatives $b = r\text{-lim}_{i\to\infty} \{(f(a) - f(v_i)) /(a - v_i)) ; i = 1, 2, 3, \ldots \}$ and $b = r\text{-lim}_{i\to\infty} \{(f(a) - f(x_i)) /(a - x_i); i = 1, 2, 3, \ldots , a = \lim_{i\to\infty} x_i \}$. Then by Lemma 2.2.9, $b = r\text{-lim}_{i\to\infty} \{(f(a) - f(x_i)) /(a - z_i)) ; i = 1, 2, 3, \ldots , a = \lim_{i\to\infty} z_i$ where z_i is either v_i or $x_i \}$.

As the sequence $\{x_i \in \boldsymbol{R}; i = 1, 2, 3, \ldots \}$ is chosen arbitrarily, Lemma 4.3.3 is proved.

Lemma 4.3.4. Any strong centered (left, right, two-sided) r-derivative of $f(x)$ at a point $a \in X$ is also a weak centered (left, right, two-sided) r-derivative of $f(x)$ at the same point for any $r \in \boldsymbol{R}^+$.

Proposition 4.3.1. If b is a weak r-derivative of any type of $f(x)$ at a and $| b - e | < k$, then e is a weak $(r + k)$-derivative of the same type of $f(x)$ at a.

Corollary 4.3.1. If $b = f\,'(a)$ and $| b - e | < k$, then e is a weak k-derivative of $f(x)$ at a.

Lemma 2.2.4 implies the following result.

Lemma 4.3.5. If $b = w_r{}^z d/_{dx} f(a)$, then $b = w_q{}^z d/_{dx} f(a)$ for any $z \in \{ct, l, r, t\}$ and any $q > r$.

Let us consider two functions $f(x)$ and $g(x)$.

Proposition 4.3.2. If $f(a) = g(a)$ and $f(x) \leq g(x)$ in some neighborhood of a, then for any weak r-derivative $b = w_r{}^z d/_{dx} f(a)$, there is a weak r-derivative $c = w_r{}^z d/_{dx}\, g(a)$ such that $b \leq c$.

Definition 4.3.2. Any weak r-derivative of a function $f(x)$ at a point $a \in X$ is called a *weak fuzzy derivative* of $f(x)$ at a and of the same type (i.e., centered, right, left, or two-sided).

It is denoted by $b = wFd^z/_{dx} f(a)$.

As we will see, not all properties of strong fuzzy derivatives or derivatives are true for weak fuzzy derivatives. However, for weak fuzzy derivatives, it is often possible to deduce somewhat weaker properties of additivity than those possessed by strong fuzzy derivatives.

Proposition 4.3.3. If the slope of the straight line that connects any point of the graph of a function $f(x)$ and the point $(a, f(a))$ is bounded (from the left, from the right) in some neighborhood of a point a from X, then $f(x)$ has, at least, one weak centered (left, right, correspondingly) derivative at this point.

Proof. Let Oa be neighborhood of a point a in which the slope of the straight line that connects any point of the graph of a function $f(x)$ and the point $(a, f(a))$ is bounded by a number k. If $c \in Oa$, then the slope of the straight line that connects points $(c, f(c))$ and $(a, f(a))$ has the form

$$\{(f(a) - f(c)) /(a - c)$$

Thus, taking a sequence $l = \{x_i \in Oa;\ i = 1, 2, 3, \ldots \}$ converging to a, we have that

$$\{(f(a) - f(x_i)) /(a - x_i) < k$$

for all elements from l. Consequently (cf. Definition 4.3.1), $0 = k\text{-lim } \{(f(a) - f(x_i)) /(a - x_i)\ ;\ i = 1, 2, 3, \ldots \}$ and $0 = \mathrm{w}_k{}^{\mathrm{ct}}\mathrm{d}/_{\mathrm{d}x}\, f(a)$.

Proposition is proved.

Remark 4.3.3. Boundedness is an essential condition for validity of Proposition 4.3.4 as the following examples show.

Example 4.3.2. Let us take the function $f(x)$ that is equal to 0 at all points of the form $k\pi$ with $k = 1, 2, 3, \ldots$ and equal to the function $\cot x$ at all other points from $\boldsymbol{R}$. At the point 0, this function has no fuzzy weak derivatives.

Example 4.3.3. Let us consider the function $f(x)$ that is equal to $1 + \sqrt{1 - x^2}$ on the unit interval $[0, 1]$ and is equal to $1 + \sqrt{1 - (x - 2)^2}$ on the interval $[1, 3]$. Being continuous, this function has no fuzzy weak derivatives at the point 1.

The construction of a weak centered (left, right, two-sided) r-derivative of $f(x)$ at a point $a \in X$ gives birth to a corresponding concept in the classical calculus, namely, of a weak centered (left, right, two-sided) derivative of a function $f(x)$ at a point $a \in X$.

Definition 4.3.3. a) A number b is called a *weak centered derivative* of the function $f(x)$ at a point $a \in X$ for a sequence $\{x_i\ ;\ x_i \in \mathrm{Dom}\, f\ ;\ x_i \neq a\ ;\ i = 1, 2, 3, \ldots \}$ converging to a if $b = \lim \{(f(a) - f(x_i)) /(a - x_i)\ ;\ i = 1, 2, 3, \ldots \}$.

b) A number b is called a *weak left derivative* of the function $f(x)$ at a point $a \in X$ for a sequences $\{x_i\ ;\ x_i \in \mathrm{Dom}\, f\ ;\ x_i < a;\ i = 1, 2, 3, \ldots \}$ converging to a if $b = \lim \{(f(a) - f(x_i)) /(a - x_i)\ ;\ i = 1, 2, 3, \ldots \}$.

c) A number b is called a *weak right derivative* of the function $f(x)$ at a point $a \in X$ for a sequences $\{x_i\ ;\ x_i \in \mathrm{Dom}\, f\ ;\ x_i > a;\ i = 1, 2, 3, \ldots \}$ converging to a if $b = \lim \{(f(a) - f(x_i)) /(a - x_i)\ ;\ i = 1, 2, 3, \ldots \}$.

d) A number b is called a *weak two-sided derivative* of the function $f(x)$ at a point $a \in X$ for sequences $\{x_i\ ;\ x_i \in \mathrm{Dom}\, f\ ;\ x_i < a;\ i = 1, 2, 3, \ldots \}$ and $\{z_i\ ;\ z_i \in \mathrm{Dom}\, f\ ;\ z_i > a;\ i = 1, 2, 3, \ldots \}$ converging to a if $b = \lim \{(f(z_i) - f(x_i)) / (z_i - x_i)\ ;\ i = 1, 2, 3, \ldots \}$.

Weak derivatives are denoted by $b = \mathrm{w}^{\mathrm{ct}}\mathrm{d}/_{\mathrm{d}x}\, f(a)$, $b = \mathrm{w}^{\mathrm{l}}\mathrm{d}/_{\mathrm{d}x}\, f(a)$, $b = \mathrm{w}^{\mathrm{r}}\mathrm{d}/_{\mathrm{d}x}\, f(a)$, and $b = \mathrm{w}^{\mathrm{t}}\mathrm{d}/_{\mathrm{d}x}\, f(a)$, correspondingly.

Remark 4.3.4. Weak derivatives of functions are special cases of extraderivatives in the sense of (Burgin, 1993; 2002) of the same functions.

Lemma 4.3.6. For any number $b \in \boldsymbol{R}$, we have $b = \mathrm{w}_0{}^{\mathrm{ct}}\mathrm{d}/_{\mathrm{d}x}\, f(a)$ if and only if $b = \mathrm{w}^{\mathrm{ct}}\mathrm{d}/_{\mathrm{d}x}\, f(a)$.

Let a be a non-isolated point of X. Then definitions imply the following result.

Corollary 4.3.2. If $b = \mathrm{stFd}^z/_{\mathrm{d}x} f(a)$, then $b = \mathrm{wFd}^z/_{\mathrm{d}x}\, f(a)$.

Proposition 4.3.4 (Local uniformity of weak fuzzy differentiation). If b is a weak centered (left, right, two-sided) r-derivative of $f(x)$ at a point a and $t \in \boldsymbol{R}$, then $t \cdot b$ is a weak centered (correspondingly, left, right, two-sided) $|t| \cdot r$-derivative of the function $t \cdot f(x)$ at the point a.

Proposition 4.3.5 is a weak counterpart of Theorem 4.2.1 proved for strong fuzzy derivatives. More exactly, it corresponds to the part b) of this theorem. At the same time, a version of the part (a) from Theorem 4.2.1 is not valid both for weak derivatives and for weak fuzzy derivatives as the following example demonstrates.

Example 4.3.4. Let us consider the following functions:

$$f(x) = \begin{cases} x & \text{for } x = \pm 1, \pm 1/2, \pm 1/3, \ldots \\ 10x & \text{otherwise} \end{cases}$$

and

$$g(x) = \begin{cases} x & \text{for } x = \pm 1/\pi, \pm 1/(2\pi), \pm 1/(3\pi), \ldots \\ 10x & \text{otherwise} \end{cases}$$

Then $1 = \mathrm{w}_0{}^{\mathrm{c}}\mathrm{d}/_{\mathrm{d}x}\, f(0)$ and $1 = \mathrm{w}_0{}^{\mathrm{c}}\mathrm{d}/_{\mathrm{d}x}\, f(0)$. However, $2 = 1 + 1$ is not a weak centered $(0+0)$-derivative, or simply, weak centered derivative (cf. Lemma 4.3.6) of the function $(f + g)(x)$ at the point 0.

However, for weak fuzzy derivatives, it is possible to prove a weaker results in comparison with Theorem 4.2.1 (a).

Let us assume that:

1) $f\colon X \rightarrow \boldsymbol{R}$ and $g\colon X \rightarrow \boldsymbol{R}$ are arbitrary real functions, $r, q \in \boldsymbol{R}^+$;
2) for any sequence $\{x_i\,;\, x_i \neq a\,;\, i = 1, 2, 3, \ldots\}$ converging to a, there is weak centered r-derivative of $f(x)$ at a and a weak centered q-derivatives of $g(x)$ at a;
3) sets $\{\, c;\, c = \mathrm{w}_r{}^{\mathrm{ct}}\mathrm{d}/_{\mathrm{d}x}\, f(a)\,\}$ of all weak centered r-derivatives of $f(x)$ at a and $\{\, b;\, b = \mathrm{w}_{\mathrm{q}}{}^{\mathrm{ct}}\mathrm{d}/_{\mathrm{d}x}\, g(a)\,\}$ of all weak centered q-derivatives of $g(x)$ at a are bounded;

   and

4) $\sup \{ c; c = w_r{}^{ct}d/_{dx} f(a)\} = u$, $\sup \{ b; b = w_r{}^{ct}d/_{dx} g(a) \} = v$.

Proposition 4.3.5. If b is a weak centered r-derivative of $f(x)$ at a and c is a weak centered q-derivative of $g(x)$ at a, then there is a number $e \in \boldsymbol{R}$ such that e is a weak centered $(r+q)$-derivative of $(f + g)(x)$ at a and $e \leq \min \{b + v; c + u \}$.

In addition, it is possible to prove the following result.

Proposition 4.3.6. If b is a weak centered (left, right, two-sided) r-derivative of a function $f(x)$ at a and c is a strong centered (correspondingly, left, right, two-sided) q-derivative of a function $g(x)$ at a, then $b \pm c$ is a weak centered (correspondingly, left, right, two-sided) $(r + q)$-derivative of the function $(f \pm g)(x)$ at the point a.

Proof. a) Let us assume that $b = w_r{}^{ct}d/_{dx} f(a)$ and $c = st_r{}^{ct}d/_{dx}\, g(a)$. By Definitions 4.2.1 and 4.3.1, it means $b = r\text{-}\lim_{i\to\infty} (f(a) - f(x_i)) / (a - x_i)$ for all sequences $\{x_i\,;\, x_i \neq a\,;\, i = 1, 2, 3, \dots \}$ converging to a and $c = q\text{-}\lim_{i\to\infty} (g(z_i) - g(x_i)) / (z_i - x_i)$ for some sequence $\{c_i\,;\, c_i \neq a\,;\, i = 1, 2, 3, \dots \}$ converging to a.

Then by Theorem 2.2.5, for the sequence $\{c_i\,;\, c_i \neq a\,;\, i = 1, 2, 3, \dots \}$, we have

$$
\begin{aligned}
b + c = (r + q)\text{-}\lim_{i\to\infty} [(f(a) - f(c_i)) / (a - c_i) + (g(a) - g(c_i)) / (a - c_i)] = \\
(r + q)\text{-}\lim_{i\to\infty} [[(f(a) - f(c_i)) + (g(a) - g(c_i))] / (a - c_i)] = \\
(r + q)\text{-}\lim_{i\to\infty} [[(f(a) + g(c_i)) - (f(a)) + g(c_i))] / (a - c_i)]
\end{aligned}
$$

It means that $b + c$ is is a weak centered $(r + q)$-derivative of the function $(f + g)(x)$ at the point a.

In a similar way, using Theorem 2.2.5, we have that $b - c$ is is a weak centered $(r + q)$-derivative of the function $(f - g)(x)$ at the point a and tb is a weak centered $|t|\cdot a$-derivative of the function $t{\cdot}f(x)$ at the point a.

The proof of the corresponding statements for left, right, and two-sided fuzzy derivatives is the same.

Proposition is proved.

Definition 4.3.4. A function $g(x)$ is called a *weak centered* (*left*, *right*, *two-sided*) *r-derivative* of a function $f(x)$ *on a set* $D \subseteq X$ if for any $d \in D$, we have $g(d) = w_r{}^{ct}\, d/_{dx} f(d)$ ($g(d) = w_r{}^{l}\, d/_{dx} f(d)$, $g(d) = w_r{}^{r}\, d/_{dx} f(d)$, and $g(d) = w_r{}^{t}\, d/_{dx} f(d)$, correspondingly).

It is denoted by $g(x) = w_r{}^{ct}d/_{dx} f(x)$, $g(x) = w_r{}^{l}d/_{dx} f(x)$, $g(x) = w_r{}^{r}d/_{dx} f(x)$, and $g(x) = w_r{}^{t}d/_{dx} f(x)$, correspondingly).

Proposition 4.3.5 implies the following result.

Corollary 4.3.3. If $h(x)$ is a weak r-derivative of $f(x)$ and $k(x)$ is a weak q-derivative of $g(x)$, then there is a function $l(x)$ such that $l(x)$ is a weak $(r+q)$-derivative of $(f+g)(x)$.

4.4. Conditional Fuzzy Derivatives

The first condition for making music is not to make a noise.
Jose Bergamin (1895-1983)

Weak and strong centered, left, right, two-sided fuzzy derivatives are naturally systematized by the constructions of weak and strong conditional fuzzy derivatives.

Let us consider some condition B for sequences of pairs of real numbers. Such condition B may be given in a *descriptive form*. For example:

$B_{tt(a)}$: a sequence $l = \{ (a_i , b_i); i = 1, 2, 3, \dots \}$ for which $\lim_{i\to\infty} a_i = a$ and all $b_i = a$.

$B_{ct(a)}$: a sequence $l = \{ (a_i , b_i); i = 1, 2, 3, \dots \}$ for which $a_i < a < b_i$ for all $i = 1, 2, 3, \dots$, $\lim_{i\to\infty} a_i = a$ and $\lim_{i\to\infty} b_i = a$.

Condition B may be given in a *model form*. For example:

B_E : Eq$\{ l_1 = \{ (a_{1i}, b_{1i}) ; i = 1, 2, 3, \dots \}, \dots , l_n = \{ (a_{ni} , b_{ni}) ; i = 1, 2, 3, \dots \} \}$.

The meaning of this model condition is: a sequence l that is equal to one of the sequences $l_1 = \{ (a_{1i}, b_{1i}) ; i = 1, 2, 3, \dots \}, \dots , l_n = \{ (a_{ni} , b_{ni}) ; i = 1, 2, 3, \dots \}$. These sequences constitute the model set E.

Condition B may be also given in an *algorithmic form*, that is, by an algorithm or procedure for calculating corresponding sequences. It is usually done in constructive or recursive mathematics. However, here we do not consider such conditions.

If a sequence $l = \{ (a_i , b_i) ; i = 1, 2, 3, \dots \}$ satisfies a condition B, it is denoted by $l \in$ B.

Remark 4.4.1. Sequences of pairs of numbers, having the form $l = \{ (a_i , b_i); i = 1, 2, 3, \dots \}$ and satisfying $a_i \leq b_i$ for all $i = 1, 2, 3, \dots$, may be interpreted as sequences of intervals. In this case, a_i is the beginning of the corresponding interval, while b_i is the end of the same interval. This demonstrates intrinsic ties between neoanalysis and interval analysis (Moore, 1966; Alefeld, and Herberger, 1983).

Let $X, Y \subseteq \boldsymbol{R}, f: X \to Y$ be a function, $b \in \boldsymbol{R}$, $r \in \boldsymbol{R}^+$, and B be a condition in some way related to a point $a \in X$. As an example of conditions related to a, we may take one of the conditions $B_{ct(a)}$ and $B_{lt(a)}$ considered above.

Definition 4.4.1. A number b is called a *strong* B-*conditional r-derivative* of a function $f(x)$ at a point $a \in X$ if $b = r\text{-}\lim_{i\to\infty} (f(z_i) - f(x_i)) / (z_i - x_i)$ for all sequences $l = \{ (x_i, z_i) ; i = 1, 2, 3, \dots \}$ that satisfy the condition B.

It is denoted by $b = \mathrm{st}_r\mathrm{Bd}/_{\mathrm{d}x} f(a)$.

Lemma 4.4.1. For any point a from $\boldsymbol{R}$ and any real function $f(x)$, we have $b = \mathrm{st}_r\mathrm{Bd}/_{\mathrm{d}x} f(a)$ if and only if $b = r$-lim E where E = $\{\{ (f(a) - f(x_i)) /(a - x_i); i = 1, 2, 3, \dots \}; \{ x_i ; i = 1, 2, 3, \dots \}$ is a sequence that converges to a and satisfies the condition B $\}$.

Lemma 4.4.2. If $d = \mathrm{st}_r\mathrm{Bd}/_{\mathrm{d}x} f(a)$, then $d = \mathrm{st}_q\mathrm{Bd}/_{\mathrm{d}x} f(a)$ for any $q > r$.

Definition 4.4.2. A number b is called a *weak* B-*conditional r-derivative* of a function $f(x)$ at a point $a \in X$ if $b = r\text{-}\lim_{i\to\infty} (f(z_i) - f(x_i)) / (z_i - x_i)$ for some sequence $l = \{ (z_i, x_i); i = 1, 2, 3, \dots \}$ that satisfies the condition B.

It is denoted by $b = \mathrm{w}_r\mathrm{Bd}/_{\mathrm{d}x} f(a)$.

When r is not specified, we call any strong (weak) B-conditional r-derivative of $f(x)$ at a point $a \in X$ a *strong* (*weak*) B-*conditional fuzzy derivative* of $f(x)$ at a point $a \in X$.

Lemma 4.4.3. $b = \text{st}_r\text{Bd}/_{\text{d}x} f(a)$ if and only if b is a unique weak r-derivative $b = \text{w}_r\text{Bd}/_{\text{d}x}$ $f(a)$ of a function $f(x)$ at a point $a \in X$.

Example 4.4.1. Let us take the condition $\text{B}_{\text{tt}(a)}$: a sequence $l = \{ (x_i , z_i) ; i = 1, 2, 3, \dots \}$ satisfies $\text{B}_{\text{tt}(a)}$ if $x_i < a < z_i$ for all $i = 1, 2, 3, \dots$, $\lim_{i\to\infty} x_i = a$ and $\lim_{i\to\infty} z_i = a$.

Proposition 4.4.1. a) Any strong $\text{B}_{\text{tt}(a)}$-conditional r-derivative of $f(x)$ at a point $a \in X$ is exactly a strong two sided r-derivative of $f(x)$ at a point $a \in X$ and *vice versa* (cf. Section 4.2).

b) Any weak $\text{B}_{\text{tt}(a)}$-conditional r-derivative of $f(x)$ at a point $a\in X$ is exactly a weak two sided r-derivative of $f(x)$ at a point $a \in X$ and *vice versa* (cf. Section 4.3).

Example 4.4.2. Let us take condition $\text{B}_{\text{lt}(a)}$: all sequences $l = \{ (x_i , z_i) ; i = 1, 2, 3, \dots \}$, for which $x_i < a$ for all $i = 1, 2, 3, \dots$, $\lim_{i\to\infty} x_i = a$ and all $z_i = a$.

Proposition 4.4.2. a) Any strong $\text{B}_{\text{lt}(a)}$-conditional r-derivative of $f(x)$ at a point $a\in X$ is exactly a strong left r-derivative of $f(x)$ at a point $a \in X$ and *vice versa* (cf. Section 4.2).

b) Any weak $\text{B}_{\text{lt}(a)}$-conditional r-derivative of $f(x)$ at a point $a\in X$ is exactly a weak left r-derivative of $f(x)$ at a point $a\in X$ and *vice versa* (cf. Section 4.3).

Example 4.4.3. Let us take condition $\text{B}_{\text{rt}(a)}$: all sequences $l = \{ (x_i , z_i) ; i = 1, 2, 3, \dots \}$, for which $a < z_i$ for all $i = 1, 2, 3, \dots$, $\lim_{i\to\infty} z_i = a$ and all $x_i = a$.

Proposition 4.4.3. a) Any strong $\text{B}_{\text{rt}(a)}$-conditional r-derivative of $f(x)$ at a point $a\in X$ is exactly a strong left r-derivative of $f(x)$ at a point $a \in X$ and *vice versa* (cf. Section 4.2).

b) Any weak $\text{B}_{\text{rt}(a)}$-conditional r-derivative of $f(x)$ at a point $a\in X$ is exactly a weak left r-derivative of $f(x)$ at a point $a \in X$ and *vice versa* (cf. Section 4.3).

Example 4.4.4. Let us take condition $\text{B}_{\text{ct}(a)}$: all sequences $l = \{ (x_i , z_i) ; i = 1, 2, 3, \dots \}$, for which $x_i \neq a$ for all $i = 1, 2, 3, \dots$, $\lim_{i\to\infty} x_i = a$, and all $z_i = a$.

Proposition 4.4.4. a) Any strong $\text{B}_{\text{ct}(a)}$-conditional r-derivative of $f(x)$ at a point $a \in X$ is exactly a strong centered r-derivative of $f(x)$ at a point $a \in X$ and *vice versa* (cf. Section 4.2).

b) Any weak $\text{B}_{\text{ct}(a)}$-conditional r-derivative of $f(x)$ at a point $a \in X$ is exactly a weak centered r-derivative of $f(x)$ at a point $a\in X$ and *vice versa* (cf. Section 4.3).

Let us assume that $r, q \in \boldsymbol{R}^+$, while $f: X\to \boldsymbol{R}$ and $g: X \to \boldsymbol{R}$ are arbitrary real functions.

Theorem 4.4.1. *a*) If b is a strong B-conditional r-derivative of a function $f(x)$ at a and c is a strong B-conditional q-derivative of a function $g(x)$ at a, then $b + c$ is a strong B-conditional $(r + q)$-derivative of the function $(f + g)(x)$ and $b - c$ is a strong B-conditional $(r + q)$-derivative of the function $(f - g)(x)$ at the point a.

b) If b is a strong B-conditional r-derivative of $f(x)$ at a point a and $t\in\boldsymbol{R}$, then $t\cdot b$ is a strong B-conditional $|t|\cdot a$-derivative of the function $t\cdot f(x)$ at the point a.

Proof. a) Let us assume that $b = \text{st}_r\text{Bd}/_{\text{d}x} f(a)$ and $c = \text{st}_r\text{Bd}/_{\text{d}x}\, g(a)$. By Definition 4.4.1, it means $b = r\text{-}\lim_{i\to\infty} (f(z_i) - f(x_i)) / (z_i - x_i)$ and $c = q\text{-}\lim_{i\to\infty} (g(z_i) - g(x_i)) / (z_i - x_i)$ for all sequences $l = \{ (x_i, z_i) ; i = 1, 2, 3, \dots \}$ that satisfy the condition B.

Then by Theorem 2.2.5, for all sequences $l = \{ (x_i, z_i) ; i = 1, 2, 3, \dots \}$ that satisfy the condition B, we have

$$b + c = (r + q)\text{-}\lim_{i\to\infty} [(f(z_i) - f(x_i)) / (z_i - x_i) + (g(z_i) - g(x_i)) / (z_i - x_i)] =$$
$$(r + q)\text{-}\lim_{i\to\infty} [[(f(z_i) - f(x_i)) + (g(z_i) - g(x_i))] / (z_i - x_i)] =$$
$$(r + q)\text{-}\lim_{i\to\infty} [[(f(z_i) + g(z_i)) - (f(x_i)) + g(x_i))] / (z_i - x_i)]$$

It means that $b + c$ is is a strong B-conditional $(r + q)$-derivative of the function $(f + g)(x)$ at the point a.

In a similar way, we have that $b - c$ is is a strong B-conditional $(r + q)$-derivative of the function $(f - g)(x)$ at the point a and tb is a strong B-conditional $|t|\cdot a$-derivative of the function $t\cdot f(x)$ at the point a.

Theorem is proved.

Remark 4.4.2. Properties of conditional derivatives essentially depend on the chosen condition B. For instance, a weak fuzzy derivative of a function $f(x)$ at a point a does not always coinside with some strong fuzzy derivative of $f(x)$ at a in the absolute case considered in Sections 4.2 and 4.3. However, for a conditional fuzzy derivative (of a function $f(x)$ at a point a), not only any conditional strong fuzzy derivative of $f(x)$ at a a conditional weak fuzzy derivative of $f(x)$ at a of the same type (cf. Lemma 4.3.4), but also any conditional weak fuzzy derivative of $f(x)$ at a a conditional strong fuzzy derivative of $f(x)$ at a of the same type. Indeed, if, for example, $b = \mathrm{w}_r{}^z \mathrm{Bd}/_{\mathrm{d}x} f(a)$ and $b = r\text{-lim}\ \{(f(a) - f(x_i)) / (a - x_i)\ ;\ i = 1, 2, 3, \dots\}$, then taking $\mathrm{B} = \{\ l = \{\ (a_i\ ,\ a\ )\ ;\ i = 1, 2, 3, \dots\ \}\ \}$ we have according to Definition 4.4.1, that $b = \mathrm{st}_r{}^z \mathrm{Bd}/_{\mathrm{d}x} f(a)$.

4.5. Extended Fuzzy Derivatives

These three have very specific things they care about,
and everything is derivative of that.
David Dukes (1945-2000)

Much more functions have weak derivatives than conventional derivatives. However, there are even functions "good" in many aspects, which do not have weak derivatives at some points. For example, the function $f(x) = \sqrt{1 - x^2}$ does not have either weak or conventional derivatives at points $x = 1$ and $x = -1$. To remedy this deficiency, we take an extended real line $\boldsymbol{R}_\infty = \boldsymbol{R} \cup \{\infty, -\infty\}$ and define weak and weak fuzzy derivatives, which take values in $\boldsymbol{R}_\infty$ instead of $\boldsymbol{R}$. We call such derivatives extended and denote them by $\mathrm{W}_r{}^z \mathrm{d}/_{\mathrm{d}x} f(a)$ where $\mathrm{z} \in \{\mathrm{ct, l, r, t}\}$.

Definition 4.5.1. An element b from $\boldsymbol{R}_\infty$ is called an *extended weak centered* (*left*, *right*, *two-sided*) *derivative* of a function $f(x)$ at a point $a \in X$ if $b \in \boldsymbol{R}$ and $b = \lim\ \{(f(a) - f(x_i)) / (a - x_i)\ ;\ i = 1, 2, 3, \dots\}$ or $b = \pm\infty$ and $b = n\text{-lim}\ \{(f(a) - f(x_i)) / (a - x_i)\ ;\ i = 1, 2, 3, \dots\}$ for all $n \in \boldsymbol{N}$ (correspondingly: and all $x_i < a$; and all $x_i > a$; $b = \lim_{i\to\infty} (f(z_i) - f(x_i)) / (z_i - x_i)$ or $\pm\infty = n\text{-lim}\ \{(f(a) - f(x_i)) / (a - x_i)\ ;\ i = 1, 2, 3, \dots\}$ for all $n \in \boldsymbol{N}$, with $z_i < a < x_i$ for all $i = 1, 2, 3, \dots$, and the sequences $\{x_i;\ i = 1, 2, 3, \dots\}$, $\{z_i;\ i = 1, 2, 3, \dots\}$ converging to a) for some sequence $\{x_i;\ x_i \neq a\ ;\ i = 1, 2, 3, \dots\}$ converging to a.

It is denoted by $b = \mathrm{W}^{\mathrm{ct}} \mathrm{d}/_{\mathrm{d}x} f(a)$ ($b = \mathrm{W}^{\mathrm{l}} \mathrm{d}/_{\mathrm{d}x} f(a)$, $b = \mathrm{W}^{\mathrm{r}} \mathrm{d}/_{\mathrm{d}x} f(a)$, and $b = \mathrm{W}^{\mathrm{t}} \mathrm{d}/_{\mathrm{d}x} f(a)$, correspondingly).

Example 4.5.1. Let us take the membership function $\mathrm{m}_{\mathrm{Q}}(x)$ of the set of rational numbers (cf. Example 2.1). At each rational point a, $\mathrm{m}_{\mathrm{Q}}(x)$ has the following extended derivatives: $0 =$

$W_r^z d/dx f(a)$, where $z \in \{ct, l, r, t\}$, $\infty = W_r^{ct} d/dx\ f(a)$, $-\infty = W_r^{ct} d/dx\ f(a)$, $\infty = W_r^{t} d/dx\ f(a)$, $-\infty = W_r^{t} d/dx f(a)$, $\infty = W_r^{l} d/dx\ f(a)$, $-\infty = W_r^{r}\ d/dx\ f(a)$. At each irrational point a, $m_Q(x)$ has the following extended derivatives: $0 = W_r^{z} d/dx\ f(a)$, where $z \in \{ct, l, r, t\}$, $\infty = W_r^{ct} d/dx$ $f(a)$, $-\infty = W_r^{ct} d/dx\ f(a)$, $\infty = W_r^{t} d/dx f(a)$, $-\infty = W_r^{t} d/dx\ f(a)$, $\infty = W_r^{r} d/dx f(a)$, $-\infty = W_r^{l}$ $d/dx f(a)$.

Lemma 4.5.1. Any extended weak left or right derivative of a function $f(x)$ at a point a is also an extended weak centered derivative of $f(x)$ at a.

Remark 4.5.1. Extended weak derivatives of functions provide only the first approximation for derivatives with infinite values. It is possible to achieve more precise approximation when we build and consider extended weak and weak fuzzy derivatives that take values in the sets of hypernumbers (Burgin, 1993a; 2002). Hypernumbers give much better representation for infinity than the conventional two symbols ∞ and $-\infty$. In a space of hypernumbers, there are many different infinite numbers, with which we can operate in a similar way to real numbers. However, in the context of neoclassical analysis, it is sufficient to consider only two infinite values: ∞ and $-\infty$.

Remark 4.5.2. The following result demonstrates that the concept of an extended weak fuzzy derivative may be reduced to two other concepts: a weak fuzzy derivative and weak extended derivative.

Lemma 4.5.1. If an element b from $\boldsymbol{R}_\infty$ is an extended weak centered (left, right, two-sided) r-derivative of a function $f(x)$ at a point $a \in X$, then either b is a number and thus, a weak centered (left, right, two-sided, correspondingly) r-derivative of $f(x)$ at $a \in X$, or b is equal to ∞ or $-\infty$ and is an extended weak centered (left, right, two-sided, correspondingly) derivative of $f(x)$ at the point $a \in X$,

Let us take some function $f: X \rightarrow \boldsymbol{R}$ and assume that X is one of the following sets: $\boldsymbol{R}$, (a, b), (a, ∞) or $(-\infty, b)$ with $a, b \in \boldsymbol{R}$. Definitions directly imply the following result, which emphasizes importance of extended derivatives.

Proposition 4.5.1. Any function $f(x)$ has both an extended weak left and an extended weak right derivatives at any point a from X.

Proof. Let us take a sequence of points x_i that converge to a from the left (i.e., all $x_i < a$) and consider the sequence $l = \{ b_i = (f(a) - f(x_i)) /(a - x_i) ; i = 1, 2, 3, \ldots \}$. If the sequence l is unbounded, then it is possible to find a subsequence $h = \{ c_i \}$ of l such that elements converge either to ∞ or to $-\infty$. In the first case, ∞ is an extended weak left derivative of $f(x)$ at the point a. In the second case, $-\infty$ is an extended weak left derivative of $f(x)$ at the point a.

When the sequence l is bounded, then (cf. Section 2.1) it is possible to find a converging subsequence $h = \{ d_i \}$ of l. The limit $b = \lim h$ is an extended weak left derivative of $f(x)$ at the point a.

A proof for an extended weak right derivative is almost the same.

Proposition is proved.

Lemma 4.5.1 and Proposition 4.5.1 imply the following result.

Corollary 4.5.1. Any function $f(x)$ has an extended weak centered derivative at any point a from X.

In a similar way to Definition 4.5.1, we define extended strong fuzzy derivatives, which take values in $\boldsymbol{R}_\infty$ instead of $\boldsymbol{R}$. We call such derivatives extended and denote them by $\mathrm{St}_r{}^{z}\mathrm{d}/_{\mathrm{d}x} f(a)$ where $z \in \{\mathrm{ct, l, r, t}\}$.

Definition 4.5.2. An element b from $\boldsymbol{R}_\infty$ is called an *extended strong centered* (*left*, *right*, *two-sided*) *r-derivative* of a function $f(x)$ at a point $a \in X$ if $b \in \boldsymbol{R}$ and $b = r\text{-lim } \{(f(a) - f(x_i)) /(a - x_i);\ i = 1, 2, 3, \dots \}$ or $b = \pm\infty$ and $b = n\text{-lim } \{(f(a) - f(x_i)) /(a - x_i)\ ;\ i = 1, 2, 3, \dots \}$ for all $n \in \boldsymbol{N}$ (and all $x_i < a$; and all $x_i > a$; $b = r\text{-lim } \{(f(z_i) - f(x_i)) / (z_i - x_i)\ ;\ i = 1, 2, 3, \dots \}$ or $\pm\infty = n\text{-lim } \{(f(a) - f(x_i)) /(a - x_i)\ ;\ i = 1, 2, 3, \dots \}$ for all $n \in \boldsymbol{N}$, with $z_i < a < x_i$ for all $i = 1, 2, 3, \dots$, and the sequences $\{x_i\ ;\ i = 1, 2, 3, \dots \}$, $\{z_i\ ;\ i = 1, 2, 3, \dots \}$ converging to a) for all sequences $\{x_i;\ x_i \neq a\ ;\ i = 1, 2, 3, \dots \}$ converging to a.

It is denoted by $b = \mathrm{St}_r{}^{\mathrm{ct}}\mathrm{d}/_{\mathrm{d}x} f(a)$ ($b = \mathrm{St}_r{}^{\mathrm{l}}\mathrm{d}/_{\mathrm{d}x} f(a)$, $b = \mathrm{St}_r{}^{\mathrm{r}}\mathrm{d}/_{\mathrm{d}x} f(a)$, and $b = \mathrm{St}_r{}^{\mathrm{t}}\mathrm{d}/_{\mathrm{d}x} f(a)$, correspondingly).

Remark 4.5.2. When $r = 0$, then extended strong centered derivative of $f(x)$ is an extended derivative of $f(x)$, that is, the conventional derivative which may take additional values ∞ and $-\infty$. This extends the scope of differentiable functions. For example, if we take the function $f(x) = \sqrt{1 - x^3}$, we see that it does not have derivatives at points $x = 1$ and $x = -1$, but it is differentiable in this extended sense: $f'(1) = -\infty$ and $f'(-1) = \infty$.

It is also possible to define extended weak fuzzy derivatives.

Proposition 4.5.1 and Lemma 2.2.1 imply the following result.

Corollary 4.5.1. Any function $f(x)$ has both an extended weak left and an extended weak right fuzzy derivatives at any point a from X.

4.6. Fuzzy Differentiable Functions

To every thing there is a season,
and a time to every purpose under the heaven …
Ecclesiastes

In a previous section we studied fuzzy derivatives of real functions, here we obtain properties of real functions that have fuzzy derivatives.

Definition 4.6.1. A real function $f(x)$ is *r-differentiable* (*from the left*, *from the right*, *from two sides*) at a point a from X if $f(x)$ has a strong centered (strong left, strong right, strong two-sided) r-derivative at a.

Theorem 4.4.1 implies the following result.

Proposition 4.6.1. *a*) If a real function $f(x)$ is r-differentiable (from the left, from the right, from two sides) at a point a from X and a real function $g(x)$ is q-differentiable (from the left, from the right, from two sides) at a point a from X, then functions $f + g$ and $f - g$ are $(r + q)$-differentiable (from the left, from the right, from two sides) at a point a from X.

b) If a real function $f(x)$ is r-differentiable (from the left, from the right, from two sides) at a point a from X and $t \in \boldsymbol{R}$, then the function $t{\cdot}f(x)$ is $|t|{\cdot}r$-differentiable (from the left, from the right, from two sides) at a point a from X.

Definition 4.6.2. A function $f(x)$ is *fuzzy differentiable* (*from the left*, *from the right*) at a point a from X if there is some number r such that $f(x)$ has a strong centered (strong left, strong right) r-derivative at a.

Remark 4.6.1. There are such functions that have no derivative at any point of $\boldsymbol{R}$ but are fuzzy differentiable at all points of $\boldsymbol{R}$. To demonstrate this, let us consider the function $f(x)$ that is defined by the following formula:

$$f(x) = \Sigma_{n=1}^{\infty} g(4^{n-1}x)/4^{n-1} \text{ where } g(x + n) = |x| \text{ for all } x \text{ with } |x| \leq 1/2.$$

It is demonstrated in (Gelbaum and Olmsted, 1964) that this function has no derivative at any point of $\boldsymbol{R}$. At the same time, it is possible to prove that 0 is a strong centered and two-sided 5-derivative of $f(x)$ at any point x from $\boldsymbol{R}$.

Theorem 4.6.1. If a function $f(x)$ is fuzzy differentiable at a point a from X, then there is such a minimal number q that $f(x)$ has a strong centered q-derivative at a, i.e., $q = \inf \{r ; f(x)$ has a strong centered r-derivative at $a\}$.

Proof. Let us consider the set FD $(f, a) = \{ r ; f(x)$ has a strong centered r-derivative at $a\}$ and the number $q = \inf$ FD (f, a). As the function $f(x)$ is fuzzy differentiable at a point a, the set FD(f, a) is not empty. If c is a number from FD (f, a), then for an arbitrary sequence l, which has the form

$$l = \{ (f(z_i) - f(x_i))/(z_i - x_i); z_i > a > x_i, i = 1, 2, 3, \dots, a = \lim_{i\to\infty} x_i = \lim_{i\to\infty} z_i\} \quad (4.10)$$

there is a point u in X such that $u = q$-lim l. The set H$(l) = \{ r ; l$ has an r-limit$\}$ is a closed ray. By the definition of a strong centered fuzzy derivative, we have FD$(f, a) = \bigcap \{$H$(l_i) ; l_t$ has the form (4.10)$\}$.

Any intersection of closed rays is a closed set or is empty. But in our case, we have FD$(f, a) \neq \varnothing$. Consequently, it is a closed ray of positive numbers. As a closed set, FD(f, a) contains q.

Theorem is proved.

Corollary 4.6.1. A function $f(x)$ has the classical derivative at a if and only if FD$(f, a) = \boldsymbol{R}^{+}$.

Corollary 4.6.2. A function $f(x)$ is differentiable on X if and only if FD$(f, a) = \boldsymbol{R}^{+}$ for all points a from X.

These results and others demonstrate that the concept of a fuzzy differentiability is a natural extension of the concept of the conventional differentiability.

Remark 4.6.2. For weak fuzzy derivatives, Theorem 4.6.1 is invalid.

Let us investigate interrelations between different types of fuzzy differentiation and continuity of functions.

Theorem 4.6.2. The following conditions are equivalent for a function $f(x)$:

(a) $f(x)$ is fuzzy differentiable at a point a from X;
(b) $f(x)$ is fuzzy differentiable from the left and from the right at a point a from X;
(c) $f(x)$ is continuous at a and is fuzzy differentiable from two-sides at a.

Proof. (a) implies (c): Let $f(x)$ be a fuzzy differentiable function at a point a from X. Then by Definition 4.2.1, there is some number r such that $f(x)$ has a strong centered r-derivative b at a. It means that for any sequence $\{x_n; x_n \in X, n = 1, 2, 3, \dots\}$, if $a = \lim_{n\to\infty} x_n$, then $b = r$-lim l where $l = \{ (f(a) - f(x_n))/(a - x_n) ; n = 1, 2, 3, \dots \}$.

By the definition of fuzzy limits (cf. Chapter 2), for any $\varepsilon > 0$, there is a natural number $m \in \omega$ such that for all $n > m$ the following inequality is valid:

$$| b - (f(a) - f(x_n))/(a - x_n)) | \leq r + \varepsilon.$$

It implies the inequality

$$| (f(a) - f(x_n))/(a - x_n) | \leq r + |b| + \varepsilon.$$

Consequently, we have

$$|f(a) - f(x_n)| \leq (r + 1 + |b|)| a - x_n|.$$

Thus, convergence of a sequence $\{x_n ; n = 1, 2, 3, \dots\}$ to the point a implies that $f(x_n) \to f(a)$.

It means that the function $f(x)$ is continuous at a. Besides, Proposition 4.2.1 implies that b is a strong two-sided r-derivative of $f(x)$ at a, i.e., $f(x)$ is fuzzy differentiable from two-sides at a, while Lemma 4.2.3 implies that b is a strong right r-derivative of $f(x)$ at a and that b is a strong left r-derivative of $f(x)$ at a.

Implications (c) $\Rightarrow$ (b) and (b) $\Rightarrow$ (a) follow from Definitions 4.2.1 and 4.2.3.

Theorem is proved.

Corollary 4.6.3. If $f(x)$ is a fuzzy differentiable function at a point $a \in X$, then $f(x)$ is continuous at a.

As another corollary, we obtain such a classical result (cf., for example, (Ross, 1996; Stewart, 2003; or Fihtengoltz, 1955)) as Theorem 4.1.1.

Let $D \subseteq X$.

Definition 4.6.3. a) A function $f(x)$ is *fuzzy differentiable* (from the left, from the right) *in* D if $f(x)$ is fuzzy differentiable (from the left, from the right) at any point a from D.

b) A function $f(x)$ is *fuzzy differentiable* (from the left, from the right) *inside* D if $f(x)$ is fuzzy differentiable (from the left, from the right) at any inner point a from D, i.e., at points a from D such that some interval $[a - k, a + k]$ with $k > 0$ is a subset of D.

Theorem 4.6.2 and the definition of continuous functions imply the following result.

Theorem 4.6.3. Any fuzzy differentiable in (inside) D function $f(x)$ is continuous in (inside) D.

In turn, Theorem 4.6.3 implies such classical result as Theorem 4.1.2.

Let us assume that $r, q \in \mathbf{R}^+$, while $f: X \to \mathbf{R}$ and $g: X \to \mathbf{R}$ are arbitrary real functions. Then Theorem 4.2.1 implies the following result.

Theorem 4.6.4. *a)* **(Global additivity of fuzzy differentiation)** If a function $f(x)$ is r-differentiable (from the right or from the left) in [inside] D and function $g(x)$ is q-differentiable(from the right or from the left) in [inside] D, then the function $(f \pm g)(x)$ is $(r + q)$-differentiable (from the right or from the left) in [inside] D.

b) **(Global uniformity of fuzzy differentiation)** If a function $f(x)$ is r-differentiable (from the right or from the left) in [inside] D and $t \in \boldsymbol{R}$, then the function $t \cdot f(x)$ is $|t| \cdot r$-differentiable (from the right or from the left) in [inside] D.

In turn, Theorem 4.6.4 implies such classical result as Theorem 4.1.6.

Connections between left and right continuity and left and right differentiability are given in the following theorem.

Theorem 4.6.5. If a function $f(x)$ is r-differentiable from the right (left) at a point $a \in X$, then $f(x)$ is continuous from the right (left) at a.

Proof is similar to the proof of implication (a) $\Rightarrow$ (c) in Theorem 4.6.2.

In turn, Theorem 4.6.5 implies such classical result as Theorem 4.1.9.

Corollary 4.6.4. If a function $f(x)$ is r-differentiable from the right (left) in (inside) D, then $f(x)$ is continuous from the right (left) in (inside) D.

Definition 4.6.4. A real function $f(x)$ is *weakly r-differentiable* (*from the left*, *from the right*, *from two sides*) at a point a from X if $f(x)$ has a weak centered (strong left, strong right, strong two-sided) r-derivative at a.

Proposition 4.6.1. If a real function $f(x)$ is weakly r-differentiable (from the left, from the right, from two sides) at a point a from X and $t \in \boldsymbol{R}$, then the function $t \cdot f(x)$ is weakly $|t| \cdot r$-differentiable (correspondingly, from the left, from the right, from two sides) at the point a.

Remark 4.6.1. At the same time, in contrast to r-differentiable functions, the part (a) of Proposition 4.6.1 can be invalid for weakly r-differentiable functions as the following example demonstrates.

Example 4.6.2. Let us consider the following functions:

$$f(x) = \begin{cases} 0 & \text{when } x = 1/2^n \text{ ; } n = 1, 2, \ldots \\ 1/x & \text{when } x \neq 1/2^n \text{ ; } n = 1, 2, \ldots \end{cases}$$

and

$$g(x) = \begin{cases} 0 & \text{when } x = 1/3^n \text{ ; } n = 1, 2, \ldots \\ 1/x & \text{when } x \neq 1/3^n \text{ ; } n = 1, 2, \ldots \end{cases}$$

By definition, both functions $f(x)$ and $g(x)$ are weakly 0-differentiable from the left at 0. However, the function $(f + g)(x)$ is not weakly r-differentiable from the left at the point 0 for any positive number r.

The concept of a weak fuzzy derivative leads us to the concept of a weakly fuzzy differentiable function.

Definition 4.6.5. A function $f(x)$ is called *weakly fuzzy differentiable* (*from the left*, *from the right*) at a point a from X if there is such number r that $f(x)$ has a weak centered (weak left, weak right) r-derivative at a.

Proposition 4.6.1 implies the following result.

Corollary 4.6.5. If a function $f(x)$ is weakly fuzzy differentiable (from the left, from the right) at a point a from X, then $t \in \boldsymbol{R}$, then the function $t \cdot f(x)$ is weakly fuzzy differentiable (correspondingly, from the left, from the right) at the point a for any $t \in \boldsymbol{R}$.

To conclude, it is necessary to remark that weak differentiability, both crisp and fuzzy, is actually weaker than their strong counterparts - differentiability and fuzzy differentiability. For instance, weak differentiability does not imply continuity of functions, weak fuzzy derivatives are not additive and so on. However, weak derivatives and weak fuzzy derivatives allow one to estimate strong fuzzy derivatives and are useful for finding extrema of functions (cf. Chapter 6).

4.7. Sets and Fuzzy Sets of Fuzzy Derivatives

What is a contemporary iconography for the spiritual?
Is it some fuzzy space?
Anish Kapoor (1954-)

In this section, we introduce and study sets, fuzzy and intuitionistic fuzzy sets of strong, weak, complete and conditional derivatives. Structure of sets, fuzzy and intuitionistic fuzzy sets of strong and weak fuzzy derivatives is described. It is possible to consider sets and fuzzy sets of strong and weak fuzzy derivatives as complete (set valued or fuzzy set valued) derivatives of functions. Conditions are found when such derivatives are fuzzy numbers.

Let us consider the following sets:

$WBFD_r(f, a)$ is the set of all weak B-conditional r-derivatives of a function $f(x)$ at a point $a \in X$;

$SBFD_r(f, a)$ is the set of all strong B-conditional r-derivatives of a function $f(x)$ at a point $a \in X$.

Definition 4.7.1. The set $WBFD_r(f, a)$ ($SBFD_r(f, a)$) is called the *complete weak* (correspondingly, *strong*) B-*conditional r-derivative* of $f(x)$ at a point a.

Particular cases of the sets $WBFD_r(f, a)$ and $SBFD_r(f, a)$ are:

- $WCFD_r(f, a)$ is the set of all weak centered r-derivatives of a function $f(x)$ at a point $a \in X$ when $B = B_{ct(a)}$ is the condition that a sequence $l = \{ (x_i , z_i) ; i = 1, 2, 3, \dots \}$ satisfies $B_{ct(a)}$ if $x_i \neq a$ for all $i = 1, 2, 3, \dots$, $\lim_{i\to\infty} x_i = a$, and all $z_i = a$.
- $WLFD_r(f, a)$ is the set of all weak left r-derivatives of a function $f(x)$ at a point $a \in X$ when $B = B_{lt(a)}$ is the condition that a sequence $l = \{ (x_i , z_i) ; i = 1, 2, 3, \dots \}$ satisfies $B_{lt(a)}$ if which $x_i < a$ for all $i = 1, 2, 3, \dots$, $\lim_{i\to\infty} x_i = a$ and all $z_i = a$.
- $WRFD_r(f, a)$ is the set of all weak right r-derivatives of a function $f(x)$ at a point $a \in X$ when $B = B_{rt(a)}$ is the condition that a sequence $l = \{ (x_i , z_i) ; i = 1, 2, 3, \dots \}$ satisfies $B_{rt(a)}$ if which $x_i > a$ for all $i = 1, 2, 3, \dots$, $\lim_{i\to\infty} x_i = a$ and all $z_i = a$.
- $WTFD_r(f, a)$ is the set of all weak two-sided r-derivatives of a function $f(x)$ at a point $a \in X$ when $B = B_{tt(a)}$ is the condition that a sequence $l = \{ (x_i , z_i) ; i = 1, 2, 3, \dots \}$ satisfies $B_{tt(a)}$ if $x_i < a < z_i$ for all $i = 1, 2, 3, \dots$, $\lim_{i\to\infty} x_i = a$ and $\lim_{i\to\infty} z_i = a$.
- $SCFD_r(f, a)$ is the set of all strong centered r-derivatives of a function $f(x)$ at a point $a \in X$ when $B = B_{ct(a)}$ is the condition that a sequence $l = \{ (x_i , z_i) ; i = 1, 2, 3, \dots \}$ satisfies $B_{ct(a)}$ if $x_i \neq a$ for all $i = 1, 2, 3, \dots$, $\lim_{i\to\infty} x_i = a$, and all $z_i = a$.

- $\mathrm{SLFD}_r(f, a)$ is the set of all strong left r-derivatives of a function $f(x)$ at a point $a \in X$ when $B = B_{lt(a)}$ is the condition that a sequence $l = \{ (x_i, z_i) ; i = 1, 2, 3, \dots \}$ satisfies $B_{lt(a)}$ if which $x_i < a$ for all $i = 1, 2, 3, \dots$, $\lim_{i\to\infty} x_i = a$ and all $z_i = a$.
- $\mathrm{SRFD}_r(f, a)$ is the set of all strong right r-derivatives of a function $f(x)$ at a point $a \in X$ when $B = B_{rt(a)}$ is the condition that a sequence $l = \{ (x_i, z_i) ; i = 1, 2, 3, \dots \}$ satisfies $B_{rt(a)}$ if which $x_i > a$ for all $i = 1, 2, 3, \dots$, $\lim_{i\to\infty} x_i = a$ and all $z_i = a$ and
- $\mathrm{STFD}_r(f, a)$ is the set of all strong two-sided r-derivatives of a function $f(x)$ at a point $a \in X$ when $B = B_{tt(a)}$ is the condition that a sequence $l = \{ (x_i, z_i) ; i = 1, 2, 3, \dots \}$ satisfies $B_{tt(a)}$ if $x_i < a < z_i$ for all $i = 1, 2, 3, \dots$, $\lim_{i\to\infty} x_i = a$ and $\lim_{i\to\infty} z_i = a$.

In what follows, $\mathrm{YXFD}_r(f, a)$ denotes one of these sets (i.e., Y may be equal to W or S, while X may be equal to C, L, R, T) and is called the *complete r-derivative* of $f(x)$ at a point $a \in X$ having type (Y,X).

Example 4.7.1. Let us consider the function $f(x) = |x|$. It does not have the classical derivative at the point 0. However, $f(x)$ has many different complete fuzzy derivatives at this point. For instance, $\mathrm{SCFD}_1(f, 0) = \{0\}$, $\mathrm{SCFD}_2(f, 0) = [-1, 1]$, and $\mathrm{SCFD}_3(f, 0) = [-2, 2]$. Besides, we have $\mathrm{SCFD}_0(f, 1) = \{1\}$ and $\mathrm{SCFD}_1(f, 1) = [0, 2]$.

Let us consider some properties of sets of fuzzy derivatives.

Lemma 4.3.4 implies the following result.

Proposition 4.7.1. $\mathrm{SBD}_r(f, a) \subseteq \mathrm{WBD}_r(f, a)$ for any condition B.

Corollary 4.7.1. $\mathrm{SXFD}_r(f, a) \subseteq \mathrm{WXFD}_r(f, a)$ when X is equal to C, L, R or T.

Theorem 4.7.1. Each complete r-derivative $\mathrm{SXFD}_r(f, a)$ with $X \in \{C, L, R, T\}$ is a bounded convex closed set, i.e., $\mathrm{SXFD}_r(f, a) = [b, c]$ for some numbers $b, c \in \boldsymbol{R}$, or $\mathrm{SXFD}_r(f, a) = \varnothing$ if f has no strong r-derivatives of the type X.

Proof. We prove the statement of the theorem for strong centered fuzzy derivatives, that is, for the set $\mathrm{SCFD}_r(f, a)$. For all types of strong derivatives the proof is similar.

By the definition and Lemma 4.2.2, $\mathrm{SCFD}_r(f, x)$ is the set of all r-limits of the set E of all sequences having form $(f(a) - f(x_i)) /(a - x_i)$, for which $a = \lim_{i\to\infty} x_i$. At the same time, by Corollary 2.18, the set $L_r(E) = \{c \in \boldsymbol{R};\ c = r\text{-}\lim E\}$ of all r-limits is a closed interval or an empty set for any set E of sequences. Consequently, $\mathrm{SXFD}_r(f, a)$ is a bounded convex closed set.

Theorem 4.7.1 is proved.

Let us also consider the following sets:

$\mathrm{EWBFD}_r(f, a)$ is the set of all extended weak B-conditional r-derivatives of a function $f(x)$ at a point $a \in X$;

$\mathrm{ESBFD}_r(f, a)$ is the set of all extended strong B-conditional r-derivatives of a function $f(x)$ at a point $a \in X$.

Definition 4.7.2. The set $\mathrm{EWBFD}_r(f, a)$ ($\mathrm{ESBFD}_r(f, a)$) is called the *complete extended weak* (correspondingly, *strong*) B-*conditional r-derivative* of $f(x)$ at a point a.

Particular cases of the sets $\mathrm{EWBFD}_r(f, a)$ and $\mathrm{ESBFD}_r(f, a)$ are:

- EWCFD$_r$(f, a) is the set of all extended weak centered r-derivatives of a function $f(x)$ at a point $a \in X$ when B = B$_{ct(a)}$ is the condition that a sequence $l = \{ (x_i , z_i) ; i = 1, 2, 3, \dots \}$ satisfies B$_{ct(a)}$ if $x_i \neq a$ for all $i = 1, 2, 3, \dots$, $\lim_{i\to\infty} x_i = a$, and all $z_i = a$.
- EWLFD$_r$(f, a) is the set of all extended weak left r-derivatives of a function $f(x)$ at a point $a \in X$ when B = B$_{lt(a)}$ is the condition that a sequence $l = \{ (x_i , z_i) ; i = 1, 2, 3, \dots \}$ satisfies B$_{lt(a)}$ if which $x_i < a$ for all $i = 1, 2, 3, \dots$, $\lim_{i\to\infty} x_i = a$ and all $z_i = a$.
- EWRFD$_r$(f, a) is the set of all extended weak right r-derivatives of a function $f(x)$ at a point $a \in X$ when B = B$_{rt(a)}$ is the condition that a sequence $l = \{ (x_i , z_i) ; i = 1, 2, 3, \dots \}$ satisfies B$_{rt(a)}$ if which $x_i > a$ for all $i = 1, 2, 3, \dots$, $\lim_{i\to\infty} x_i = a$ and all $z_i = a$.
- EWTFD$_r$(f, a) is the set of all extended weak two-sided r-derivatives of a function $f(x)$ at a point $a \in X$ when B = B$_{tt(a)}$ is the condition that a sequence $l = \{ (x_i , z_i) ; i = 1, 2, 3, \dots \}$ satisfies B$_{tt(a)}$ if $x_i < a < z_i$ for all $i = 1, 2, 3, \dots$, $\lim_{i\to\infty} x_i = a$ and $\lim_{i\to\infty} z_i = a$.
- ESCFD$_r$(f, a) is the set of all extended strong centered r-derivatives of a function $f(x)$ at a point $a \in X$ when B = B$_{ct(a)}$ is the condition that a sequence $l = \{ (x_i , z_i) ; i = 1, 2, 3, \dots \}$ satisfies B$_{ct(a)}$ if $x_i \neq a$ for all $i = 1, 2, 3, \dots$, $\lim_{i\to\infty} x_i = a$, and all $z_i = a$.
- ESLFD$_r$(f, a) is the set of all extended strong left r-derivatives of a function $f(x)$ at a point $a \in X$ when B = B$_{lt(a)}$ is the condition that a sequence $l = \{ (x_i , z_i) ; i = 1, 2, 3, \dots \}$ satisfies B$_{lt(a)}$ if which $x_i < a$ for all $i = 1, 2, 3, \dots$, $\lim_{i\to\infty} x_i = a$ and all $z_i = a$.
- ESRFD$_r$(f, a) is the set of all extended strong right r-derivatives of a function $f(x)$ at a point $a \in X$ when B = B$_{rt(a)}$ is the condition that a sequence $l = \{ (x_i , z_i) ; i = 1, 2, 3, \dots \}$ satisfies B$_{rt(a)}$ if which $x_i > a$ for all $i = 1, 2, 3, \dots$, $\lim_{i\to\infty} x_i = a$ and all $z_i = a$.
  and
- ESTFD$_r$(f, a) is the set of all extended strong two-sided r-derivatives of a function $f(x)$ at a point $a \in X$ when B = B$_{tt(a)}$ is the condition that a sequence $l = \{ (x_i , z_i) ; i = 1, 2, 3, \dots \}$ satisfies B$_{tt(a)}$ if $x_i < a < z_i$ for all $i = 1, 2, 3, \dots$, $\lim_{i\to\infty} x_i = a$ and $\lim_{i\to\infty} z_i = a$.

In what follows, EYXFD$_r$(f,a) denotes one of these sets (i.e., Y may be equal to W or S, while X may be equal to C, L, R, T) and is called the *complete extended r-derivative* of $f(x)$ at a point $a \in X$ having type (Y,X).

Proposition 4.7.2. ESXFD$_r$(f, a) $\subseteq$ EWXFD$_r$(f, a) when X is equal to C, L, R or T.

Theorem 4.7.2. The following conditions are equivalent for all X $\in$ {C, L, R, T}:

1. a function $f(x)$ has a strong fuzzy derivative at a point a of the type X;
2. the sets EWXFD$_r$(f, a) are bounded for all $r \geq 0$;
3. there is such $t \geq 0$ that the sets SXFD$_r$(f, a) are non-empty for all $r \geq t$;
4. the set EWXFD$_0$(f, a) is bounded;
5. the sets WXFD$_r$(f, a) and EWXFD$_r$(f, a) are equal, non-empty and bounded for all $r \geq 0$;
6. the sets WXFD$_0$(f, a) and EWXFD$_0$(f, a) are equal, non-empty and bounded.

Proof. 1 → 2. Let $d = \mathrm{st}_q{}^{x}\mathrm{d}/_{\mathrm{d}x}\, f(a)$ for some number $q \geq 0$. If $b \in \mathrm{EWXFD}_r(f, a)$ for some number $r \geq 0$, then $|\, d - b \,| \leq r + q$ by Proposition 4.2.4. Consequently, the diameter of the set $\mathrm{EWXFD}_r(f, a)$ is not larger than $2(r + q)$. Thus, this set is bounded.

Besides, any set $\mathrm{WXFD}_r(f, a)$ is not empty because any bounded sequence of real numbers contains a convergent subsequence.

2 → 3 because $t \geq 0$.

3 → 4 because r may be equal to 0.

4 → 1. Let us assume that the set $\mathrm{EWXFD}_0(f, a)$ is bounded. Then the set $\mathrm{WXFD}_0(f, a)$ is non-empty and bounded by Proposition 4.5.1. Consequently, the diameter of the set $\mathrm{WXFD}_0(f, a)$ is some positive number d and there is some number $b \in \boldsymbol{R}$ such that the distance from b to any element from $\mathrm{WXFD}_0(f, a)$ is less or equal to $q = d/2$. By the definition, we have $b = \mathrm{st}_q{}^{x}\mathrm{d}/_{\mathrm{d}x}\, f(a)$.

3 ↔ 5 because $\mathrm{WXFD}_r(f, a)$ is bounded and $\mathrm{WXFD}_r(f, a)$ and $\mathrm{EWXFD}_r(f, a)$ are equal if and only if $\mathrm{EWXFD}_r(f, a)$ is bounded.

4 ↔ 6 because $\mathrm{WXFD}_0(f, a)$ is bounded and $\mathrm{WXFD}_0(f, a)$ and $\mathrm{EWXFD}_0(f, a)$ are equal if and only if $\mathrm{EWXFD}_0(f, a)$ is bounded.

Theorem 4.7.2 is proved because any equivalence relation is transitive and symmetric (cf Appendix A).

Theorem 4.7.2 provides for the following criterion of fuzzy differentiability.

Proposition 4.7.3. A function $f(x)$ is fuzzy differentiable (from the left, from the right) at a point a from X if and only if the set $\mathrm{EWCFD}_0(f, a)$ ($\mathrm{EWLFD}_0(f, a)$, $\mathrm{EWRFD}_0(f, a)$, correspondingly) is non-empty and bounded.

Proposition 4.7.4. The set $\mathrm{EWCFD}_0(f, a)$ consists of a single point from $\boldsymbol{R}$ if and only if the classical derivative $f'(a)$ exists.

Proof. Sufficiency. Let the classical derivative $f'(a)$ of the function $f(x)$ at a point a exists. Then by Proposition 4.2.3, $f'(a)$ is equal to the strong centered 0-derivative $\mathrm{st}_0{}^{ct}\mathrm{d}/_{\mathrm{d}x}\, f(a)$ of $f(x)$ at a. As this strong centered 0-derivative is unique, the set $\mathrm{ESCFD}_0(f, a)$ consists of a single point, say, b. Then by Lemma 4.3.1, b is a unique weak centered 0-derivative of the function $f(x)$ at the point a, i.e., the set $\mathrm{EWCFD}_0(f, a)$ also consists of a single point.

Necessity. Let the set $\mathrm{EWCFD}_0(f, a)$ consists of a single point from $\boldsymbol{R}$. Then the set $\mathrm{SCFD}_0(f, a)$ consists of a single point as $\mathrm{SXFD}_0(f, a) \subseteq \mathrm{EWXFD}_0(f, a)$ by Proposition 4.7.1 and the element from $\mathrm{EWCFD}_0(f, a)$ is not equal to infinity. Then by Proposition 4.2.3, the classical derivative $f'(a)$ of $f(x)$ at a exists.

Proposition is proved.

Corollary 4.7.2. The set $\mathrm{ESCFD}_0(f, a)$ consists of a single point if and only if the classical derivative $f'(a)$ exists.

From Theorem 2.5.1, we obtain the following result.

Proposition 4.7.5. The set $\mathrm{YXFD}_r(f, a)$ is a union of closed intervals for all X equal to C, L, R, or T and Y equal to W or S.

Remark 4.7.1. While any set $\mathrm{SXFD}_r(f, a)$ with $\mathrm{X} \in \{\mathrm{C, L, R, T}\}$ always is equal to one closed interval in $\boldsymbol{R}$, sets $\mathrm{WXFD}_r(f, a)$ with $\mathrm{X} \in \{\mathrm{C, L, R, T}\}$ can consist of any finite or even infinite quantity of closed intervals in $\boldsymbol{R}$. It is demonstrated by the following examples.

Example 4.7.2. Let us take $X = [0, 1]$ and consider the following function:

$$f(x) = \begin{cases} 0 & \text{when } x = 1/2^n \text{ ; n = 1, 2, ...} \\ x & \text{when } x \neq 1/2^n \text{ ; } n = 1, 2, \ldots \end{cases}$$

Then $\mathrm{WRFD}_0(f, 0) = [0,0]\cup[1,1]$ and $\mathrm{WRFD}_{1/3}(f, 0) = [-1/3, 1/3]\cup[2/3, 4/3]$, while $\mathrm{SRFD}_0(f, 0) = \varnothing$.

Example 4.7.3. Let us take $X = [0, 1]$ and consider the following function:

$$f(x) = \begin{cases} 0 & \text{when } x \neq 1/k^n;\ n = 1, 2, \ldots ;\ k = 1, 2, \ldots \\ x & \text{when } x = 1/k^n;\ n = 1, 2, \ldots ;\ k = 1, 2, \ldots \end{cases}$$

Then $\mathrm{WRFD}_0(f, 0) = \bigcup_{k=0}^{\infty} [-k, k] = \boldsymbol{R}$ and $\mathrm{WRFD}_{1/4}(f, 0) = \bigcup_{m=0}^{\infty} [m - 1/4, m + 1/4] = \boldsymbol{R}$, while $\mathrm{SRFD}_0(f, 0) = \varnothing$.

Lemmas 4.2.4 and 4.3.5 imply the following result.

Proposition 4.7.6. If $r \leq p$, then $\mathrm{SBD}_r(f, a) \subseteq \mathrm{SBD}_p(f, a)$ and $\mathrm{WBD}_r(f, a) \subseteq \mathrm{WBD}_p(f, a)$.

Corollary 4.7.3. If $r \leq p$, then $\mathrm{YXFD}_r(f, a) \subseteq \mathrm{YXFD}_p(f, a)$ where X is equal to C, L, R or T and Y is equal to S or W.

Let us assume that B is a condition on sequences closed with respect to subsequences. As any bounded sequence of real numbers contains a convergent subsequence (cf. Section 2.1), we have the following result.

Proposition 4.7.7. If $\mathrm{WBD}_r(f, a) \neq \varnothing$ for some $r \geq 0$, then $\mathrm{WBD}_0(f, a) \neq \varnothing$.

Proof. If $\mathrm{WBD}_r(f, a) \neq \varnothing$, then there is a sequence $l = \{ (x_i, z_i) ; i = 1, 2, 3, \ldots \}$ such that it converges to $b = r\text{-lim } \{(f(z_i) - f(x_i)) /(z_i - x_i) ; i = 1, 2, 3, \ldots \}$, satisfies the condition B and $b = r\text{-lim } \{(f(z_i) - f(x_i)) /(z_i - x_i) ; i = 1, 2, 3, \ldots \}$ for some number b. Taking the sequence $h = \{ (f(z_i) - f(x_i)) / (z_i - x_i) ; i = 1, 2, 3, \ldots \}$ that defines the weak derivative b, we can find in it a converging to some number c subsequence $k = \{ (f(u_i) - f(v_i)) / (u_i - v_i) ; i = 1, 2, 3, \ldots \}$. This subsequence corresponds to the subsequence $t = \{ (u_i, v_i) ; i = 1, 2, 3, \ldots \}$ of the sequence $l = \{ (x_i, z_i) ; i = 1, 2, 3, \ldots \}$. As the condition B is closed with respect to subsequences, t also satisfies the condition B. As any subsequence of a converging sequence converges to the same limit, both sequences $\{x_i ; i = 1, 2, 3, \ldots \}$ and $\{ z_i ; i = 1, 2, 3, \ldots \}$ converge to a. Consequently, $c = \mathrm{w}_0\mathrm{Bd}/_{\mathrm{d}x} f(a)$ and thus, $\mathrm{WBD}_0(f, a) \neq \varnothing$.

Proposition is proved.

Corollary 4.7.4. If $\mathrm{WXFD}_r(f, a) \neq \varnothing$ for some $r \geq 0$, then $\mathrm{WXFD}_0(f, a) \neq \varnothing$ where X is equal to C, L, R or T.

Corollary 4.7.5. If $r \leq p$, then $\mathrm{WBD}_p(f, a) \neq \varnothing$ if and only if $\mathrm{WBD}_r(f, a) \neq \varnothing$.

Indeed, if $\mathrm{WBD}_r(f, a) \neq \varnothing$, then by Proposition 4.7.6, $\mathrm{WBD}_p(f, a) \neq \varnothing$. If $\mathrm{WBD}_p(f, a) \neq \varnothing$, then by Proposition 4.7.7, $\mathrm{WBD}_0(f, a) \neq \varnothing$. Consequently, by Proposition 4.7.6, $\mathrm{WBD}_r(f, a) \neq \varnothing$.

Corollary 4.7.6. If $r \leq p$, then $\mathrm{WXFD}_p(f, a) \neq \varnothing$ if and only if $\mathrm{WXFD}_r(f, a) \neq \varnothing$ where X is equal to C, L, R or T.

Let $\mathrm{WBD}_0(f, a) \neq \varnothing$.

Proposition 4.7.8. If b is a weak (strong) B-conditional r-derivative of $f(x)$ at x, then $\mathbf{d}(b, \mathrm{WBD}_0(f,a)) \leq r$.

Let $\mathrm{WXFD}_0(f, a) \neq \varnothing$.

Corollary 4.7.7. If b is a weak (strong) r-derivative of $f(x)$ at x, then $\mathbf{d}(b, \mathrm{WXFD}_0(f, a)) \leq r$.

Let $\mathrm{WBD}_k(f, a) \neq \varnothing$.

Proposition 4.7.9. If b is a weak (strong) B-conditional r-derivative of $f(x)$ at x, then $\mathbf{d}(b, \mathrm{WBD}_k(f, a)) \leq r - k$ where $r - k = r - k$ when $r \leq k$, otherwise $r - k = 0$.

Let $\mathrm{WXFD}_0(f, a) \neq \varnothing$.

Corollary 4.7.8. If b is a weak (strong) r-derivative of $f(x)$ at x, then $\mathbf{d}(b, \mathrm{WXFD}_k(f,a)) \leq r - k$ where $r - k = r - k$ when $r \leq k$, otherwise $r - k = 0$.

Sets $\mathrm{YXFD}_r(f, a)$ define complete global r-derivatives $\mathrm{YXFD}_r f$ of the function $f(x)$ on its domain X. Each set $\mathrm{YXFD}_r f$ is a binary relation on X, and namely, $\mathrm{YXFD}_r f = \{(x, z);\ a \in X,\ z \in \mathrm{YXFD}_r(f, a)\}$.

Proposition 4.7.10. The set $\mathrm{YXFD}_r f$ is a closed subset of $\boldsymbol{R}$ for all numbers $r \geq 0$.

Theorem 4.2.1 implies the following result.

Theorem 4.7.3. The set $\mathrm{SXFD}(f, a) = \bigcup_{r \geq 0} \mathrm{SXFD}_r(f, a)$ of all strong fuzzy derivatives of a function $f(x)$ at a is a real linear space.

Corollary 4.7.9. $\mathrm{SXFD}_{r+q}(f + g, a) \supseteq \mathrm{SXFD}_r(f, a) \oplus \mathrm{SXFD}_q(g, a)$ where $\oplus$ is the Minkowski sum (cf. Appendix B) of two sets of numbers.

Remark 4.7.2. However, we cannot substitute the equality for the inclusion in Corollary 4.7.9 as the following example demonstrates.

Example 4.7.4. Let us consider the following functions

$$f(x) = \begin{cases} 0 & \text{if } x \text{ is a rational number;} \\ 1 & \text{if } x \text{ is an irrational number} \end{cases}$$

and

$$g(x) = \begin{cases} 0 & \text{if } x \text{ is a rational number;} \\ -1 & \text{if } x \text{ is an irrational number.} \end{cases}$$

Then $\mathrm{SXFD}_0(f, a) = \varnothing$ and $\mathrm{SXFD}_0(g, a) = \varnothing$, while $\mathrm{SXFD}_{0+0}(f + g, a) = \mathrm{SXFD}_0(f + g, a) = \{0\}$.

All sets $\mathrm{YXFD}_r(f, a)$ and $\mathrm{YBD}_r(f, a)$ with Y equal to W or S and X equal to C, L, R, or T may be naturally combined into fuzzy sets.

Complete weak and strong B-conditional r-derivatives $\mathrm{WBD}_r(f, a)$ and $\mathrm{SBD}_r(f, a)$ define complete local B-conditional derivatives $\mathrm{WBD}(f, a)$ and $\mathrm{SBD}(f, a)$ of $f(x)$ at a point $a \in X$. These derivatives are fuzzy sets, for which $\mathrm{WBD}_r(f, a)$ and $\mathrm{SBD}_r(f, a)$ are their α-level sets, correspondingly.

Definition 4.7.3. The *complete local strong* B-*conditional derivative* $\mathrm{SBD}(f, a)$ (*complete local weak* B-*conditional derivative* $\mathrm{WBD}(f, a)$) is a fuzzy subset of $\boldsymbol{R}$ in the sense of Zadeh that has the form SBD(f, a) = ($\boldsymbol{R}$, μs_{BD} , [0,1]) (correspondingly, WBD(f, a) = ($\boldsymbol{R}$, μw_{BD}, [0,1]) where the membership function μs_{BD} is defined by the equality

$$\mu s_{BD}(z) = 1/(1 + \mathrm{ms}(a, z)),$$

where ms(a,z) = inf $\{r\in\boldsymbol{R}^+;\ z \in \mathrm{SBD}_r(f, a)\}$ and the membership function μw_{BD} is defined by the equality $\mu w_{BD}(z) = 1/(1 + \mathrm{mw}(a, z))$, where $\mathrm{mw}(a, z) = \inf \{r\in\boldsymbol{R}^+;\ z \in \mathrm{WBD}_r(f, a)\}$.

Their particular cases are:

1. The complete local weak centered derivative WCFD(f, a) = ($\boldsymbol{R}$, μw_{CFD} , [0,1]) of $f(x)$ at a point $a\in X$;
2. The complete local weak left derivative WLFD(f, a) = ($\boldsymbol{R}$, μw_{LFD} , [0,1]) of $f(x)$ at a point $a \in X$;
3. The complete local weak right derivative WRFD(f, a) = ($\boldsymbol{R}$, μw_{RFD} , [0,1]) of $f(x)$ at a point $a \in X$;
4. The complete local weak two-sided derivative WTFD(f, a) = ($\boldsymbol{R}$, μw_{TFD} , [0,1]) of $f(x)$ at a point $a \in X$;
5. The complete local strong centered derivative SCFD(f, a) = ($\boldsymbol{R}$, μs_{CFD} , [0,1]) of $f(x)$ at a point $a \in X$;
6. The complete local strong left derivative SLFD(f, a) = ($\boldsymbol{R}$, μs_{LFD} , [0,1]) of $f(x)$ at a point $a \in X$;
7. The complete local strong right derivative SRFD(f, a) = ($\boldsymbol{R}$, μs_{RFD} , [0,1]) of $f(x)$ at a point $a \in X$;
8. The complete local weak two-sided derivative STFD(f, a) = ($\boldsymbol{R}$, μs_{TFD} , [0,1]) of $f(x)$ at a point $a \in X$.

Remark 4.7.6. In the case of fuzzy strong centered, right, left, and two-sided derivatives, we can take the operation *min* instead of the operation *inf*. This minimum exists by Theorem 2.5.1.

Let us consider a set of sequences E = $\{ l_j = \{ (a_{ji}); i = 1, 2, 3, \dots \}; j = 1, 2, 3, \dots \}$. A sequence $l = \{ (b_j); j = 1, 2, 3, \dots \}$ is a *diagonalization* of E if $b_j = a_{jt}$ for all $j = 1, 2, 3, \dots$ and if $b_j = a_{jt}$, $b_k = a_{kq}$ and $j < k$, then $t < q$.

Let us assume that condition B is stable with respect to diagonalization, i.e., for any set E of sequences, which satisfy condition B, any its diagonalization also satisfies this condition.

Proposition 4.7.11. The set SBD$_r$(f, a) is the α-level set (cf., Section 1.3.1) of the fuzzy set SBD(f, a) where $\alpha = 1/(1 + r)$.

Proof. 1. Let p be an arbitrary number from the α-level set of the fuzzy set SBD(f, a). It means that $\mu s_{BD}(p) = \beta \geq \alpha$. Then $\beta = 1/(1 + q)$ and $p \in \mathrm{SBD}_{q+\varepsilon}(f, a)$ for all numbers $\varepsilon \in \boldsymbol{R}^{++}$. Utilizing the diagonalization process, we can show that $p \in \mathrm{SBD}_q(f, a)$.

By the initial condition from Proposition 4.7.11, we have $\alpha = 1/(1 + r)$. Thus, the inequality $\beta \geq \alpha$ implies the inequality $q < r$. By Lemma 4.2.4, $p \in \mathrm{SBD}_r(f, a)$. As p is an arbitrary number from the α-level set of the fuzzy set SBD(f, a), all elements from the α-level set of the fuzzy set SBD(f, a) belong to the set SBD$_r$(f, a).

2. Let t be an arbitrary number from the set SBD$_r$(f, a). Then $\mathrm{ms}(a,t) \leq r$ and $\mu s_{BD}(t) = 1/(1 + \mathrm{ms}(a, t) \geq \alpha = 1/(1 + r)$. Consequently, t belongs to the α-level set of the fuzzy set

SBD(f, a). As t is an arbitrary number from the set SBD$_r$(f, a), all elements from the set SBD$_r$(f, a), belong to the α-level set of the fuzzy set SBD(f, a).

Proposition is proved.

Corollary 4.7.10. The set SXFD$_r$(f, a) is the α-level set of the fuzzy set SXFD(f, a), where $\alpha = 1/(1 + r)$.

For weak fuzzy derivatives, we have a more complicated result.

Proposition 4.7.12. The closure of the set WBD$_r$(f, a) is the α-level set of the fuzzy set WBD(f, a) where $\alpha = 1/(1 + r)$.

Proof is similar to the proof of Proposition 4.7.11.

Corollary 4.7.11. The closure of the set WXFD$_r$(f, a) is the α-level set of the fuzzy set WXFD(f,a) where $\alpha = 1/(1 + r)$.

By Proposition 4.7.12, closures of sets WBD$_r$(f, a) are α-level sets of the fuzzy set WBD$_r$(f, a) and by Proposition 4.7.12, sets SBD$_r$(f, a) are α-level sets of the fuzzy set SBD$_r$(f, a). According to Definition 4.7.3, we have $\alpha = 1/(1 + r)$. Consequently, sets YXFD$_r$(f, a) are α-level sets of the fuzzy sets YXFD(f, a), correspondingly, where $\alpha = 1/(1 + r)$, Y may be equal to W or S and X may be equal to C, L, R, or T.

Remark 4.7.7. It is possible to consider fuzzy set derivatives: SBD(f, a) = ($\boldsymbol{R}$, ms(a, z), $\boldsymbol{R}^+$) and WBD(f, a) = ($\boldsymbol{R}$, mw(a, z), $\boldsymbol{R}^+$), which are fuzzy sets in the sense of Cai Wen. However, in this case, the classical derivative $f\,'(a)$ of $f(x)$ at a belongs to these fuzzy sets with the zero extent. This contradicts our intuition that fuzzy derivatives are extensions of the classical derivative and consequently, the classical derivative is the most typical and must have the highest level of membership.

Lemma 4.7.1. $\mu_{SBD}(z) = 1$ for the fuzzy set SCFD(f, a) if and only if z is the classical derivative $f\,'(a)$ of $f(x)$ at a.

If we take the join for all points $a \in X$ of all complete conditional derivatives YBD(f, a) with Y $\in$ {S, W}, we obtain the complete global fuzzy derivative YBD f of $f(x)$ on X having type Y. Here YBD(f, a) is a fuzzy binary relation on $\boldsymbol{R}$, and namely, YBD f = ($\boldsymbol{R}^2$, μ_{YBD} , [0,1]) where the membership function μ_{YBD} is defined on $\boldsymbol{R}^2$ by the equality $\mu(x, z) = 1/(1 + \text{m}(x, z))$. The result of Theorem 4.7.1 provides for correctness of the definition of fuzzy sets YBD(f, a) for all $a \in \boldsymbol{R}$, as well as of the fuzzy set YBD f. By Lemma 4.71, $\mu_{YBD}(z) = \mu_{YBD}(a, z) = 1$ if and only if $z = f\,'(a)$ at the point a. Consequently, fuzzy sets YXFD(f, a) and YXFD f are fuzzy set derivatives of crisp (ordinary) functions related to similar constructions that were introduced by Kalina (1997; 1998; 1999).

Sets YXFD$_r$(f, a) define the complete fuzzy derivative YXFD(f, a) of the function $f(x)$ at a point $a \in X$ having type (Y,X). It is called also a complete local fuzzy derivative of f. Each YXFD(f, a) is a fuzzy subset of $\boldsymbol{R}$, and namely, YXFD(f, a) =($\boldsymbol{R}$, μ_x , [0,1]) where the membership function μ_x is defined by the equality $\mu_x(z) = 1/(1 + \text{m}(a, z))$ where $\text{m}(a, z) = \min \{r \in \boldsymbol{R} \,;\, z \in \text{YXFD}_r(f, a)\}$. This minimum exists by Theorem 4.7.1.

If we take the join of all complete fuzzy derivatives YXFD(f, a), we obtain the complete global fuzzy derivative YXFD f of the function $f(x)$ on X having type (Y,X). Here YXFD(f, a) is a fuzzy binary relation on $\boldsymbol{R}$, and namely, YXFD f = ($\boldsymbol{R}^2$, μ , [0,1]) where the membership function μ is defined on $\boldsymbol{R}^2$ by the equality $\mu(x, z) = 1/(1 + \text{m}(x, z))$. The result of Theorem 4.7.1 provides for correctness of the definition of fuzzy sets YXFD(f, a) for all $a \in \boldsymbol{R}$, as well as of the fuzzy set YXFD*f*. By Lemma 4.7.1, $\mu_x(z) = \mu(a, z) = 1$ if and only if $z = f'(a)$ at the point a.

Here we can see in an explicit form how investigation of ordinary functions involves construction of fuzzy sets and relations.

Remark 4.7.3. Complete fuzzy derivatives do not possess many properties of ordinary derivatives as well as of other (strong centered, left, right, two-sided etc.) fuzzy derivatives. For example, let us take $f(x) = |x|$ and $g(x) = - |x|$. Then $f + g$ is the function identically equal to zero. All its derivatives are also equal to zero at all points. Consequently, $\mu_0(0) = 1$ for $f + g$.

At the same time, the value of the membership function $\mu_0(0)$ for the sum of any pair of fuzzy sets YXFD(f,0) and YXFD(g,0) is equal to ½. Thus, the complete fuzzy derivative of the sum $(f + g)(x)$ is not equal to the sum of the complete fuzzy derivatives of $f(x)$ and of $g(x)$. However, it is well known that the conventional differentiation is a linear operator (Dieudonné, 1960) and the same is true for all kinds of strong (strong centered, left, right, two-sided etc.) fuzzy derivatives (cf. Theorems 4.2.1 and 4.4.1 and Corollary 4.2.2).

Theorem 4.7.3 implies the following result.

Corollary 4.7.12. The set SXFD(f) = $\bigcup_{r\geq 0}$ SXFD$_r$(f) of all fuzzy derivatives of $f(x)$ is a real linear space.

The conditional r-derivatives introduced in the previous section define the complete strong local B-conditional r-derivative SBD$_r$(f, a) of $f(x)$ at a point $a \in X$. Namely, SBD$_r$(f, a) = $\{ z \in \boldsymbol{R}$; z is a strong local B-conditional r-derivative of $f(x)$ at a point $a\}$.

By the definitions, all fuzzy sets SBD(f, a) and WBD(f, a), as well as all YXFD(f, a) with X $\in$ {C, L, R, T} and Y $\in$ {S, W} are fuzzy numbers in the sense of (Averkin *et al.*, 1986). As the definition of a fuzzy number in (Averkin *et al.*, 1986) is the most general (cf. Section 1.3.1), it is interesting to know when these fuzzy sets are fuzzy numbers in a more restricted sense.

Proposition 4.7.13. The fuzzy set SCFD(f, a) is normal if and only if the function $f(x)$ has the classical derivative $f'(a)$ at a point a.

Theorem 4.7.4. A complete local strong derivative SCFD(f, a) is a fuzzy number in the sense of (Furukawa, 1996) if and only if the function $f(x)$ has the classical derivative $f'(a)$ at the point a.

By Theorem 4.7.1, each set SXFD$_r$(f, a) with X $\in$ {C, L, R, T} is a bounded convex closed set, i.e., SXFD$_r$(f, a) = $[b,c]$ for some numbers b, $c \in \boldsymbol{R}$, or SXFD$_r$(f, a) = $\varnothing$ if f has no strong r-derivatives of the type X. Consequently, all its α–level sets of the fuzzy set SXFD(f, a) are convex. Then Lemma 4.7.1 implies the following result.

Theorem 4.7.5. Each complete local strong derivative SXFD(f, a) with X $\in$ {C, L, R, T} is a convex fuzzy set,

Theorem 4.7.5 and Lemma 4.7.1 imply the following result.

Theorem 4.7.6. A complete local strong derivative SCFD(f, a) is a fuzzy number in the sense of (Zimmermann, 2001) if and only if the function $f(x)$ has the classical derivative $f'(a)$ at the point a.

Definition 4.7.4. The *complete local* B-*conditional derivative* BD(f, a) is a tetrad that has the form BD(f, a) = ($\boldsymbol{R}$, μ_{BD} , ν_{BD} , [0,1]) where the function μ_{BD} is defined by the equality $\mu_{BD}(z) = 1/(1 + \text{ms}(a, z))$, where $\text{ms}(a, z) = \inf \{r \in \boldsymbol{R}^+;\ z \in \text{SBD}_r(f, a)\}$and the non-membership function ν_{BD} is defined by the equality $\nu_{BD}(z) = 1 - [1/(1 + \text{mw}(a, z))] = \text{mw}(a, z)/(1 + \text{mw}(a, z))$, where $\text{mw}(a, z) = \inf \{r \in \boldsymbol{R}^+; z \in \text{WBD}_r(f, a)\}$.

As all fuzzy sets WCFD(f, a), WRFD(f, a), WLFD(f, a), WTFD(f, a), SCFD(f, a), SRFD(f, a), SLFD(f, a), STFD(f, a), and their membership functions are particular cases of the fuzzy sets WBD(f, a) and SBD(f, a) and their membership functions, the complete local B-conditional derivative BD(f, a) gives, as its particular cases, the complete local centered derivative CFD((f, a), the complete local left derivative LFD((f, a), the complete local right derivative RFD((f, a), and the complete local two-sided derivative TFD((f, a).

Lemma 4.7.2. $\nu_A(z) \leq \mu_A(z)$ for such $z \in \boldsymbol{R}$ that $1 \geq \text{ms}(a, z)$ and $\nu_A(z) \geq \mu_A(z)$ for such $z \in \boldsymbol{R}$ that $1 \leq \text{mw}(a, z)$.

Really, let us assume that $1 \leq \text{mw}(a, z)$. Then

$$1 + \text{mw}(a, z) \leq \text{mw}(a, z) + \text{mw}(a, z)\cdot\text{ms}(a, z)$$

because $1 \leq \text{mw}(a,z)$ and $\text{mw}(a,z) \leq \text{ms}(a,z)$. Consequently,

$$1 + \text{mw}(a, z) \leq \text{mw}(a, z)\cdot(1 + \text{ms}(a, z))$$

and

$$1/(1 + \text{ms}(a, z)) \leq \text{mw}(a, z)/(1 + \text{mw}(a, z)).$$

This means that $\nu_A(z) \geq \mu_A(z)$.

If we take $1 \geq \text{ms}(a, z))$, then

$$1 + \text{mw}(a, z) \geq \text{mw}(a, z) + \text{mw}(a, z)\cdot\text{ms}(a, z)$$

because $1 \geq \text{ms}(a,z)$ and $\text{mw}(a,z) \leq \text{ms}(a,z)$. Consequently,

$$1 + \text{mw}(a, z) \geq \text{mw}(a, z)\cdot(1 + \text{ms}(a, z))$$

and

$$1/(1 + \text{ms}(a, z)) \geq \text{mw}(a, z)/(1 + \text{mw}(a, z)).$$

This means that $\nu_A(z) \leq \mu_A(z)$.

Lemma is proved.

Proposition 4.7.14. The complete local B-conditional derivative BD((f, a) is an intuitionistic fuzzy set.

Proof. We may assume that μ_{BD} is the membership function and ν_{BD} is the non-membership function of the intuitionistic fuzzy set BD((f, a). To have a correct definition, we need only to prove that $0 \leq \mu_A(z) + \nu_A(z) \leq 1$ for any $z \in \boldsymbol{R}$ because $\mu_A(z)$, $\nu_A(z) \leq 1$ for any $z \in \boldsymbol{R}$. Really, by the definitions of weak and strong fuzzy derivatives, we have $\text{ms}(a, z) \geq \text{mw}(a, z)$ for all a and z from $\boldsymbol{R}$. This implies

$$1/(1 + \mathrm{ms}(a, z)) \leq 1/(1 + \mathrm{mw}(a, z)).$$

Consequently,

$$1 - 1/(1 + \mathrm{ms}(a, z)) \geq 1 - 1/(1 + \mathrm{mw}(a, z)).$$

Then

$$1 = 1/(1 + \mathrm{ms}(a, z)) + (1 - [1/(1 + \mathrm{ms}(a, z))]) \geq$$
$$1/(1 + \mathrm{ms}(a, z)) + (1 - 1/(1 + \mathrm{mw}(a, z))) = \mu_A(z) + \nu_A(z).$$

In addition, $0 \leq \mu_A(z) + \nu_A(z)$ because $0 \leq \mu_A(z)$ and $0 \leq \nu_A(z)$.

Proposition is proved.

Corollary 4.7.13. The complete local centered derivatives CFD((f, a), right derivatives RFD((f, a), left derivatives LFD((f, a), and two-sided derivatives TFD((f, a) are intuitionistic fuzzy sets.

By the definitions, all fuzzy sets SBD((f, a) and WBD((f, a), as well as all YXFD((f, a) with X $\in$ {C, L, R, T} and Y $\in$ {S, W} are intuitionistic fuzzy numbers.

When a function $f(x)$ has the classical derivative $f'(a)$ at a point a, the fuzzy sets of strong derivatives SCFD((f, a), SRFD((f, a), SLFD((f, a), STFD((f, a), and WCFD((f, a), WRFD((f, a), WLFD((f, a), WTFD((f, a), as well as the intuitionistic fuzzy sets CFD((f, a), RFD((f, a), LFD((f, a), and TFD((f, a), have relatively simple structure as the following results demonstrate.

Let $f'(a) = b$.

Proposition 4.7.15. a) SCFD((f, a) = SRFD((f, a) = SLFD((f, a) = STFD((f, a) = WCFD((f, a) = WRFD((f, a) = WLFD((f, a) = WTFD((f, a) = ($\boldsymbol{R}$, μ_D , [0,1]) where $\mu_D(z) = 1/(1+ | z - b |)$.

b) CFD((f, a) = RFD((f, a) = LFD((f, a) = TFD((f, a) = ($\boldsymbol{R}$, μ_D , ν_D , [0,1]) where $\mu_D(z) = 1/(1+ | z -b |)$ and $\nu_D = 1 - \mu_D(z)$ for all $z \in \boldsymbol{R}$.

Chapter 5

MONOTONE AND FUZZY MONOTONE FUNCTIONS

Not to go back, is somewhat to advance...
Alexander Pope (1688-1744)

In this chapter, weak fuzzy derivatives and weak derivatives are applied to a study of monotone and fuzzy monotone functions. Monotone functions defined in the Appendix have many good properties. However, when we get function values from measurement or computation, it is impossible to tell precisely whether the function is monotone in the exact mathematical sense. This brings us to the concept of a fuzzy monotone function studied in this chapter. In such a function, monotonicity may be slightly violated.

Different properties of monotone and fuzzy monotone functions are obtained. Observations are made when it is possible to ignore some deviations from the conventional monotonicity and still have properties of monotone functions. Thus, some properties of fuzzy monotone functions are the same or at least similar to the properties of the monotone functions, while other properties differ in many aspects from those in the standard theory. It is demonstrated that in a broader context of fuzzy limits and derivatives, it is possible to extend many results of the classical mathematical analysis that are related to monotone functions. Many classical results are obtained as direct corollaries of statements proved in this chapter.

Such a transition to a fuzzy context provides for completion of some basic results of the classical mathematical analysis. For instance, it is known that if a differentiable function $f(x)$ has a non-negative derivative, then $f(x)$ is an increasing function. The converse of this statement is not true as there are increasing functions that do not have derivatives at some points. However, here we prove that a function $f(x)$ is increasing if and only if any extended weak derivative of $f(x)$ is larger than or equal to zero (Theorem 5.5). This gives a complete criterion because, as it is demonstrated in Proposition 4.5.1 and Corollary 4.5.1, any real function has an extended weak derivative at any point.

Moreover, neoclassical analysis does not only bring new results, which complete their classical analogues, but also produces deeper insights and a better understanding of the classical theory.

Let X be a subset of $\boldsymbol{R}$ and $f: X \to \boldsymbol{R}$ be an arbitrary function. We know from Proposition 4.5.1 that $f(x)$ has an extended weak derivative at any point.

Theorem 5.1. If any extended weak derivative $b = W^{ct}d/_{dx} f(a)$ is larger than zero for all $a \in X$, then the function $f(x)$ is increasing on X.

Proof. Let us suppose that the condition concerning extended weak derivatives is valid but $f(x)$ is not an increasing function on X. Then there are such elements $c, d \in \boldsymbol{R}$, for which $c < d$ and $f(c) > f(d)$. We consider separately several cases.

1) There is a sequence $\{ c_i ; i = 1, 2, 3, \ldots\}$ such that $c < c_i < d$ for all $i = 1, 2, 3, \ldots, f(c) \geq f(c_i)$ also for all $i = 1, 2, 3, \ldots$, and $\lim_{i\to\infty} c_i = c$. Then we have $0 \geq (f(c) - f(c_i)) / (c - c_i)$ for all $i = 1, 2, 3, \ldots$. Consequently, the extended weak derivative $u = \lim_{i\to\infty} (f(c) - f(c_i)) / (c - c_i)$ is less than 0 or equal to 0 at the point c. It is possible that $u = -\infty$. However, in any case, we come to a contradiction to our supposition that any weak derivative $b = w^{ct}d/_{dx} f(a)$ is larger than zero for all $a \in X$. This excludes a possibility for the case (1).
2) There is a sequence $\{ c_i ; i = 1, 2, 3, \ldots\}$ such that $c < c_i < d$ for all $i = 1, 2, 3, \ldots, f(c_i) \geq f(d)$ for all $i = 1, 2, 3, \ldots$, and $\lim_{i\to\infty} c_i = d$. Then we have $0 \geq (f(d) - f(c_i)) / (d - c_i)$ for all $i = 1, 2, 3, \ldots$. Consequently, the extended weak derivative $v = \lim_{i\to\infty} (f(d) - f(c_i)) / (d - c_i) \leq 0$ at the point d. It is possible that $v = -\infty$. However, in any case, we come to a contradiction to our supposition. This excludes a possibility for the case (2).
3) There are no such sequences. Let us take the number $e = \inf \{ x ; c < x < d \text{ and } f(c) > f(x) \}$.
   (a) It is possible that $e = c$. In this situation, the definition of infimum implies that there is a sequence $\{ c_i ; i = 1, 2, 3, \ldots \}$ such that $c < c_i < d$ for all $i = 1, 2, 3, \ldots$, $f(c) > f(c_i)$ for all $i = 1, 2, 3, \ldots$, and $\lim_{i\to\infty} c_i = c$. Thus we come to the case (1), which contradicts to the assumption.
   (b) Let $e > c$. In this situation, the definition of infimum implies that there is a sequence $\{ c_i ; i = 1, 2, 3, \ldots\}$ such that $c < c_i < d$ for all $i = 1, 2, 3, \ldots$, $f(c_i) \geq f(c) > f(e)$ for all $i = 1, 2, 3, \ldots$, and $\lim_{i\to\infty} c_i = e$. Thus, we come to a case similar to the case (2). The same reasoning shows that in this case we also have a contradiction to the assumption.

As all possible cases imply contradictions to the assumption, Theorem 5.1 is proved.

Let a function $f(x)$ has an extended strong derivative $St^{ct}d/_{dx} f(a)$ at all points a from X.

Corollary 5.1. If any extended strong derivative $b = St^{ct}d/_{dx} f(a)$ exists and is larger than zero for all $a \in X$, then the function $f(x)$ is increasing on X.

When the classical derivative of a function at a point exists, then all extended weak derivatives at this point coincide with the classical derivative. Consequently, Theorem 5.1 gives us the following classical result for differentiable functions (cf., for example, (Marsden and Weinstein, 1981; Ross, 1996)).

Theorem 5.2. If the classical derivative $f'(x)$ exists and is larger than zero on X, then the function $f(x)$ is increasing on X.

Theorem 5.1 and Lemmas 4.3.6 and 4.3.7 imply the following result.

Corollary 5.2. If for some $r \in \boldsymbol{R}^+$, all extended weak fuzzy derivatives $b = \mathrm{W}_r{}^{\mathrm{ct}}\mathrm{d}/_{\mathrm{d}x} f(a)$ are larger than zero for all $a \in X$, then the function $f(x)$ is increasing on X.

Let for some $r \in \boldsymbol{R}^+$, a function $f(x)$ has an extended strong r-derivative $\mathrm{St}_r{}^{\mathrm{ct}}\mathrm{d}/_{\mathrm{d}x} f(a)$ at all points a from X.

Corollary 5.3. If for some $r \in \boldsymbol{R}^+$, all extended strong r-derivatives $b = \mathrm{St}_r{}^{\mathrm{ct}}\mathrm{d}/_{\mathrm{d}x} f(a)$ exist and are larger than zero for all $a \in X$, then the function $f(x)$ is increasing on X.

Remark 5.2. The condition from Theorem 5.1 is not necessary for an arbitrary function to be increasing as the example of the function $f(x) = x^3$ shows. This function is strictly increasing, but all its extended weak derivatives at 0 are equal to 0.

Theorem 5.3. If any extended weak derivative $b = \mathrm{W}^{\mathrm{ct}}\mathrm{d}/_{\mathrm{d}x} f(a)$ is smaller than zero for all $a \in X$, then the function $f(x)$ is decreasing on X.

Proof is similar to the proof of Theorem 5.1.

Let a function $f(x)$ has an extended strong derivative $\mathrm{St}^{\mathrm{ct}}\mathrm{d}/_{\mathrm{d}x} f(a)$ at all points a from X.

Corollary 5.4. If any extended strong derivative $b = \mathrm{St}^{\mathrm{ct}}\mathrm{d}/_{\mathrm{d}x} f(a)$ exists and is smaller than zero for all $a \in X$, then the function $f(x)$ is decreasing on X.

As its direct corollary, Theorem 5.3 gives us the following classical result for differentiable functions (cf., for example, (Marsden and Weinstein, 1981; Ross, 1996)).

Theorem 5.4. If the classical derivative $f'(x)$ exists and is smaller than zero on X, then the function $f(x)$ is decreasing on X.

Theorem 5.3 and Lemmas 4.3.6 and 4.3.7 imply the following result.

Corollary 5.5. If for some $r \in \boldsymbol{R}^+$, all extended weak fuzzy derivatives $b = \mathrm{W}_r{}^{\mathrm{ct}}\mathrm{d}/_{\mathrm{d}x} f(a)$ are smaller than zero for all $a \in X$, then the function $f(x)$ is decreasing on X.

Let for some $r \in \boldsymbol{R}^+$, a function $f(x)$ has an extended strong r-derivative $\mathrm{St}_r{}^{\mathrm{ct}}\mathrm{d}/_{\mathrm{d}x} f(a)$ at all points a from X.

Corollary 5.6. If for some $r \in \boldsymbol{R}^+$, all extended strong r-derivatives $b = \mathrm{St}_r{}^{\mathrm{ct}}\mathrm{d}/_{\mathrm{d}x} f(a)$ exist and are smaller than zero for all $a \in X$, then the function $f(x)$ is decreasing in X.

These results imply corresponding results for locally monotone functions.

Definition 5.1. A function $f(x)$ is *increasing* (*strictly increasing*, *decreasing*, *strictly decreasing*) *at a point* $a \in X$, if there is a neighborhood Oa of the point a such that the function $f(x)$ is increasing (strictly increasing, decreasing, strictly decreasing, correspondingly) in Oa.

Increasing functions and decreasing at a point a functions are called *monotone at the point* a functions. Strictly increasing functions and strictly decreasing at a point a functions are called *strictly monotone at the point* a functions. The corresponding properties of functions are called *local monotonicity* and *local strict monotonicity*, respectively.

Example 5.1. Functions x and x^3 are increasing and monotone at the point 0, while functions $|x|$ and x^2 are not increasing and not monotone at the point 0.

Corollary 5.7. If any extended weak derivative $b = \mathrm{W}^{\mathrm{ct}}\mathrm{d}/_{\mathrm{d}x} f(c)$ is larger (smaller) than zero for all c from some neighborhood Oa of the point a, then the function $f(x)$ is increasing (decreasing) at a.

Definition 5.2. A binary relation $U \subseteq \boldsymbol{R} \times \boldsymbol{R}$ ($U \subseteq \boldsymbol{R}_\infty \times \boldsymbol{R}_\infty$) is called *continuous* at a point a from $\boldsymbol{R}$ if for any $\varepsilon \in \boldsymbol{R}^{++}$, there is $\delta \in \boldsymbol{R}^{++}$ such that

for any $x \in \boldsymbol{R}$, $|x - a| < \delta$ implies

$\forall\, z, u \in \boldsymbol{R}\, ((u \in U(a) \,\&\, z \in U(x)) \rightarrow |z - u| < \varepsilon)$.

Definition 5.3. A binary relation $U \subseteq \boldsymbol{R} \times \boldsymbol{R}$ ($U \subseteq \boldsymbol{R}_\infty \times \boldsymbol{R}_\infty$) is called *strictly continuous* at a point a from $\boldsymbol{R}$ if for any $\varepsilon \in \boldsymbol{R}^{++}$, there is $\delta \in \boldsymbol{R}^{++}$ such that for any $x \in \boldsymbol{R}$, $|x - a| < \delta$ implies $D(U(x)) < \varepsilon$) where $D(U(x))$ is the diameter of the set $U(x)$.

Lemma 5.1. A binary relation U is strictly continuous at a point a if and only if it is continuous at a.

Proof. 1. Let a binary relation $U \subseteq \boldsymbol{R} \times \boldsymbol{R}$ ($U \subseteq \boldsymbol{R}_\infty \times \boldsymbol{R}_\infty$) be strictly continuous at a point a from $\boldsymbol{R}$ and ε be some number from $\boldsymbol{R}^{++}$. Then there is $\delta \in \boldsymbol{R}^{++}$ such that for any $x \in \boldsymbol{R}$, $\rho(x, a) < \delta$ implies $D(U(x)) < (1/3)\varepsilon$) where $D(U(x))$ is the diameter of the set $U(x)$. Let us take $\eta = \min \{(1/3)\varepsilon , \delta\}$, such $x \in \boldsymbol{R}$ that $\rho(x, a) < \eta$, and consider two points $u \in U(a)$ and $z \in U(x)$. Then $\rho(z, u) \leq \rho(z, x) + \rho(x, a) \leq D(U(x)) + \eta$ because $\rho(z, x) \leq D(U(x))$. This implies $\rho(z, u) <(1/3)\varepsilon + (1/3)\varepsilon < \varepsilon$ when $\rho(x, a) < \eta$. Thus, the relation U satisfies Definition 5.1, i.e., the relation U is continuous at a point a.

2. Let a binary relation $U \subseteq \boldsymbol{R} \times \boldsymbol{R}$ ($U \subseteq \boldsymbol{R}_\infty \times \boldsymbol{R}_\infty$) be continuous at a point a from $\boldsymbol{R}$ and ε be some number from $\boldsymbol{R}^{++}$. Then there is $\delta \in \boldsymbol{R}^{++}$ such that for any $x \in \boldsymbol{R}$, $|x - a| < \delta$ implies $\forall\, z, u \in \boldsymbol{R}\, ((u \in U(a) \,\&\, z \in U(x)) \rightarrow |z - u| < (1/5)\varepsilon$). If $v \in U(x)$, then $|z - v| = |z - u + u - a + a - v| \leq |z - u| + |u - a| + |v - a| < (1/5)\varepsilon + |\, u - a| + |\, v - a|$. As $|\, x - a| < \delta$ and $u \in U(a)$, then by definition of continuity at a, we have $|\, u - a| < (1/5)\varepsilon$. As $v \in U(x)$ and $a \in U(a)$, then by definition of continuity at a, we have $|\, v - a| < (1/5)\varepsilon$. Thus, $|\, z - v| <(1/5)\varepsilon + (1/5)\varepsilon + (1/5)\varepsilon < \varepsilon$.

Lemma 5.1 is proved.

Any function is a binary relation (see Appendix A). Thus, it is interesting what is the connection between continuity of functions and continuity of binary relations. The following result given an answer to this question.

Lemma 5.2. Any function is continuous at a point a if and only if it is continuous at a as a binary relation.

The set $\mathrm{WD}(f, x)$ of all weak derivatives of a function $f(x)$ defines the binary relation $\mathbf{WD}(f, x) = \{ (a, b);\ b = \mathrm{W}^{ct}\mathrm{d}/_{\mathrm{d}x}\, f(a),\ a \in \boldsymbol{R} \}$ in $\boldsymbol{R}$. Let us assume that $\mathbf{WD}(f, x)$ is a continuous relation at a point a. Then Corollary 5.7 makes it possible to prove the following result.

Proposition 5.1. If any extended weak derivative $b = \mathrm{W}^{ct}\mathrm{d}/_{\mathrm{d}x}\, f(a)$ is larger (smaller) than zero, then the function $f(x)$ is increasing (decreasing) at the point a.

Proof. If any extended weak derivative $b = \mathrm{W}^{ct}\mathrm{d}/_{\mathrm{d}x}\, f(a)$ of $f(x)$ at a is larger than zero, then continuity of the relation $\mathbf{WD}(f, x)$ implies that any extended weak derivative $y = \mathrm{W}^{ct}\mathrm{d}/_{\mathrm{d}x}\, f(c)$ of $f(x)$ at c is larger than zero for all c from some neighborhood $\mathrm{O}a$ of the point a. By Corollary 5.7, $f(x)$ is increasing at a.

Similar arguments show that a function $f(x)$ is decreasing at a, when any extended weak derivative $b = \mathrm{W}^{ct}\mathrm{d}/_{\mathrm{d}x}\, f(a)$ is smaller than zero. This concludes the proof of Proposition 5.1.

When the classical derivative of a function at a point exists, then all extended weak derivatives at this point coincide with the classical derivative. Consequently, Proposition 5.1 implies the following result.

Proposition 5.2. If the classical derivative $f'(a)$ exists and is larger (smaller) than zero, then the function $f(x)$ is increasing(decreasing) at the point a.

Lemma 5.3. If any extended weak derivative $b = W^{ct}d/_{dx}\, f(a)$ of a function $f(x)$ at a point a from X is larger than or equal to zero, then for any $\varepsilon \in \boldsymbol{R}^{++}$, there is a neighborhood $O_\varepsilon a$ of the point a such that the following condition (K) is true:

$$\forall\, x \in O_\varepsilon a\, (\, (x > a \rightarrow f(x) - f(a) > - \varepsilon(\, x - a\,)\,)$$

and

$$(\, x < a \rightarrow f(a) - f(x) > - \varepsilon(\, a - x\,)\,)).$$

Proof. If the condition (K) is not true, there are a number $k \in \boldsymbol{R}^{++}$ and a sequence $\{\, c_i\,;\, i = 1, 2, 3, \ldots\,\}$ such that $a < c_i$ for all numbers $i = 1, 2, 3, \ldots$, $\lim_{i\to\infty} c_i = a$, and $f(c_i) - f(a) < -k(c_i - a\,)$, or there is a sequence $\{\, a_i\,;\, i = 1, 2, 3, \ldots\,\}$ such that $a > a_i$ for all $i = 1, 2, 3, \ldots$, $\lim_{i\to\infty} a_i = a$, and $f(a) - f(a_i) < -\, k(\, a - a_i)$. Consequently, either $(f(c_i) - f(a))/\,(\, c_i - a\,) < -\, k$ for all $i = 1, 2, 3, \ldots$, and extended weak derivative $b = W^{ct}d/_{dx} f(a) = \lim_{i\to\infty}(f(c_i) - f(a))/\,(\, c_i - a\,) \leq -\, k < 0$ or $(f(a) - f(a_i))\,/\,(\, a - a_i\,) < -\, k$ for all $i = 1, 2, 3, \ldots$, and extended weak derivative $d = W^{ct}d/_{dx}\, f(a) = \lim_{i\to\infty}(\, f(a) - f(a_i))/\,(\, a - a_i\,) \leq -\, k < 0$. It is possible that b or/and d is equal to $-\infty$. This contradicts our assumption and thus, concludes the proof.

In a similar way, we can prove the following result.

Lemma 5.4. If any extended weak derivative $b = W^{ct}d/_{dx} f(a)$ of a function $f(x)$ at a point a from X is less than or equal to zero, then for any $\varepsilon \in \boldsymbol{R}^{++}$, there is a neighborhood $O_\varepsilon a$ of the point a such that the following condition (K) is true:

$$\forall\, x \in O_\varepsilon a\, (\, (x > a \rightarrow f(x) - f(a) < \varepsilon(\, x - a\,)\,)$$

and

$$(\, x < a \rightarrow f(a) - f(x) > \varepsilon(\, a - x\,)\,)).$$

Conditions from Lemmas 5.3 and 5.4 bring us to the following concepts of fuzzy increasing and fuzzy decreasing functions.

Let $d \in \boldsymbol{R}^{+}$.

Definition 5.4. A function $f: X \rightarrow \boldsymbol{R}$ is called:

1. *d-increasing* on X if $a < b$ implies $f(a) \leq f(b) + d(b - a)$ for all a and b from X;
2. *d-decreasing* on X if $a < b$ implies $f(a) \geq f(b) - d(b - a)$ for all a and b from X;
3. *strictly d-increasing* on X if $a < b$ implies $f(a) < f(b) + d(b - a)$ for all a and b from X;
4. *strictly d-decreasing* on X if $a < b$ implies $f(a) > f(b) - d(b - a)$ for all a and b from X.

If a function $f(x)$ is d-increasing (d-decreasing) on X for some d, then $f(x)$ is called *fuzzy increasing* (*fuzzy decreasing*) on X. Both classes, fuzzy increasing functions and fuzzy decreasing functions are called *fuzzy monotone*. Strictly fuzzy increasing functions and strictly fuzzy decreasing functions are called *strictly fuzzy monotone*.

Example 5.2. Functions $x + (½)\sin x$ and $x^3 + \sin^2 x$ are 1-increasing and 1-monotone, while they are not increasing and not monotone.

Informally, when a function is fuzzy increasing it means that it allows some deviations from being increasing, that is, from growing all time. Functions that we can observe in reality, either in nature and society or in virtual reality of computers, do not often allow one to verify whether they are increasing or not. It is possible to make 1000 observations and the results will implicate that the function is increasing. However, we get one more result and it violates the strict condition of growth. Usually such results are discarded, as they can be results of noise, mistakes in measurement procedure or violation of initial conditions. However, in many cases, this is not true. Besides, it is useful to able to tolerate some deviations. For instance, some historians of science suggest that if Kepler had modern means for measuring positions of celestial bodies (planets and the Sun), he would never discover his laws of motion.

The concepts of fuzzy increasing and decreasing functions allows us to neglect small violations of monotonicity conditions and the parameter d gives us a measures to what extent we allow non-monotonicity.

It is also necessary to remark that the parameter d is scalable because it does not depend on the distance between points.

An important feature of fuzzy monotonicity is that a 0-increasing function is simply an increasing function and a 0-decreasing function is simply a decreasing function.

Definition 5.4 directly implies the following result.

Lemma 5.5. If a function $f(x)$ is d-increasing (d-decreasing) on X, then for any $\varepsilon \in \boldsymbol{R}^{++}$, the function $f(x)$ is strictly $(d + \varepsilon)$-increasing ($(d + \varepsilon)$-decreasing) on X.

From global monotonicity, we come to local monotonicity.

Definition 5.5. A function $f: X \to \boldsymbol{R}$ is called:

1. *d-increasing* at a point a if there is a neighborhood Oa of the point a such that $f(x)$ is d-increasing on Oa;
2. *d-decreasing* at a point a if there is a neighborhood Oa of the point a such that $f(x)$ is d-decreasing on Oa;
3. *strictly d-increasing* at a point a if there is a neighborhood Oa of the point a such that $f(x)$ is strictly d-increasing on Oa;
4. *strictly d-decreasing* at a point a if there is a neighborhood Oa of the point a such that $f(x)$ is strictly d-decreasing on Oa;

If a function $f(x)$ is d-increasing (d-decreasing) at a point a for some d, then $f(x)$ is called *fuzzy increasing* (*fuzzy decreasing*) at the point a. Both classes, fuzzy increasing at a point a functions and fuzzy decreasing at a point a functions are called *fuzzy monotone* at the point a. Strictly fuzzy increasing at a point a functions and strictly fuzzy decreasing at a point a functions are called *strictly fuzzy monotone* at the point a.

Example 5.3. Functions x^2, $x \cdot \sin x$ and $x^3 \cdot \sin x$ are ε-increasing and ε-monotone at the point 0 for any $\varepsilon > 0$, while they are not increasing and not monotone at the point 0.

An important feature of local fuzzy monotonicity is that a 0-increasing at a point function is simply an increasing at this point function and a 0-decreasing at a point function is simply a decreasing at this point function.

Let k be an arbitrary non-negative real number.

Proposition 5.3. If a function $f(x)$ defined in a convex set X of real numbers is [strictly] k-increasing (k-decreasing) at any point a from X, then the function $f(x)$ is [strictly] k-increasing (k-decreasing) on X.

Proof. Let us suppose that $f(x)$ is k-increasing at any point a from a convex set X of real numbers and consider two elements $c, d \in X$. As the set X is convex, the interval $[c, d]$ is a subset of the set X. Let us also consider a system of all neighborhoods Oa for all points a from the interval $[c, d]$ such that $f(x)$ is k-increasing in each neighborhood Oa.

Interval is a compact space. Consequently (Kuratowsky, 1966), it is possible to choose a finite number of these neighborhoods which cover $[c, d]$. Let it be neighborhoods Oc_i , which correspond to the finite sequence of points $c_1 < c_2 < \dots < c_t$ in $[c, d]$. This means that $[c, d] \subseteq \bigcup_{i=1}^{t} Oc_i$. It is possible to suppose that this system of points is minimal as we can eliminate any reducible point. Because the sequence $c_1 < c_2 < \dots < c_t$ is minimal, all Oc_j are open intervals and they cover the closed interval $[c, d]$, there are points d_j that belong to the intersections $Oc_j \cap Oc_{j+1}$ for all $j = 1, 2, 3, \dots , t - 1$.

It is possible that $c = c_1$. If it is not so, $c \in Oc_1$ because $c \in \bigcup_{i=1}^{t} Oc_i$ and if $c \in Oc_j$ with $j > 1$, then $Oc_1 \subseteq Oc_j$ and it is possible to eliminate the neighborhood Oc_1 and the point c_1 contrary to our assumption that the sequence $c_1 < c_2 < \dots < c_t$ is minimal. It is also possible that $d = c_t$. Otherwise, $d \in Oc_t$ because $c \in \bigcup_{i=1}^{t} Oc_i$.

By the construction of the neighborhood Oc_1 and mononicity of $f(x)$ at c_1 , we have the inequality $f(c_1) - f(c) \geq - d(c_1 - c)$ as $c \leq c_1$ and $c \in Oc_1$. This implies $f(c) - f(c_1) \leq d(c_1 - c) = (r/10)(c_1 - c)$.

Because the sequence $c_1 < c_2 < \dots < c_t$ is minimal, all $O_\varepsilon c_j$ are open intervals and they cover the closed interval $[c, d]$, we conclude that there are points d_j that belong to the intersections $O_\varepsilon c_j \cap O_\varepsilon c_{j+1}$ for all $j = 1, 2, \dots , t - 1$.

By the construction of the neighborhoods Oc_i , by mononicity of $f(x)$ at all points c_i and as $d_i \in Oc_i$ for all $i = 1, 2, \dots , t$, we have $f(c_i) - f(d_i) \leq (r/10)(c_{i+1} - c_i)$ and as $d_i \in Oc_{i+1}$ for all $i = 1, 2, \dots , t$, we have $f(d_i) - f(c_{i+1}) \leq (r/10)(c_{i+1} - c_i)$.

Besides, $f(c_t) - f(d) \leq (r/10)(d - c_t)$ because $d \in Oc_t$ and this neighborhood Oc_t satisfies condition of monotonicity for $f(x)$.

Addition of all inequalities that we have derived gives us

$$\begin{gathered} f(c) - f(d) = (f(c) - f(c_1)) + (f(c_1) - f(d_1)) + (f(d_1) - f(c_2)) \\ + (f(c_2) - f(d_2)) + (f(d_2) - f(c_3)) + \dots \\ + (f(c_{t-1}) - f(d_{t-1})) + (f(d_{t-1}) - f(c_t)) + (f(c_t) - f(d)) = \\ f(c) - f(d) \leq k(c_1 - c) + k(d_1 - c_1) + k(c_2 - d_1) + \\ k(d_2 - c_2) + k(c_3 - d_2) + \dots \\ + k(c_t - d_{t-1}) + k(c_t - d_{t-1}) + k(d - c_t) = \\ k(c_1 - c + c_2 - c_1 + c_3 - c_2 + \dots + c_t - c_{t-1} + d - c_t) = k(d - c). \end{gathered}$$

Thus, we come to the conclusion that $(f(c) - f(d)) \leq k(d - c)$ and consequently, $f(c) \leq f(d))$ $+ k(d - c)$. As the points c and d are arbitrary, this concludes the proof that the function $f(x)$ is k-increasing on X.

Proofs for strictly increasing, decreasing and strictly decreasing functions are similar to the proof for increasing functions. Thus, the proposition is proved.

Corollary 5.8. If a function $f(x)$ defined in a convex set X of real numbers is [strictly] increasing (decreasing) at any point a from X, then the function $f(x)$ is [strictly] increasing (decreasing) on X.

Note that any convex set in $\boldsymbol{R}$ with a finite diameter is equal to one of the intervals $[c, d]$, (c, d), $(c, d]$, and $[c, d)$ where c and d are any real numbers.

Remark 5.4. Convexity of the set X is necessary for validity of Proposition 5.3 and Corollary 5.8 as the following example demonstrates.

Example 5.1. Let $X = (0, 1) \cup (2, 3)$. We define $f(x) = x$ when $0 < x < 1$ and $f(x) = x - 10$ when $2 < x < 3$. Then $f(x)$ is locally increasing at all points of X, but it is not increasing on X,

Lemmas 5.1, 5.3 and 5.4 imply the following result.

Proposition 5.4. If any extended weak derivative $b = W^{ct}d/_{dx} f(a)$ of a function $f(x)$ at a point a from X is larger (smaller) than or equal to zero, then for any $\varepsilon \in \boldsymbol{R}^{++}$, the function $f(x)$ is strictly ε-increasing (ε-decreasing) at the point a.

Lemma 5.6. If $d < c$ and a function $f(x)$ is (strictly) d-increasing (d-decreasing) on X, then the function $f(x)$ is (strictly) c-increasing (c-decreasing) on X.

Theorem 5.5. a) A function $f(x)$ is increasing on X if and only if any extended weak derivative $b = W^{ct}d/_{dx} f(a)$ is larger than or equal to zero for all $a \in X$.

b) A function $f(x)$ is decreasing on X if and only if any extended weak derivative $b =$ $W^{ct}d/_{dx} f(a)$ is less than or equal to zero for all $a \in X$.

Proof. a) *Necessity* can be proved in the same way as for differentiable increasing functions or is directly deduced from Theorem 5.3 because if the extended weak derivative b $= W^{ct}d/_{dx} f(a)$ is less than zero for some point $a \in X$, then there is a sequence of points in X such that $f(x)$ is decreasing on elements of this sequence. This implies that $f(x)$ is not an increasing function on X.

Sufficiency. Let us suppose that the condition about extended weak derivatives is valid but $f(x)$ is not an increasing function on X. Then there are such elements $c, d \in \boldsymbol{R}$, for which c $< d$ and $f(c) > f(d)$. We take the number $r = (f(c) - f(d))/ (d - c)$, put $\varepsilon = r/10$ and consider the system of all neighborhoods $O_\varepsilon a$ for all points a from the interval $[c, d]$. Neighborhoods $O_\varepsilon a$ are defined in Lemma 5.3 for the chosen $\varepsilon = r/10$. It is possible to do this because all extended weak derivatives $b = W^{ct}d/_{dx} f(a)$ are larger or equal to zero for all $a \in X$.

Interval is a compact space. Consequently (Kuratowsky, 1966), it is possible to choose a finite number of these neighborhoods which cover $[c, d]$. Let it be neighborhood $O_\varepsilon c_i$, which correspond to the finite sequence of points $c_1 < c_2 < \ldots < c_t$. This means that $[c, d] \subseteq \bigcup_{i=1}^{t}$ $O_\varepsilon c_i$. It is possible to suppose that this system of points is minimal as we can eliminate any reducible point.

It is possible that $c = c_1$. If it is not so, $c \in O_\varepsilon c_1$ because $c \in \bigcup_{i=1}^{t} O_\varepsilon c_i$ and if $c \in O_\varepsilon c_j$ with $j > 1$, then $O_\varepsilon c_1 \subseteq O_\varepsilon c_j$ and it is possible to eliminate the neighborhood $O_\varepsilon c_1$ and the point c_1, while the sequence $c_1 < c_2 < \dots < c_t$ is minimal. It is also possible that $d = c_t$. Otherwise, $d \in O_\varepsilon c_t$.

Because the sequence $c_1 < c_2 < \dots < c_t$ is minimal, all $O_\varepsilon c_j$ are open intervals and they cover the closed interval $[c, d]$, we conclude that there are points d_j that belong to the intersections $O_\varepsilon c_j \cap O_\varepsilon c_{j+1}$ for all $j = 1, 2, \dots, t - 1$.

By the construction of the neighborhood $O_\varepsilon c_1$, we have (cf. Lemma 5.3) the inequality $f(c_1) - f(c) \geq - \varepsilon(c_1 - c)$ as $c \leq c_1$. This implies $f(c) - f(c_1) \leq \varepsilon(c_1 - c) = (r/10)(c_1 - c)$.

By the construction of the neighborhoods $O_\varepsilon c_i$, mononicity of $f(x)$ at c_i and as $d_i \in O_\varepsilon c_i$ for all $i = 1, 2, \dots, t$, we have $f(c_i) - f(d_i) \leq (r/10)(c_{i+1} - c_i)$ and as $d_i \in O_\varepsilon c_{i+1}$ for all $i = 1, 2, \dots, t$, we have $f(d_i) - f(c_{i+1}) \leq (r/10)(c_{i+1} - c_i)$.

Besides, $f(c_t) - f(d) \leq (r/10)(d - c_t)$ because $d \in O_\varepsilon c_t$ and this neighborhood $O_\varepsilon c_t$ satisfies condition from Lemma 5.3.

Addition of all inequalities that we have obtained gives us

$$\begin{gathered} f(c) - f(d) = (f(c) - f(c_1)) + (f(c_1) - f(d_1)) + (f(d_1) - f(c_2)) \\ + (f(c_2) - f(d_2)) + (f(d_2) - f(c_3)) + \dots \\ + (f(c_{t-1}) - f(d_{t-1})) + (f(d_{t-1}) - f(c_t)) + (f(c_t) - f(d)) = \\ f(c) - f(d) \leq (r/10)(c_1 - c) + (r/10)(d_1 - c_1) \\ + (r/10)(c_2 - d_1) + (r/10)(d_2 - c_2) + (r/10)(c_3 - d_2) + \dots \\ + (r/10)(c_t - d_{t-1}) + (r/10)(c_t - d_{t-1}) + (r/10)(d - c_t) = \\ (r/10)(c_1 - c + c_2 - c_1 + c_3 - c_2 + \dots + c_t - c_{t-1} + d - c_t) = (r/10)(d - c). \end{gathered}$$

This bring us to the conclusion that $(f(c) - f(d)) / (d - c) \leq r/10$, while we know that $(f(c) - f(d)) / (d - c) = r$. This contradiction concludes the proof.

Proof of the part b) is similar to the proof of the part a) and is based on Lemma 5.4.

Theorem is proved.

Remark 5.3. Theorems 5.1 and 5.3 follow from Theorem 5.5, but they have simpler independent proofs.

Let a function $f(x)$ has an extended strong derivative $c = \mathrm{St}^{ct}d/_{dx} f(a)$ at all points a from X.

Corollary 5.9. A function $f(x)$ is increasing on X if and only if any extended strong derivative $d = \mathrm{St}^{ct}d/_{dx} f(a)$ is larger or equal to zero for all $a \in X$.

Theorem 5.5 and Lemmas 4.3.6 and 4.3.7 imply the following result.

Corollary 5.9. A function $f(x)$ is increasing on X if and only if for some $r \in \boldsymbol{R}^+$, all extended weak fuzzy derivatives $b = \mathrm{W}_r{}^{ct}d/_{dx} f(a)$ are larger than zero for all $a \in X$.

Let for some $r \in \boldsymbol{R}^+$, a function $f(x)$ has an extended strong r-derivative $c = \mathrm{St}_r{}^{ct}d/_{dx} f(a)$ at all points a from X.

Corollary 5.10. A function $f(x)$ is increasing on X if and only if all extended strong r-derivatives $b = \mathrm{St}_r{}^{ct}d/_{dx} f(a)$ are larger than zero for all $a \in X$.

Corollary 5.11. A function $f(x)$ is increasing at a point a if and only if any extended weak derivative $b = W^{ct}d/_{dx}f(a)$ is larger than zero.

Definition 5.5. A point a is called a *weak critical point* of a function $f(x)$ if $0 = W^{ct}d/_{dx}$ $f(a)$.

Definition 5.5. A point a is called a *semi-critical point* of the function $f(x)$ if all extended weak centered derivatives $W^{ct}d/_{dx} f(a)$ of $f(x)$ at a are equal to zero.

Lemma 5.7. If a function $f(x)$ is continuous at a and a is a semi-critical point of the function $f(x)$, then a is a critical point of the function $f(x)$.

Theorem 5.6. A function $f(x)$ is strictly increasing on X if and only if:

(1) any extended weak derivative $b = W^{ct}d/_{dx} f(a)$ is larger or equal to zero for all $a \in X$;

and

(2) X does not contain a closed interval of weak critical points of $f(x)$, i.e., an interval at each point x of which zero is an extended weak derivative $W^{ct}d/_{dx}f(x)$ of the function $f(x)$ at x.

Proof. Necessity of the condition (1) follows from Theorem 5.5 as any strictly increasing function is increasing. At the same time, violation of the condition (2) means that there are many points in X at which $f(x)$ has the same value. In this case, $f(x)$ is not strictly increasing.

Sufficiency. Let us suppose that conditions (1) and (2) are valid but $f(x)$ is not a strictly increasing function on X. By Theorem 5.5, $f(x)$ is an increasing function on X. If $f(x)$ is not strictly increasing, then there are two different points $c, d \in X$ for which $f(c) = f(d)$ because Condition (1) implies $f(c) \leq f(d)$.

Let us consider an arbitrary point $a \in X$, for which $c < a < d$ is valid. If $f(a) > f(c) = f(d)$, then $f(d) - f(a) < 0$ and $f(x)$ is not an increasing function. If $f(a) < f(c) = f(d)$, then $f(a) - f(c) < 0$ and $f(x)$ is not an increasing function. Thus, we have only one option for the value $f(a)$, namely, $f(a) = f(c) = f(d)$. As a is an arbitrary point from the interval $[c, d]$, $f(x)$ is a constant function on this interval. This invalidates the condition (2), and concludes the proof of the theorem

Let a function $f(x)$ has an extended strong derivative $St^{ct}d/_{dx} f(a)$ exists at all points a from X.

Corollary 5.12. A function $f(x)$ is strictly increasing on X if and only if:

(1) any extended strong derivative $d = St^{ct}d/_{dx} f(a)$ is larger or equal to zero for all $a \in X$;

and

(2) X does not contain a closed interval of critical points of the function $f(x)$.

Corollary 5.13. A function $f(x)$ is strictly increasing on X if and only if:

(1) any extended weak derivative $b = W^{ct}d/_{dx} f(a)$ is larger or equal to zero for all $a \in X$;

and

(2) X does not contain a closed interval of semi-critical points of the function $f(x)$, i.e., an interval on which all extended weak derivative $W^{ct}d/_{dx} f(x)$ are identically equal to zero.

Corollary 5.14. A function $f(x)$ is strictly increasing at a point a if and only if any extended weak derivative $b = W^{ct}d/_{dx} f(a)$ is larger than zero and there does not exist an interval of critical points of the function $f(x)$ such that it contains a.

Let a function $f(x)$ has an extended strong derivative $St^{ct}d/_{dx} f(a)$ at all points a from X.

Corollary 5.15. A function $f(x)$ is decreasing on X if and only if any extended strong derivative $d = St^{ct}d/_{dx} f(a)$ is smaller than or equal to zero for all $a \in X$.

Corollary 5.16. A function $f(x)$ is decreasing on X if and only if for some $r \in \boldsymbol{R}^+$, all extended weak fuzzy derivatives $b = W_r{}^{ct}/_{dx} f(a)$ are smaller than or equal to zero for all $a \in X$.

Let for some $r \in \boldsymbol{R}^+$, a function $f(x)$ has an extended strong r-derivative $St_r{}^{ct}/_{dx} f(a)$ at all points a from X.

Corollary 5.17. A function $f(x)$ is decreasing on X if and only if all extended strong r-derivatives $b = St_r{}^{ct}/_{dx} f(a)$ are smaller than or equal to zero for all $a \in X$.

Corollary 5.18. A function $f(x)$ is decreasing at a point a if and only if any extended weak derivative $b = W^{ct}/_{dx} f(a)$ is smaller than or equal to zero.

Lemma 5.8. A function $f(x)$ is increasing if and only if it is strictly d-increasing for any $d > 0$ and is decreasing if and only if $f(x)$ is strictly d-decreasing for any $d > 0$.

This and Theorems 5.5 and 5.6 give us the following result.

Theorem 5.7. A function $f(x)$ is strictly d-increasing (d-decreasing) on X for all $d > 0$ if and only if any extended weak derivative $b = W^{ct}/_{dx} f(a)$ is larger (less) than or equal to zero for all $a \in X$.

Corollary 5.19. A differentiable function $f(x)$ is strictly d-increasing (d-decreasing) on X for all $d > 0$ if and only if its derivative is larger (less) than or equal to zero for all $a \in X$.

Theorem 5.8. A function $f(x)$ is strictly decreasing on X if and only if:

(1) any extended weak derivative $b = W^{ct}d/_{dx} f(a)$ is smaller or equal to zero for all $a \in X$;

and

(2) X does not contain a closed interval of weak critical points of $f(x)$, i.e., an interval on which the extended weak derivative $W^{ct}d/_{dx} f(x)$ is identically equal to zero.

Proof is similar to the proof of Theorem 5.6.

Let a function $f(x)$ has an extended strong derivative $St^{ct}/_{dx} f(a)$ at all points a from X.

Corollary 5.20. A function $f(x)$ is strictly decreasing on X if and only if:

(1) any extended strong derivative $d = St^{ct}/_{dx} f(a)$ is smaller or equal to zero for all $a \in X$;

and

(2) X does not contain a closed interval of critical points of the function $f(x)$.

Corollary 5.21. A function $f(x)$ is strictly decreasing on X if and only if:

(1) any extended weak derivative $b = \mathrm{W}^{\mathrm{ct}}/_{\mathrm{d}x} f(a)$ is smaller than or equal to zero for all $a \in X$;

and

(2) X does not contain a closed interval of semi-critical points of the function $f(x)$, i.e., an interval on which all extended weak derivative $\mathrm{W}^{\mathrm{ct}}/_{\mathrm{d}x} f(x)$ are identically equal to zero.

Corollary 5.22. A function $f(x)$ is strictly decreasing at a point a if and only if any extended weak derivative $b = \mathrm{W}^{\mathrm{ct}}/_{\mathrm{d}x} f(a)$ is smaller than or equal to zero and there does not exist an interval of critical points of the function $f(x)$ such that it contains a.

Let us consider an r-differentiable at a point a function $f(x)$.

Lemma 5.9. For any r-derivative $v = \mathrm{St}_r{}^{\mathrm{ct}}/_{\mathrm{d}x} f(a)$, we have $v > k$, then any extended weak derivative $v = \mathrm{W}^{\mathrm{ct}}/_{\mathrm{d}x} f(a)$ is larger than k.

Let $r, k, h \in \boldsymbol{R}^{+}$.

Theorem 5.9. a) If a function $f(x)$ is r-differentiable in (a, b) and for any point u from (a, b), $v = \mathrm{St}_r{}^{\mathrm{ct}}\mathrm{d}/_{\mathrm{d}x} f(u)$ implies $v > -k$, then $f(x)$ is k-increasing in (a, b).

b) If a function $f(x)$ is r-differentiable in (a, b) and for any point u from (a, b), $v = \mathrm{St}_r{}^{\mathrm{ct}}/_{\mathrm{d}x} f(u)$ implies $v < h$, then $f(x)$ is h-decreasing in (a, b).

Proof. Let us consider the function $g(x) = f(x) + kx$. Then $\mathrm{St}_r{}^{\mathrm{ct}}/_{\mathrm{d}x}\, g(u) = \mathrm{St}_r{}^{\mathrm{ct}}/_{\mathrm{d}x} f(u) + k$ for any point u from (a, b). Thus, $\mathrm{St}_r{}^{\mathrm{ct}}/_{\mathrm{d}x}\, g(u) > 0$ for all points u from (a, b). Then by Lemma 5.9, any extended weak derivative $v = \mathrm{W}_r{}^{\mathrm{ct}}/_{\mathrm{d}x} f(a)$ is larger than 0. By Theorem 5.5, $g(x)$ is an increasing function in (a, b), i.e., for any points c and d from (a, b), if $c < d$, then $g(c) \leq g(d)$. This gives the inequality $f(c) + kc \leq f(d) + kd$. Consequently, $f(c) \leq f(d) + k(c - d)$. It means that $f(x)$ is k-increasing in (a, b) and part (a) is proved.

Proof of part (b) is similar.

Chapter 6

FUZZY MAXIMA AND MINIMA OF REAL FUNCTIONS

The true test of a first-rate mind is the ability
to hold two contradictory ideas at the same time.
F. Scott Fitzgerald (1896-1940)

In this chapter, we apply neoclassical analysis to problems of optimization, with the emphasis on the mathematical context of these problems. Fuzzy derivatives are used to a study of maxima and minima, as well as fuzzy maxima and minima of real functions. When we do not make a distinction between a maximum and minimun or when we want to comprise them in a common class, we call it an extremum. Fuzzy maxima (minima) are points where function values are larger (respectively, smaller) up to some constant r than all other values. These, larger, values are called r-maxima (respectively, r-minima) of the corresponding function. As in the case of conventional maxima and minima, there are local and global r-maxima and r-minima.

There are several reasons to study fuzzy extrema. In some cases, a maximum or minimum of a function exists, but is not achievable in a constructive way (for example, it an irrational number). It makes necessary to find a good approximation to such maximum or minimum. This approximation is a kind of a fuzzy maximum or fuzzy minimum. In addition, there are situations when exact maximum or minimum of a function does not exist, but it is possible to find a point where the function value plus some constant r is larger (respectively, smaller) than all other values. For instance, let us consider the function $f(x)$ that is equal to x in the interval (0, 1) and equal zero in all other points. This function does not have a maximum, but for any r, it has many r-maxima.

We consider here criteria of extrema for conventional, crisp functions, taking into account uncertainties that emerge from limit operations. Different conditions for maxima and minima of real functions are obtained. Some of them are the same or at least similar to the conditions for the differentiable functions, while others differ in many aspects from those for the standard differentiable functions. Many classical results are obtained as direct corollaries of propositions for fuzzy derivatives, which are proved in this section. Such results as the Fuzzy Fermat theorem, Fuzzy Rolle theorem, and Fuzzy Mean Value theorem are proved. These results provide sound theorctical base for computational methods of optimization that achieves a better correlation between mathematical analysis and numerical analysis than traditional methods. In addition, application of fuzzy derivatives allows one to utilize analytical methods of calculus in a much more voluminous extent of situations. The classical Fermat theorem, Rolle theorem, and Mean Value theorem become direct corollaries of the new theorems obtained here.

6.1. MAXIMA AND MINIMA OF FUNCTIONS

For since the fabric of the universe is most perfect
and the work of a most wise Creator,
nothing at all takes place in the universe in which some
rule of maximum or minimum does not appear.
Leonard Euler (1707–1783)

Maxima and minima of functions are very important. Many problems that we encounter in our life are optimization problems when it is necessary to find a maximum or minimum of some function or, at least, points that are close to such maximum or minimum.

Let $X, Y \subseteq \boldsymbol{R}, f: X \rightarrow Y$ be a real function, $b \in \boldsymbol{R}$, and $r \in \boldsymbol{R}^{+}$.

Definition 6.1.1. A point b (point a) is called a *point of a global*, or *absolute*, *maximum* (correspondingly, *minimum*) of the function $f(x)$ if for any point $x \in X$, we have $f(b) \geq f(x)$ (correspondingly, $f(x) \geq f(a)$), while the value $f(b)$ (correspondingly, value $f(a)$) is called a *global*, or *absolute*, *maximum* (correspondingly, *minimum*) [or *r-maximum* (*r-minimum*) *value*] of the function $f(x)$.

It is denoted by $f(b) = \max \{ f(x); x \in X \}$ and $f(a) = \min\{ f(x); x \in X \}$.

Example 6.1.1. The point 0 is a point of a global minimum of the function $f(x) = x^2$.

Example 6.1.2. The point 0 is a point of a global minimum and 1 is a point of a global maximum of the function $f(x) = x$ in the interval [0, 1].

Definition 6.1.2. A point b is called a *point of a local*, or *relative*, *maximum* (*minimum*) of the function $f(x)$ if there is a neighborhood Ob of the point b such that for any point $x \in Ob$, we have $f(b) \geq f(x)$ ($f(x) \geq f(b)$), while the value $f(b)$ is called a *local*, or *relative*, *maximum* (*minimum*) [or *maximum* (*minimum*) *value*] of the function $f(x)$.

Note that a neighborhood of a point a in a real line $\boldsymbol{R}$ is an interval (open in the case of an open neighborhood and closed in the case of a closed neighborhood) interval that contains a.

Example 6.1.3. The point 1/3 is a point of a global minimum and -1/3 is a point of a global maximum of the function $f(x) = x^3 - x$.

Maxima and minima are called by the common name *extrema* of functions. Points of (global and local) maxima and minima are called *points of* (global and local) *extrema* or *critical points* of functions.

Definition 6.1.3. A point a is called a *critical point* of a function $f(x)$ if the derivative of $f(x)$ at a is equal to 0 or does not exist.

The classical result of the calculus is the following *Fermat theorem*.

Theorem 6.1.1 (Fermat theorem). If a function $f(x)$ is differentiable on an open interval (a, b) and a is a point of a local maximum or minimum of $f(x)$, then $f'(a) = 0$, i.e., a is a critical point.

Here we do not give a proof, as this result is a direct corollary of its fuzzy counterpart proved in Section 6.3.

Let $X \subseteq \boldsymbol{R}$ be the domain of a function $f(x)$ and $a \in X$.

Definition 6.1.4. The *optimality domain* DO($f(a)$) of the value $f(a)$ is the largest interval $I \subseteq X$ such that $a \in I$ and $f(a)$ is either a global maximum or global minimum of the function $f(x)$ in I.

Proposition 6.1.1. Any value $f(a)$ with $a \in X$ has the optimality domain DO($f(a)$).

Proof. For the value $f(a)$, there are two options: either is a local optimum (that is, maximum or minimum) or not. In the second case, DO($f(a)$) = $[a, a]$. In the first case, let us assume that $f(a)$ is a local maximum. When $f(a)$ is a local minimum, all considerations are the same.

Then there is an open interval (d, b) such that $a \in (d, b)$ and $f(a) \geq f(x)$ for all $x \in (d, b)$. This shows that if we have an increasing sequence of open intervals such that $f(a)$ is a local maximum in each of them, then the union of these intervals is an open interval and $f(a)$ is a local maximum in it. Thus, there is a maximal interval with these properties. Let it be (d, b), $(d, b]$, $[d, b)$ or $[d, b]$. We consider the case when it is (d, b) as all other cases are similar and treated in the same way. Here maximal means that (d, b) does not belong to any other interval with these properties. We show that (d, b) is the largest such interval.

Indeed, if there is another interval (c, e) that is not included in (d, b) and $f(a)$ is a local maximum in it, then the union $(d, b) \cup (c, e)$ is an interval because the point a belongs to both intervals (d, b) and (c, e). Besides, this new interval is larger than (d, b) and $f(a)$ is a local maximum in it. This contradict maximality of the interval (d, b) and shows that (d, b) is the largest and thus, it is the optimality domain DO($f(d)$) of a value $f(d)$.

Proposition is proved.

Two more classical results of the calculus are the *Rolle Theorem* and *Mean Value Theorem*.

Theorem 6.1.2 (the Rolle Theorem). If a function $f(x)$ is differentiable on (a, b), continuous on $[a, b]$ and $f(a) = f(b)$, then there is [at least one] point $c \in [a, b]$ such that $f'(a) = 0$.

Theorem 6.1.3 (the Mean Value Theorem). If a function $f(x)$ is differentiable on (a, b) and continuous on $[a, b]$, then there is [at least one] point $c \in [a, b]$ such that $f'(a) = (f(b) - f(a)) / (b - a)$.

Here we do not give proofs of these theorems, as their results are direct corollaries of their fuzzy counterparts proved in Section 6.3.

We know that any differentiable function is continuous (cf. Section 4.1). The Mean Value Theorem allows us to obtain a stronger result.

Let $f(x)$ be a differentiable function in an interval $[a, b]$.

Theorem 6.1.4. If $f'(x) \leq k$ in $[a, b]$, then $f(x)$ is uniformly (r, kr)-continuous in $[a, b]$ for any $r \in \boldsymbol{R}^+$.

Proof. Let us take two points c and d in $[a, b]$ such that $0 < d - c < r$. As $f(x)$ is a differentiable function in the interval $[a, b]$, it is differentiable in the interval $[c, d]$ and by the Mean Value Theorem, there is a point e such that $c \leq e \leq d$ and $((f(d) - f(c))/(d - c)) = f'(e)$. Thus, we have

$$f(d) - f(c) = (d - c) \cdot f'(e) \leq k(d - c) < kr$$

Theorem is proved as c and d are arbitrary points in $[a, b]$ with $0 < d - c < r$.

Corollary 6.1.1. If $f'(x) \leq k$ in $[a, b]$, then $f(x)$ is (r, kr)-continuous in $[a, b]$ for any $r \in \boldsymbol{R}^+$.

If we put r equal to 0, we see that Theorem 6.1.4 implies that any differentiable in a closed interval function is uniformly continuous because any uniformly (0, 0)-continuous function is uniformly continuous (cf. Section 3.5) and Corollary 6.1.1 implies that any differentiable function is continuous because any (0, 0)-continuous function is continuous (cf. Sections 3.3 and 3.4).

At the same time, not every continuous function is (r, kr)-continuous for given k and r. For instance, the function $f(x) = 100x$ is continuous but it is not (r, kr)-continuous for $k = 5$ and $r = 1$.

6.2. Fuzzy Maxima and Minima of Functions

Improvization certainly is the touch-stone of spirit.
Jean Baptiste Moliere (1622-1673)

Let $X, Y \subseteq \boldsymbol{R}$, $f: X \rightarrow Y$ be a real function, $b \in \boldsymbol{R}$, and $r \in \boldsymbol{R}^+$.

Definition 6.2.1. A point b (point a) is called a *point of a global* (also called *absolute*) *r-maximum* (*r-minimum*) of the function $f(x)$ if for any point $x \in X$, we have $f(b) + r \geq f(x)$ ($f(x) + r \geq f(a)$), while the value $f(b)$ (value $f(a)$) is called a *global*, or *absolute*, *r-maximum* (*r-minimum*) [or *r-maximum* (*r-minimum*) *value*] of the function $f(x)$.

It is denoted by $f(b) = r\text{-max}\ \{ f(x); x \in X \}$ and $f(a) = r\text{-min}\{ f(x); x \in X \}$.

Example 6.2.1. Let us consider the following function:

$$f(x) = \begin{cases} x^2 & \text{when } 0 \leq x < 1; \\ 0 & \text{otherwise.} \end{cases}$$

Then 0.9 is a point of a global 0.2-maximum of $f(x)$ and 0.99 is a point of a global 0.02-maximum of $f(x)$.

Example 6.2.2. Let us consider the following function:

$$f(x) = \begin{cases} x^2 & \text{when } 0 < x \leq 1; \\ 1 & \text{otherwise.} \end{cases}$$

Then 0.1 is a point of a global 0.01-minimum of $f(x)$ and 0.01 is a point of a global 0.0001-minimum of $f(x)$.

Example 6.2.3. Let us consider the following functions:

$$g(x) = \begin{cases} x & \text{when } 0 \leq x \leq 1; \\ 2 - x & \text{when } 1 \leq x \leq 2. \end{cases}$$

and the function f: $[0, \infty] \rightarrow [0, 2]$ that is defined by the formula $f(x) = (2 - (1/n))$ for all $n = 1$, 2, 3, ... when $2n \leq x \leq 2(n + 1)$ and $f(x) = g(x)$ when $0 \leq x \leq 2$.

Then 101 is a point of a global 0.02-maximum of $f(x)$ and 201 is a point of a global 0.01-maximum of $f(x)$ because $2 - (1/100) > 2 - (1/n) + (1/50)$ and $2 - (1/200) > 2 - (1/n) + (1/100)$ for any $n = 1, 2, 3, \ldots$

Example 6.2.4. Let us take the function $f(x) = \sin x$ with $x \in [0, \pi]$, i.e., f maps the interval $[0, \pi]$ into the interval $[0, \pi]$. We know that $\pi/2$ is the point of the global maximum of $f(x)$. However, computers and calculators allow us to consider only rational points. So, in computations, it is necessary to take some rational points as points of maxima and minima. For simplicity, we assume that it is possible to consider only points $(0.1)k$ with $k = 0, 1, 2, \ldots, 31$ that belong to the interval $[0, \pi]$. These points do not contain the point of the global maximum of $f(x)$. However, 1.6 is the point of a global 0.0005-maximum of $f(x)$, while 1.5 and 1.6 are points of a global 0.003-maximum of $f(x)$.

In the considered case, the discretization step is equal to 0.1. Computers and modern calculators have much smaller discretization step than 0.1, but it always exists. As a consequence, in many cases of computations, it is possible to consider only points of fuzzy extrema.

Remark 6.2.1. In Example 6.2.4, the point of maximum exists. So, it is possible to consider points 1.6 or 1.5 as approximations to this point. However, when the values of a function are obtained by measurement, it is possible that the function does not have the maximum (minimum) and we can also find only points of fuzzy maximum (minimum).

Definition 6.2.1 implies the following results.

Lemma 6.2.1. A point b is a point of a global maximum (minimum) of the function $f(x)$ if and only if b is a point of a global 0-maximum (0-minimum) of the function $f(x)$.

Lemma 6.2.2. If $r \leq p$, then any global r-maximum (r-minimum) of the function $f(x)$.is a global p-maximum (p-minimum) of the function $f(x)$.

Proposition 6.2.1. If $|f(x)| < r$ for all points $x \in X$, then any point $a \in X$ is a point of a global r-maximum and a point of a global r-minimum of the function $f(x)$.

Remark 6.2.2. In general, the concepts of a global r-maximum and r-minimum have sense only for sufficiently small values of the number $r \in \boldsymbol{R}^+$ as Proposition 6.2.1 demonstrates. However, in most cases, we may achieve exact extrema only in the realm of abstract mathematics with its idealized real numbers. In computations, we, as a rule, can get only r-maximum or r-minimum for some r because when a global extremum is an irrational number or even a very small rational number, numerical computations cannot reach this number. Analogous computations also cannot give the exact value of an extremum due to the noise in signal transmission and the finite precision of any measurement as measurement is necessary to display the result of an analogous computation.

Remark 6.2.3. Proposition 6.2.1 also demonstrates that if there no restriction on the number r, the concepts of a local r-maximum and r-minimum, in some theoretical (but not practical) sense, degenerate for continuous functions. Really, if we take a continuous function $f(x)$ and some number $r \in \boldsymbol{R}^+$, then in any sufficiently small neighborhood any continuous function is bounded above and below by r. As a result, any number becomes a local r-maximum and r-minimum of the function $f(x)$. However, when the number r is small in comparison with a given neighborhood, these concepts make sense and are important even for continuous functions.

Definition 6.2.2. A point b is called a *point of a local*, or *relative*, *r-maximum* (*r-minimum*) of a function $f(x)$ if there is a neighborhood Ob of the point b such that for any point $x \in$ Ob, we have $f(b) + r \geq f(x)$ ($f(x) + r \geq f(b)$), while the value $f(b)$ is called a *local*, or *relative*, *r-maximum* (*r-minimum*) [or *r-maximum* (*r-minimum*) *value*] of the function $f(x)$.

Example 6.2.5. Let us consider the following function:

$$f(x) = \begin{cases} x^2 & \text{when } 0 \leq x < 1; \\ x - 1 & \text{otherwise.} \end{cases}$$

Then 0.9 is a point of a local 0.2-maximum of f and 0.99 is a point of a local 0.02-maximum of f.

Example 6.2.6. Let us take the function $f(x) = \sin x$ with arbitrary x from $\boldsymbol{R}$. We know that $\pi/2$ is the point of the local maximum of $f(x)$. However, computers and calculators allow us to consider only rational points. So, in computations, it is necessary to take some rational points as points of maxima and minima. For simplicity, we assume that it is possible to consider only points $(0.1)k$ with $k = 0, 1, 2, \ldots$. These points do not contain the point of local maxima and minima of the function $f(x)$. However, 1.6 is the point of a local 0.0005-maximum of $f(x)$, while 1.5 and 1.6 are points local 0.003-maximum of $f(x)$.

Lemma 6.2.3. Any point of a global r-maximum (r-minimum) of a function $f(x)$ is a point of a local r-maximum (r-minimum) of the same function.

Lemma 6.2.4. A point b is a point of a local maximum (minimum) of a function $f(x)$ if and only if b is a point of a local 0-maximum (0-minimum) of the function $f(x)$.

Lemma 6.2.5. If $r \leq p$, then any point of a local r-maximum (r-minimum) of a function $f(x)$ is a point of a local p-maximum (p-minimum) of the function $f(x)$.

Proposition 6.2.2. If d is a local r-maximum (r-minimum) of $f(x)$ in (a, b), z is a point from (a, b) and $| f(z) - f(d)| < q$, then z is a local $(r + q)$-maximum (respectively, $(r + q)$-minimum) of $f(x)$ in (a, b).

Proof. At first, we consider the case of a local r-maximum. As d is a local r-maximum of $f(x)$ in (a, b), then there is a neighborhood Od of the point d such that for any point $x \in$ Od, we have $f(d) + r \geq f(x)$. Then $f(z) + r + q \geq f(d) + r \geq f(x)$ for any point $x \in$ Od. Thus, by Definition 6.2.2, the point z is a local $(r + q)$-maximum (respectively, $(r + q)$-minimum) of $f(x)$ in (a, b).

Proposition is proved.

In such a way, we come to the concepts of relative local r-maximum and r-minimum of a function, as well as to the concepts of points of a relative local r-maximum and relative local r-minimum of a function.

Let $k, r, q \in \boldsymbol{R}^{++}$, $b \in X$, and assume that h: $\boldsymbol{R}^{+} \rightarrow \boldsymbol{R}^{+}$ is some function.

Definition 6.2.3. A point b is called a *point of a relative* [*relative with respect to h*] *local k-maximum* (*k-minimum*) of the function $f(x)$ if there is a number $p \in \boldsymbol{R}^{++}$ such that for any number $0 < q < p$, there is a number $r \in \boldsymbol{R}^{++}$ such that for any point $x \in (b - q, b + q)$, we have $f(b) + r \geq f(x)$ ($f(x) + r \geq f(b)$) and $(r/q) < k$ [$(r/q) < h(k)$], while the value $f(b)$ is called a *relative* [*relative with respect to h*] *local k-maximum* (*k-minimum*) of the function $f(x)$.

The latter condition from Definition 6.2.3 implies that the rate of the function value convergence to an extremum is not less than the rate of the argument convergence or in the case that is relative with respect a function h, the rate of the function value convergence to an extremum is not less than the value of some function of the rate of the argument convergence. For example, if arguments of the function $f(x)$ converge exponentially, then the values of the function $f(x)$ have to converge, at least exponentially. This condition is related to the characteristic property of asymptotic series (cf., for example, (Wong, 1989)).

Theorem 6.2.1. Any bounded function (theoretically) reaches its global r-maximum and r-minimum for any $r \in \boldsymbol{R}^{++}$.

Proof. Let $f: X \to \boldsymbol{R}$ be a bounded function. Then there are such numbers u and v that $u = \sup \{ f(x) \, ; x \in X \}$ and $v = \inf \{ f(x) \, ; x \in X \}$. If these numbers belong to the range of $f(x)$, then u is a global maximum and v is a global minimum of $f(x)$. Then by Lemmas 6.2.1 and 6.2.2, u is a global r-maximum and v is a global r-minimum of f for any $r \in \boldsymbol{R}^{++}$. Thus, if both numbers u and v belong to the range of $f(x)$, then the statement of the theorem is true.

Let us consider the case when numbers u and v do not belong to the range of $f(x)$ and take an arbitrary number $r \in \boldsymbol{R}^{++}$. Then by the definitions of supremum and infimum, there are such points a and b from X that $u - f(a) < r$ and $f(b) - v < r$. As $u \geq f(x)$ and $f(x) \geq v$ for any $x \in X$, we have $f(a) + r \geq f(x)$ and $f(x) + r \geq f(b)$ for any $x \in X$. Consequently, u is a global r-maximum and v is a global r-minimum of $f(x)$.

The case when one of two numbers u and v does not belong to the range of $f(x)$ and the second belongs is considered in a similar way.

Theorem is proved.

This result allows us to get a new criterion of boundedness.

Corollary 6.2.1. A function $f(x)$ is bounded on X if and only if $f(x)$ reaches its global r-maximum and r-minimum for any $r \in \boldsymbol{R}^{++}$.

Proof. Sufficiency is proved in Theorem 6.2.1, and we need to prove only necessity.

Necessity. Let us suppose that $f(x)$ reaches its global r-maximum and r-minimum for any $r \in \boldsymbol{R}^{++}$ and consider points a and b from X, for which a is a global 1-maximum and b is a global 1-minimum of $f(x)$. Then by Definition 6.2.1, $f(a) + 1 \geq f(x)$ and $f(x) + 1 \geq f(b)$ for all x from X. Consequently, the values of $f(x)$ on X are bounded by the number $f(a) + 1$ from above and by the number $f(b) - 1$ from below. Thus, $f(x)$ is bounded on X.

Corollary is proved.

It is proved (Section 3.2) that any fuzzy continuous in a closed interval function is bounded. Consequently, Theorem 6.2.1 implies the following result.

Theorem 6.2.2 (the Second Fuzzy Weierstrass Theorem). Any fuzzy continuous in a closed interval function $f(x)$ reaches its global r-maximum and r-minimum for any $r \in \boldsymbol{R}^{++}$.

For continuous functions, we have a stronger result - the Second Weierstrass Theorem (cf. Section 3.1.2).

Let $f(x)$ be a continuous function, $r \geq 0$ and $a \in I \subseteq X$.

Lemma 6.2.6. If $f(a)$ is a global r-maximum (r-minimum) in I, then $f(a)$ is a global r-maximum (r-minimum) in the closure of I in X.

Indeed, if J is the closure of I in $\boldsymbol{R}$ and $c \in J$, then (cf. Appendix C and (Kelly, 1957)) topology of the real line $\boldsymbol{R}$ implies that $c = \lim_{i\to\infty} c_i$ for some sequence $l = \{a_i \in I; i = 1, 2, 3, \ldots\}$. Besides, the closure K of I in X is the intersection of J and X. Thus, if $f(a) + r \geq f(c_i)$ for all $i = 1, 2, 3, \ldots$ and c belongs to the closure of I in X, $c = \lim_{i\to\infty} c_i \in X$, then $c = \lim_{i\to\infty} c_i$ and $f(a) + r \geq f(c)$, i.e., $f(a)$ is a global r-maximum of $f(x)$ in the closure of I in X as c is an arbitrary point from the closure of I in X.

The case for a global r-minimum of $f(x)$ in the closure of I in X. is treated in the same way.

Lemma is proved.

Corollary 6.2.2. If $f(a)$ is a global maximum (minimum) in I, then $f(a)$ is a global maximum (minimum) in the closure of I in X.

Lemma 6.2.6 also implies the following result.

Corollary 6.2.3. If $f(x)$ is a continuous function, then the optimality domain DO($f(a)$)) of a value $f(a)$ is closed in X.

Remark 6.2.4. If $f(x)$ is not continuous, then Corollary 6.2.3 and Lemma 6.2.2 can be incorrect as the following example demonstrates.

Example 6.2.7. Let us consider the following function (cf. Fig. 6.1):

$$f(x) = \begin{cases} 1 - x^2 \text{ when } -1 < x < 1; \\ |x| + 1 \text{ when } x \leq -1 \text{ and } 1 \leq x. \end{cases}$$

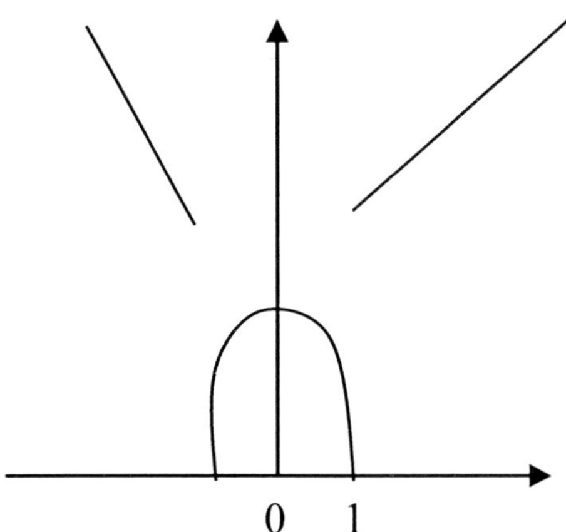

Figure 6.1. The graph of the function $f(x)$.

In this case, DO($f(0)$)) of a value $f(0)$ is equal to (-1, 1).

Definition 6.2.4. The *r-optimality domain* rDO($f(a)$)) of the value $f(a)$ is the largest interval $I \subseteq X$ such that $a \in I$ and $f(a)$ is either a global r-maximum or global r-minimum in I.

Proposition 6.1.1 implies the following result.

Theorem 6.2.3. a) Any value $f(z)$ where $z \in X$ has the r-optimality domain rDO($f(z)$)) for any $r \geq 0$.

b) When z is an inner point of X, then the length of rDO($f(z)$)) is larger than 0 for some $r \geq 0$ if and only if $f(x)$ is fuzzy continuous at the point z.

c) If $f(x)$ is a continuous function and z is an inner point of X, then the length of rDO($f(z)$)) is larger than 0 for any $r > 0$.

Corollary 6.2.4. rDO($f(a)$)) $\subseteq (r + q)$DO($f(z)$)) if $|f(z) - f(a)| < q$.

When we have two even closely situated functions, it is impossible to judge where are extremums of one of them by extremums of the other.

Example 6.2.8. Let us consider functions $f(x) = 0.001\sin x$ and $g(x) = 0$ for all values of x. These function are very close to one another in the natural metric in the function space **d**(f,

g) = sup {| $f(x)$ - $g(x)$|; $x \in X$ }. Actually, **d**(f, g) = 0.001. At the same time, the function $f(x)$ has maxima only at points $\pi/2 + 2k\pi$, while any real number is a point of maximum for the function $g(x)$.

The advantage of fuzzy maxima and minima is that they allow us to find relations between maxima and minima of different functions. This is especially useful when a function is approximated by another one.

Let us consider functions $f(x)$ and $g(x)$ defined in a set $X \subseteq \boldsymbol{R}$ and the number $k = \sup \{ | f(x) - g(x) | ; x \in X \}$.

Theorem 6.2.4. a) If $b = r\text{-max}\{g(x); x \in X\}$, then $b = (r + 2k)\text{-max}\{ f(x); x \in X\}$ and any point of global r-maximum of $g(x)$ in X is a point of global $(r + 2k)$-maximum of $f(x)$ in X.

b) If $a = r\text{-min}\{g(x); x \in X\}$, then $a = (r + 2k)\text{-min}\{ f(x); x \in X\}$ and any point of global r-minimum of $g(x)$ in X is a point of global $(r + 2k)$-minimum of $f(x)$ in X.

Proof. Let $b = r\text{-max}\{g(x); x \in X\}$, $g(u) = b$, and $c \in X$. Then

$$g(u) + r \geq g(c) \text{ or } | g(u) - g(c)| \leq r.$$

This implies

$$| f(u) - f(c)| = | f(u) + g(u) - g(u) + g(c) - g(c) - f(c)| \leq$$
$$| f(u) + g(u)| + | g(c) - g(u)| + | - g(c) - f(c)| \leq k + r + k$$

Consequently, $| f(u) - f(c)| \leq r + 2k$ and $f(u) + r + 2k \geq f(c)$. It means that $b = (r + 2k)\text{-max}\{ f(x); x \in X\}$.

Statement b) is proved in a similar way.

Corollary 6.2.3. $\max\{ g(x); x \in X\} = k\text{-max}\{ f(x); x \in X\}$ and $\min\{ g(x); x \in X\} = k\text{-min}\{ f(x); x \in X\}$.

Remark 6.2.5. Numbers r and $2k$ give exact estimate for fuzzy maximums of the function $f(x)$ as the following example demonstrates.

Example 6.2.8. Let us consider the following functions defined in the interval [0, 4]

$$g(x) = \begin{cases} x & \text{when } 0 < x \leq 1; \\ 2 - x & \text{when } 1 < x \leq 2; \\ x - 2 & \text{when } 2 < x \leq 3; \\ 4 - x & \text{when } 3 < x \leq 4 \end{cases}$$

and

$$f(x) = \begin{cases} 3x - 1 & \text{when } 0 < x \leq 1; \\ 5 - 3x & \text{when } 1 < x \leq 2; \\ 3x - 7 & \text{when } 2 < x \leq 3; \\ 11 - 3x & \text{when } 3 < x \leq 4 \end{cases}$$

Then $k = \sup \{ \mid f(x) - g(x)| ; x \in X \} = 1$. At the same time, 1 is a point of the global 1-maximum of $g(x)$ as $g(x) = 0$, $3 = 1 + 2k$, and 1 is the global maximum of $g(x)$, while 1 is a point of the global 3-maximum of $f(x)$ and it is not the global r-maximum of $f(x)$ for any $r < 3$.

6.3. CRITERIA FOR FUZZY MAXIMA AND MINIMA

The greatest obstacle in finding the truth
is not the lie itself, but
that which seems to be the truth.
Arthur Schopenhauer (1788-1860)

Strict and fuzzy maxima and minima of functions are related to strict and fuzzy critical points.

Definition 6.3.1. A point a is called an *r-critical point* of the function $f(x)$ if $0 = \text{st}_r{}^t\text{d}/_{\text{d}x} f(a)$.

Example 6.3.1. Let us consider the function $|x|$. We have that 0 is a 1-critical point of this function as $0 = \text{st}_1{}^t\text{d}/_{\text{d}x} |x|(0)$.

Definition 6.3.2. A point a is called a *weak r-critical point* of the function $f(x)$ if $0 = \text{w}_r{}^t\text{d}/_{\text{d}x} f(a)$.

Example 6.3.2. Let us consider the function $f(x) = |x|$. We have that 0 is a weak 0-critical point of this function as $0 = \text{w}_0{}^t\text{d}/_{\text{d}x} |x|(0)$, while 0.1 is a weak 0.1-critical point of this function as $0.1 = \text{w}_{0.1}{}^t\text{d}/_{\text{d}x} |x|(0)$.

Definition 6.3.3. A point a is called a *weak critical point* of the function $f(x)$ if $0 = \text{w}^t\text{d}/_{\text{d}x} f(a)$.

Example 6.3.4. 0 is a weak critical point of the function $|x|$.

Let X be an open subset of $\boldsymbol{R}$.

Theorem 6.3.1 (The Fuzzy Fermat Theorem). If a function $f(x)$ is r-differentiable on X and a is a point of a local maximum or minimum of $f(x)$, then $0 = \text{w}_0{}^t\text{d}/_{\text{d}x} f(a)$ and $0 = \text{st}_q{}^t\text{d}/_{\text{d}x} f(a)$ with $q \leq 2r$.

Proof. We consider only the case of a local maximum because the case of a local minimum is proved in a similar way. At first, let us note that by Theorem 4.6.2 and Propositions 4.2.1 and 4.2.2, the function $f(x)$ is continuous in X and has strong centered, two-sided, left and right r-derivatives at any point a from X.

Let a be a point of a local maximum of $f(x)$ on X. As X is an open subset of $\boldsymbol{R}$, there is a neighborhood $\text{O}a = [a - k, a + k]$ of the point a that is a subset of X and $f(a) \geq f(x)$ for all points x from $\text{O}a$. When there is an interval $[e, d]$ that is a subset of $\text{O}a$, $d > a > e$, and $f(a) = f(x)$ for all points x from $[e, d]$, then the statement of the theorem is true by the definitions of strong and weak two-sided r-derivatives.

When there is no intervals $[e, d]$ such that it is a subset of $\text{O}a$ and $f(a) = f(x)$ for all points x from $[e, d]$, then there is an interval $[e, d]$ such that it is a subset of $\text{O}a$, $d > a > e$, and $f(a) > f(d)$ or (and) $f(a) > f(e)$. We consider only the case when $f(a) > f(d)$ as the case when $f(a) >$

$f(e)$ is considered in a similar way. By the properties of intervals, there is a sequence $l = \{ z_i; i = 1, 2, 3, \dots \}$ of points from $[e, d]$ such that $z_i < z_{i+1} < a$ for all $i = 1, 2, 3, \dots$ and $\lim_{i\to\infty} z_i = a$. By the definition of a local maximum, $f(a) \geq f(z_i)$ for all $i = 1, 2, 3, \dots$ and we can take a sequence $l = \{ z_i; i = 1, 2, 3, \dots \}$ such that $f(z_i) \leq f(z_{i+1})$ for all $i = 1, 2, 3, \dots$. In addition, $\lim_{i\to\infty} f(z_i) = f(a)$ because $f(x)$ is continuous at a.

As $f(a) > f(d)$, there is a number $n \in \omega$ such that $f(d) < f(z_i)$ when $i > n$ because $\lim_{i\to\infty} f(z_i) = f(a)$. Changing, if necessary, the sequence $l = \{ z_i ; i = 1, 2, 3, \dots \}$, we may assume that $f(d) < f(z_1)$. By the choice of points, we have $f(d) < f(z_i) \leq f(a)$ for all $i = 1, 2, 3, \dots$. Thus, by the intermediate value theorem for continuous functions (cf., Chapter 3) for each z_i , there is a point x_i such that $a < x_i < d$ and $f(x_i) = f(z_i)$. Consequently, $0 = \lim \{(f(z_i) - f(x_i)) /(z_i - x_i) ; i = 1, 2, 3, \dots \} = \mathrm{w}_0{}^{\mathrm{t}}\mathrm{d}/_{\mathrm{d}x} f(a)$. Properties of fuzzy derivatives imply $0 = \mathrm{w}_r{}^{\mathrm{t}}\mathrm{d}/_{\mathrm{d}x} f(a)$ when $0 < r$.

As $f(x)$ is r–differentiable at the point a, it has a strong centered r-derivative at the point a. Let $u = \mathrm{st}_r{}^{\mathrm{ct}}\mathrm{d}/_{\mathrm{d}x} f(a)$. Then by Proposition 4.2.1, $u = \mathrm{st}_r{}^{\mathrm{t}}\mathrm{d}/_{\mathrm{d}x} f(a)$. It means that $u = r\text{–}\lim_{i\to\infty} (f(z_i) - f(x_i)) /(z_i - x_i)$. Properties of fuzzy limits imply that $|u - w | \leq r$ for any weak two-sided derivative w of $f(x)$ at a. Consequently, $| u - 0| \leq r$ because we have proved that 0 is a weak two-sided derivative of $f(x)$ at a. Lemma 2.2.8 implies that $0 = \mathrm{st}_q{}^{\mathrm{t}}\mathrm{d}/_{\mathrm{d}x} f(a)$ with $q \leq 2r$.

Theorem is proved.

It is possible to show that we cannot take $q \leq 2r$ in Theorem 6.3.1.

Let us consider the following function

$$f(x) = \begin{cases} 3x & \text{when } x > 0 \\ -x & \text{when } x \leq 0 \end{cases}$$

This function has the global and thus, a local, minimum at the point 0. Besides, $1 = \mathrm{st}_2{}^{\mathrm{t}}\mathrm{d}/_{\mathrm{d}x} f(a)$ and $0 = \mathrm{st}_3{}^{\mathrm{t}}\mathrm{d}/_{\mathrm{d}x} f(a)$. It is possible to show that 0 is not a strong two-sided derivative $\mathrm{st}_q{}^{\mathrm{t}}\mathrm{d}/_{\mathrm{d}x} f(a)$ for any $q < 3$.

Corollary 6.3.1. If $f(x)$ is r-differentiable on X and a is a point of a local maximum or minimum of $f(x)$, then:

$$0 = \mathrm{w}_q{}^{\mathrm{r}}\mathrm{d}/_{\mathrm{d}x} f(a),\ 0 = \mathrm{w}_q{}^{\mathrm{l}}\mathrm{d}/_{\mathrm{d}x} f(a), \text{ and } 0 = \mathrm{w}_q{}^{\mathrm{ct}}\mathrm{d}/_{\mathrm{d}x} f(a) \text{ with } q \leq 2r.$$

$$0 = \mathrm{st}_q{}^{\mathrm{r}}\mathrm{d}/_{\mathrm{d}x} f(a),\ 0 = \mathrm{st}_q{}^{\mathrm{l}}\mathrm{d}/_{\mathrm{d}x} f(a), \text{ and } 0 = \mathrm{st}_q{}^{\mathrm{ct}}\mathrm{d}/_{\mathrm{d}x} f(a) \text{ with } q \leq 2r.$$

Corollary 6.3.2. If $f(x)$ is fuzzy differentiable on X and a is a point of a local maximum or minimum of $f(x)$, then:

$$0 = \mathrm{w}_r{}^{\mathrm{t}}\mathrm{d}/_{\mathrm{d}x} f(a) \text{ and } 0 = \mathrm{st}_q{}^{\mathrm{t}}\mathrm{d}/_{\mathrm{d}x} f(a) \text{ for some } r, q \in \boldsymbol{R}^+.$$

$$0 = w_r{}^{r}d/_{\mathrm{d}x} f(a),\ 0 = \mathrm{w}_r{}^{\mathrm{l}}\mathrm{d}/_{\mathrm{d}x} f(a), \text{ and } 0 = \mathrm{w}_r{}^{\mathrm{ct}}\mathrm{d}/_{\mathrm{d}x} f(a) \text{ for some } r \in \boldsymbol{R}^+.$$

$$0 = \text{st}_q{}^{\text{r}}\text{d}/\text{d}x f(a),\ 0 = \text{st}_q{}^{\text{l}}\text{d}/\text{d}x f(a), \text{ and } 0 = \text{st}_q{}^{\text{ct}}\text{d}/\text{d}x f(a) \text{ for some } q \in \boldsymbol{R}^+.$$

Remark 6.3.1. Theorem 6.3.1 and Corollaries 6.3.1 and 6.3.2 give necessary conditions for local maxima and minima of functions. As in the classical case, these conditions are useful for testing points in the domain of a function in order to find a local extremum.

Remark 6.3.2. Theorem 6.3.1 and Corollaries 6.3.1 and 6.3.2 also give conditions for finding fuzzy local maxima and minima of fuzzy differentiable functions because as it is demonstrated in Chapter 4, any r-differentiable function is continuous and fuzzy local maxima and minima of continuous functions are approximations of their local maxima and minima.

Remark 6.3.3. For fuzzy continuous functions, the results of Theorem 6.3.1 and Corollary 6.3.1 are invalid even for global extrema as the following example demonstrates.

Example 6.3.5. Let us consider the following function (cf. Fig. 6.2):

$$f(x) = \begin{cases} x & \text{if } x \in [0, 1] \text{ and } x \neq 1 - (1/2n); \\ 1 + (1/2n) & \text{if } x = 1 - (1/2n). \end{cases}$$

This function has the global maximum $1 + (1/2n)$ at the point $1 - (1/2n)$. However, $f(x)$ is not fuzzy differentiable at this point.

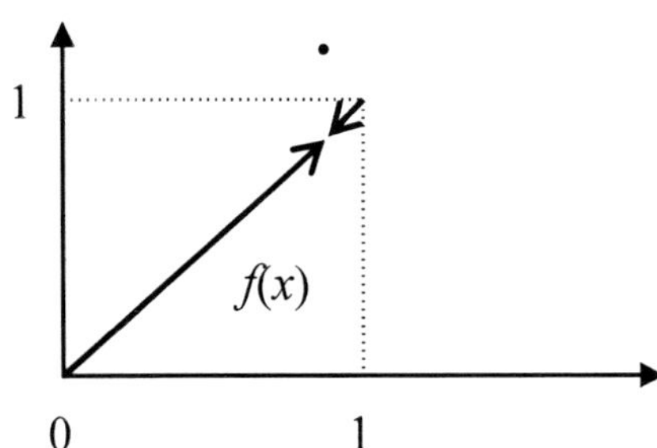

Figure 6.2. A function that is not differentiable at the point of global maximum

The function $f(x)$ is (1/n)-continuous and the point $1 - (1/2n)$ is the global maximum of $f(x)$ on the interval (0, 1). However, $1 - (1/2n)$ is neither r-critical nor weak r-critical point for any $r < ½$.

Remark 6.3.4. In a general case, Theorem 6.3.1 is valid only for two-sided fuzzy derivatives as the following example demonstrates.

Example 6.3.6. Let us consider the function $f(x)$ that is equal to $1 + \sqrt{1 - x^2}$ on the unit interval [0, 1] and is equal to $1 + \sqrt{1 - (x - 2)^2}$ on the interval [1, 2]. Being continuous, this function has no weak centered derivatives at the point 1. At the same time, 1 is the point of global maximum of the function $f(x)$.

Theorem 6.3.1 implies such classical result as the Fermat Theorem (Theorem 6.1.1).

Theorem 6.3.2 (Fuzzy Rolle Theorem). If a function $f(x)$ is r-differentiable on $[a, b]$ and $f(a) = f(b)$, then there is [at least one] point $c \in [a, b]$ such that $0 = \mathrm{w}_0{}^{\mathrm{t}}\mathrm{d}/_{\mathrm{d}x} f(c)$ and $0 = \mathrm{st}_q{}^{\mathrm{t}}\mathrm{d}/_{\mathrm{d}x} f(c)$ with $q \leq 2r$.

Proof. As it is proved in Chapter 4, fuzzy differentiability implies that the function $f(x)$ is continuous on $[a, b]$. A continuous function on a closed interval reaches its maximum and minimum (cf. Theorem 3.1.2). Let us take a point c in which $f(x)$ reaches its maximum. There three possibilities for this: 1) $c = a$; 2) $c = b$; and 3) $a < c < b$. If $c = a$ or $c = b$, then we have $f(a) = f(b) \geq f(x)$ for all $x \in [a, b]$. It gives us two possibilities: either $f(a) = f(x)$ for all $x \in [a, b]$ or there is a point $c \in [a, b]$ such that $f(a) > f(c)$. In the first case, the function $f(x)$ is identically equal to $f(a)$ on $[a, b]$ and the statement of the theorem is evidently true.

In the second case, there is a point $d \in [a, b]$ in which $f(x)$ reaches its minimum. There are three possibilities for this: 1) $d = a$; 2) $d = b$; and 3) $a < d < b$. If $d = a$ or $d = b$, then the function $f(x)$ is identically equal to $f(a)$ on $[a, b]$ and the statement of the theorem is evidently true.

If $a < d < b$. Then d is a point of a local minimum in the open interval (a, b) and the statement of the theorem follows from Theorem 6.3.1.

To conclude the proof, we have consider the case (3) when $a < c < b$. Then c is a point of a local maximum in the open interval (a, b) and the statement of the theorem follows from Theorem 6.3.1.

Theorem is proved.

Corollary 6.3.3. If a function $f(x)$ is r-differentiable on (a, b), continuous on $[a, b]$ and $f(a) = f(b)$, then there is [at least one] point $c \in [a, b]$ such that $0 = \mathrm{w}_r{}^{\mathrm{t}}\mathrm{d}/_{\mathrm{d}x} f(c)$ and $0 = \mathrm{st}_q{}^{\mathrm{t}}\mathrm{d}/_{\mathrm{d}x} f(c)$ with $q \leq 2r$.

Theorem 6.3.2 also implies such classical result as the *Rolle Theorem* (Theorem 6.1.2).

Theorem 6.3.3 (the Fuzzy Mean Value Theorem). If a function $f(x)$ is r-differentiable on $[a, b]$, then there is [at least one] point $c \in [a, b]$ such that $(f(b) - f(a)) / (b - a) = \mathrm{w}_0{}^{\mathrm{t}}\mathrm{d}/_{\mathrm{d}x} f(c)$ and $(f(b) - f(a)) / (b - a) = \mathrm{st}_q{}^{\mathrm{t}}\mathrm{d}/_{\mathrm{d}x} f(c)$ with $q \leq 2r$.

Proof. Let us consider the function $g(x) = f(x) - l(x)$ where $l(x) = [(f(b) - f(a)) / (b - a)]x + [(bf(a) - af(b)) / (b - a)]$. Then we have

$$l(a) = [(f(b) - f(a)) / (b - a)]a + [(bf(a) - af(b)) / (b - a)] =$$
$$[(af(b) - af(a) + bf(a) - af(b)) / (b - a)] =$$
$$[(bf(a) - af(a)) / (b - a)] = f(a)$$

and

$$l(b) = [(f(b) - f(a)) / (b - a)]b + [(bf(a) - af(b)) / (b - a)] =$$
$$[(bf(b) - bf(a) + bf(a) - af(b)) / (b - a)] =$$
$$[(bf(b) - af(b)) / (b - a)] = f(b)$$

Thus, $l(a) = f(a)$, $l(b) = f(b)$. In addition, $l'(c) = (f(b) - f(a)) / (b - a)$ for all c from (a, b). By the definition of a strong fuzzy derivative, we have $(f(b) - f(a)) / (b - a) = \mathrm{st}_q{}^{\mathrm{t}}\mathrm{d}/_{\mathrm{d}x}\, l(c)$ for any $q \in \boldsymbol{R}^+$ and for all c from (a, b). By the definition of a weak fuzzy derivative, we have $(f(b) - f(a)) / (b - a) = u = \mathrm{w}_0{}^{\mathrm{t}}\mathrm{d}/_{\mathrm{d}x}\, l(c)$ for any $r \in \boldsymbol{R}^+$ and for all c from (a, b).

So, we have $g(a) = g(b) = 0$. By the Fuzzy Rolle's Theorem, then there is [at least one] point $c \in [a, b]$ such that $0 = \mathrm{w}_0{}^{\mathrm{t}}\mathrm{d}/_{\mathrm{d}x}\, g(c)$ and $0 = \mathrm{st}_q{}^{\mathrm{t}}\mathrm{d}/_{\mathrm{d}x}\, g(c)$ with $q \leq 2r$.

As it is demonstrated in Chapter 4, weak and strong fuzzy derivatives are additive. Then by Proposition 4.3.8, we have $0 + u = u = (f(b) - f(a)) / (b - a) = \mathrm{w}_{0+0}{}^{\mathrm{t}}\mathrm{d}/_{\mathrm{d}x}\, [g(c) + l(c)] = \mathrm{w}_0{}^{\mathrm{t}}\mathrm{d}/_{\mathrm{d}x}\, f(c)$ as $0 = \mathrm{w}_0{}^{\mathrm{t}}\mathrm{d}/_{\mathrm{d}x}\, g(c)$ and any weak derivative $u = \mathrm{w}_0{}^{\mathrm{t}}\mathrm{d}/_{\mathrm{d}x}\, l(c)$ is equal to $[(f(b) - f(a)) / (b - a)]$. Thus, $(f(b) - f(a)) / (b - a) = \mathrm{w}_0{}^{\mathrm{t}}\mathrm{d}/_{\mathrm{d}x} f(c)$.

In a similar way, by Theorem 4.2.1, as $0 = \mathrm{st}_q{}^{\mathrm{t}}\mathrm{d}/_{\mathrm{d}x}\, g(c)$ and $u = [(f(b) - f(a)) / (b - a)] = \mathrm{st}_0{}^{\mathrm{t}}\mathrm{d}/_{\mathrm{d}x}\, l(c)$, we have $0 + u = u = \mathrm{st}_{q+0}{}^{\mathrm{t}}\mathrm{d}/_{\mathrm{d}x}\, [g(c) + l(c)] = \mathrm{st}_q{}^{\mathrm{t}}\mathrm{d}/_{\mathrm{d}x} f(c)$. Thus, $(f(b) - f(a)) / (b - a) = \mathrm{st}_q{}^{\mathrm{t}}\mathrm{d}/_{\mathrm{d}x}\, f(c)$ with $q \leq 2r$.

Theorem is proved.

Corollary 6.3.4. If a function $f(x)$ is r-differentiable on (a, b) and continuous on $[a, b]$, then there is [at least one] point $c \in [a, b]$ such that $(f(b) - f(a)) / (b - a) = \mathrm{w}_r{}^{\mathrm{t}}\mathrm{d}/_{\mathrm{d}x} f(c)$ and $(f(b) - f(a)) / (b - a) = \mathrm{st}_q{}^{\mathrm{t}}\mathrm{d}/_{\mathrm{d}x} f(c)$ with $q \leq 2r$.

Theorem 6.3.3 also implies such a classical result as the *Mean Value Theorem* (Theorem 6.1.3).

The Fuzzy Mean Value Theorem allows us to extend the result of Theorem 6.1.4 from differentiable functions to the scope of fuzzy differentiable functions.

Let us take an r-differentiable on $[a, b]$ function $f(x)$.

Theorem 6.3.4. If for any point c from $[a, b]$, some number $r \in \boldsymbol{R}^+$, and any $u = \mathrm{w}_r{}^{\mathrm{t}}\mathrm{d}/_{\mathrm{d}x}\, f(c)$, we have $u \leq k$, then $f(x)$ is uniformly (q, kq)-continuous in $[a, b]$ for any $q \in \boldsymbol{R}^+$.

Proof. Let us take two points c and d in $[a, b]$ such that $0 < d - c < q$. As $f(x)$ is a differentiable function in the interval $[a, b]$, it is differentiable in the interval $[c, d]$. By the Fuzzy Mean Value Theorem, there is a point e such that $c \leq e \leq d$ and $u = (f(b) - f(a)) / (b - a) = \mathrm{w}_r{}^{\mathrm{t}}\mathrm{d}/_{\mathrm{d}x} f(c)$. Thus, we have

$$f(d) - f(c) = (d - c) \cdot u \leq k(d - c) < kq$$

Theorem is proved as c and d are arbitrary points in $[a, b]$ with $0 < d - c < q$.

Theorem 6.3.4 and Lemma 3.5.2 imply the following result.

Corollary 6.3.5. If for any point c from $[a, b]$, some number $r \in \boldsymbol{R}^+$, and any $u = \mathrm{w}_r{}^{\mathrm{t}}\mathrm{d}/_{\mathrm{d}x}\, f(c)$, we have $u \leq k$, then $f(x)$ is (q, kq)-continuous in $[a, b]$ for any $q \in \boldsymbol{R}^+$.

Theorem 6.3.4 and Lemma 4.3.1 imply the following result.

Corollary 6.3.6. If for any point c from $[a, b]$, some number $r \in \boldsymbol{R}^+$, and any $u = \mathrm{st}_{2r}{}^{\mathrm{t}}\mathrm{d}/_{\mathrm{d}x} f(c)$, we have $u \leq k$, then $f(x)$ is uniformly (q, kq)-continuous in $[a, b]$ for any $q \in \boldsymbol{R}^+$.

Corollary 6.3.6 and Lemma 3.5.2 imply the following result.

Corollary 6.3.7. If for any point c from $[a, b]$, some number $r \in \boldsymbol{R}^+$, and any $u = \mathrm{st}_{2r}{}^{\mathrm{t}}\mathrm{d}/_{\mathrm{d}x} f(c)$, we have $u \leq k$, then $f(x)$ is (q, kq)-continuous in $[a, b]$ for any $q \in \boldsymbol{R}^+$.

Let $p \in \boldsymbol{R}^+$

Definition 6.3.4. A function $f(x)$ is called *p-stable* in a set X if $(f(b) - f(a)) \leq p$ for any points a and b from X.

This is a fuzzification of the concept of a constant function as the following result demonstrates.

Lemma 6.3.6. A 0-stable function is constant.

Example 6.3.7. The function $f(x) = \sin x$ is 2-stable.

Example 6.3.8. The function $f(x) = (\sin x)(\cos x)$ is 1-stable.

Real p-stable functions are closely related to fuzzy continuous functions. Namely, p-stability is a kind (r, q)-continuity (cf. Chapter 3) as the following lemma demonstrates.

Lemma 6.3.7. A function $f(x)$ is p-stable on an interval $[a, b]$ if and only if f is (r, p)-continuous on X where $r = b - a$.

Real p-stable functions are also related to functions with finite fluctuation, which are studied in (Burgin and Glushchenko, 1997)

Lemma 6.3.8. A function $f(x)$ has the finite fluctuation on a set X if and only if it is p-stable on X for some $p \in \boldsymbol{R}^+$.

Let $r \in \boldsymbol{R}^+$.

Theorem 6.3.5 (Fuzzy Stability Theorem). If a function $f(x)$ is r-differentiable on (a, b) and $0 = \mathrm{st}_q{}^{\mathrm{t}}\mathrm{d}/_{\mathrm{d}x} f(x)$ for some $q \in \boldsymbol{R}^+$ and all x from (a, b), then $f(x)$ is p-stable where $p = q(b - a)$.

Proof. By the definition of a strong fuzzy derivative, if $b = \mathrm{st}_q{}^{\mathrm{t}}\mathrm{d}/_{\mathrm{d}x} f(x)$, then $|b| < q$. Consequently, $| f(x) - f(z) | < q(z - x) < q(b - a)$ when $a \leq x < z \leq b$. This means that $f(x)$ is p-stable with $p = q(b - a)$.

Remark 6.3.5. In Theorem 6.3.5, it is reasonable to assume that $r < q$. In this case, it is possible that $0 \neq \mathrm{st}_r{}^{\mathrm{t}}\mathrm{d}/_{\mathrm{d}x} f(x)$ for some (or even for all) x from (a, b). At the same time, for each x from the interval (a, b), we have $a = \mathrm{st}_r{}^{\mathrm{t}}\mathrm{d}/_{\mathrm{d}x} f(x)$ for some number a because the function $f(x)$ is r-differentiable in (a, b).

When p is sufficiently small, Theorem 6.3.4 states that the function $f(x)$ is almost a constant in (a, b). In particular, we have the following result.

Corollary 6.3.8. If a function $f(x)$ is r-differentiable on (a, b) and $0 = \mathrm{st}_0{}^{\mathrm{t}}\mathrm{d}/_{\mathrm{d}x} f(x)$ for some $q \in \boldsymbol{R}^+$ and all x from (a, b), then $f(x)$ is a constant function.

Corollary 3.8 implies the following classical result.

Corollary 6.3.9 (cf., for example, (Ross, 1996)). If a function $f(x)$ is differentiable on (a, b) and $f'(x) = 0$ for all x from (a, b), then $f(x)$ is constant.

Let $r \in \boldsymbol{R}^+$, $f(x)$ and $g(x)$ be r-differentiable on (a, b) functions and $\mathbf{d}(f, g) = \sup \{ | f(x) - g(x) | ; x \in (a, b) \}$. Then Theorem 6.3.5 implies the following result.

Theorem 6.3.6 (Fuzzy Identity Theorem). If $0 = \mathrm{st}_q{}^{t}\mathrm{d}/_{\mathrm{d}x}(f - g)(x)$ for some $q \in \boldsymbol{R}^+$ and all x from (a, b), then for any number c from (a, b), we have $\mathbf{d}(f, h) \leq p$ where $p = q(b - a)$ and $h(x) = g(x) + f(c) - g(c)$.

When p is sufficiently small, Theorem 6.3.6 states that in the interval (a, b) the function $f(x)$ is almost equal to the function $g(x)$ plus a constant, i.e., the difference in (a, b) between $f(x)$ and $g(x)$ plus some constant is very small.

Theorem 6.3.6 implies the following classical result.

Corollary 6.3.10 (cf., for example, (Ross, 1996)). If functions $f(x)$ and $g(x)$ are differentiable on (a, b) and $f'(x) = g'(x)$ for all x from (a, b), then $f(x) = g(x) + c$ for some number $c \in \boldsymbol{R}^+$.

Chapter 7

FUZZY INTEGRATION

Convergence connotes a meeting,
an integration, a harmony ...
James Robbins Kidd (1915-1982)

Integral calculus is one of two central parts of the calculus and analysis. The main concept of integral calculus is the concept of an integral. The main goal of this chapter is to extend classical constructions of integrals in order to make these constructions more flexible and more relevant to real life conditions where data are obtained from measurement and computation.

Conventional integration that is used for finding areas and volumes, for solving a variety of physical problems, and for finding probability is defined in mathematics by a limit process. The main idea of fuzzy integration is to use fuzzy limits instead of conventional limits. What does it give for integration? The main advantage is that it essentially extends the scope of integrable functions. For instance, it is known that the Lebesgue integral exists for a much wider range of functions than the Riemann integral. However, as it is demonstrated in this chapter, the fuzzy Riemann integral exists for a much wider range of functions than the Lebesgue integral. Thus, the neoclassical approach presented here allows one to integrate functions that have neither Riemann nor Lebesgue nor gauge integral.

At the same time, fuzzy integrals are natural extensions of classical integrals and as such, preserve many important properties of classical integrals. For instance, fuzzy integrability implies boundedness of a function. In addition, the neoclassical approach allows researchers to build better models for numerical integration when the result is not given by a formula but is computed.

In this chapter, we develop a theory of fuzzy integration based on the theory of fuzzy limits presented in Chapter 2. In Section 7.1, we introduce the fuzzy Riemann integral. We construct the fuzzy Riemann integral using three different approaches developed by Riemann, Darboux, and Daniell in the traditional setting and based on the Riemann and Darboux sums, and integrals of step functions, respectively. It is proved that all three approaches give the same concept called the fuzzy Riemann integral. It is also demonstrated (Theorem 7.1.1) that there are fuzzy Riemann integrals in many cases when the Riemann integral and even the

Lebesgue integral do not exist. For completeness, we also define classical Riemann, Darboux, and Daniell integrals.

In Section 7.2, various properties of the fuzzy Riemann integral that are useful for analytical and numerical evaluation of integrals are studied. These properties allow one to get respective properties of the Riemann integral as direct corollaries. Many properties of fuzzy and conventional integrals are similar. For instance, additivity of the Riemann integral is transformed to shifted additivity (cf. Appendix) of the fuzzy Riemann integral and uniformity to shifted uniformity. At the same time, other properties of fuzzy and conventional integrals are essentially different. For instance, the Riemann integral has a unique value (if any) for any interval of integration, while the fuzzy Riemann integral can have different values for the same interval of integration.

Applications of fuzzy integration provide for better utilization of analytical methods of calculus in a much more voluminous extent of practical situations. In addition, fuzzy integration helps to achieve better correlation between mathematical analysis and numerical analysis.

It is necessary to remark that the term "fuzzy integration" allows different interpretations. It may be integration of fuzzy functions (cf., for example, (Matloka, 1987; Zhenyuan and Guoli, 1988; Nanda, 1989)). It may also be integration with fuzzy measures (cf. (Sugeno, 1974; Murofushi and Sugeno, 1989; Grabisch, Murofushi, and Sugeno, 2000). Besides, it may be information integration based on fuzzy logic (cf., for example, (Chen, et al, 2001). Here we consider fuzzy integration in the context of neoclassical analysis, i.e., integration based on the concept of a fuzzy limit.

7.1. THE FUZZY RIEMANN, DARBOUX, AND DANIELL INTEGRALS

The experimental verification of a theory
concerning any natural phenomenon
generally rests on the result of integration.
Joseph William Mellor (1873-1938)

There are several ways to construct the Riemann integral. Correspondingly, there are even more ways to construct the fuzzy Riemann integral. Here we consider some of them, starting with the most popular approach to the Riemann integral.

Let us consider two real numbers $a < b$ and a bounded real function f: $[a, b] \to \mathbf{R}$. The interval $[a, b]$ is divided into n subintervals $[x_i, x_{i+1}]$ by choosing points $a = x_0 < x_1 < x_2 < \dots < x_i < x_{i+1} < \dots < x_n = b$. The system of intervals $\mathrm{P} = \{[x_{i-1}, x_i]; i = 1, 2, 3, \dots, n\}$ with $a = x_0$ and $x_n = b$ is called a *partition* of the interval $[a, b]$.

A partition $\mathrm{Q} = \{[z_{j-1}, z_j]; j = 1, 2, 3, \dots, m\}$ of the interval $[a, b]$ is called a *subpartition* of the partition $\mathrm{P} = \{[x_{i-1}, x_i]; i = 1, 2, 3, \dots, n\}$ of the interval $[a, b]$ if $m > n$ and each point x_i is equal to some point z_j.

Traditionally, the length of the interval $\mathrm{I}_i = [x_{i-1}, x_i]$ is denoted by $\Delta x_i = x_i - x_{i-1}$. This allows us to define the *norm of the partition* P that is equal to $||\mathrm{P}|| = \max \{\Delta x_i ; i = 1, 2, 3, \dots n\}$. A sequence $L = \{ \mathrm{P}_i ; i = 1, 2, 3, \dots \}$ of partitions is *decreasing* if the inequality $||\mathrm{P}_i|| > ||\mathrm{P}_{i+1}||$ for norms is true for all $i = 1, 2, 3, \dots$

In each interval $[x_{i-1}, x_i]$, some number c_i is selected ($i = 1, 2, 3, \ldots n$). Then the sum $\sum_{i=1}^{n} f(c_i)\, \Delta x_i$ (or otherwise denoted by $\sum_{I_i \in P} f(c_i)\, \Delta x_i$) where $I_i = [x_{i-1}, x_i]$ is an interval from the partition P = $\{[x_{i-1}, x_i];\ i = 1, 2, 3, \ldots, n\}$) is called the *Riemann sum* corresponding to the partition P. Thus, we can see that a Riemann sum depends on several parameters: the function $f(x)$, partition P, and choice of points c_i.

Let us consider the set E_f of sequences of Riemann sums of the function $f(x)$ corresponding to all decreasing sequences $L = \{P_i;\ i = 1, 2, 3, \ldots\}$ of partitions P_i in which the norm $\|P_i\|$ tends to 0.

Definition 7.1.1. (any course of the calculus, cf., for example, (Ribenboim, 1964; Ross, 1996)). When the limit lim E_f exists, it is called the *Riemann integral* or *definite Riemann integral* of the function $f(x)$ over the interval $[a, b]$ and is denoted by

$$\int_a^b f(x)\, dx \qquad \text{or} \qquad \int_a^b f dx$$

In this case, the function $f(x)$ is called the *integrand*, the numbers a and b are called the *bounds of integration*, and the interval $[a, b]$ is called the *interval of integration.*

By the definition of a limit (cf. Section 2.1), we have the following result.

Lemma 7.1.1. The Riemann integral of a function $f(x)$ over the interval $[a, b]$ is a unique number when it exists.

In similar way, we define the fuzzy Riemann integral.

Let $r \in \boldsymbol{R}^+$.

Definition 7.1.2. A number $u = r$-lim E_f is called a *Riemann r-integral* or *definite Riemann r-integral* of the function $f(x)$ over the interval $[a, b]$ and is denoted by

$$u = r\text{-}\int_a^b f(x)\, dx$$

As in the classical case, the function $f(x)$ is called the *integrand*, the numbers a and b are called the *bounds of integration*, and the interval $[a, b]$ is called the *interval of integration.*

When there is, at least, one number $u = r\text{-}\int_a^b f(x)\, dx$, then the function $f(x)$ is called *Riemann r-integrable* over the interval $[a, b]$.

In a general case, the number $u = r\text{-}\int_a^b f(x)\, dx$ is called a *fuzzyRiemann integral* or *definite fuzzy Riemann integral* of the function $f(x)$ over the interval $[a, b]$.

Example 7.1.1. Let us consider the characteristic function $\chi_Q(x)$ of the set $\boldsymbol{Q}$ of rational numbers, i.e., $\chi_Q(x) = 1$ when $x \in \boldsymbol{Q}$ and $\chi_Q(x) = 0$ otherwise. Then

$$½ = ½\text{-}\int_0^1 \chi_Q(x)dx,\ 1 = 1\text{-}\int_0^1 \chi_Q(x)dx, \text{ and } 0 = 1\text{-}\int_0^1 \chi_Q(x)dx\,.$$

Remark 7.1.1. When the Riemann integral $\int_a^b f(x)dx$ exists, the equality $u = r\text{-}\int_a^b f(x)dx$, means that the distance from u to $\int_a^b f(x)dx$ is not larger than r or that u is an approximation of the integral $\int_a^b f(x)dx$ with the precision r.

Remark 7.1.2. In contrast to the Riemann integral $\int_a^b f(x)dx$, the Riemann r-integral of a function $f(x)$ over the interval $[a, b]$ is not, as a rule, unique.

Properties of fuzzy limits and Riemann integrals imply the following result.

Proposition 7.1.1. $0\text{-}\int_a^b f(x)dx$ is the Riemann integral $\int_a^b f(x)dx$ of the function $f(x)$ over the interval $[a, b]$.

However, there are functions that do not have the Riemann integral but have fuzzy Riemann integrals.

Example 7.1.2. Let us consider the characteristic function $\chi_{IR}(x)$ of the set $\boldsymbol{IR}$ of irrational numbers, i.e., $\chi_{IR}(x) = 1$ when $x \in \boldsymbol{IR}$ and $\chi_{IR}(x) = 0$ otherwise. This function does not have the Riemann integral over the interval [0, 1]. However, it has fuzzy Riemann integrals over the interval [0, 1]:

$$½ = ½\text{-} \int_0^1 \chi_{IR}(x)dx,\ 1 = 1\text{-} \int_0^1 \chi_{IR}(x)dx, \text{ and } 0 = 1\text{-} \int_0^1 \chi_{IR}(x)dx .$$

It is possible to show that many functions that have neither Riemann nor even Lebesgue integral have fuzzy Riemann integrals.

Example 7.1.3. Let us consider the characteristic function $\chi_A(x)$ of a nonmeasurable set A in the interval [0, 1], i.e., $\chi_A(x) = 1$ when $x \in A$ and $\chi_A(x) = 0$ otherwise. It is possible to find examples of nonmeasurable sets in (Gelbaum and Olmsted, 1964). According to the theory of Lebesgue integration (cf., for example, (Saks, 1964)), this function has neither Riemann nor Lebesgue integral. Nevertheless, it has fuzzy Riemann integrals:

$$½ = ½\text{-} \int_0^1 \chi_A(x)dx,\ 1 = 1\text{-} \int_0^1 \chi_A(x)dx, \text{ and } 0 = 1\text{-} \int_0^1 \chi_A(x)dx .$$

The following result shows to what extent fuzzy integration extends the scope of integration.

Theorem 7.1.1. Any bounded in a closed interval function has a fuzzy Riemann integral over this interval.

Proof. Let $f(x)$ be a bounded in the interval $[a, b]$ function. It means that there is a real number C such that $|f(x)| < C$ for all $x \in [a, b]$.

Let us take a partition P = $\{[x_{i-1}, x_i]; i = 1, 2, 3, \dots, n\}$ of the interval $[a, b]$ and a Riemann sum $\Sigma_{i=1}^n f(c_i)\,\Delta x_i$ for the function $f(x)$ corresponding to the partition P. Then we have

$$|\,\Sigma_{i=1}^n f(c_i)\,\Delta x_i\,| \le \Sigma_{i=1}^n |f(c_i)|\,\Delta x_i \le \Sigma_{i=1}^n C\Delta x_i = C\Sigma_{i=1}^n \Delta x_i = C(b-a).$$

Taking $r = C(b - a)$, we have $|\,\Sigma_{i=1}^n f(c_i)\,\Delta x_i\,| < r + \varepsilon$ for any $\varepsilon > 0$. Thus, if we have a decreasing sequence $L = \{\, P_j\,;\, j = 1, 2, 3, \dots \}$ of partitions of the interval $[a, b]$ and the corresponding sequence $l = \{\, a_{P_j}\,;\, j = 1, 2, 3, \dots \}$ of Riemann sums where $\Sigma_{i=1}^n f(c_i)\,\Delta x_i = a_P$ is Riemann sum of the function $f(x)$ corresponding to a partition P = $\{[x_{i-1}, x_i]; i = 1, 2, 3, \dots, n\}$, then, by the definition of an r-limit, we have $0 = r\text{-}\lim l$ as $|0 - a_{P_j}| < r + \varepsilon$ for any $\varepsilon > 0$ and all $j = 1, 2, 3, \dots$.

Consequently, $0 = r\text{-}\int_a^b f(x)\,dx$.

Theorem is proved.

Remark 7.1.3. In many cases, it is possible to get a more exact (in comparison to Theorem 7.1.1) estimate r for fuzzy Riemann integration of the function $f(x)$ over the interval $[a, b]$ (cf. Example 7.1.1). However, in a general case, the estimate r from Theorem 7.1.1 cannot be improved as Example 7.1.4 demonstrates.

Example 7.1.4. Let us consider the function

$$m(x) = \begin{cases} 1 & \text{if } x \text{ is a rational number,} \\ -1 & \text{otherwise.} \end{cases}$$

Then by Definition 7.1.2,

$$0 = 1\text{-}\int_0^1 m(x)dx \text{ and for any } c \ne 0, \text{ we have } c \ne 1\text{-}\int_0^1 \chi_A(x)dx\,.$$

In a similar way, for any number $r < 1$, we have $0 \ne r\text{-}\int_0^1 \chi_A(x)dx$.

However, properties of fuzzy limits imply that $c = (1 + |c|)\text{-}\int_0^1 \chi_A(x)dx$.

In a similar way, we have $0 = 5\text{-}\int_0^1 m(x)dx$ and $c = (3 + |c|)\text{-}\int_0^1 \chi_A(x)dx$.

We already know (cf. Chapter 3) that any fuzzy continuous function on a closed interval is bounded. This and Theorem 7.1.1 give the following result.

Corollary 7.1.1. Any fuzzy continuous function on a closed interval has a fuzzy Riemann integral over this interval.

However, to show that a function has a fuzzy integral is only the first step. What is more important is to find the estimate of fuzziness of this integral, that is, the number r. This brings us to the following problem:

Is any r-continuous function Riemann r-integrable?

In Chapter 3, we obtained a criterion for boundedness of functions in a closed interval (Theorem 3.4.2). Here we give another criterion for boundedness of functions in a closed interval.

Theorem 7.1.2. A function has a fuzzy Riemann integral over a closed interval if and only if it is bounded in this interval.

Proof. Theorem 7.1.1 shows that we need to prove only necessity of function boundedness for a possibility of fuzzy Riemann integration.

Let a function $f(x)$ has the Riemann r-integral over an interval $[a, b]$. By Definition 7.1.2, all sequences of Riemann sums of the function $f(x)$ corresponding to all decreasing sequences L of partitions of $[a, b]$ r-converge to the same number. By Theorem 2.2.2, all Riemann sums corresponding to these partitions are bounded by the same number.

At the same time, when the function $f(x)$ is unbounded in $[a, b]$, it is possible to build a decreasing sequence L of partitions of $[a, b]$ such that corresponding Riemann sums of the function $f(x)$ will converge to ∞ when $f(x)$ is unbounded above and to $-\infty$ when $f(x)$ is unbounded below.

To do this, let us consider the first case when $f(x)$ is unbounded above and bounded below in $[a, b]$. We can take the a decreasing sequence $L = \{ \mathrm{P}_j ; j = 1, 2, 3, \dots \}$ of partitions of the interval $[a, b]$ where the partition $\mathrm{P}_n = \{[x_{i-1, n}, x_{i, n}]; i = 1, 2, 3, \dots , n\}$ consists of 2^n equal parts of the interval $[a, b]$. By definition, L is a decreasing sequence of partitions with $\| \mathrm{P}_n \| = [(b - a)/2^n]$ and consequently, $\| \mathrm{P}_n \| \to 0$ when $n \to \infty$. As $f(x)$ is unbounded above in $[a, b]$, we can correspond to each partition P_n a point a_n from $[a, b]$ such that $f(a_n) > 2^{2n}$. As $f(x)$ is bounded below in $[a, b]$, there is a number d such that $f(x) > d$ for all $x \in [a, b]$.

Let us consider the Riemann sum $S_n = \Sigma_{i=1}^{n} f(c_i)\, \Delta x_i$ for the function $f(x)$ corresponding to the partition P_n and take one of the points c_i (just the point c_i such that a_n belongs to $[x_{j-1, n}, x_{j, n}]$) equal to a_n. By the construction of P_n, we have $\Delta x_i = 2^{-n}$ for all $i = 1, 2, 3, \dots , n$. Besides, $f(c_i) > d$ for all $i = 1, 2, 3, \dots , n$. Thus, we have

$$\Sigma_{i=1}^{n} f(c_i)\, \Delta x_i = \Sigma_{i \neq j} f(c_i)\, \Delta x_i + f(a_n)\, \Delta x_i >$$
$$d(b - a - 2^{-n}) + 2^{-n} \cdot f(a_n) > d(b - a - 2^{-n}) + 2^{-n} \cdot 2^{2n} =$$
$$d(b - a - 2^{-n}) + 2^n$$

As $2^n \to \infty$, $2^{-n} \to 0$ when $n \to \infty$ and the number b is fixed, we see that the Riemann sums S_n for the function $f(x)$ corresponding to the partitions P_n tend to infinity when $n \to \infty$. Thus, $f(x)$ is not fuzzy integrable.

Similar considerations show that when $f(x)$ is bounded above and unbounded below in $[a, b]$, we can find a decreasing sequence $L = \{ \mathrm{P}_j ; j = 1, 2, 3, \dots \}$ of partitions of the interval $[a, b]$ such that the corresponding Riemann sums S_n for the function $f(x)$ corresponding to the partitions P_n tend to $-\infty$ when $n \to \infty$. Thus, $f(x)$ is also not fuzzy integrable.

The last possibility is that $f(x)$ is unbounded above and unbounded below in $[a, b]$. Let us take, as before, the a decreasing sequence $L = \{ \mathrm{P}_j ; j = 1, 2, 3, \dots \}$ of partitions of the interval $[a, b]$ where the partition $\mathrm{P}_n = \{[x_{i-1, n}, x_{i, n}]; i = 1, 2, 3, \dots , n\}$ consists of 2^n equal parts of the interval $[a, b]$. Building the Riemann sum $S_n = \Sigma_{i=1}^{n} f(c_i)\, \Delta x_i$ for the function $f(x)$ corresponding to the partition P_n, we take the numbers $d_{i, n} = \sup \{ f(x); x \in [x_{i-1, n}, x_{i, n}] \}$ for all $i = 1, 2, 3, \dots , n$. The numbers $d_{i, n}$ exist because $f(x)$ is defined for all point from the interval $[a, b]$. Then we can find $d_n = \min \{ d_{i, n} ; i = 1, 2, 3, \dots , n \}$. By the definition of supremum, for each pair i and n, there is a point $c_{i,n} \in [x_{i-1,n}, x_{i,n}]$ such that $0 < d_{i,n} - f(c_{i,n}) < 2^{-n}$. At the same time, as $f(x)$ is unbounded above in $[a, b]$, we can find a point a_n from $[a, b]$ such that $f(a_n) > 2^{2n}(|d_n| + 1)(b - a)$. We take these points $c_{i,n}$ and a_n as arguments for $f(x)$ in the corresponding intervals $[x_{i-1,n}, x_{i,n}]$.

Thus, for n such that $b - a > 2^{-n}$, we have

$$\begin{gathered}\Sigma_{i=1}^{n} f(c_i)\, \Delta x_i = \Sigma_{i \neq j} f(c_{i,n})\, \Delta x_i + f(a_n)\, \Delta x_i > \\ (d_n - 2^{-n})(b - a - 2^{-n}) + 2^{-n} \cdot f(a_n) > \\ (d_n - 2^{-n})(b - a - 2^{-n}) + 2^{-n} \cdot 2^{2n}(d_n + 1)(b - a) > \\ d_n \cdot (b - a - 2^{-n}) + 2^{-n} \cdot 2^{2n} \cdot |d_n| \cdot (b - a) = \\ d_n \cdot (b - a - 2^{-n}) + 2^{n} \cdot |d_n| \cdot (b - a) = \\ d_n \cdot (b - a) - d_n \cdot (2^{-n}) + 2^{n} |d_n| \cdot (b - a) > \\ 2d_n \cdot (b - a) + 2^{n} |d_n| \cdot (b - a) > \\ (2^{n} - 2) \cdot |d_n|\, (b - a)\end{gathered}$$

As $2^n \to \infty$, $2^{-n} \to 0$ when $n \to \infty$ and the number b is fixed, we see that the Riemann sums S_n for the function $f(x)$ corresponding to the partitions P_n tend to infinity when $n \to \infty$. Thus, $f(x)$ is not fuzzy integrable.

We have demonstrated that unbounded in $[a, b]$ functions cannot have the fuzzy Riemann integral over the interval $[a, b]$. Thus, assumption of integrability shows that $f(x)$ is a bounded in $[a, b]$ function.

Theorem is proved.

This shows that fuzzy Riemann integrability and fuzzy continuity are equivalent properties for functions in a closed interval. Namely, Theorems 3.4.2 and 7.1.2 imply the following result.

Theorem 7.1.3. A function has a fuzzy Riemann integral over a closed interval if and only if it is fuzzy continuous in this interval.

It is interesting to remark that for continuous functions, Cauchy was the first to build what we call now the Riemann integral. It is called the Riemann integral because Riemann extended this construction to arbitrary functions. However, taking the fuzzy Riemann integral, we see that it is sufficient to define it only for fuzzy continuous functions because it would be impossible to directly extend this construction to larger classes of functions.

Another construction of a definite integral is developed based on Darboux sums. Namely, the lower Darboux sum of a function $f(x)$ over the interval $[a, b]$ is

$$\Sigma_{i=1}^{n} m(f_i)\, \Delta x_i$$

where $m(f_i) = \inf \{f(x); x \in [x_{i-1}, x_i]\}$.

The upper Darboux sum of a function $f(x)$ over the interval $[a, b]$ is

$$\Sigma_{i-1}^{n} M(f_i)\, \Delta x_i$$

where $M(f_i) = \sup \{f(x); x \in [x_{i-1}, x_i]$.

Darboux sums are also denoted by $\sum_{I_i \in P} m(f_i)\, \Delta x_i$ and $\sum_{I_i \in P} M(f_i)\, \Delta x_i$ where $I_i = [x_{i-1}, x_i]$ is an interval from the partition $P = \{[x_{i-1}, x_i];\ i = 1, 2, 3, \ldots, n\}$).

Let us consider sets UDE_f and LDE_f of upper and lower Darboux sums.

Definition 7.1.3 (cf., for example, (Ross, 1996)). The infimum $\inf UDE_f$ (supremum $\sup LDE_f$) is called the *upper* (*lower*) *Darboux integral* or *definite upper* (*lower*) *Darboux integral* of the function $f(x)$ over the interval $[a, b]$ and is denoted by

$$UD\int_a^b f(x)\, dx \qquad (\ LD\int_a^b f(x)\, dx,\ \text{respectively}\ )$$

Definitions imply the following result.

Lemma 7.1.2. $LD\int_a^b f(x)\, dx \leq UD\int_a^b f(x)\, dx$.

In some cases, upper and lower Darboux integrals coincide.

Definition 7.1.4. (cf., for example, (Ross, 1996)). When $\inf UDE_f = \sup LDE_f$, then it is called the *Darboux integral* or *definite Darboux integral* of the function $f(x)$ over the interval $[a, b]$ and is denoted by

$$D\int_a^b f(x)\, dx$$

Theorem 7.1.4 (cf., for example, (Ross, 1996)). Riemann and Darboux integrals coincide.

We do not give a proof of this result because it is a direct corollary of Theorem 7.1.5, Proposition 7.1.1 and Lemma 7.1.3.

In a similar way, we define the fuzzy Darboux integral. To do this, we need concepts of fuzzy (or approximate) supremum and infimum defined in Section 2.2.

Remark 7.1.4. The situation with fuzzy infima and suprema is different from the situation with fuzzy limits. There are many cases, when the limit of a sequence does not exist but this sequence has many fuzzy limits (cf. Section 2.2). At the same time, as the set $\boldsymbol{R}$ of all real numbers is a complete metric space (Kelly, 1957), supremum $\sup X$ and infimum $\inf X$ exist for any subset X of $\boldsymbol{R}$. So, we may ask a question why we need fuzzy infima and suprema if conventional infima and suprema always exist. The answer is that in many cases, such infima and suprema are only theoretical constructions. Computations cannot give such values. For instance, π as a number can be only denoted by a symbol, but it is impossible to write it as a decimal number. In a decimal (or, e.g., binary) form, we can obtain only approximations of π. The situation is similar even with such a simple number as $\sqrt{2}$. When we have a result of a measurement, then it is impossible even to make a distinction between some fuzzy limits. Thus, in reality, we often can work only with fuzzy infima and suprema but not with their conventional prototypes - infima and suprema. Fuzzy infima and suprema are approximations of conventional infima and suprema.

However, when we are working with incomplete spaces, fuzzy infima and suprema are important by themselves.

Definition 7.1.5. A partition P = $\{[x_{i-1}, x_i]; i = 1, 2, 3, \dots, n\}$ of the interval $[a, b]$ is called a *refinement* of a partition Q = $\{[z_{j-1}, z_j]; j = 1, 2, 3, \dots, m\}$ of the interval $[a, b]$ if each interval $[x_{i-1}, x_i]$ consists of one or several intervals $[z_{j-1}, z_j]$.

For instance, it is possible that $[x_0, x_1] = [z_0, z_1]$ and $[x_1, x_2] = [z_1, z_2] \cup [z_2, z_3] \cup [z_3, z_4]$. Another example: the partition Q = {[0, ¼], [¼, ½], [½, ¾], [¾, 1] } of the interval [0, 1] is a refinement of the partition P = {[0, ½], [½, 1] } of the interval [0, 1].

Lemma 7.1.3. If partition P is a refinement of a partition Q, then

$$\sum_{I_i \in P} m(f_i)\, \Delta x_i \leq \sum_{I_i \in Q} m(f_i)\, \Delta x_i$$

and

$$\sum_{I_i \in Q} M(f_i)\, \Delta x_i \leq \sum_{I_i \in P} M(f_i)\, \Delta x_i$$

Definition 7.1.6. A number $u = r\text{-infUDE}_f$ ($u = r\text{-supLDE}_f$) is called an *upper* (*lower*) *Darboux r-integral* or *definite upper* (*lower*) *Darboux r-integral* of the function $f(x)$ over the interval $[a, b]$ and is denoted by

$$u = \mathrm{UD}r\text{-}\int_a^b f(x)\mathrm{d}x \qquad \left(u = \mathrm{LD}r\text{-}\int_a^b f(x)\mathrm{d}x, \text{ respectively}\right)$$

For fuzzy upper and lower Darboux integrals, the result of Lemma 7.1.1 can be incorrect. For instance, if $f(x) = 1$ when $x \in [0, 1]$, then

$$\mathrm{UD}\int_0^1 f(x)\,\mathrm{d}x = \mathrm{LD}\int_0^1 f(x)\,\mathrm{d}x = 1$$

but

$$-1 = \mathrm{UD}1\text{-}\int_0^1 f(x)\,\mathrm{d}x \text{ and } 1 = \mathrm{LD}1\text{-}\int_0^1 f(x)\,\mathrm{d}x$$

Remark 7.1.5. When the upper Darboux integral $\mathrm{UD}\int_a^b f(x)\mathrm{d}x$ exists, the equality $u = \mathrm{UD}r\text{-}\int_a^b f(x)\mathrm{d}x$, means that the distance from u to $\mathrm{UD}\int_a^b f(x)\mathrm{d}x$ is not larger than r or that u is an approximation of the integral $\mathrm{UD}\int_a^b f(x)\mathrm{d}x$ with the precision r. The same is true for the lower Darboux integral.

Definition 7.1.7. If for a number u, we have $u = r\text{-infUDE}_f$ and $u = r\text{-supLDE}_f$, then u is called a *Darboux r-integral* or *definite Darboux r-integral* of the function $f(x)$ over the interval $[a, b]$ and is denoted by

$$u = \mathrm{D}r\text{-}\int_a^b f(x)\,\mathrm{d}x$$

Remark 7.1.6. When the Darboux integral $\mathrm{D}\int_a^b f(x)\mathrm{d}x$ exists, the equality $u = \mathrm{D}r\text{-}\int_a^b f(x)\mathrm{d}x$, means that the distance from u to $\mathrm{D}\int_a^b f(x)\mathrm{d}x$ is not larger than r or that u is an approximation of the integral $\mathrm{D}\int_a^b f(x)\mathrm{d}x$ with the precision r.

Lemma 7.1.4. $\mathrm{D}0\text{-}\int_a^b f(x)\,\mathrm{d}x$ is the Darboux integral of the function $f(x)$ over the interval $[a, b]$.

Theorem 7.1.5. $u = \mathrm{D}r\text{-}\int_a^b f(x)\,\mathrm{d}x$ if and only if $u = r\text{-}\int_a^b f(x)\,\mathrm{d}x$, that is, fuzzy Riemann and Darboux integrals coincide.

Proof. a) We need to show that if $u = \mathrm{D}r\text{-}\int_a^b f(x)\mathrm{d}x$, then $u = r\text{-}\int_a^b f(x)\,\mathrm{d}x$. Let $u =$

$\mathrm{D}r\text{-}\int_a^b f(x)\,\mathrm{d}x$, then $u = r\text{-}\mathrm{infUDE}_f$ and $u = r\text{-}\mathrm{supLDE}_f$. By Definition 2.2.6, it means that $u + r$ is larger than or equal to any sum $\Sigma_{i=1}^{n} m(f_i)\,\Delta x_i$ and for any $\varepsilon \in \boldsymbol{R}^{++}$, there is a sum $\Sigma_{i=1}^{n} m(f_i)\,\Delta x_i = v$ such that $|\,u - v| < r + \varepsilon$. By Definition 2.2.7, it means that $u + r$ is smaller than or equal to any sum $\Sigma_{i=1}^{n} M(f_i)\,\Delta x_i$ and for any $\varepsilon \in \boldsymbol{R}^{++}$, there is a sum $\Sigma_{j=1}^{m} M(f_j)\,\Delta x_j = w$ such that $|\,u - w| < r + \varepsilon$. Thus, it possible to take a sequence $H = \{\mathrm{P}_t\,;\, t = 1, 2, 3, \ldots\}$ of partitions such that the sequence $\{\, c_t = \Sigma_{i=1}^{n} m(f_{ti})\,\Delta x_{ti}\,;\, t = 1, 2, 3, \ldots\}$ of the corresponding lower Darboux sums r-converges to u and a sequence $K = \{\mathrm{P}_q\,;\, q = 1, 2, 3, \ldots\}$ of partitions such that the sequence $\{\, c_j = \Sigma_{i=1}^{n} M(f_i)\,\Delta x_{ji}\,;\, t = 1, 2, 3, \ldots\}$ of the corresponding upper Darboux sums r-converges to u. Taking a common refinement of the sequences H and K, that is a sequence of partitions that are mutual refinements of partitions from H and K, we see that they have the same property, namely, the corresponding upper and lower Darboux sums r-converge to u. Thus, it is possible to assume that $K = H$.

Let us consider a decreasing sequence $L = \{\mathrm{P}_j\,;\, j = 1, 2, 3, \ldots\}$ of partitions in which the norm $\|\mathrm{P}_i\,\|$ tends to 0 and corresponding Riemann sums $u_{\mathrm{P}j} = \Sigma_{i=1}^{n} f(c_{ji})\,\Delta x_{ji}$ where $\mathrm{P}_j = \{[x_{j,i-1}\,, x_{ji}\,];\, i = 1, 2, 3, \ldots, n\}$. By the definition of upper and lower sums, we have

$$c_j = \Sigma_{i=1}^{n} m(f_{ii})\,\Delta x_{ji} \leq u_{\mathrm{P}j} = \Sigma_{i=1}^{n} f(c_{ji})\,\Delta x_{ji} \leq \Sigma_{i=1}^{n} M(f_{ji})\,\Delta x_{ji} = b_j \qquad (7.1)$$

At first, let us suppose that $L = K = H$. Then by Theorem 2.2.10, we have $u = r\text{-}\lim_{j\to\infty} u_{\mathrm{P}j}$ as $u = r\text{-}\lim_{j\to\infty} c_j$ and $u = r\text{-}\lim_{j\to\infty} b_j$. If $L \neq H$, we take a new sequence L_1 that is the common refinement of the sequences L and H. The inequality (7.1) holds for the corresponding

Riemann sums. So, $u = r\text{-}\lim_{j\to\infty} u_{P_l}$ where $L_1 = \{P_l ; l = 1, 2, 3, \dots\}$. The sequence $\{ u_{P_j} ; j = 1, 2, 3, \dots \}$ is a subsequence of the sequence $\{ u_{P_l} ; l = 1, 2, 3, \dots \}$. Thus, by Lemma 2.2.3, we have $u = r\text{-}\lim_{j\to\infty} u_{P_j}$.

As the decreasing sequence $L = \{P_j ; j = 1, 2, 3, \dots \}$ of partitions in which the norm $\|P_i\|$ tends to 0 was arbitrary, we have

$$u = r\text{-}\int_a^b f(x)\,dx.$$

b) We need to demonstrate that if $u = r\text{-}\int_a^b f(x)\,dx$, then $u = Dr\text{-}\int_a^b f(x)\,dx$. Let $u = r\text{-}\int_a^b f(x)\,dx$. It means that for any decreasing sequence $L = \{P_j ; j = 1, 2, 3, \dots \}$ of partitions in which the norm $\|P_j\|$ tends to 0 and corresponding Riemann sums $u_{P_j} = \Sigma_{i=1}^{n} f(c_{ji})\,\Delta x_{ji}$, $u = r\text{-}\lim_{j\to\infty} u_{P_j}$.

This allows us to show that $u + r$ is larger than or equal to any sum $\Sigma_{i=1}^{n} m(f_i)\,\Delta x_i$ and $u - r$ is smaller than or equal to any sum $\Sigma_{i=1}^{n} M(f_i)\,\Delta x_i$. Really, let us assume that $u - r > \Sigma_{i=1}^{n} M(f_i)\Delta x_i = d$ for some partition $P = \{ I_i ; i = 1, 2, 3, \dots , n\}$ and take a decreasing sequence $L = \{P_j ; j = 1, 2, 3, \dots \}$ of refinements of P in which the norm $\|P_j\|$ tends to 0. If we denote $\Sigma_{I_j\in Q} M(f_j)\,\Delta x_j$ by d_j , then by Lemma 7.1.2, we have $d \geq d_1 \geq d_2 \geq \dots \geq d_j \geq \dots$ There are two possible situations: either we have infinitely many different elements in this sequence or elements d_t become equal to some d_j for all $t = j + 1, j + 2, j + 3, \dots$

In the first situation, we have an infinite sequence of Darboux sums $\Sigma_{I_j\in Q} M(f_j)\,\Delta x_j$ less than $u - r$. If $c = u - r - d$, then by the definition of $M(f_j)$, there are points $c_j \in I_j$ such that $M(f_j) - f(c_j) < (\frac{1}{4} c)/(b - a)$ for all $j = 1, 2, 3, \dots$ Consequently,

$$\Sigma_{I_j\in Q} M(f_j)\,\Delta x_j - \Sigma_{I_j\in Q} f(c_j)\,\Delta x_j <$$
$$[(\tfrac{1}{4} c)/(b - a)]\,\Sigma_{I_j\in Q}\Delta x_j =$$
$$[(\tfrac{1}{4} c)/(b - a)](b - a) = \tfrac{1}{4} c.$$

Let us denote the sum $\Sigma_{I_j\in Q} f(c_j)\,\Delta x_j$ by u_j for all $j = 1, 2, 3, \dots$. Then the distance all these elements u_j to u is larger than $r + \frac{1}{2} c$. Indeed, when $u_j \leq d_j$, we have

$$|\, u - u_j \,| \geq |\, u - d_j \,| > |\, u - d \,| = r + c,$$

and when $u_j > d_j$, we have

$$|\, u - u_j \,| = |\, u - d_j \,| - |\, u_j - d_j \,| >$$
$$|\, u - d \,| - |\, u_j - d_j \,| = r + c - |\, u_j - d_j \,| >$$
$$r + c - \tfrac{1}{4} c = r + \tfrac{3}{4} c > r + \tfrac{1}{2} c$$

At the same time, by the definition of the Riemann r-integral of $f(x)$ over $[a, b]$, almost all elements u_j belong to the ε-neighborhood of u for any ε. As this assumption is violated for $\varepsilon = \frac{1}{4}$, we come to a contradiction. Thus, our assumption is invalid.

The second situation is possible only if $d_j = \int_a^b f(x)\, dx$ because in this case, all Riemann and Darboux sums start coinciding when elements of partitions become sufficiently small. Then the number u cannot be an r-limit of Riemann sums because its distance to the Riemann integral of $f(x)$ over $[a, b]$ is larger than r. This contradicts our assumption.

Theorem is proved.

Another construction of a definite integral is based on step functions and the general approach to integration developed in (Burgin, 1995a).

Let $[a, b]$ be a finite or infinite (e.g., $[a, \infty)$ or $(-\infty, \infty)$) interval of real numbers.

Definition 7.1.8. A function h: $[a, b] \to \boldsymbol{R}$ is called a *step function* if there is a partition P $= \{[x_{i-1}, x_i]; i = 1, 2, 3, \ldots n\}$ of the interval $[a, b]$ such that h is constant, i.e., equal to some number c_i, inside each interval $[x_{i-1}, x_i]$ of this partition. At the boundary point x_i, the value $h(x_i)$ is equal either to c_i or to c_{i+1} $(i = 1, 2, 3, \ldots n)$.

The definite integral $\int_a^b h(x)\, dx$ for a step function $h(x)$ is defined in a natural way, taking into account that the area of a rectangle R is equal to $l \cdot w$ where l is the length of R and w is the width of R. Namely, if $h(x) = c_i$ in the interval Δx_i, $i = 1, 2, \ldots, n$, then we have

$$\int_a^b h(x)\, dx = \Sigma_{i=1}^{n} c_i \cdot \Delta x_i$$

Note that the definite integral of a step function on a partition always exists.

It is possible to define the same step function in $[a, b]$ with respect to different partitions of the interval $[a, b]$.

Lemma 7.1.5. The definite integral of a step function in $[a, b]$ does not depend on partitions of the interval $[a, b]$ with respect to which this function is considered.

Now let us consider a general function f: $[a, b] \to \boldsymbol{R}$ and define for this function the definite integral, utilizing integrals of step functions that are already defined. When the function $f(x)$ is bounded in $[a, b]$, we can always find two step functions $h(x)$ and $k(x)$ such that $h(x) \leq f(x) \leq k(x)$ for all x in the interval $[a, b]$. These functions are considered as step boundaries of the function $f(x)$. Namely, $h(x)$ is called a *lower step boundary* and $k(x)$ an *upper step boundary* of the function $f(x)$.

Taking a common refinement of partitions of the interval $[a, b]$ that correspond to step functions $h(x)$ and $k(x)$ and using the comparison property and Lemma 7.1.5, we conclude that

$$\int_a^b h(x)\, dx \leq \int_a^b k(x)\, dx$$

Lemma 7.1.6. For any lower Darboux sum $\Sigma_{i=1}^{n} m(f_i)\, \Delta x_i$ of a function $f(x)$ over the interval $[a, b]$, there is a step function $h(x)$ such that $h(x) \leq f(x)$ for all x from $[a, b]$ and

$$\Sigma_{i=1}^{n} m(f_i)\ \Delta x_i = \int_a^b h(x)\ dx$$

Lemma 7.1.7. For any step function $h(x)$ such that $h(x) \leq f(x)$ for all x from $[a, b]$, there is a lower Darboux sum $\Sigma_{i=1}^{n} m(f_i)\ \Delta x_i$ of a function $f(x)$ over the interval $[a, b]$ such that

$$\int_a^b h(x)\ dx \leq \Sigma_{i=1}^{n} m(f_i)\ \Delta x_i$$

Lemma 7.1.8. For any upper Darboux sum $\Sigma_{i=1}^{n} M(f_i)\ \Delta x_i$ of a function $f(x)$ over the interval $[a, b]$, there is a step function $h(x)$ such that $f(x) \leq k(x)$ for all x from $[a, b]$ and

$$\Sigma_{i=1}^{n} M(f_i)\ \Delta x_i = \int_a^b k(x)\ dx$$

Lemma 7.1.9. For any step function $k(x)$ such that $f(x) \leq k(x)$ for all x from $[a, b]$, there is an upper Darboux sum $\Sigma_{i=1}^{n} M(f_i)\ \Delta x_i$ of a function $f(x)$ over the interval $[a, b]$ such that

$$\int_a^b k(x)\ dx \geq \Sigma_{i=1}^{n} M(f_i)\ \Delta x_i$$

Let us consider the sets UBS_f and LBS_f all the upper and lower step boundaries of the function $f(x)$ and the sets UBI_f and LBI_f of their integrals.

Definition 7.1.9. The infimum $\inf UBI_f$ (supremum $\sup LBI_f$) is called the *upper* (*lower*) *Daniell integral* or *definite upper* (*lower*) *Daniell integral* of the function $f(x)$ over the interval $[a, b]$ and is denoted by

$$UDE \int_a^b f(x)\ dx \qquad (LDE \int_a^b f(x)\ dx, \text{ respectively})$$

Lemma 7.1.10. $LDE \int_a^b f(x)\ dx \leq UDE \int_a^b f(x)\ dx$.

Definition 7.1.10. When $\inf UBI_f = \sup LBI_f$, then this number is called the *Daniell integral* or *definite Daniell integral* of the function $f(x)$ over the interval $[a, b]$ and is denoted by

$$DE \int_a^b f(x)\ dx$$

Lemma 7.1.11. For any interval $[a, b]$ and function $f(x)$, we have

$$\int_a^b f(x)\ dx \ \leq\ DE \int_a^b f(x)\ dx$$

Theorem 7.1.6. Riemann and Daniell integrals coincide.

We do not give a proof of this theorem as it is a direct corollary of Theorem 7.1.7.

It is necessary to remark that if the function $f(x)$ is not bounded there may be no lower or upper step approximations of $f(x)$: only bounded functions allow Riemann integration. However, it is possible to introduce improper integrals where the limit is taken two times.

Besides, not all bounded functions have the Riemann integral even over a bounded interval. At the same time by Theorem 7.1.1, any bounded in an interval function has a fuzzy Riemann integral over this interval. This allows one to essentially extend the scope of improper integration, taking fuzzy integrals where the function is bounded and then going to a fuzzy limit in the whole domain.

Definition 7.1.11. A number $u = r\text{-infUDI}_f$ ($u = r\text{-supLDI}_f$) is called an *upper* (*lower*) *Daniell r-integral* or *definite upper* (*lower*) *Daniell r-integral* of the function $f(x)$ over the interval $[a, b]$ and is denoted by

$$u = \text{UDE}r\text{-}\int_a^b f(x)\,\text{d}x \qquad (u = \text{LDE}r\text{-}\int_a^b f(x)\,\text{d}x, \text{ respectively})$$

Remark 7.1.7. When the upper Daniell integral $\text{UDE}\int_a^b f(x)\text{d}x$ exists, the equality

$u = \text{UDE}r\text{-}\int_a^b f(x)\text{d}x$, means that the distance from u to $\text{UD}\int_a^b f(x)\text{d}x$ is not larger than r or that u is an approximation of the integral $\text{UD}\int_a^b f(x)\text{d}x$ with the precision r. The same is true for the lower Daniell integral.

Definition 7.1.12. If for a number u, we have $u = r\text{-infUDI}_f$ and $u = r\text{-supLDI}_f$, then u is called a *Daniell r-integral* or *definite Daniell r-integral* of the function $f(x)$ over the interval $[a, b]$ and is denoted by

$$u = \text{DE}r\text{-}\int_a^b f(x)\,\text{d}x$$

Remark 7.1.8. When the Daniell integral $\text{DE}\int_a^b f(x)\text{d}x$ exists, the equality $u = \text{DE}r\text{-}\int_a^b f(x)\text{d}x$, means that the distance from u to $\text{DE}\int_a^b f(x)\text{d}x$ is not larger than r or that u is an approximation of the integral $\text{DE}\int_a^b f(x)\text{d}x$ with the precision r.

Lemma 7.1.12. $\text{DE}0\text{-}\int_a^b f(x)\,\text{d}x$ is the Daniell integral of the function $f(x)$ over the interval $[a, b]$.

Theorem 7.1.7. $u = \text{DE}r\text{-}\int_a^b f(x)\,\text{d}x$ if and only if $u = r\text{-}\int_a^b f(x)\,\text{d}x$, that is, the fuzzy Riemann and Daniell integrals coincide.

Remark 7.1.9. It is possible to ask why we need three different constructions to define actually the same concept of the Riemann integral and three different constructions to define the fuzzy Riemann integral. The reason is that these constructions provide different opportunities in different cases. For instance, in some situations, Riemann sums exist, while Darboux sums and boundary step functions do not exist.

Let us consider the following function.

$$f(x) = \begin{cases} 1/x & \text{when } 0 < x \leq 1; \\ 1 & \text{when } x = 0. \end{cases}$$

This function has a Riemann sum for any partition of the interval [0, 1]. At the same time, $f(x)$ has neither Darboux sums nor upper step functions for this interval.

A distinction between the constructions of the Darboux and Daniell integrals is that Darboux sums involve such non-constructive operations as infimum and supremum. In contrast to this, it possible to take only constructive step functions when we build the Daniell integral. Here the meaning of the term *constructive* allows different interpretations:

1. *Constructive* means what is possible to exactly measure.
2. *Constructive* means computable by a computer.
3. *Constructive* means recursively computable, e.g., recursively computable, or simply, recursive real numbers (cf. (Rice, 1951; Freund, 1983)).
4. *Constructive* means inductively computable, e.g., inductively computable real numbers (cf. (Burgin, 2003b)).

7.2. Properties of the Fuzzy Riemann Integral

Science is an integral part of culture.
Stephen Jay Gould (1941–2002)

At first, we present a property for the fuzzy Riemann integral, and then from it, we derive the corresponding property for the Riemann integral.

(1) Fuzzy (Shifted) Additivity Property with Respect to Functions

Theorem 7.2.1. If a function $f(x)$ is r-integrable and a function $g(x)$ is q-integrable over the same interval $[a, b]$, then their sum is $(r + q)$-integrable and we have:

if $u = r\text{-}\int_a^b f(x)\, dx$ and $v = q\text{-}\int_a^b g(x)\, dx$, then $u + v = (r + q)\text{-}\int_a^b (f(x) + g(x))dx$ and $u - v = (r + q)\text{-}\int_a^b (f(x) - g(x))dx$

Proof. By Definition 7.1.2, $u = r\text{-lim } E_f$ and $v = r\text{-lim } E_g$ where E_f and E_g are sets of all sequences of Riemann sums of functions $f(x)$ and $g(x)$ corresponding to all decreasing sequences $L = \{P_i ; i = 1, 2, 3, \dots \}$ of partitions P_i of the interval $[a, b]$ in which the norm $||P_i||$ of partitions tends to 0. Thus, if E_{f+g} and E_{f-g} are sets of all sequences of Riemann sums of the functions $f(x) + g(x)$ and $f(x) - g(x)$ corresponding to all decreasing sequences $L = \{P_i ; i = 1, 2, 3, \dots \}$ of partitions in which the norm $||P_i||$ tends to 0, then by definition, we have $E_{f+g} = E_f + E_g$ and $E_{f-g} = E_f - E_g$ where operations on sequences are performed element-wise. By properties of fuzzy limits (cf. Section 2.6), $u = r\text{-lim } E$ and $v = q\text{-lim } D$, then $u + v = (r + q)\text{-lim } (E + D)$ and $u - v = (r + q)\text{-lim } (E - D)$. Consequently, $u + v = (r + q)\text{-lim } E_{f+g}$ and $u - v = (r + q)\text{-lim } E_{f-g}$. By Definition 7.1.2, this gives us the necessary statement.

When $r = q = 0$, Theorem 7.2.1 and Proposition 7.1.1 imply the following well-known property of the Riemann integral.

(1a) ***Additivity Property with Respect to Functions***

If two functions $f(x)$ and $g(x)$ are integrable over the same interval $[a, b]$, then their sum is also integrable and we have:

$$\int_a^b f(x)\,dx + \int_a^b g(x)\,dx = \int_a^b (f(x) + g(x))dx$$

(2) Fuzzy (Shifted) Uniformity Property

Theorem 7.2.2. If a function $f(x)$ is r-integrable over an interval $[a, b]$, then for any real number c, the function $f(x)$ is $|c|r$-integrable over the interval $[a, b]$ and we have the following property:

$$\text{if} \quad u = r\text{-}\int_a^b f(x)\,dx, \quad \text{then} \quad cu = |c|r\text{-}\int_a^b cf(x)\,dx$$

Proof. By Definition 7.1.2, $u = r\text{-lim } E_f$. Thus, if E_{cf} is the set of all sequences of Riemann sums of the function $cf(x)$ corresponding to all decreasing sequences $L = \{P_i ; i = 1, 2, 3, \dots \}$ of partitions in which the norm $||P_i||$ tends to 0, then $E_{cf} = cE_f$. By Theorem 2.6.2, if $u = r\text{-lim } E$ and c is a real number, then $cu = |c|r\text{-lim } cE$. Consequently, $cu = |c|r\text{-lim } E_{cf}$. By Definition 7.1.2, this gives us the necessary statement.

When $r = q = 0$, Theorem 7.2.2 and Proposition 7.1.1 imply the following well-known property of the Riemann integral.

(2a) ***Uniformity Property***

If a function $f(x)$ is integrable over an interval $[a, b]$, then for any real number c, we have:

$$\int_a^b cf(x)\,\mathrm{d}x = c\cdot\int_a^b f(x)\mathrm{d}x$$

(Fuzzy) additivity and uniformity, i.e., Theorems 7.2.1 and 7.2.2, imply the following linearity property.

(3) Fuzzy (Shifted) Linearity Property

Theorem 7.2.3. If a function $f(x)$ is r-integrable and a function $g(x)$ is q-integrable over the same interval $[a, b]$, then for any real numbers c and d, their sum is $(|c|r + |d|q)$-integrable over the interval $[a, b]$ and we have:

$$\text{if } u = r\text{-}\int_a^b f(x)\,\mathrm{d}x \text{ and } v = q\text{-}\int_a^b g(x)\,\mathrm{d}x\text{, then } cu + dv = (|c|r + |d|q)\text{-}\int_a^b (cf(x) + dg(x))\mathrm{d}x$$

As a corollary, we have the following classical result.

(3a) *Linearity Property*

If two functions $f(x)$ and $g(x)$ are integrable over the same interval $[a, b]$, then any linear combination of the two is also integrable over the interval $[a, b]$ and for any real number c and d, we have:

$$c\cdot\int_a^b f(x)\,\mathrm{d}x + d\cdot\int_a^b g(x)\,\mathrm{d}x = \int_a^b (c\cdot f(x) + d\cdot g(x))\mathrm{d}x$$

Proofs of the following properties are similar to the proofs of properties 1 – 3.

(4) Fuzzy Reversibility Property

Theorem 7.2.4. If a function $f(x)$ is r-integrable over the interval $[a, b]$ and

$$u = r\text{-}\int_a^b f(x)\,\mathrm{d}x\text{, then } -u = r\text{-}\int_b^a f(x)\,\mathrm{d}x$$

When $r = 0$, Theorem 7.2.4 and Proposition 7.1.1 imply the following well-known property of the Riemann integral.

(4a) *Reversibility Property*

Provided the function is integrable over the interval $[a, b]$, we have:

$$\int_a^b f(x)\mathrm{d}x = - \int_b^a f(x)\mathrm{d}x$$

(5) Fuzzy (Shifted) Additivity Property with Respect to the Integration Interval

Theorem 7.2.5. Given three numbers a, b and c (whose order is unimportant), and provided a function $f(x)$ is r-integrable over the interval $[a, b]$, and is q-integrable over the interval $[b, c]$, then the function $f(x)$ is $(r + q)$-integrable over the interval $[a, c]$, and we have:

$$\text{if } u = r\text{-}\int_a^b f(x)\,\mathrm{d}x \text{ and } v = q\text{-}\int_b^c f(x)\,\mathrm{d}x \text{, then } u + v = (r + q)\text{-}\int_a^c f(x)\,\mathrm{d}x$$

When $r = q = 0$, Theorem 7.2.5 and Proposition 7.1.1 imply the following well-known property of the Riemann integral.

(5a) ***Additivity Property with Respect to the Integration Interval***

Given three numbers a, b and c (whose order is unimportant), a function $f(x)$, and provided the function is integrable over each of the intervals considered, we have:

$$\int_a^c f(x)\mathrm{d}x = \int_a^b f(x)\mathrm{d}x + \int_b^c f(x)\,\mathrm{d}x$$

(6) Fuzzy Monotonicity Property

Theorem 7.2.6. If functions $f(x)$ and a function $g(x)$ are r-integrable over the same interval $[a, b]$, $f(x) > g(x) + 2r/k$ for all x from $[a, b]$ where $k = b - a$, then for

$$c = r\text{-}\int_a^b f(x)\mathrm{d}x \text{ and } d = r\text{-}\int_a^b g(x)\mathrm{d}x\text{, we have } c \geq d.$$

Corollary 7.2.1. If $f(x) > 0$ for all x from $[a, b]$ and

$$c = r\text{-}\int_a^b f(x)\mathrm{d}x \text{ , then } c \geq -2r.$$

Corollary 7.2.2. If $f(x) > 2r/k$ for $b - a = k$ and all x from $[a, b]$ and

$$c = r\text{-}\int_a^b f(x)\mathrm{d}x \text{ , then } c \geq 0.$$

When $r = q = 0$, Theorem 7.2.6 and Proposition 7.1.1 imply the following well-known property of the Riemann integral.

(6a) ***Monotonicity (Comparison) Property***

If two functions $f(x)$ and $g(x)$ are integrable over the same interval $[a, b]$ (with $a < b$), if $f(x) \leq g(x)$ for all x in $[a, b]$ then:

$$\int_a^b f(x)\, \mathrm{d}x \leq \int_a^b g(x)\, \mathrm{d}x$$

Corollary 7.2.3. a) If $f(x) \geq 0$ for all x from $[a, b]$, then:

$$\int_a^b f(x)\mathrm{d}x \geq 0.$$

b) If $f(x) \geq 0$ for all x from $[a, b]$, $f(c) > 0$ for some c from $[a, b]$, and $f(x)$ is a continuous function, then:

$$\int_a^b f(x)\mathrm{d}x > 0$$

Remark 7.2.1. In contrast to the case of the conventional Riemann integral, it is possible that $f(x)$ is a continuous function and $f(x) > 0$ for all x from $[a, b]$, but

$$v = r\text{-}\int_a^b f(x)\mathrm{d}x \text{ and } v < 0.$$

Example 7.2.1. Let us consider the function that is identically equal to ¼ .

Then $\frac{1}{4} = \int_0^1 f(x)\, \mathrm{d}x$, $-\frac{1}{2} = 1\text{-}\int_0^1 f(x)\, \mathrm{d}x$, and $-\frac{1}{2} = v < 0$.

(7) Fuzzy Absolute Value Property

Theorem 7.2.7. If $u = r\text{-}\int_a^b f(x)\, \mathrm{d}x$, then the absolute value $|f(x)|$ is also fuzzy

b

integrable over the interval $[a, b]$, and if $v = r\text{-}\int_a |f(x)|\, dx$, then $|u| < v + 2r$.

Remark 7.2.2. It is possible that $v = r\text{-}\int_a^b |f(x)|\, dx$, $u = r\text{-}\int_a^b f(x)\, dx$, and $v < u$.

Example 7.2.2. Let us consider the following function:

$$f(x) = \begin{cases} 1 & \text{when } x \text{ is irrational;} \\ 0 & \text{when } x \text{ is rational.} \end{cases}$$

Then $0 = 1\text{-}\int_0^1 |f(x)|\, dx$, $1 = |1\text{-}\int_0^1 f(x)\, dx|$, and $0 = v < 1 = u$.

When $r = q = 0$, Theorem 7.2.7 and Proposition 7.1.1 imply the following well-known property of the Riemann integral.

(7a) ***Absolute Value Property***

If a function $f(x)$ is integrable over an interval $[a, b]$, then the absolute value $|f(x)|$ is also integrable over the interval $[a, b]$ and the following inequality holds (with $a < b$):

$$\int_a^b |f(x)|\, dx \geq \left| \int_a^b f(x)\, dx \right|$$

Thus, we see that in some cases, properties of the fuzzy Riemann integral and conventional Riemann integral are the same (cf., for example, Theorem 7.2.4). In other cases, corresponding properties of the fuzzy Riemann integral and conventional Riemann integral are similar (cf. for example, Theorems 7.2.1 – 7.2.3), while there are cases when corresponding properties are essentially different (cf. for example, Remarks 7.1.2, 7.2.1 and 7.2.2).

Chapter 8

FUZZY DYNAMICAL SYSTEMS

Sometimes one creates a dynamic impression
by saying something, and sometimes one creates as significant
an impression by remaining silent.
Dalai Lama Tenzin Gyatso (1935-)

In this chapter, we consider applications of neoclassical analysis to dynamical systems. The main goal here is to reconsider classical results from the theory of dynamical systems and ergodic theory, making them appropriate for treating systems about which we have imprecise, vague, uncertain or incomplete information.

The main interest here is the Poincaré recurrence theorem. It is one of the basic results of ergodic theory (cf., for example, (Sinai, 1977; Aoki and Hiraide, 1994; Katok and Hasselblatt, 1997)). It has many applications in mathematical physics. A physical interpretation of this theorem states that if a mechanical system governed by Newton's laws has a fixed total energy that restricts its dynamics to bounded subsets of its phase space, then the system will eventually return as closely as we choose to any given initial set of states. In the kinetic theory of gases, the system will eventually return to any neighborhood of a given initial set of molecular positions and velocities.

There are other applications. For instance, the Poincaré Recurrence Theorem explains why, according to the contemporary cosmological model, the Universe is pulsating under the indicated initial conditions (Gurovich and Fridman, 1968). In (Furstenberg, 1981) relations between the Poincaré recurrence and number theory are considered.

However, the Poincaré recurrence theorem is inherently classical and hence many branches of modern physics, such as quantum mechanics or the theory of gauge fields, do not provide necessary conditions for application of this theorem, as well as many other results of ergodic theory.

Some researchers suggest (cf., for example, (Bugajski, 1995)) that "quantum" in physical theories may be interpreted as "fuzzy", i.e., that the most characteristic feature of quantum observables is their "fuzziness." Thus, to model quantum systems, we need fuzzy dynamical systems.

Another cause of imprecision in physics is measurement. Any real measurement is never absolutely precise but gives only approximate results. So, the data that we process are not absolutely precise. This inaccuracy leads to the inaccuracy in the result of data processing. The problem is to estimate the resulting inaccuracy.

Two more aspects that demand fuzzy models are incompleteness of information in control processes and finite precision of computations. For instance, computing points of a trajectory of a system, we can find these points only with a finite precision.

As we know, imprecision and fuzziness is much larger in social and behavioral sciences than in natural sciences. This precludes valid application of the Poincaré recurrence theorem in social and behavioral sciences.

All this shows that constructions and methods developed in the classical theory of dynamical systems are only approximations to what exists in reality. In many situations, such approximations have been giving a sufficiently adequate representation of studied phenomena. However, scientists and, especially, engineers have discovered many cases in which such methods did not work because classical approach is too rough, using abstractions that are too far from reality. Fuzzy dynamical systems give an appropriate framework for capturing system uncertainty and imprecision.

We study qualitative properties of dynamical systems related to recurrent points. According to the basic Poincaré recurrence theorem, the set of recurrent points has full measure in any subset of the classical dynamical system space. In general, this property can be invalid for fuzzy dynamical systems. In this chapter, we consider fuzzy dynamical systems of two types.

Dynamical systems of the first type are based on fuzzy measures in the same way as conventional dynamical systems are based on classical measures.

Dynamical systems of the second type allow one to use not only measure preserving mappings as it is done in the classical case, but also mappings that preserve the measure only to some extent.

These less restricted conditions are more relevant to real life than classical conditions for dynamical systems. Indeed, when we make measurement, there are no means to get absolutely precise results. Consequently, it is impossible to test all conditions of classical measures, as well as to assure that some transformation completely preserves a given measure.

We start our exposition with classical dynamical systems in Section 8.1. In Section 8.2, elements of the theory of fuzzy measures are presented to make our exposition more complete. In Section 8.3, fuzzy dynamical systems are introduced. Section 8.4 contains a study of recurrent points in fuzzy dynamical systems, explicating differences and similarities between classical and fuzzy dynamical systems.

According to the basic Poincaré recurrence theorem, the set of recurrent points, i.e., points that in the system evolution return to the same part from which they started, has full measure in any subset of the classical dynamical system space. In general, this property is invalid for fuzzy dynamical systems. We give some examples of this situation. However, it is demonstrated (Theorem 8.4.1) that for fuzzy dynamical systems of the first type, the Poincaré property of recurrent points remains true if the fuzzy measure defined in the system space satisfies some, rather weak, additional conditions. Examples show that these conditions are essential for validity of the result. It is also proved (Corollary 4.3) that infinitely recurrent points, i.e., points that in the system evolution return infinitely many times to the same part from which they started, have the same property. At the same time, for fuzzy dynamical systems of the second type, the Poincaré property of recurrent points is invalid in general, and it is possible (Theorem 8.4.2) to find only sufficiently big subsets of recurrent points. In contrast to fuzzy dynamical systems of the first type, infinitely recurrent points of fuzzy dynamical systems of the second type do not have even a weaker property that have recurrent points, namely, the measure of the set of infinitely recurrent points can be equal to zero.

Fuzzy Poincaré recurrence theorems allow us to explain various phenomena in social, cognitive and individual practice. For instance, if we analyze history of science as the dynamics of scientific knowledge, we can see many recurrent ideas and approaches.

Let us look at the idea of atomic structure of matter (cf., for example, (Lindsay, 1972)). The beginning was in ancient Greece. Two Attic philosophers Leucippus (fifth century B.C.E.) and Democritus (circa 460-370 B.C.E.) taught that the whole universe consists of a void and a very large number of invisible and indivisible particles that were called *atoms*. Probably this idea emerged even earlier and came to them from the Ionian philosopher Anaxagoras (circa 500-428 B.C.E.). Later Epicurus (341-270 B.C.E.) adopted the atomic idea and taught it to his students, while the great Roman poet Lucretius (99-55 B.C.E.) explained the atomic idea in his famous poem "De rerum natura" (On the nature of things).

However, the atomic idea found mighty opponents. One of the most influential was Aristotle (384-332) who refused to accept these ideas. The influence of Aristotle was so great that for a long time, the majority of philosophers and physicists, when physics became a separate science, also refused to accept the atomic idea at all. Only in the 17th century, this idea returned to the scientific discourse. The main protagonists of this theory were Pierre Gassendi (1592-1655) and Robert Boyle (1627-1691). In the 1660s, Boyle proposed a corpuscular theory of matter to explain behavior of gases. According to this theory, there was only one fundamental element, all corpuscles of it were identical and different substances were constructed by combining the corpuscles in different ways. Then the 18th century was barren of further progress in the domain of the atomic theory, so far as physics is concerned.

In 1808, the atomic theory was again resurrected by John Dalton (1766-1844). He discovered a law of partial pressures of gases, describing how gases of equal volume contribute pressures in nearly integer ratios. He concluded that these were ratios of atomic weights which were a characteristic of indivisible atoms. This would also explain chemical composition and the nature of the chemical elements. Amedeo Avogadro (1776-1856) developed the molecular theory and his law that all gases at the same temperature, pressure and volume contain the same number of molecules even though their weights are different. James Clerk Maxwell (1831-1879) and Ludvig Boltzmann (1844-1906) went on to explain the laws of thermodynamics through the statistical physics of molecular motion. The atomic theory was having unprecedented success in explaining a wide variety of physical phenomena and due to many supporting discoveries, became domineering in the 19th century, although disagreement continued, with eminent physicists like Ernst Mach (1838-1916) and Wilhelm Ostwald (1853-1932) denouncing the atomic theory.

Examples of other recurrent ideas in physics are the concept of void/vacuum, corpuscular structure of light, and idea of relativity. Examples of recurrent ideas in mathematics are reduction of geometry to arithmetic/algebra, axiomatic approach, and integration. It is possible to explain all these phenomena in cognition by the fuzzy Poincaré recurrence theorem.

Thus, we can observe recurrence predicted by fuzzy versions of the Poincaré recurrence theorem not only for physical systems but also in such domains of human culture as science, mathematics, economics, etc.

It is useful to know that it is possible to develop many directions of classical theory of dynamical systems, for example, symbolic dynamics and chaos theory (Robinson, 1995), in the fuzzy context constructed in this Chapter.

It is also necessary to remark that other researchers studied other types of fuzzy dynamical systems and their properties (cf., for example, (Kloeden, 1982; Bugajski, 1995; Dumitrescu, 1995; Friedman and Sandler, 1996; 1999; Dumitrescu, Hloiu, and Dumitrescu, 2000)).

8.1. Dynamical Systems

The greatest obstacle in finding the truth
Is not the lie itself,
But what seems to be the truth.
Arthur Schopenhauer (1788-1860)

We begin with the classical concept of a dynamical system.

All systems in physics evolve in time and this evolution is governed by some set of physical laws. These systems can range from relatively simple, such as a clock pendulum, to extremely complex, like the complete climate system. A dynamical system can be in a definite state and its evolution is a sequence of state changes, i.e., the system goes from one state to another in this process. The evolution rule of the dynamical system is a fixed rule that describes what future states follow from the current state.

In a mathematical model, states are usually presented by collections of real or complex numbers, or more generally, by sets of points in an appropriate state space. Changes in the state of the system correspond to changes in the numbers that represent states. When the state space is such a geometrical space as a manifold, points are represented by their coordinates. A general mathematical definition of a dynamical system is based on the concept of a measure. To rigorously introduce this concept, we need several other mathematical concepts, a more detailed exposition of which can be found in (Kolmogorov and Fomin, 1999) or in (Shilov and Gurevich, 1978).

Definition 8.1.1. A system $\mathbf{B}$ of sets is called a *set ring* if it satisfies the following conditions:

(F1) $A, B \in \mathbf{B}$ implies $A \cap B \in \mathbf{B}$.

(F2) $A, B \in \mathbf{B}$ implies $A \Delta B \in \mathbf{B}$ where $A \Delta B = (A \setminus B) \cup (B \setminus A)$.

Proposition 8.1.1. For any set ring $\mathbf{B}$, we have $\varnothing \in \mathbf{B}$ and $A, B \in \mathbf{B}$ implies $A \cup B$, $A \setminus B \in \mathbf{B}$.

Indeed, $A \Delta A = \varnothing$, $A \setminus B = A \setminus (A \cap B) = A \Delta (A \cap B)$, and $A \cup B \in \mathbf{B}$ as $(A\Delta B)\Delta(A\cap B) = (((A \setminus B) \cup (B \setminus A)) \setminus (A \cap B))) \cup ((A \cap B) \setminus ((A \setminus B) \cup (B \setminus A))) = ((A \setminus B) \cup (B \setminus A)) \cup (A \cap B) = A \cup B$.

It is possible to find a more detailed proof of this result in (Kolmogorov and Fomin, 1999).

Let $\mathbf{P}(X)$ be the set of all subsets of a set X.

Definition 8.1.2. A set ring $\mathbf{B} \subseteq \mathbf{P}(X)$ with a unit element, i.e., an element E from $\mathbf{B}$ such that for any A from $\mathbf{B}$, we have $A \cap E = A$, is called a *set algebra.*

Definition 8.1.3. A set algebra $\mathbf{B}$ is called a *set σ-algebra* it satisfies the following condition:

$$\text{if } A_n \in \mathbf{B} \text{ for all } i, j, n = 1, \dots, m, \dots, \text{ then } A = \bigcup_{n=1}^{\infty} A_n \in \mathbf{B}.$$

Example 8.1.1. A Borel field of sets from $\mathbf{P}(X)$ where X is an interval in $\boldsymbol{R}$ (in particular, it can be the whole $\boldsymbol{R}$) is the least set σ-algebra that contains all intervals from X.

Let $\mathbf{P}(X)$ be the set of all subsets of a set X and $\mathbf{B}$ be a set ring from $\mathbf{P}(X)$, in particular, $\mathbf{B} \subseteq \mathbf{P}(X)$.

Definition 8.1.4. A *measure* on $\mathbf{B}$ in X is a function μ: $\mathbf{B} \to \boldsymbol{R}^+$ that assigns to each set from $\mathbf{B}$ a positive real number and satisfies the following condition:

(M1) for any $A = \bigcup_{n=1}^{m} A_n$ from $\mathbf{B}$ with $A_i \cap A_j = \varnothing$ and $A_n \in \mathbf{B}$ for all $i, j, n = 1, \dots, m$ with $i \neq j$, we have

$$\mu(A) = \Sigma_{n=1}^{m} \mu(A_n)$$

Often the system $\mathbf{B}$ is called the *algebra of measurable sets* with respect to the measure μ.

Lemma 8.1.1. $\mu(\varnothing) = 0$.

Indeed, the equality $\varnothing = \varnothing + \varnothing$ implies by (M1), that $\mu(\varnothing) = 2\mu(\varnothing)$, i.e., $\mu(\varnothing) = 0$.

An important property of measures is σ-additivity.

Definition 8.1.5. A measure μ in X is called *σ-additive* if the following condition is satisfied:

(M2) for any $A = \bigcup_{n=1}^{\infty} A_n$ from $\mathbf{B}$ with $A_i \cap A_j = \varnothing$ for all $i \neq j$ and $A_n \in \mathbf{B}$ for all $n = 1, \dots, m, \dots$, we have

$$\mu(A) = \Sigma_{n=1}^{\infty} \mu(A_n)$$

In many sources, the term measure means a σ-additive measure.

Let us take a set X with a measure μ and an algebra $\mathbf{B}$ of its measurable subsets.

Definition 8.1.6. The system $(X, \mathbf{B}, g)$ is called a *measurable space.*

Definition 8.1.7. A mapping T: $X \to X$ is called an *endomorphism* of the measurable space $(X, \mathbf{B}, \mu)$ if for any A from $\mathbf{B}$, $T^{-1}(A)$ also belongs to $\mathbf{B}$ and

$$\mu(A) = \mu(T^{-1}(A)) \tag{8.1}$$

where $T^{-1}(A)$ is the inverse image of A.

Example 8.1.2. If k is a real number and μ is the Lebesgue measure on the real line, then the mapping $T(x) = x + k$ is an endomorphism of the real line $\boldsymbol{R}$.

Now we can give a general mathematical definition of a dynamical system (cf., for example, (Aoki and Hiraide, 1994) or (Katok and Hasselblatt, 1997)).

Definition 8.1.8. A *dynamical system* in a space X with a measure m consists of a measurable space (X, **A**, m), some semigroup R with the operation $\circ$, which is called *time* in the space (X, **A**, m), and a *parametric system* $\mathbf{T}_R = \{ T^i ; i \in R \}$ of endomorphisms of the space (X, **A**, m), i.e., $T^i (T^j (x)) = T^{i \circ j}(x)$ for any $i, j \in R$ and $x \in X$.

Such system may be fuzzy in any component. The basic set X can be a fuzzy set. The field **A** of sets from X can be fuzzy. The measure m can be fuzzy. The temporal semigroup R can be fuzzy. Endomorphisms T^i in the space (X, **A**, m) can be a fuzzy. Finally, the composition of these endomorphisms can be also fuzzy. As a result, we obtain different types of fuzzy dynamical systems.

Here we consider two types of fuzziness: dynamical systems in spaces with a fuzzy measure (fuzzy dynamical systems of the first type) and dynamical systems with fuzzy endomorphisms (fuzzy dynamical systems of the second type).

8.2. Fuzzy Measures

There is beauty in everything, but not everybody sees it.
Confucius

Sugeno (1974) introduced the concept of a fuzzy measure.

Let **P**(X) be the set of all subsets of a set X and **A** be a Borel field of sets from **P**(X), specifically, $\mathbf{A} \subseteq \mathbf{P}(X)$.

Definition 8.2.1. A *fuzzy measure* on **A** in X (in the sense of Sugeno) is a function g: $\mathbf{A} \to [0,1]$ that assigns a number in the unit interval [0,1] to each set from A so that the following conditions are valid:

(FM1) $X \in \mathbf{A}$, $g(\emptyset) = 0$, and $g(X) = 1$, i.e., the function g is normed.

(FM2) the function g is monotone, i.e., for any A and B from A, the inclusion $A \subseteq B$ implies $g(A) \leq g(B)$.

(FM3) For any non-decreasing sequence $A_1 \subseteq A_2 \subseteq \dots \subseteq A_n \subseteq A_{n+1} \subseteq \dots$ of sets from **A**, the following equality is valid

$$g(\bigcup_{n=1}^{\infty} A_n) = \lim_{n\to\infty} g(A_n) \quad (8.2)$$

(FM4) For any non-increasing sequence $A_1 \supseteq A_2 \supseteq \dots \supseteq A_n \supseteq A_{n+1} \supseteq \dots$ of sets from **A**, the following equality is valid

$$g(\bigcap_{n=1}^{\infty} A_n) = \lim_{n\to\infty} g(A_n)) \quad (8.3)$$

If $A \in \mathbf{A}$, then the value g(A) is called the fuzzy measure of the set A.

Then this definition was improved further through elimination of the condition (FM4) (Sugeno, 1977; Zimmermann, 2001).

Definition 8.2.2. A *fuzzy measure* on **A** in X (in the sense of Zimmermann) is a function g: **A** $\rightarrow [0,1]$ that assigns a number in the unit interval $[0,1]$ to each set from **A** so that the following conditions are valid:

(FM1) $g(\varnothing) = 0$ and $g(X) = 1$.

(FM2) For any A and B from **A**, the inclusion $A \subseteq B$ implies $g(A) \leq g(B)$.

(FM3) For any non-decreasing sequence $A_1 \subseteq A_2 \subseteq \ldots \subseteq A_n \subseteq A_{n+1} \subseteq \ldots$ of sets from **A**, the following equality is valid

$$g(\bigcup_{n=1}^{\infty} A_n) = \lim_{n\to\infty} g(A_n) \,) \tag{8.4}$$

Bannon (1981) demonstrated that many measures with infinite universe studied by different researchers, such as probability measures, belief functions, plausibility measures, and so on, are fuzzy measures in the sense of Sugeno.

We take even more general definition of a fuzzy measure based on (Klir and Wang, 1993).

Let **P**(X) be the set of all subsets of a set X and **B** be an algebra of sets from **P**(X), in particular, **B** $\subseteq$ **P**(X).

Definition 8.2.3. A *fuzzy measure* on **B** in X (in the sense of Klir and Wang) is a function g: **B** $\rightarrow$ $\boldsymbol{R}^+$ that assigns to each set from **B** a positive real number, is monotone (FM2):

(FM1) $g(\varnothing) = 0$.

(FM2) For any A and B from **B**, the inclusion $A \subseteq B$ implies $g(A) \leq g(B)$.

In what follows, we call g simply a fuzzy measure and call **B** the *algebra of fuzzy measurable sets* with respect to the fuzzy measure g.

Definition 8.2.4. A fuzzy measure g on **B** in X is called *bounded* if $X \in$ **B**.

Definition 8.2.5. A bounded fuzzy measure g on **B** in X is called *normed* if g: **B** $\rightarrow [0,1]$ and $g(X) = 1$.

Definition 8.2.6. Two bounded fuzzy measures g and m are called equivalent if they have the same algebra **B** of fuzzy measurable sets and there is a one-to one mapping e: $[0,1] \rightarrow [0,1]$ such that for any set A from **B**, we have

$$g(A) \leq e(m(A))$$

Lemma 8.2.1. Any bounded fuzzy measure is equivalent to a normed fuzzy measure.

Popular examples of fuzzy measures are possibility, belief and plausibility measures.

Example 8.2.1. Possibility theory is based on possibility measures.

Let us consider some set X and its power set **P**(X).

Definition 8.2.7. A *possibility measure* in X is a partial function Pos: **P**(X) $\rightarrow [0,1]$ that is defined on a subset **A** from **P**(X) and satisfies the following axioms (cf. Zadeh, 1978; Higashi and Klir, 1982; Zimmermann, 2001):

(Po1) $\varnothing, X \in$ **A**, $Pos(\varnothing) = 0$, and $Pos(X) = 1$.

(Po2) For any A and B from **A**, the inclusion $A \subseteq B$ implies $Pos(A) \leq Pos(B)$.

(Po3) For any system $\{ A_i ; i \in I \}$ of sets from **A**,

$$Pos(\bigcup_{i \in I} A_i) = \sup_{i \in I} Pos(A_i) \text{)} \quad (8.5)$$

A possibility measure is a fuzzy measure in the sense of Definition 8.2.3, but it is not always a fuzzy measure in the sense of Definitions 8.2.1 and 8.2.2 (Puri and Ralesky, 1982).

Example 8.2.2. Possibility is also described by a more general class of measures (cf. (Oussalah, 2000; Zadeh, 1978)).

Definition 8.2.8. A *quantitative possibility measure* in X is a function P: $\mathbf{P}(X) \to [0,1]$ that satisfies the following axioms:

(Po1) $\varnothing, X \in \mathbf{A}$, $P(\varnothing) = 0$, and $P(X) = 1$;
(Po2a) For any A and B from **A**, $P(A \cup B) = \max \{P(A), P(B)\}$.

A quantitative possibility measure is a fuzzy measure in the sense of Definition 8.2.3.

Example 8.2.3. Dual to a quantitative possibility measure is a necessity measure (Oussalah, 2000).

Definition 8.2.9. A *quantitative necessity measure* in X is a function N: $\mathbf{P}(X) \to [0,1]$ that satisfies the following axioms:

(Ne1) $N(\varnothing) = 0$, and $N(X) = 1$.
(Ne2) For any A and B from $\mathrm{P}(X)$, $N(A \cap B) = \min \{N(A), N(B)\}$.

A quantitative necessity measure is a fuzzy measure in the sense of Definition 8.2.3.

Possibility and necessity measures are important in support logic programming (Baldwin, 1986). It uses fuzzy measures for reasoning under uncertainty and approximate reasoning in expert systems, based on the logic programming style.

Definition 8.2.10. A fuzzy measure g in X is called:

a) *additive* if for any A and B from **B**, the equality $A \cap B = \varnothing$ implies
$$g(A \cup B) = g(A) + g(B) \text{)} \quad (8.6)$$
b) *super-additive* if for any A and B from **B**, the equality $A \cap B = \varnothing$ implies
$$g(A \cup B) \geq g(A) + g(B) \text{)} \quad (8.7)$$
c) *subadditive* if for any A and B from **B**, the equality $A \cap B = \varnothing$ implies
$$g(A \cup B) \leq g(A) + g(B) \text{)} \quad (8.8)$$

Let g be an additive fuzzy measure.

Lemma 8.2.2. If $g(A) = 0$, then $g(A \cup B) = g(B)$.

Proposition 8.2.1. A fuzzy measure is a measure if and only if it is additive.

Indeed, on the one hand, the equality (8.6) is a particular case of the condition (M1). Thus, by Lemma 8.1.1, any measure is an additive fuzzy measure.

On the other hand, if the equality (8.6) is true, then by induction, we can prove the condition (M1). Thus, any additive fuzzy measure is a measure.

Let k be a non-negative real number.

Definition 8.2.11. A fuzzy measure g in X is called k-additive if for any A and B from **B**, the equality $A \cap B = \varnothing$ implies

$$g(A \cup B) \geq (1/2) \cdot (1 + k) \cdot (g(A) + g(B)) \;) \qquad (8.9)$$

Lemma 8.2.3. Any super-additive fuzzy measure g is k-additive for any $k \leq 2$.

If we take a non-decreasing sequence $A_1 \subseteq A_2 \subseteq \ldots \subseteq A_n \subseteq A_{n+1} \subseteq \ldots$ of sets from X, then the sequence of numbers $g(A_1) \subseteq g(A_2) \subseteq \ldots \subseteq g(A_n) \subseteq g(A_{n+1}) \subseteq \ldots$ is non-decreasing, bounded and thus, has the limit $\lim_{n\to\infty} g(A_n)$.

Definition 8.2.12. A fuzzy measure g in X is called *continuous* if for any non-decreasing sequence $A_1 \subseteq A_2 \subseteq \ldots \subseteq A_n \subseteq A_{n+1} \subseteq \ldots$ of sets from **B**, the union $\bigcup_{n=1}^{\infty} A_n$ belongs to **B** and the following equality is valid

$$g(\bigcup_{n=1}^{\infty} A_n) = \lim_{n\to\infty} g(A_n) \qquad (8.10)$$

Proposition 8.2.2. An additive fuzzy measure is continuous if and only if it is σ-additive.

Remark 8.2.1. Additivity of g is essential for validity of Proposition 8.2.2 as the following example demonstrates.

Example 8.2.4. Let us consider the set $X = [0, 1]$ and the set algebra **B** generated by all intervals of the form $[a, b]$, $[a, b)$, $(a, b]$ or (a, b) with $a \geq 0$ and $b \leq 1$. For any of these intervals I, we define $g(I) = b - a$ when $b - a > 1/8$ and $g(I) = 0$ when $b - a \leq 1/8$. Any set B from the algebra **B** is a union of a finite number of intervals because if A and B are intervals, then $A \cap B$ and $A \setminus B$ are intervals (may be empty) and $A \Delta B = (A \setminus B) \cup (B \setminus A)$ is either an interval or a union of two intervals. If $B \in$ **B**, $a = \inf B$ and $b = \sup B$, then B is equal to one of the intervals $[a, b]$, $[a, b)$, $(a, b]$ or (a, b) minus some finite set $\{I_1, I_2, \ldots, I_t\}$ of subintrvals from (a, b). Then $g(I) = 0$ when $b - a \leq 1/8$ and $g(I) = (b - a) - \Sigma_{i=1}^{t} g(I_i)$ when $b - a > 1/8$. For instance, $g([3/10, 4/10]) = 0$, $g([3/7, 4/7]) = 1/7$, $g([6/32, 7/32] \cup [8/32, 9/32]) = 0$, $g([6/32, 7/32] \cup [18/32, 19/32]) = (19/32 - 6/32) - (18/32 - 7/32) = 13/32 - 11/32 = 2/32 = 1/16$, and $g([6/32, 7/32] \cup [11/32, 12/32]) = (12/32 - 6/32) - 0 = 6/32 = 3/16$.

It is possible to show that the function g is a continuous fuzzy measure in [0, 1]. However, it is not additive as, for example, $g([3/7, 4/7]) = 1/7 > g([6/14, 7/14]) + g([7/14, 8/14]) = 0 + 0 = 0$, and thus, g is not a measure and not σ-additive.

Let r be a non-negative real number.

Definition 8.2.13. A fuzzy measure g in X is called *r-continuous* if for any non-decreasing sequence $A_1 \subseteq A_2 \subseteq \ldots \subseteq A_n \subseteq A_{n+1} \subseteq \ldots$ of sets from **B**, the union $\bigcup_{n=1}^{\infty} A_n$ belongs to **B** and the following equality is valid

$$g(\bigcup_{n=1}^{\infty} A_n) = r\text{-}\lim_{n\to\infty} g(A_n) \qquad (8.11)$$

Because the difference between fuzzy measures of any two sets from X cannot be larger than 1 when the fuzzy measure is normed, we have the following result.

There are r-continuous fuzzy measures that are not continuous.

Example 8.2.5. Let us consider the set $X = [0, 1]$ and the set algebra **B** generated by all intervals of the form $[a, b]$, $[a, b)$, $(a, b]$ or (a, b) with $a \geq 0$ and $b \leq 1$. For any of these intervals I, we define $g(I) = b - a$ when $b - a \geq 1/8$ and $g(I) = 0$ when $b - a \leq 1/8$. Any set B from the algebra **B** is a union of a finite number of intervals because if A and B are intervals, then $A \cap B$ and $A \setminus B$ are intervals (may be empty) and $A \Delta B = (A \setminus B) \cup (B \setminus A)$ is either an interval or a union of two intervals. If $B \in \mathbf{B}$, $a = \inf B$ and $b = \sup B$, then B is equal to one of the intervals $[a, b]$, $[a, b)$, $(a, b]$ or (a, b) minus some finite set $\{I_1, I_2, \dots, I_t\}$ of subintrvals from (a, b). Then $g(I) = 0$ when $b - a < 1/8$ and $g(I) = (b - a) - \Sigma_{i=1}^{t} g(I_i)$ when $b - a > 1/8$. For instance, $g([3/10, 4/10]) = 0$, $g([3/8, 4/8]) = 1/8$, $g([6/32, 7/32] \cup [8/32, 9/32]) = 0$, $g([6/32, 7/32] \cup [18/32, 19/32]) = (19/32 - 6/32) - (18/32 - 7/32) = 13/32 - 11/32 = 2/32 = 1/16$, and $g([6/32, 7/32] \cup [11/32, 12/32]) = (12/32 - 6/32) - 0 = 6/32 = 3/16$.

It is possible to show that the function g is a (1/8)-continuous fuzzy measure in [0, 1]. However, it is not continuous as if we take $A_n = [1/4, 3/8 - 1/(n + 10))$, then $\bigcup_{n=1}^{\infty} A_n = [1/4, 3/8]$ belongs to **B**, $g(\bigcup_{n=1}^{\infty} A_n) = (1/8)\text{-}\lim_{n\to\infty} g(A_n)$ but $g(\bigcup_{n=1}^{\infty} A_n) = 1/8 \neq \lim_{n\to\infty} g(A_n)$ as all $g(A_n) = 0$.

Proposition 8.2.3. An arbitrary normed fuzzy measure g in X is 1-continuous.

Definition 8.2.14. A fuzzy measure g in X is called *uniformly fuzzy continuous* if it is r-continuous for some non-negative real number $r < 1$.

Definition 8.2.15. A fuzzy measure g in X is called *fuzzy continuous* if for any non-decreasing sequence $A_1 \subseteq A_2 \subseteq \dots \subseteq A_n \subseteq A_{n+1} \subseteq \dots$ of sets from **B** there is a non-negative real number $r < 1$ such that the union $\bigcup_{n=1}^{\infty} A_n$ belongs to **B** and the following equality is valid

$$g(\bigcup_{n=1}^{\infty} A_n) = r\text{-}\lim_{n\to\infty} g(A_n) \tag{8.12}$$

In addition to ordinary measures, fuzzy measures encompass many kinds of measures introduced and studied by different researchers. Here are some examples.

Example 8.2.6. Beliefs play an important role in people's behavior. To study beliefs by methods of fuzzy set theory, the concept of belief measure was introduced and studied (Shafer, 1976; Banon, 1981).

Definition 8.2.16. A *belief measure* in X is a partial function Bel: $\mathbf{P}(X) \to [0,1]$ that is defined on a subset **A** from $\mathbf{P}(X)$ and satisfies the following axioms:

(Be1) $\varnothing, X \in \mathbf{A}$, $Bel(\varnothing) = 0$, and $Bel(X) = 1$.

(Be2) For any system $\{ A_i ; i = 1, 2, \dots, n \}$ of sets from **A** and any n from $\boldsymbol{N}$,

$Bel(A_1 \cup \dots \cup A_i) \geq \Sigma_{i=1}^{n} Bel(A_i) - \Sigma_{i<j} Bel(A_i \cap A_j) + \dots + (-1)^{n+1} Bel(A_1 \cap \dots \cap A_i)$

Axiom (Be2) implies that belief measure is a super-additive fuzzy measure as for $n = 2$ and arbitrary subsets A and B from X, we have

$$Bel(A \cup B) \geq Bel(A) + Bel(B) - Bel(A \cap B)\,) \tag{8.13}$$

Axiom (Be2) also implies the following property

$$0 \leq Bel(A) + Bel(\overline{A}\,) \leq 1) \tag{8.14}$$

where $\overline{A}$ is a complement of A.

For each set $A \in \mathbf{P}(X)$, the number $Bel(A)$ is interpreted as the degree of belief (based on available evidence) that a given element x of X belongs to the set A.

Another interpretation treats subsets of X as answers to a particular question. It is assumed that some of the answers are correct, but we do not know with full certainty which ones they are. Then the number $Bel(A)$ estimates our belief that the answer A is correct.

Example 8.2.7. Plausibility measures are related to belief measures.

Definition 8.2.17. A *plausibility measure* in X is a partial function Pl: $\mathbf{P}(X) \to [0,1]$ that is defined on a subset $\boldsymbol{A}$ from $\mathbf{P}(X)$ and satisfies the following axioms (Banon, 1981; Shafer, 1976):

(Pl1) $\varnothing, X \in \boldsymbol{A}$, $Pl(\varnothing) = 0$, and $Pl(X) = 1$.

(Pl2) For any system $\{ A_i ; i = 1, 2, \dots , n \}$ of sets from $\boldsymbol{A}$ and any n from $\boldsymbol{N}$,

$$Pl(A_1 \cap \dots \cap A_i) \leq \Sigma_{i=1}^{n} Pl(A_i) - \Sigma_{i<j} Pl(A_i \cup A_j) + \dots + (-1)^{n+1} Pl(A_1 \cup \dots \cup A_i)) \quad (8.15)$$

Axiom (Pl2) implies that plausibility measure is a subadditive fuzzy measure as for $n = 2$ and arbitrary subsets A and B from X, we have

$$Pl(A \cap B) \leq Pl(A) + Pl(B) - Pl(A \cup B)) \quad (8.16)$$

Transforming this inequality, we have

$$Pl(A \cup B) \leq Pl(A \cup B) + Pl(A \cap B) \leq Pl(A) + Pl(B)) \quad (8.17)$$

Belief measure and plausibility are *dual measures* as for any belief measure $Bel(A)$, $Pl(A) = 1 - Bel(\overline{A})$ is a plausibility measure and for any plausibility measure $Pl(A)$, $Bel(A) = 1 - Pl(\overline{A})$ is a belief measure.

Example 8.2.8. When Axiom (Be2) for belief measures is replaced with a stronger axiom

$$Bel(A \cup B) = Bel(A) + Bel(B) \text{ whenever } A \cap B = \varnothing \quad (8.18)$$

we obtain a special type of belief measures, the *classical probability measures* (sometimes also referred to as *Bayesian belief measures*).

In a similar way, it is possible to obtain some special kinds of probability measures, for instance, those that are studied in (Dempster, 1967).

It is interesting to consider dynamical systems in spaces with a belief, plausibility or possibility measure. These systems allow one to model mental processes, cognition, and information processing in intelligent systems. For example, it is possible to consider data- and knowledge bases as dynamical systems with a belief measure and study their behavior. Such a belief measure can reflect user beliefs in correctness and validity of data, as well as user beliefs in truth and groundedness of knowledge systems.

8.3. Dynamical Systems with Fuzzy Measures and Fuzzy Transformations

The universe is dynamic.
When we are creative,
we are the most alive and in touch with it.
Brad Dourif (1950-)

Let us take a set X with a fuzzy measure g and an algebra **B** of its fuzzy measurable subsets.

Definition 8.3.1. The system (X, **B**, g) is called a *fuzzy measurable space*.

Definition 8.3.2. A mapping T: $X \to X$ is called an *endomorphism* of the fuzzy measurable space (X, **B**, g) if for any A from **B**, $T^{-1}(A)$ also belongs to **B** and

$$g(A) = g(T^{-1}(A))) \tag{8. 19}$$

where $T^{-1}(A)$ is the inverse image of A.

Example 8.3.1. If k is a real number, then the mapping $T(x) = x + k$ is an endomorphism of the real line $\boldsymbol{R}$.

Let r be a non-negative real number.

Definition 8.3.3. A mapping T: $X \to X$ is called an *r-endomorphism* of X if it is into X, for any A from **B**, $T^{-1}(A)$ also belongs to **B** and

$$| g(A) - g(T^{-1}(A))| \leq r \tag{8.20}$$

Example 8.3.2. If k is a real number, then the mapping $T(x) = (1 - r)x$ is an r-endomorphism of the interval [0, 1].

Remark 8.3.1. This construction is close to the concept of Ulam (1964) of an approximate homomorphism of a universal algebra. For instance, in the category of groups, one can consider a mapping f from a group A to a group B such that $f(xy)$ not necessarily equals to $f(x)f(y)$, but must be "close" to $f(x)f(y)$. Such approximate homomorphisms were also studied by Farah (1998; 2000).

Definition 8.3.2. A mapping T: $X \to X$ is called a fuzzy endomorphism of X if it is an r-endomorphism of X for some $r \geq 0$.

Definition 8.3.3. A mapping T: $X \to X$ is called an automorphism (r-automorphism, fuzzy endomorphism) of X if it is a one-to-one endomorphism (r-endomorphism, fuzzy endomorphism, correspondingly) of X.

Example 8.3.3. The mapping $T(x) = - x$ is an automorphism of the interval [-1, 1].

Let (X, **B**, g) be a fuzzy measurable space, R be some semigroup with the operation $\circ$, which we call time in the space (X, **B**, g), and $\mathbf{T}_R = \{ T^i ; i \in R \}$ be a parametric system of endomorphisms of (X, **B**, g), i.e., $T^i(T^j(x)) = T^{i \circ j}(x)$ for any $i, j \in R$ and $x \in X$

Definition 8.3.4. A structure ((X, **B**, g), R, $\mathbf{T}_R$) is called a *fuzzy dynamical system of the first type*.

Let (X, **B**, m) be a measurable space, R be some semigroup with the operation °, which we call time in the space (X, **B**, m), and $\mathbf{T}_R = \{ T^i ; i \in R \}$ be a parametric system of r-endomorphisms of (X, **B**, m), i.e., $T^i(T^j(x)) = T^{i \circ j}(x)$ for any $i, j \in R$ and $x \in X$

Definition 8.3.5. A structure ((X, **B**, m), R, $\mathbf{T}_R$) is called a *fuzzy dynamical system of the second type.*

Example 8.3.4. Let (X, **B**, m) be the interval [0, 1] with the Lebesgue measure m on the algebra **B** of Borel sets. Then the mapping $T(x) = x^2$ is a (1/4)-endomorphism and ((X, **B**, m), $\boldsymbol{Z}$, $\mathbf{T}_Z$) is a fuzzy dynamical system of the second type with time $\mathbf{T}_Z = \{ T^i ; i \in \boldsymbol{Z} \}$.

8.4. Recurrent Points in Fuzzy Dynamical Systems

To confine our attention to terrestrial matters
would be to limit the human spirit.
Stephen Hawking (1942-)

Let $A \subseteq X$ and $T: X \to X$ be a mapping.

Definition 8.4.1. A point a from X is called *recurrent* with respect to A and T if $T^n(a)$ belongs to A for some natural number $n > 0$.

Let $A \subseteq X$, R be a partially ordered semigroup with 0, and $\mathbf{T}_R = \{ T^i ; i \in R \}$ be a parametric system of endomorphisms of (X, **B**, g).

Definition 8.4.2. A point a from X is called *recurrent* with respect to A and $\mathbf{T}_R$ if $T^i(a)$ belongs to A for some $i \in R$ with $i > 0$.

Remark 8.4.1. There is a close connection between recurrent points, fuzzy/approximate fixed points (Dugungji and Granas, 1982; Burgin, 1987a) and fuzzy/approximate periodic points. This connection is explicated by introduction of recurrence time for a point x with respect to a set A that contains this point.

Definition 8.4.3. The *recurrence time* TR(x, A) of x with respect to A is a sequence of integers (t_1, t_2, ... , t_i, ...) such that $T^{t_i}(x) \in A$ for all $i = 1, 2, \ldots$

For any point in a dynamical system with discrete positive time, there is the least recurrent time.

Let $f: X \to Y$. and $k \in \boldsymbol{R}^+$.

Definition 8.4.4. A point $x \in X$ is called a *k-fuzzy fixed point* for a mapping f if

$$| f(x) - x| \leq k$$

Proposition 8.4.1. A point $x \in X$ is a k-fuzzy fixed point for a mapping T^n if there is a set $A \subseteq X$ such that n belongs to some recurrence time of x with respect to A and $Diam(A) \leq k$.

It is possible to prove a similar result for fuzzy/approximate periodic points.

For fuzzy dynamical systems of the first type with discrete countable time, the Poincaré recurrence theorem is not valid, in general, as it is demonstrated by the following example.

Example 8.4.1. Let X = [-1, 1], **A** be a Borel algebra of sets in X, m be the Lebesgue measure on X, and $D = \{ \pm 1/n ; n = 2, 3, \ldots \}$. As a countable set, $D \in$ **A**. We define the fuzzy measure g by the following rules:

if $A \in \mathbf{A}$ and $A \cap D \neq \varnothing$, then $g(A) = (10/21)\cdot(m(A) + (1/10))$;
if $A \in \mathbf{A}$ and $A \cap D = \varnothing$, then $g(A) = (10/21)\cdot m(A)$.

These rules imply that $g(D) = g(C) = 1/21$ for any $C \subseteq D$. In addition, both conditions FM1a and FM2 from Definition 8.2.3 are satisfied. So, g is a fuzzy measure. In addition, $g(X) = 1$, i.e., the fuzzy measure g is normed. As a result, we obtain a fuzzy measurable space (X, $\boldsymbol{A}$, g).

We define on X the following transformation:

$$T(x) = \begin{cases} \frac{1}{2} & \text{if } x = -\frac{1}{2}; \\ x & \text{if } x \notin D; \\ (1/n+1) & \text{if } x = \pm 1/n \end{cases}$$

As the transformation T maps D into itself, T is an endomorphism of X. As a result, we obtain a fuzzy dynamical system $((X, \boldsymbol{A}, g), R, \mathbf{T}_Z)$ of the first type. This system has discrete time $\boldsymbol{Z}$.

However, for any interval $[1/n, \frac{1}{2}]$ with $k > 2$, the measure g of non-recurrent points is not zero but is equal to 1/21.

At the same time, when we take fuzzy dynamical systems with k-additive fuzzy measure and discrete countable time, the result of the Poincaré recurrence theorem remains true.

Let R be the additive group $\boldsymbol{Z}$ of all integer numbers, k be a non-negative real number, and $\mathbf{D} = ((X, \mathbf{B}, g), R, \mathbf{T}_Z)$ be a fuzzy dynamical system of the first type with discrete time $\boldsymbol{Z}$.

Theorem 8.4.1 (Generalized Poincaré Recurrence Theorem). For any set A from $\mathbf{B}$ in a fuzzy dynamical system $\mathbf{D}$ of the first type with a k-additive fuzzy measure g, the measure of the set of non-recurrent with respect to A and $\mathbf{T}_Z$ points in A is equal to zero, or, in other words, almost all points in A are recurrent.

Proof. Given a subset A of a set X, let us consider the set Y of all non-recurrent with respect to A and $\mathbf{T}_Z$ points in A. Then by the definition $Y = A \cap ((\bigcap_{n=1}^{\infty} T^{-n}(X \setminus A))$. Really, if $x \in \bigcap_{n=1}^{\infty} T^{-n}(X \setminus A)$ and x is recurrent with respect to A and $\mathbf{T}_Z$, then $T^m(x) \in A$ for some $m > 0$. Besides, $x \in T^{-m}(X \setminus A)$. Consequently, $T^m(x) \in T^m(T^{-m}(X \setminus A)) = X \setminus A$. This contradicts to our supposition that $T^m(x) \in A$ and thus, the point x is not recurrent and $x \in Y$. As x is an arbitrary point from $\bigcap_{n=1}^{\infty} T^{-n}(X \setminus A)$, we have $\bigcap_{n=1}^{\infty} T^{-n}(X \setminus A) \subseteq Y$.

If $x \notin T^{-m}(X \setminus A)$ for some $m > 0$, then $x \in T^{-m}(A)$. Consequently, $T^m(x) \in A$, i.e., x is a recurrent point. Thus, $x \in T^{-m}(X \setminus A)$ for all $m > 0$ and we have proved that $Y = A \cap ((\bigcap_{n=1}^{\infty} T^{-n}(X \setminus A))$.

In addition, $Y \cap T^{-n}(Y) = \varnothing$ for all $n > 0$. Really, if $x \in Y$, then $T^n(x) \notin A$ for all $n > 0$. However, if $x \in T^{-n}(Y)$, then $T^n(x) \in Y$ and $Y \subseteq A$, i.e., $T^n(x) \in A$.

Moreover, $T^{-n}(Y) \cap T^{-m}(Y) = \varnothing$ when $m \neq n$. Really, if $0 < m < n$, then $T^{-n}(Y) \cap T^{-m}(Y) = T^{-m}(Y \cap T^{-(n-m)}(Y) = \varnothing$ as $Y \cap T^{-k}(Y) = \varnothing$ for all $k > 0$.

Now let us suppose that $g(Y) = a > 0$. As g is a k-additive fuzzy measure, the transformation T preserves measure, and $Y \cap T^{-1}(Y) = \varnothing$, we have $g(Y \cup T^{-1}(Y)) \geq (1/2) \cdot (1 + k)(g(Y) + g(T^{-1}(Y))) = (1 + k)(g(Y) = (1 + k)a$. This implies $g(T^{-2}(Y) \cup T^{-3}(Y))) = g(T^{-2}(Y \cup T^{-1}(Y))) = g(Y \cup T^{-1}(Y)) \geq (1 + k)a$. Then by the same arguments, we have $g(Y \cup T^{-1}$

$(Y) \cup T^{-2}(Y) \cup T^{-3}(Y))) \geq (1 + k)^2 a$. By induction, we prove that $g(\bigcup_{n=0}^{r} T^{-n}(Y))) \geq (1 + k)^m a$ when $r = 2^m - 1$.

As $k > 0$ and $a > 0$, there is such m for which $g(\bigcup_{n=0}^{r} T^{-n}(Y)) > 1$. Because $\bigcup_{n=0}^{r} T^{-n}(Y) \subseteq X$ and $g(X) = 1$, monotonicity of g yields a contradiction. Thus, our assumption cannot be true and $g(Y) = 0$.

Theorem is proved.

Any additive fuzzy measure g is a k-additive fuzzy measure with $k = 1$. This implies the following result.

Corollary 8.4.1. For any measurable set A in a fuzzy dynamical system **D** of the first type with an additive fuzzy measure, the measure of the set of non-recurrent with respect to A and $\mathbf{T}_Z$ points in A is equal to zero or almost all points from A are recurrent.

Remark 8.4.2. For fuzzy dynamical systems of the first type in which the measure is not additive, two conclusions of Corollary 8.4.1 are not always equivalent as it is demonstrated by the following example.

Example 8.4.2. Let $X = [-1, 1]$, A be a Borel field of sets in X, m be a Lebesgue measure on X, and $D = \{ \pm 1/n \,;\, n = 2, 3, \dots \}$. As a countable set, D is an element of A. We define the fuzzy measure g by the following rules:

if $A \subseteq (1/3\ ,\ ½)$, then $g(A) = (m(A) - (1/10))$;

for all other sets $A \in A$, $g(A) = m(A)$.

Both conditions from Definition 8.2.3 are satisfied. As a result, we obtain a fuzzy measurable space $(X, \boldsymbol{A}, g)$.

We define on X the following transformation:

$$T(x) = \begin{cases} ½ & \text{if } x = -½; \\ x & \text{if } x \notin D; \\ (1/n+1) & \text{if } x = \pm 1/n \end{cases}$$

As the transformation T maps D into itself, T is an endomorphism of X. As a result, we obtain a fuzzy dynamical system $((X, \boldsymbol{A}, g), R, \mathbf{T}_Z)$ of the first type. This system has discrete time $\boldsymbol{Z}$.

In addition, for any set $\boldsymbol{A} \in A$, the measure g of non-recurrent points is equal to zero, but the measure g of all recurrent points in the segment $[1/3\ ,\ ½]$ is not equal to $g([1/3\ ,\ ½]) = 1/6$.

Corollary 8.4.2 (Poincaré Recurrence Theorem). For any measurable set A in a dynamical systems with an σ-additive measure, the measure of the set of non-recurrent with respect to A and $\mathbf{T}_Z$ points in A is equal to zero or almost all points from A are recurrent.

Definition 8.4.5. A point a from X is called *infinitely* (*finitely*) *recurrent* with respect to A and $\mathbf{T}_R$ if $T^i(a)$ belongs to A for an infinite (finite) number of $i \in R$ with $i > 0$.

Corollary 8.4.3. For any set A from **B** in a fuzzy dynamical system **D** of the first type with a k-additive continuous fuzzy measure g, the measure of the set of non-infinitely recurrent with respect to A and $\mathbf{T}_Z$ points in A is equal to zero.

Proof. Given a subset A of a set X, let us consider the set Y of all point from A that are not infinitely recurrent with respect to A and $\mathbf{T}_Z$ points in A. If x belongs to Y, then there is some number p such that for all $k > p$, $T^k(a)$ does not belong to A. By the definition, it means that x

is non-recurrent with respect to A and $\mathbf{T}^p{}_Z$. By the Generalized Poincaré Recurrence Theorem the measure of the set Y_p of all such points is equal to zero. As $Y = \bigcup_{p=1}^{\infty} Y_p$, for all n, $A_n \subseteq A_{n+1}$, and the measure in **D** is continuous, the measure of the set Y is equal to zero.

Corollary 8.4.3 is proved.

Corollary 8.4.4. For any measurable set A in a fuzzy dynamical system **D** of the first type with an additive continuous fuzzy measure, almost all points in A are infinitely recurrent with respect to A and $\mathbf{T}_Z$.

Let us consider a space X with a measure (fuzzy measure) m, a non-negative real number k, some property P of elements from X, and a subset A of X.

Definition 8.4.6. We say that *k-almost all* elements from A have property P if $m(A) - m(A_P) \leq k$ were $A_P = \{ x \in A \, ; x \text{ has the property } P \}$.

Let R be the additive group $\boldsymbol{Z}$ of all integer numbers and $\mathbf{D} = ((X, \boldsymbol{A}, \text{g}), R, \mathbf{T}_Z)$ be a fuzzy dynamical system of the second type with an additive fuzzy measure g and discrete time $\boldsymbol{Z}$.

Theorem 8.4.2 (Fuzzy Poincaré Recurrence Theorem). For any set A from $\boldsymbol{A}$ in a fuzzy dynamical system **C** of the second type with an r-endomorphism T, the measure of the set of non-recurrent with respect to A and $\mathbf{T}_Z$ points in A is less than or equal to r, or, in other words, r-almost all points in A are recurrent.

Proof. Let us consider a subset A of the set X and the set Y of all non-recurrent with respect to $\mathbf{T}_Z$ points in A. As it is demonstrated in the proof of Theorem 8.4.1, $Y = A \cap ((\bigcap_{n=1}^{\infty} T^{-n}(X \setminus A))$. In addition, $Y \cap T^{-n}(Y) = \varnothing$ for all $n > 0$. Moreover, $T^{-n}(Y) \cap T^{-m}(Y) = \varnothing$ when $m \neq n$. Really, if $0 < m < n$, then $T^{-n}(Y) \cap T^{-m}(Y) = T^{-m}(Y \cap T^{-(n-m)}(Y) = \varnothing$ as $Y \cap T^{-k}(Y) = \varnothing$ for all $k > 0$.

The measure g is additive, the transformation T is an r-endomorphism, and $Y \cap T^{-1}(Y) = \varnothing$. So, we have $\text{g}(Y \cup T^{-1}(Y)) \geq 2\text{g}(Y) - r$ because $\text{g}(T^{-1}(Y)) \geq \text{g}(Y) - r$. By induction, $\text{g}(\bigcup_{n=0}^{p} T^{-n}(Y)) \geq (p+1)\text{g}(Y) - pr$ for an arbitrary $p = 1, 2, \ldots$.

Now let us assume that $\text{g}(Y) = a > r$. Then $\text{g}(\bigcup_{n=0}^{p} T^{-n}(Y)) \geq (p+1)a - pr > p(a - r)$. As $a - r$ is a fixed positive number and p tends to infinity, for some p we have $\text{g}(\bigcup_{n=0}^{p} T^{-n}(Y)) > p(a - r) > 1$. This is a contradiction because all considered sets are subsets of X, $\text{g}(X) = 1$, and g satisfies condition (2) from Definition 2.3.

Equivalence of the two conclusions of the theorem follows from Lemma 2.1.

Theorem is proved.

For $r = 0$, we have Poincaré Recurrence Theorem for the case when the measure is only additive.

Corollary 8.4.5. For any measurable set A in a dynamical systems, the measure of the set of non-recurrent with respect to A and $\mathbf{T}_Z$ points in A is equal to zero or almost all points from A are recurrent.

However, the statement of Corollary 8.4.4 about existence of infinite recurrent points for dynamical systems of the first type, as well as for classical dynamical systems, is not true for dynamical systems of the second type. It is demonstrated by the following example.

Example 8.4.3. Let $X = [0, 1]$, A be a Borel field of sets in X, m be a Lebesgue measure on X, and $T(x) = (9/10)x$. The mapping $T(x)$ is a $(1/10)$-endomorphism of X as $m(A) - m(TA) \leq 1/10$. As a result, we obtain a fuzzy dynamical system $((X, A, m), \boldsymbol{Z}, \mathbf{T}_Z)$ of the second type with the discrete time $\boldsymbol{Z}$.

Let us consider the set $A = [8/10, 1]$ and the set NRA of all non-recurrent points from A. Then $m(\text{NR}A) = (8/9) - (8/10) = 4/45 < 1/10$. This correlates with the statement of Fuzzy Poincaré Recurrence Theorem. However, any point from A returns to A at most two times. Moreover, for any set $B = [a, b]$ with $0 < a < b < 1$, there are no infinitely recurrent points.

Thus, we have extended the scope of application of the Poincaré recurrence theorem, proving its more advanced versions for fuzzy dynamical systems. Some extended versions of the Poincaré recurrence theorem are somewhat weaker than the initial result of Poincaré but examples show that this is an intrinsic in the fuzzy context and cannot be remedied. With weaker assumptions it is futile, as a rule, to hope to get the same strong results, although weaker results are still useful and more realistic in practice.

Chapter 9

CONCLUSION

A journey of a thousand miles begins with a single step.
Confucius

Thus, we have seen how neoclassical analysis extends the scope of the classical calculus and analysis, brings new results and completes several basic classical results, as well as produces deeper insights and a better understanding of the classical theory. It is possible to compare the development of neoclassical analysis with introduction of irrational numbers, which make the space of all numbers complete and increase power of mathematical methods.

In Conclusion, we compare neoclassical analysis with such related areas as interval analysis (Section 9.1) and fuzzy set theory (Section 9.2). In the last section of this chapter, open problems and directions for further research are formulated.

9.1. COMPARISON OF INTERVAL ANALYSIS AND NEOCLASSICAL ANALYSIS

He who knows others is learned.
He who knows himself is wise
Lao-Tzu

It is necessary to state that although the approach of neoclassical analysis has common features with interval analysis, these two directions have essential differences.

Distinctive features:

1. Interval analysis works with intervals, while in a way similar to computers, neoclassical analysis works with individual numbers. The latter approach better correlates with contemporary situation in the information processing field because computers still work with numbers and not with fuzzy sets or intervals.
2. Interval analysis gives estimation (as an interval) for all possible approximations reflecting limits for all appropriate approximations, while neoclassical analysis gives

estimation (as a number) for an individual point reflecting to what extent this point is a good approximation of what we need.

3. In interval analysis, it is necessary to have precise operations for the ends of intervals, while in neoclassical analysis, it is possible to use imprecise operations.
4. Methods of interval analysis are different from methods of the classical calculus, while methods of neoclassical analysis are rather close to methods of the classical calculus
5. Interval analysis is dealing with approximations, assuming that there is always one absolutely correct result. Neoclassical analysis also works with approximations, but at the same time, it considers situations when there are no absolutely correct results but only approximately correct ones.
6. Interval analysis is treating all kinds of imprecision related to computing. Neoclassical analysis deals only with situations of uncertainty, imprecision, and vagueness that emerge in limit processes.
7. Objects of interval analysis are crisp (classical) sets, while objects of neoclassical analysis (fuzzy limits, fuzzy continuous functions, fuzzy derivatives, etc.) form fuzzy sets.
8. Neoclassical analysis introduces such mathematical objects as, for instance, fuzzy continuous functions, which have no analogs in interval analysis.

At the same time, neoclassical analysis and interval analysis have a lot in common.

Common features:

1. Both interval analysis and neoclassical analysis reflect uncertainty and imprecision.
2. According to international standards, a measurement result is complete only when accompanied by a quantitative estimation of its uncertainty and imprecision. The uncertainty and imprecision measure is required in order to decide if the result is adequate for its intended purpose and to ascertain if the result is consistent with other similar results. Over the years, many different approaches to evaluating and expressing the uncertainty and imprecision of measurement results have been developed and used. Both interval analysis and neoclassical analysis give mathematical foundations for operating with measures of uncertainty and imprecision. In neoclassical analysis, such a measure is given by the number r in r-limits, r-continuity, and r-derivatives, while the measure in interval analysis is the length of an interval.
3. Performing an operation (analytical, computational or measurement) and evaluating the result, e.g., finding a limit in neoclassical analysis or adding intervals in interval analysis, both interval analysis and neoclassical analysis base their actions on a distance criterion. For instance, the concept of a fuzzy continuity is defined using distances between values of the independent variable and distances between values of the function, while intervals in interval analysis are defined based on the distance between exact and approximate solutions. Definition of the metric continuity measure also utilizes the distance from a given function to a continuous function. At the same time, intervals in interval analysis represent possible distance between the

correct result and results distorted by errors or bounded by restrictions that have computers and calculators.

4. Results of operations, such as limits, derivatives or values of integrals, (and even initial data) in neoclassical analysis can be integrated into a form of intervals similar to intervals in interval analysis. For instance (cf. Section 2.5), r-limits of a sequence form an interval. Each continuous function determines an interval (or more exactly, a ball, which can be considered as a multidimensional version of a one-dimensional interval) of (fuzzy) r-continuous functions.

9.2. Comparison of Fuzzy Set Theory and Neoclassical Analysis

He who knows, does not speak.
He who speaks, does not know.
Lao-Tzu

Neoclassical analysis has many similarities and dissimilarities with fuzzy set theory.

Distinctive features:

1. In comparison with fuzzy set theory, neoclassical analysis does not go that far in changing mathematics: while fuzzy set theory changes all mathematical structures and concepts to their fuzzy replicas, neoclassical analysis preserves basic conventional mathematical objects, such as sets and functions. In particular, fuzzy set theory works with fuzzy sets (e.g., fuzzy numbers are also fuzzy sets), while neoclassical analysis works with conventional numbers or spaces, such as metric spaces.
2. The standard fuzzy set theory gives a precise estimation (as a value of the membership function) of similarity/closeness, while neoclassical analysis is satisfied with some approximate estimate (as a number).
3. Fuzzy set theory works with precise concepts reflecting imprecise objects (numbers, functions, sets, etc.), while neoclassical analysis works with imprecise concepts (like fuzzy limits or fuzzy continuous functions) reflecting precise objects (numbers, functions, sets, etc.).
4. Methods of fuzzy set theory often are different from methods of the classical calculus, while methods of neoclassical analysis are rather close to methods of the classical calculus
5. Fuzzy set theory assumes that the membership function gives an exact estimation of fuzziness. Estimates in neoclassical analysis, such as the number r in r-limits, r-continuity and r-derivatives, are not necessarily exact. As a rule, they often reflect absence of exact knowledge even about the estimate.
6. Fuzzy set theory tries to deal with all kinds of imprecision and fuzziness. Neoclassical analysis deals only with situations of uncertainty, imprecision, and vagueness that emerge in limit processes.

At the same time, neoclassical analysis and fuzzy set theory have a lot in common.

Common features:

1. Both fuzzy set theory and neoclassical analysis are developed to reflect uncertainty, imprecision, and vagueness.
2. Performing an operation (analytical, computational or measurement) and evaluating the result, both fuzzy set theory and neoclassical analysis can build their estimates on metric criteria. For instance, the concept of a fuzzy limit is defined using distances between numbers or between elements in metric spaces, while the membership function in a fuzzy set can be defined based on the distance (closeness) between a property of an object and some standard property. Definition of the metric continuity measure (cf. Section 3.2) also utilizes the distance from a given function to a continuous function. At the same time, it is possible to interpret values of the fuzzy set membership function as values of some (often informal) metric.
3. Performing an operation (analytical, computational or measurement) and evaluating the result, both fuzzy set theory and neoclassical analysis take into account to what extent the result satisfies some condition (graduality and granularity of properties).
4. Results of operations (and even initial data) in neoclassical analysis are naturally integrated into the form of fuzzy sets, which are objects from fuzzy set theory. For instance, fuzzy limits of a sequence form a fuzzy set (cf. Section 2.5). Fuzzy derivatives form a fuzzy set (cf. Section 4.7). The same is true for fuzzy continuous functions.
5. Both fuzzy set theory and neoclassical analysis work with approximations and at the same time, they consider situations when there are no absolutely correct results but only approximately correct ones. For instance (cf. Section 2.7), there are situations when fuzzy limits of divergent sequences exist. There are fuzzy sets that are not normalized, i.e., in which the membership function is always less than 1. It means that there are no objects that completely satisfy all conditions that define this fuzzy set.

9.3. Further Development of Neoclassical Analysis: Directions for Research and Open Problems

To every thing there is a season,
and a time to every purpose under the heaven:
A time to break down and time to build up;
A time to cast away stones and time to gather stones together; ...
Ecclesiastes

History of mathematics and science teaches us that when a theory is built, it generates new problems and opens new directions for further research. This is true for neoclassical analysis. In this section, we list some of such problems and directions for further research. We begin with the most general problems.

Analysis of function of one variable is only the first step in functional analysis. After building neoclassical analysis of function of one variable, it is necessary to consider functions of several variables.

Problem 1. *Develop neoclassical analysis for multivariable and vector-valued functions.*

In particular, we have the following problem.

Problem 2. *Develop fuzzy differential and integral calculi for functions in* $\boldsymbol{R}^n$.

The next step is to analyze functions on infinite dimensional spaces.

Problem 3. *Develop neoclassical analysis in infinite dimensional spaces.*

In particular, we have the following problem.

Problem 4. *Develop fuzzy differential calculus and integral calculi for functions in Hilbert spaces.*

Problem 5. *Develop neoclassical analysis on conventional and infinite-dimensional manifolds.*

It is necessary to remark that multivariable and vector-valued functions and mappings of infinite dimensional spaces, i.e., operators, are functions in metric spaces. Thus, the results of neoclassical analysis obtained for metric spaces are true for multivariable and infinite dimensional cases. However, real multivariable functions and operations in spaces of these functions (integration and differentiation) have their specific properties, as well as operators in infinite dimensional spaces and operations in spaces of these functions (integration and differentiation) have their specific properties. Thus, all these cases deserve separate development in the context of neoclassical analysis.

It would also be natural to extend fuzzy analysis to more general contexts.

Problem 6. *Develop a non-standard fuzzy differential and integral calculi.*

Problem 7. *Develop fuzzy differential and integral calculi for extrafunctions.*

It is an interesting problem to combine neoclassical approach with interval analysis and fuzzy set theory, finding what uncertainty due to limit procedures is added to the initial uncertainty. In fuzzy set theory, there are means for differentiation of fuzzy functions. In some sense, it is an exact differentiation of fuzzy functions because it is based on a fuzzification the standard concept of a limit. However, using the theory of fuzzy limits, it is possible to elaborate new methods for differentiation of fuzzy functions and relations. Consequently we have the following problem.

Problem 8. *Construct a theory of fuzzy differentiation for fuzzy functions.*

Such a theory might be obtained by synthesizing methods from fuzzy set theory with constructions from neoclassical analysis. The first stage in doing this is connected with the following problem.

Problem 9. *Elaborate a theory of fuzzy limits for sequences of fuzzy numbers and fuzzy functions.*

Fuzzy metric and normed spaces provide a good base for the development of neoclassical analysis.

Problem 10. *Develop fuzzy continuity for fuzzy numerical functions and fuzzy mappings of (fuzzy) metric spaces.*

Fuzzy continuous functions are also related to roughly continuous functions studied in (Pawlak, 1995). It brings us to the following problem.

Problem 11. *Develop fuzzy continuity for rough functions.*

One more direction for development is complex analysis, which studies complex functions.

Problem 12. *Develop neoclassical analysis for complex functions*, i.e., *build neoclassical complex analysis.*

Problem 13. *Investigate specific properties of the complete local and global fuzzy derivatives of ordinary complex functions.*

Some more problems that are also important for the further development of neoclassical analysis.

Problem 14. *Study fuzzy continuity of binary relations.*

In the area of fuzzy differentiation, it would be interesting to do the following.

Problem 15. *Build fuzzy derivatives for discontinuous functions and functions defined on discrete sets.*

Problem 16. *Build fuzzy derivatives as a fuzzy linear part of the function deviation* (cf. Gateaux derivative).

Problem 17. *Build theory of fuzzy differential equations.*

Problem 18. *Build theory of fuzzy differential inclusions.*

In the area of fuzzy integration, it would be interesting to do the following.

Problem 19. *Build indefinite fuzzy integrals as fuzzy antiderivatives and find their properties.*

Problem 20. *Build indefinite fuzzy integrals as inverse operators to fuzzy differentiation and find their properties.*

Problem 21. *Build improper fuzzy integrals, i.e., integrals of unbounded functions or over unbounded intervals of integration, and find their properties.*

It is demonstrated (cf., Chapter 8) that any bounded function has a fuzzy integral over any interval where this function is bounded. However, more important is to find the estimate of fuzziness of this integral, that is, the number r. This issue brings us to the next problem.

Problem 22. *Is any r-continuous function Riemann r-integrable over a finite interval?*

For $r = 0$, the answer is positive because any continuous function is Riemann integrable over a finite interval (cf. Section 7.1).

Problem 23. *Find conditions when a function $f(x)$ is Riemann r-integrable.*

Problem 24. *Find conditions when a function $f(x)$ is Riemann r-integrable but is not q-integrable for any $q < r$.*

More general problems are related to the theory of fuzzy Lebesgue integration and fuzzy gauge integration, as well as to finding relations between these theories and fuzzy Riemann integration. As it is demonstrated in this paper, the scope of fuzzy Riemann integration encompasses the scope of the traditional (crisp) Lebesgue integration. At the same time, the scope of the traditional Lebesgue integration encompasses the scope of the traditional Riemann integration. So, it is interesting to know whether the scope of fuzzy Riemann coincides with the scope of fuzzy Lebesgue integration.

Problem 25. *Build the fuzzy Lebesgue integral and find its properties. Is there a difference between the fuzzy Lebesgue and Riemann integrals?*

Problem 26. *Build the gauge fuzzy integrals and find their properties.*

Problem 27. *Build nonstandard fuzzy Riemann and Lebesgue integrals and find their properties.*

The main principle of neoclassical analysis is to take concepts and constructions from the classical calculus and to consider their extensions so that new concepts and constructions are close, in some sense, to the classical ones. Here we considered closeness build on distance. For a localized to a point construction, such as, for example, convergence at a point, this is the most natural way. When a construction for a set (for an interval for one-dimensional case or a domain in a metric space) is taken, it is possible also to use distance as a measure of closeness. It gives, as a rule several ways to build neoclassical constructions from the corresponding classical ones. For example, there are strong and weak fuzzy derivatives. However, there are other ways to define closeness. One more natural way is based on the concept of a measure. A general principle of analysis is that it possible, as a rule, to ignore sets of zero measure. Thus, it is interesting to develop neoclassical analysis in this direction. Actually, many elements and constructions of this theory already exist. There are approximate limits, convergence, continuity, and derivatives.

Approximate limits and approximately continuous functions were first introduced by Denjoy in his work (1915) on derivatives. Later they were utilized by Khinchin who introduced concept of an approximate derivative (1916) and Denjoy (1916) in the study of the Lebesgue and Denjoy–Khinchin integrals. A function $f(x)$ is approximately continuous if and only if it is continuous in the density topology. Later different authors studied and utilized approximately continuous functions (cf., for example, (Saks, 1964)). These functions have many good properties. For example, they have the Darboux property and belong to the first Baire class. Moreover, any bounded approximately continuous function is a derivative of some function.

It is also necessary to mention that, as we know, the Lebesgue integral disregards sets of measure zero. So, it may be taken as an appropriate construction for integration in this branch of neoclassical analysis.

One more important direction for research is the development of a nonarchimedean neoclassical analysis.

It is also possible to formulate some problems of neoclassical analysis applications.

Problem 28. *Develop symbolic dynamics and chaos theory in the fuzzy context of neoclassical analysis.*

The following problem is related to the Ulam's problem for approximate homomorphisms (Ulam, 1964).

Problem 29. *Given a class of spaces with measure m, is it possible to k_r-approximate every r-endomorphism by a strict endomorphism, where k_r is a constant absolute for the class of spaces considered?*

Thus, we see that neoclassical analysis has generated various interesting problems and opened new directions for further research.

APPENDIX

DENOTATIONS AND BASIC CONCEPTS

What memory has in common with art
is the knack for selection, the taste for detail...
Joseph Brodsky (1940 – 1996)

Some mathematical concepts, in spite of being basic and extensively used, have different interpretation in different books. In a similar way, different authors use dissimilar notations and/or names for the same things, as well as the same notation and/or name for distinct things. For this reason and to make our exposition self-consistent, we give here definitions and denotation for basic mathematical concepts that are used in this book.

A. SETS, FUNCTIONS, AND GENERAL STRUCTURES

Science may set limits to knowledge,
but should not set limits to imagination.
Bertrand Russell (1872 - 1970)

$\varnothing$ is the *empty set*.

If X is a set, $r \in X$ means that r belongs to X or r is a member of X.

If X and Y are sets, $Y \subseteq X$ means that Y is a *subset* of X, i.e., Y is a set such that all elements from Y belong to X.

The *union* $Y \cup X$ of two sets Y and X is the set that consists of all elements from Y and from X.

The *intersection* $Y \cap X$ of two sets Y and X is the set that consists of all elements that belong both to Y and to X.

The *difference* $Y \setminus X$ of two sets Y and X is the set that consists of all elements that belong to Y but does not belong to X.

If X is a set, then 2^X is the *power set* of X, which consists of all subsets of X. The *power set* of X is also denoted by $\mathbf{P}(X)$.

If X and Y are sets, then $X \times Y = \{(x, y); x \in X, y \in Y\}$ is the *direct* or *Cartesian product* of X and Y, in other words, $X \times Y$ is the set of all pairs (x, y), in which x belongs to X and y belongs to Y.

Y^X is the set of all mappings from X into Y.

$X^n = \underbrace{X \times X \times \ldots X \times X}_{n}$ is the *direct* or *Cartesian power* of X.

A fundamental structure of mathematics is *function*. However, functions are special kinds of binary relations between two sets.

A *binary relation* T between sets X and Y (also called a *correspondence* from X to Y) is a subset of the direct product $X \times Y$. As $X \times Y$ is the set of all pairs (x, y), in which x belongs to X and y belongs to Y, a binary relation T is a set of some of these pairs (x, y). Binary relations are also called *multivalued functions* (*multivalued mappings* or *multivalued maps*).

There are several important types of binary relations.

A *preorder* (also called a *quasiorder*) on a set X is a binary relation Q on X that satisfies the following axioms:

1. Q is reflexive, i.e., $x\text{Q}x$ for all x from X.
2. Q is transitive, i.e., $x\text{Q}y$ and $y\text{Q}z$ imply $x\text{Q}z$ for all x, y, and z from X.

A *partial order* is a preorder that satisfies the following additional axiom:

3. Q is antisymmetric, i.e., $x\text{Q}y$ and $y\text{Q}x$ imply $x = y$ for all x and y from X.

A *strict partial order* is a preorder that is not reflexive, is transitive and satisfies the following additional axiom:

4. Q is asymmetric, i.e., only one relation $x\text{Q}y$ or $y\text{Q}x$ is valid for all x and y from X.

An *equivalence* on a set X is a binary relation Q on X that is reflexive, transitive and satisfies the following additional axiom:

5. Q is symmetric, i.e., $x\text{Q}y$ implies $y\text{Q}x$ for all x and y from X.

A *function* (also called a *mapping* or *map* or *total function* or *total mapping*) f from X to Y is a binary relation between sets X and Y in which there are no elements from X that are corresponded to more than one element from Y and any element from X is corresponded to some element from Y. Often total functions are also called everywhere defined functions. Traditionally, the element $f(a)$ is called the image of the element a and denotes the value of f on the element a from X. At the same time, the function f is also denoted by $f: X \rightarrow Y$ or by $f(x)$. In the latter formula, x is a variable and not a concrete element from X.

A *partial function* (or *partial mapping*) f from X to Y is a binary relation between sets X and Y in which there are no elements from X which are corresponded to more than one element from Y. Thus, any function is also a partial function. Sometimes, when the domain of a partial function is not specified, we call it simply a function because any partial function is a total function on its domain.

A *multivalued function* (or *multivalued mapping*) f from X to Y is any binary relation between sets X and Y.

$f(x) \equiv a$ means that $f(x)$ is equal a at all points where $f(x)$ is defined.

Two important concepts of mathematics are the domain and range of a function. However, there is some ambiguity the first of them. Namely, there are two distinct meanings in current mathematical usage for this concept. In the majority of mathematical areas, including the calculus and analysis, the term "domain of f" is used for the set of all values x such that $f(x)$ is defined. However, some mathematicians (in particular, category theorists), however, consider the domain of a function $f: X \rightarrow Y$ to be X, irrespective of whether $f(x)$ exists for all x in X. To eliminate this ambiguity, we suggest the following terminology consistent with the current practice of calculus teaching.

If T is a binary relation between sets X and Y, then the set X is called the *domain* of T (it is denoted by Dom T) and Y is called the *codomain* of T (it is denoted by Codom T). The *range* Rg T of the relation T is the set of all elements from Y to which, at least, one element from X is related by T, or formally, Rg $T = \{ y ; \exists x \in X ((x, y) \in T) \}$. The *domain of definition* DDom T of the relation T is the set of all elements from X that related by T to, at least, one element from Y is or formally, DDom $T = \{x ; \exists y \in Y ((x, y) \in T) \}$.

Thus, for a (total) function $f: X \rightarrow Y$, its *domain* Dom f coincides with its *domain of definition* DDom f, while the *range* Rg f the set of all elements from Y that are images of elements from X.

For a partial function $f(x)$, its *domain of definition* DDom f is the set of all elements for which $f(x)$ is defined.

Taking two mappings (or functions) $g: X \rightarrow Y$ and $f: Y \rightarrow Z$, it is possible to build a new mapping (function) $f \circ g: X \rightarrow Z$ (also denoted by $fg: X \rightarrow Z$) that is called the *composition* or *superposition* of mappings (functions) f and g and defined by the rule $f \circ g(x) = f(g(x))$ for all x from X. The function $f \circ g(x)$ is also called the *composite function*. The composition of functions is a special case of the *composition* of binary relations. Namely, if T is a binary relation between sets X and Y and Q is a binary relation between sets Y and Z, then their composition (also called product) is the binary relation $Q \circ T$ between X and Z defined in the following way: $Q \circ T = \{ (x, z); x \in X, z \in Z$ and there is $y \in Y$ such that $(x, y) \in T$ and $(y, z) \in Q \}$.

Let us consider function the domain and range of which are partially ordered sets.

A function (mapping) $f: X \rightarrow Y$ is *increasing* if $a < b$ implies $f(a) \leq f(b)$ for all a and b from X.

A function $f: X \rightarrow Y$ is *decreasing* on X if $a < b$ implies $f(a) \geq f(b)$ for all a and b from X.

A function $f: X \rightarrow Y$ is *strictly increasing* on X if $a < b$ implies $f(a) < f(b)$ for all a and b from X.

A function $f: X \rightarrow Y$ is *strictly decreasing* on X if $a < b$ implies $f(a) > f(b)$ for all a and b from X.

Increasing functions and decreasing functions are called *monotone* functions. Strictly increasing functions and strictly decreasing functions are called *strictly monotone* functions.

The corresponding properties of functions are called *monotonicity* and *strict monotonicity*, respectively.

For instance, functions x and x^3 are increasing and monotone, while functions $|x|$ and x^2 are not increasing and not monotone.

For any set X, $\chi_X(x)$ is its *characteristic function* (also called *membership function*) of a set X if $\chi_X(x)$ is equal to 1 when $x \in X$ and is equal to 0 when $x \notin X$, and $C_X(x)$ is its partial characteristic function if $C_X(x)$ is equal to 1 when $x \in X$ and is undefined when $x \notin X$.

If $f: X \to Y$ is a function and $Z \subseteq X$, then the *restriction* $f|_Z$ of f on Z is the function defined only for elements from Z and $f|_Z(z) = f(z)$ for each element z from Z.

if U is a correspondence of a set X to a set Y (binary relation on X), i.e. $U \subseteq X \times Y$, then $U(x) = \{y \in Y; (x, y) \in U\}$ and $U^{-1}(y) = \{x \in X; (x, y) \in U\}$.

Let X be a set. An *integral operation* W on the set X is a mapping that given a subset of X, corresponds to it an element from X, and for any $x \in X$, $W(\{x\}) = x$.

Examples of integral operation are: sums, products, minimums, maximums, infima, suprema, integrals, taking the first element from a given subset, taking the sum of the first and second elements from a given subset, and so on.

Examples of finite integral operations defined for numbers are: sums, products, minimums, maximums, average, weighted average, taking the first element from a given subset, and so on.

As a rule, integral operations are partial, that is, they correspond numbers only to some subsets of X.

Proposition A.1. Any binary operation in X generates a finite ordinal integral operation on X.

It is possible to read more about integral operations and their applications in (Burgin and Karasik, 1976; Burgin 2004d).

A *multiset* is similar to a set, but can contain indiscernible elements or different copies of the same elements. It is possible to read more about multisets in (Aigner, 1979; Knuth, 1997; 1998).

A *named set* (also called a *fundamental triad*) **X** is a triad (X, f, I) where X called the *support* of **X** is some entity, I called the *component of names* (or *set of names* or *reflector*) of **X** is some entity, and f called the *naming correspondence* (or *reflection*) of the named set **X** is a correspondence between X and I. The most popular type of named sets is a named set **X** = (X, f, I) in which X and I are sets and f consists of connections between their elements. When these connections are set theoretical, i.e., they are binary relations between X and Y, and each connection between elements is represented by a pair (x, a) where x is an element from X and a is its name from I, we have a *set theoretical named set* or binary relation. Note that even when X and I are sets, the naming correspondence f is not always a binary relation between X and I or a function from X into I. For instance, f can be an algorithm, such as a Turing machine or a computer program, that converts elements from X into elements from I. Bourbaki in their fundamental monograph (1960) also represent binary relations in a form of a triad (named set). It is possible to read more about named sets in (Burgin 1990; 1992; 2004c).

It is possible to find much more material about sets, relations and functions, for example, in (Fraenkel and Bar-Hillel, 1958; Bourbaki, 1960). There is a good exposition of set theory from the perspective of analysis in (Kolmogorov and Fomin, 1999).

B. NUMBERS AND NUMERICAL FUNCTIONS

Everything is a number.
Pythagoras

N is the *set of all natural numbers* 1, 2, ..., n, ...

ω is the *sequence of all natural numbers*.

$\boldsymbol{N}_0$ is the *set of all whole numbers* 0, 1, 2, ..., n, ...

Z is the *set of all integer numbers*.

Q is the *set of all rational numbers*.

R is the *set of all real numbers* or *reals*. The geometric form of the set ***R*** is called the *real line*.

$\boldsymbol{R}^{+}$ is the *set of all non-negative real numbers*.

$\boldsymbol{R}^{++}$ is the *set of all positive real numbers*.

C is the *set of all complex numbers*.

∞ is the *positive infinity*, - ∞ is the *negative infinity*. Usually, these element are added to the set ***R***. The new set is denoted by $\boldsymbol{R}_\infty = \boldsymbol{R} \cup \{\infty, -\infty\}$. In $\boldsymbol{R}_\infty$, the element ∞ is larger than any real number and the element - ∞ is smaller than any real number. It possible to extend arithmetical operations to $\boldsymbol{R}_\infty$. Namely, we have $c + \infty = \infty + c = c \cdot \infty = \infty \cdot c = \infty + \infty = \infty \cdot \infty = \infty$ and $c + (-\infty) = -\infty + c = -\infty$ for any real number c, while $c \cdot (-\infty) = (-\infty) \cdot c = -\infty$ for any positive real number c, and $c \cdot (-\infty) = (-\infty) \cdot c = \infty$ for any negative real number c.

If a is a real number, then $|a|$ or $||a||$ denotes its *absolute value* (also called the *modulus* of a). Thus, $|a| = a$ when a is non-negative and $|a| = -a$ when a is negative. Two more important constructions are the *integral part* $[a]$ (also called the *integral value* and denoted by $\lfloor a \rfloor$) of a, which is equal to the largest integer number that is less than a. The real number a is also corresponded to $]a[$ (also denoted by $\lceil a \rceil$), which is equal to the least integer number that is larger than a. For instance, $[1.2] = 1$, $[7.99] = 7$ and $[-1.2] = -2$. The difference $a - [a]$, also denoted by a mod 1, or $\{a\}$, is called the *fractional part* of a. All these constructions define real-valued functions $|x|$, $[x]$, $\{x\}$, and $]x[$ where the function $]x[$ ($\lceil x \rceil$), is sometimes called the *ceiling function* and the function $[x]$ ($\lfloor a \rfloor$) is sometimes called the *floor function*.

A set X of numbers is called *closed* if it contains limits of all sequences with elements in X.

If $a \leq b$, then the *closed interval* $[a, b]$ of real numbers consists of all real numbers d that satisfy the condition $a \leq d \leq b$. The *open interval* (a, b) of real numbers consists of all real numbers d that satisfy the condition $a < d < b$. The *midpoint* of the *interval* $[a, b]$ or (a, b) is the point $c = a + ½ (b - a)$.

A set X of numbers is called *connected* if it is not a union of two closed subsets of X.

A set X of real numbers is called *convex* if the condition $a, b \in X$ implies that the whole interval $[a, b]$ belongs to X.

Properties of real numbers imply the following result.

Proposition B.1. Any closed convex set X of real numbers is an interval.

Proposition B.2. Any sequence of closed intervals of real numbers such that any interval is a subset of the previous one has a non-empty intersection.

It means that ***R*** is a complete metric space (cf. Appendix C).

When X is a set of real numbers, then the *supremum* (also called the least upper bound) of X denoted by sup X (or by LUB X) is the least number a (when such a number exists) that is larger than or equal to any number in X. When there is no such a number, the supremum sup $X = \infty$.

An important property of sup X is that either it is the largest element of X or sup X is not equal to any element of X and for any $\varepsilon \in \mathbf{R}^{++}$, there is an element x from X such that $| a - x| < \varepsilon$.

The *infimum* of X (also called the greatest lower bound) denoted by inf X (or by GLB X) is the number a (when such a number exists) that is smaller than or equal to any number in X and for any $\varepsilon \in \mathbf{R}^{++}$, there is an element x from X such that $| a - x| < \varepsilon$. When there is no such a number, the infimum inf $X = - \infty$.

An important property of inf X is that either it is the least element of X or inf X is not equal to any element of X and for any $\varepsilon \in \mathbf{R}^{++}$, there is an element x from X such that $| a - x| < \varepsilon$.

As the space $\mathbf{R}$ is a complete metric space (cf. Appendix C), the following result is true.

Proposition B.3. Any set X of real numbers has the infimum inf X and supremum sup X, which may be infinite.

Proposition B.4. Any bounded set X of real numbers has the finite infimum inf X and finite supremum sup X.

For rational numbers, these properties are not true in a general case.

A sequence $l = \{a_i ; i = 1, 2, 3, \ldots\}$ is called *monotone*, or *monotonous*, if either $a_i \leq a_{i+1}$ for all $i = 1, 2, 3, \ldots$ or $a_i \geq a_{i+1}$ for all $i = 1, 2, 3, \ldots$.

A sequence $l = \{a_i ; i = 1, 2, 3, \ldots\}$ is called *strictly monotone*, or *strictly monotonous*, if either $a_i < a_{i+1}$ for all $i = 1, 2, 3, \ldots$ or $a_i > a_{i+1}$ for all $i = 1, 2, 3, \ldots$.

A numerical function $f(x)$ is called:

a) *bounded above* at a point a if there are a number M and an interval $[b, c]$ such that a belongs to this interval and
$f(x) < M$ for all x from the interval $[b, c]$.
b) *bounded below* at a point a if there are a number m and an interval $[b, c]$ such that a belongs to this interval and
$f(x) > m$ for all x from the interval $[b, c]$.
c) *bounded* at a point a if it is a bounded both above and below at a point a.
d) *bounded above* (*below*) *in a set* $Z \subseteq X$ if there is a number M such that $f(x) < M$ ($f(x) > M$) for all $x \in Z$;
e) *bounded above* (*below*) if it is bounded above (below) in its domain X.
f) *bounded in a set* $Z \subseteq X$ if it is a bounded both above and below *in a set* Z.
g) *bounded* if it is bounded in X.

Axioms for operations with real and complex numbers:

1. Commutativity of addition: $a + b = b + a$;
2. Associativity of addition: $(a + b) + c = a + (b + c)$;
3. Commutativity of multiplication: $a \cdot b = b \cdot a$;
4. Associativity of multiplication: $(a \cdot b) \cdot c = a \cdot (b \cdot c)$;

5. Distributivity of multiplication with respect to addition: $a \cdot (b + c) = a \cdot b + a \cdot c$;
6. Zero is a neutral element with respect to addition: $a + 0 = 0 + a = a$;
7. One is a neutral element with respect to multiplication: $a \cdot 1 = 1 \cdot a = a$.

Let $A = \{ a_1 , a_2 , a_3 , \ldots , a_n \}$ and $C = \{ c_1 , c_2 , c_3 , \ldots , c_m \}$ be two sets of numbers. Then the Minkowski sum $A \oplus C$ of sets A and C is the set that consists of all sums $a_i + c_j$ where $i = 1, 2, 3, \ldots, n$ and $j = 1, 2, 3, \ldots , m$. In a general case, if X and Y are sets of real numbers, then $X \oplus Y = \{ x + y ; x \in X \text{ and } y \in Y \}$.

The Minkowski difference $X \ominus Y$ of the sets X and Y is the set that consists of all differences $a - b$ where $a \in X$ and $b \in Y$, i.e., $A \ominus Y = \{ a - b ; a \in A \text{ and } b \in Y \}$.

Some properties of addition and subtraction of number sets are the same as properties of addition and subtraction of numbers. For instance, addition of number sets is commutative, i.e., for any sets $A, B \subseteq \boldsymbol{R}$, we have

$$A \oplus B = B \oplus A$$

and associative, i.e., for any sets A, B, C $\subseteq$ **R**, we have

$$(A \oplus B) \oplus C = A \oplus (B \oplus C)$$

At the same time other properties are different. For instance, for any number $a \in \boldsymbol{R}$, we have $a - a = 0$, while $A \ominus A \neq \{0\}$ for any set $A \subseteq \boldsymbol{R}$ that contains more than one element, e.g., $[0, 1] \ominus [0, 1] = [-1, 1]$.

A function with $\boldsymbol{R}$ as its codomain is called a *real function* or a *real-valued function.*

A function with $\boldsymbol{R}$ as its domain is called a *function* with one *real variable.*

A function with $\boldsymbol{C}$ as its codomain is called a *complex function* or a *complex-valued function.*

A function with $\boldsymbol{C}$ as its domain is called a *function* with one *complex variable.*

Operations with numbers induce similar operations with functions:

Addition of functions $(f + g)(x) = f(x) + g(x)$.

Subtraction of functions $(f - g)(x) = f(x) - g(x)$.

Multiplication of functions $(f \cdot g)(x) = f(x) \cdot g(x)$.

Scalar multiplication of functions $(k \cdot f)(x) = k{\cdot}g(x)$ by a number k.

It is possible to find a rigorous exposition of the theory of natural, rational and real numbers, for example, in (Ross, 1996) and a more extended exposition of the theory of real numbers in (Dieudonné, 1960; Fihtengoltz, 1955).

C. Topological, Vector, Metric, and Normed Spaces

Space is the breath of art.
Frank Lloyd Wright (1867-1959)

A *topology* in a set X is a system $\mathbf{O}(X)$ of subsets of X that are called open subsets and satisfy the following axioms:

T1. $X \in \mathbf{O}(X)$ and $\varnothing \in \mathbf{O}(X)$.
T2. For all A, B , if $A, B \in \mathbf{O}(X)$, then $A \cap B \in \mathbf{O}(X)$.
T3. For all A_i, $i \in I$, if all $A_i \in \mathbf{O}(X)$, then $\bigcup_{i \in I} A_i \in \mathbf{O}(X)$.

A set X with a topology in it is called a *topological space*.

Topology in a set can be also defined by a system of neighborhoods of points from this set. In this case, a set is *open* in this topology if it contains a standard neighborhood of each of its points. For instance, if a is a real number and $t \in \boldsymbol{R}^{++}$, then an open interval $\mathrm{O}_t a = \{ x \in \boldsymbol{R};\ a - t < x < a + t \}$ is a standard neighborhood of a.

One more way to define topology in a set is to use the closure operation (Kuratowski, 1966).

If X is a subset of a topological space, then $\mathrm{Cl}(X)$ denotes the *closure* of the set X, i.e., the least closed set that contains X.

A point a is an *isolated point* in a topological space X if the set that consists of this point is open. There are no non-trivial sequences that converge to an isolated point.

A function $f: X \to \boldsymbol{R}$ defined in a topological space X is *bounded* at a point $a \in X$ if there exists a number q and a neighborhood $\mathrm{O}a$ of the point a such that for any x from $\mathrm{O}a$ the inequality $|f(a) - f(x)| < q$ is valid.

Let u be a point from a topological space X and assume that spaces X and Y are ordered. A function (mapping) $f: X \to Y$ is called:

1. *increasing* at a point u from X if there is a neighborhood Ou of the point u such that for any points a and b from Ou, $a < b$ always implies $f(a) \leq f(b)$;
2. *decreasing* at a point u from X if there is a neighborhood Ou of the point u such that for any points a and b from Ou, $a < b$ always implies $f(a) \geq f(b)$;
3. *strictly increasing* at a point u from X if there is a neighborhood Ou of the point u such that for any points a and b from Ou, $a < b$ always implies $f(a) > f(b)$;
4. *strictly decreasing* at a point u from X if there is a neighborhood Ou of the point u such that for any points a and b from Ou, $a < b$ always implies $f(a) < f(b)$.

In many interesting cases, topology is defined by a metric.

A *metric* in a set X is a mapping $\mathbf{d}: X \times X \to \boldsymbol{R}^+$ that satisfies the following axioms:

M1. $\mathbf{d}(x, y) = 0$ if and only if $x = y$,

i.e., the distance between an element and itself is equal to zero, while the distance between any different elements is equal to a positive number.

M2. $\mathbf{d}(x, y) = \mathbf{d}(y, x)$ for all $x, y \in X$,
i.e., the distance between x and y is equal to the distance between x and y.

M3. The triangle inequality:

$$\mathbf{d}(x, y) \leq \mathbf{d}(x, z) + \mathbf{d}(z, y) \text{ for all } x, y, z \in X,$$

i.e., the distance from x through z to y is never shorter than going directly from x to y, or the shortest distance between any two points is a straight line

A set X with a metric $\mathbf{d}$ is called a *metric space*. The number $\mathbf{d}(x, y)$ is called the distance between x and y in the metric space X.

For instance, in the set $\boldsymbol{R}$ of all real numbers, the distance $\mathbf{d}(x, y)$ between numbers x and y usually is the absolute value $| x - y|$, i.e., $\mathbf{d}(x, y) = | x - y|$. This metric defines the following topology in $\boldsymbol{R}$. If a is a point from $\boldsymbol{R}$, then a standard neighborhood of a has the form $O_r(a) = \{ y; | a - y| < r \text{ with } r \in \boldsymbol{R}^{++}\}$. A set is open in this topology if it contains a standard neighborhood of each of its points.

It is possible to extend a given distance between points to a distance between sets and from a point to a set. If $A, B \subseteq X$ and $x \in X$, then $\mathbf{d}(x, A) = \inf \{ \mathbf{d}(x, y); y \in A \}$ and $\mathbf{d}(A, B) = \inf \{ \mathbf{d}(x, y); x \in A, y \in B \}$. When $X = \boldsymbol{R}$, we have $\mathbf{d}(x, A) = \inf \{ \mathbf{d}(x, y); y \in A \}$ and $\mathbf{d}(A, B) = \inf \{ \mathbf{d}(x, y); x \in A, y \in B \}$.

If $A \subseteq X$, then the *diameter* of A is equal to $\text{Diam}(A) = \sup \{ \mathbf{d}(x, y); x, y \in A \}$. When $X = \boldsymbol{R}$, we have $\text{Diam}(A) = \sup \{ |x - y |; x, y \in A \}$.

A metric space X is also a topological space. As a neighborhood of a point a from X, we can take an open ball $B_r (a) = \{x; \mathbf{d}(x, a) < r, r \in \boldsymbol{R}^{+}\}$. Then topology in X is defined by the system of all such neighborhoods.

A sequence $l = \{a_i ; i = 1, 2, 3, \dots\}$ of elements from a metric space (e.g., real numbers) is called a *Cauchy sequence* if for any real number $\varepsilon > 0$ there is such $n \in \omega$ that for any $i, j \geq n$, we have $\mathbf{d}(a_j, a_i) < \varepsilon$.

Mappings of metric spaces that preserve the metric structure are called continuous. Namely, if $f: X \rightarrow Y$ is a mapping of a metric space X with a metric $\mathbf{d}_X$ into a metric space Y, then f is called:

a) *continuous* if for any real number $\varepsilon > 0$, there is a real number $\delta > 0$ such that $\mathbf{d}_X(x, z) < \delta$ implies $\mathbf{d}_Y(f(x), f(z)) < \varepsilon$ for any $x, z \in X$;
b) *bounded at a point a* if there is a number M and a real number $\delta > 0$ such that $\mathbf{d}_X(x, z) < \delta$ implies $\mathbf{d}_Y(f(a), f(x)) < M$ for all x from X.
c) *bounded in a set* $Z \subseteq X$ if there is a number M such that $\mathbf{d}_Y(f(a), f(x)) < M$ for all $x, z \in Z$;
d) *bounded* if it is bounded in its domain X.

Note that this definition of boundedness coincides for numerical functions with the definition of boundedness given in Appendix B.

A metric space is called *complete* if any Cauchy sequence has a limit.

Proposition C.1. The set $\boldsymbol{R}$ of all real numbers is complete.

Proposition C.2. In a complete metric space, any sequence of closed subsets such that any set is a subset of the previous one has a non-empty intersection.

An important property of sets in metric spaces compactness, which allows mathematicians to prove many useful theorems. Fréchet introduced the term *compact* in 1906 in the form of sequential compactness.

A subset B of a metric space is called *sequentially compact* if every sequence in it has a convergent subsequence. Later compact sets were defined in more general topological spaces. Namely, a subset B of a topological space is called *compact* if each of its open covers, i.e., a family of open sets such that B is a subset of unions of these sets, has a finite subcover, i.e., a finite number of those open sets already cover B.

It is proved that for metric spaces both definitions give the same concept.

It is possible to find much more material about metric and topological structures, for example, in (Bers, 1957; Kelly, 1957; Kuratowski, 1966; 1968; Kolmogorov and Fomin, 1999).

$\Sigma_{i=1}{}^{n} c_i$ denotes the sum $c_1 + c_2 + c_3 + \dots + c_n$.

If $l = \{a_i \in M;\ i = 1, 2, 3, \dots\}$ is a sequence, and $f\colon M \to L$ is a mapping, then $f(l) = \{f(a_i);\ i = 1, 2, 3, \dots\}$.

$a = r\text{-}\lim l$ means that a number a is an r-limit of a sequence l.

A set L is called a *linear space* or a *vector space over the field* $\boldsymbol{R}$ of real numbers if it has two operations:

- *addition*: $L \times L \to L$ denoted by $x + y$, where x and y belong to L, and
- *scalar multiplication*: $\boldsymbol{R} \times L \to L$ denoted by ax, where a belong to $\boldsymbol{R}$ and x belong to L,

satisfying the following axioms:

1. Addition is associative:
   For all x, y, z from L, we have $x + (y + z) = (x + y) + z$.
2. Addition is commutative:
   For all x, y from L, we have $x + y = y + x$.
3. Addition has an identity element:
   There exists an element $\mathbf{0}$ from L, called the *zero vector*, such that $x + \mathbf{0} = x$ for all x from L.
4. Addition has an inverse element:
   For any from L, there exists an element z from L, called the additive inverse of x, such that $x + z = \mathbf{0}$.
5. Scalar multiplication is distributive over addition in L:
   For all numbers a from $\boldsymbol{R}$ and vectors y, $\mathbf{w}$ from L, we have $a\,(y + \mathbf{w}) = a\,y + a\,\mathbf{w}$.

6. Scalar multiplication is distributive over addition in $\boldsymbol{R}$:
   For all numbers a, b from $\boldsymbol{R}$ and any vector y from L, we have $(a + b)\, y = a\, y + b\, y$.
7. Scalar multiplication is compatible with multiplication in $\boldsymbol{R}$:
   For all numbers a, b from $\boldsymbol{R}$ and any vector y from L, we have $a\,(b\, y) = (ab)\, y$.
8. The number 1 is an identity element for scalar multiplication:
   For all vectors $\boldsymbol{x}$ from L, we have $1\boldsymbol{x} = \boldsymbol{x}$.

In a similar way, a *linear space over the field* $\boldsymbol{C}$ of complex numbers or over any field $\boldsymbol{F}$ is defined.

Vectors $\boldsymbol{x}_1$, $\boldsymbol{x}_2$, ... , $\boldsymbol{x}_n$ from L are called *linearly dependent* in L if any there is an equality $\Sigma_{i=1}{}^{n}\, a_i\, \boldsymbol{x}_i = 0$ where a_i are elements from $\boldsymbol{R}$ (from $\boldsymbol{F}$, in a general case) and not all of them are equal to 0. When there are no such an equality, vectors $\boldsymbol{x}_1$, $\boldsymbol{x}_2$, ... , $\boldsymbol{x}_n$ are called *linearly independent*.

A system B of linearly independent vectors from L is called a *basis* of L if any element $\boldsymbol{x}$ from L is equal to a sum $\Sigma_{i=1}{}^{n}\, a_i\, \boldsymbol{x}_i$ where n is some natural number, $\boldsymbol{x}_i$ are elements from B and a_i are elements from $\boldsymbol{R}$ (from $\boldsymbol{F}$, in a general case).

The number of elements in a basis is called the *dimension* of the space L. It is proved that all bases of the same space have the same number of elements. The number of elements in a basis is called the *dimension* of the space L.

The space $\boldsymbol{R}$ is a one-dimensional vector (linear) space over itself. The space $\boldsymbol{R}^n$ is an n-dimensional vector (linear) space over $\boldsymbol{R}$.

Let us consider a mapping f: $L \to H$ where L and H are vector (linear) spaces over $\boldsymbol{R}$. The mapping f is called:

1. *additive* if $f(a + b) = f(a) + f(b)$ for all $a, b \in L$.
2. *subadditive* if $f(a + b) \leq f(a) + f(b)$ for all $a, b \in L$.
3. *superdditive* if $f(a + b) \geq f(a) + f(b)$ for all $a, b \in L$.
4. *shifted additive* if $f(a) + f(b) = g(f(a + b))$ for some mapping g: $H \to H$ and all $a, b \in L$.
5. *uniform* if $f(ka) = k{\cdot}f(a)$ for all $a \in L$ and $k \in \boldsymbol{R}$.
6. *shifted uniform* if $g(f(ka)) = k{\cdot}f(a)$ for some mapping g: $H \to H$, all $a \in L$ and $k \in \boldsymbol{R}$.
7. *linear* if it is uniform and additive, i.e., $f(ka + hb) = k{\cdot}f(a) + h{\cdot}f(b)$ for all $a, b \in L$ and $k, h \in \boldsymbol{R}$.
8. *shifted linear* if $k{\cdot}f(a) + h{\cdot}f(b) = g(f(ka + hb))$ for some mapping g: $H \to H$, all $a, b \in L$ and $k, h \in \boldsymbol{R}$.

Shifted additivity, uniformity and linearity are also kinds of *fuzzy additivity*, *uniformity* and *linearity*.

It is possible to introduce a natural metric in the space $\boldsymbol{R}^n$. Namely, if $\boldsymbol{x}, \boldsymbol{y} \in \boldsymbol{R}^n$, $\boldsymbol{x} = \Sigma_{i=1}{}^{n} a_i\, \boldsymbol{x}_i$, and $\boldsymbol{y} = \Sigma_{i=1}{}^{n}\, b_i\, \boldsymbol{x}_i$ where $\boldsymbol{x}_i$ are elements from a basis B of the space $\boldsymbol{R}^n$, then $\mathbf{d}(\boldsymbol{x}, \boldsymbol{y}) = \sqrt{(a_1 - b_1)^2 + (a_2 - b_2)^2 + \ldots + (a_n - b_n)^2}$. It is called the *Euclidean metric* in the space $\boldsymbol{R}^n$. The space $\boldsymbol{R}^n$ with the Euclidean metric is called an *Euclidean space* and often denoted by $\boldsymbol{E}^n$. In the metric space $\boldsymbol{R}$, the distance between a and b is defined as $\mathbf{d}(a, b) = |\, a - b|$.

Another natural metric in $\mathbf{R}^n$ is the Manhattan distance, where the distance between any two points, or vectors, is the sum of the distances between corresponding coordinates, i.e., $\mathbf{d}(x, y) = |a_1 - b_1| + |a_2 - b_2| + \ldots + |a_n - b_n|$.

It is possible to consider any non-empty set X as a metric space with the distance $\mathbf{d}(x, y) = 1$ for all x not equal to y and $\mathbf{d}(x, y) = 0$ otherwise. It is a discrete metric.

A *norm* in a linear space L over the field $\mathbf{R}$ is a mapping $||\ ||: L \to \mathbf{R}^+$ that satisfies the following axioms:

N1. $||x|| = 0$ if and only if $x = 0$,
i.e., the zero vector has zero length, while any other vector has a positive length.

N2. For any positive number a from $\mathbf{R}$, we have $||ax|| = a||x||$,
i.e., multiplying a vector by a positive number has the same effect on the length.

N3. The triangle inequality:

$$||x + y|| \leq ||x|| + ||y||$$

i.e., the norm of a sum of vectors is never less than the sum of their norms.

This implies:

$$||x|| - ||y|| \leq ||x + y||$$

A linear space L with a norm is called a *normed linear space* or simply, a *normed space*.

The space $\mathbf{R}^n$ is a normed linear space with the following norm: if $\mathbf{x} \in \mathbf{R}^n$ and $\mathbf{x} = \Sigma_{i=1}^{n} a_i \mathbf{x}_i$ where $\mathbf{x}_i$ are elements from a basis B of the space $\mathbf{R}^n$, then $||\mathbf{x}|| = \sqrt{a_1^2 + a_2^2 + \ldots + a_n^2}$.

Proposition C.3. Any normed linear space is a metric space.

Indeed, we can define $\mathbf{d}(\mathbf{x}, \mathbf{y}) = ||\ \mathbf{x} - \mathbf{y}\ ||$ and check that all axioms of metric are valid for this distance.

Another natural metric in a normed linear space is the British Rail metric (also called the Post Office metric or the SNCF metric) on a normed vector space, given by $\mathbf{d}(\mathbf{x}, \mathbf{y}) = ||\mathbf{x}|| + ||\mathbf{y}||$ for distinct vectors $\mathbf{x}$ and $\mathbf{y}$, and $\mathbf{d}(\mathbf{x}, \mathbf{x}) = 0$.

Proposition C.4. a. The set of all real functions is a linear space.
b. The set of all bounded real functions is a normed linear space.

It is possible to find much more material about algebraic structures, for example, in (Kurosh, 1963; 1974; Van der Varden, 1971; Giles, 2000).

D. LOGIC

Pure mathematics is, in its way, the poetry of logical ideas.
Albert Einstein (1879-1955)

If $A = \{a_i;\ i \in I\}$ is an infinite set, then the expression "a predicate P(x) is *true for almost all* elements from A (or *almost all* elements from A have a property P)" means that P(x) can be untrue only for a finite number of elements from A. For example, if $A = \omega$, then almost all elements of A are bigger than 10, or another example is that conventional convergence of a sequence l to x means that any neighborhood of x contains almost all elements from l. We have an important property of this concept.

Proposition D.1. If almost all elements from A have a property P and almost all elements from A have a property Q, then almost all elements from A have both properties P and Q.

Indeed, if A_P is the subset of all elements from A that have the property P and A_Q is the subset of all elements from A that have the property Q, then the intersection $A_P \cap A_Q$ consists of all elements from A that have both properties P and Q. At the same time, $A \backslash A_P$ is finite as almost all elements from A have a property P and $A \backslash A_Q$ is finite as almost all elements from A have a property Q. Consequently, the set $A \backslash (A_P \cap A_Q) = (A \backslash A_P) \cup (A \backslash A_Q)$ is also finite. It means that almost all elements from A have both properties P and Q.

This is a consequence of the statement that the union of two finite sets is a finite set.

If P and Q are two statements, then

P $\rightarrow$ Q (also denoted by P $\Rightarrow$ Q) means that P implies Q

and

P $\leftrightarrow$ Q (also denoted P $\Leftrightarrow$ Q) means that P and Q are equivalent, i.e., P implies Q and Q implies P.

The logical symbol $\forall$ means "for any".

The logical symbol $\exists$ means "there exists".

One of the most important mathematical ways of reasoning that we will use is the, so-called, *proof from contradiction*. The essence of this proof is that trying to prove that some system (object) A has a property P, we make an assumption that A does not have this property P. Then we show that this contradicts the initial conditions. This allows us to conclude that our assumption was not true and due to the Principle or Law of Excluded Middle, the system (object) A has the property P.

The *Principle* or *Law of Excluded Middle*, or in Latin, *tertium non datur* (a third is not given) is formulated in traditional logic as "A has a property B or A does not have a property B".

It is possible to find much more material on mathematical logic and logic structures, for example, in (Mendelson, 1977).

REFERENCES AND SOURCES FOR ADDITIONAL READING

Adlassnig, K. (1986) Fuzzy set theory in medical diagnosis, *IEEE Transactions on Systems, Man and Cybernetics*, v. 16 , No. 2 , pp. 260 - 265.

Aigner, M. *Combinatorial Theory*, Springer Verlag, 1979.

Albrecht A, and Skordis C. (2000) Phenomenology of a realistic accelerating universe using only planck-scale physics, *Phys. Rev. Lett.*, v. 6, No. 84(10), pp. 2076-2079..

Alefeld, G., and Herberger, J. *Introduction to Interval Computations*, London, Academic Press, 1983.

Amanton, L. Sadeg, B. and Saad-Bouzefrane, S. (2001) Disconnection Tolerance in Soft Real-Time Mobile Databases, in *Proceedings of the ISCA* 16*th International Conference "Computers and their Applications"*, Seattle, pp. 98-101.

An, P.T. (2003) Piecewise constant roughly convex functions, *Journal of Optimization Theory and Applications*, v. 117, pp. 415–438.

An, P.T. and Hai N.N. (2004) δ-Convexity in Normed Linear Spaces, *Numerical Functional Analysis and Optimization*, v. 25, No. 5& 6, pp. 407 - 422.

Antosik P., Mikulinski J., and Sikorski R. *Theory of distributions*, Elsevier, Amsterdam, 1973.

Aoki, N. and Hiraide, K. *Topological Theory of Dynamical Systems: Recent Advances*, North-Holland, Amsterdam/London/New York, 1994.

Artstein, Z. (1974) On the calculus of closed set-valued functions, *Indiana University Mathematics J.*, v. 24, pp. 433-441.

Artstein, Z. and Burns, J.A. (1975) Integration of compact set-valued functions, *Pacific J. of Mathematics*, v. 58, pp. 297-307.

Atanassov, K. T. (1983) Intuitionistic fuzzy sets, VII *ITKR's Session,* Sofia (deposed in Central Sci.-Technical Library of Bulg. Acad. of Sci., 1697/84) (in Bulgarian).

Atanasov, K. (1986) Intuitionistic Fuzzy Sets, *Fuzzy Sets and Systems*, v. 20, No. 1, pp. 87-96.

Atanasov, K. *Intuitionistic Fuzzy Sets*: *Theory and Applications*, Physica-verlag, 1999.

Atkinson, R.L., Atkinson, R.C., Smith E.E., and Bem, D.J. *Introduction to Psychology*, Harcourt Brace Jovanovich, Inc., San Diego/New York/Chicago, 1990.

Aubin, J.-P. and Cellina, A. *Differential inclusions*, Spinger-Verlag, Grundlehren der math. Wiss., Vol. 264, 1984.

Aubin, J.-P. and Frankowska, H. *Set-valued analysis*, Birkhauser, Boston, MA, 1990.

Aumann, R. (1965) Integrals of Set-Valued Functions, *Journal of Mathematical Analysis and Applications*, v. 12, pp. 1-12.

Averbukh, V. I. and Smolyanov, O. G. (1967) The theory of differentiation in linear topological spaces, *Russ Math Surv*, v. 22, No. 6, pp. 201-258.

Averbukh, V. I. and Smolyanov, O. G. (1968) The various definitions of the derivative in linear topological spaces, *Russ Math Surv*, v. 23, No. 4, pp. 67-113.

Averkin, A.N., Batyrshin, I.Z., Blishun, A.F., Silov, V.B., and Tarasov, V.B. (1986) *Fuzzy Sets in Models of Control and Artificial Intelligence*, Moscow, Nauka.

Axer, H., Südfeld, D., van Keyserlingk, D. G. and Berks, G. (2001) Fuzzy sets in human anatomy, *Artificial Intelligence in Medicine*, v. 21, No. 1-3, pp. 147-152.

Baez, S., Balachandran, A. P., Vaidya, S. and Ydri, B. (2000) Monopoles and Solitons in Fuzzy Physics, *Commun. Math. Phys.*, v. 208, pp.787-798.

Balachandran, A. P. and Vaidya, S. (2001) Instantons and Chiral Anomaly in Fuzzy Physics, *Int. J. Mod. Phys.*, v. A16, pp. 17-40.

Baldwin, J.F. (1986) Support Logic Programming, *Int. J. of Intelligent Systems*, v. 1, pp. 73-104.

Banon, G. (1981) Distinction between several subsets of fuzzy measures, *Fuzzy Sets and Systems*, v. 5, pp. 291-305.

Barán B., Rojas A., Brítez D., and Barán L., Measurement and Analysis of Poverty and Welfare using Fuzzy Sets, International Conference *Systemics, Cybernetics and Informatics SCI'99*, Orlando, Florida, 1999.

Barth, E., Zetzsche, C. and Krieger, G. (1998) Curvature measures in visual information processing, *Open Systems and Information Dynamics*, v. 5, pp. 25-39.

Bartle, R.G. *Modern Theory of Integration*, American Mathematical Society, Providence, Rhode Island, 2001.

Běhounek, L. and Cintula, P. (2005) Fuzzy Class Theory as Foundations for Fuzzy Mathematics, in "*Fuzzy Logic, Soft Computing and Computational Intelligence: Eleventh International Fuzzy Systems Association World Congress*", Tsinghua University Press/Springer, Beijing, pp.1233-1238.

Běhounek, L. and Cintula, P. (2005a) Fuzzy class theory, *Fuzzy Sets and Systems*, v. 154, pp. 34–55..

Bekenstein, J.D. (1981) Universal upper bound on the entropy-to-energy ratio for bounded systems, *Phys. Rev. D, Part. Fields*, v. 23, pp. 287-298.

Bell, J.L. *Set Theory: Boolean-Valued Models and Independence Proofs*, Third Edition, Clarendon Press, Oxford, 2005.

Bers, L. *Topology*, Courant Institute of Mathematical Sciences, New York, 1957.

Billingsley, P. *Ergodic Theory and Information*, John Wiley & Sons, New York, 1965.

Blehman, I.I., Myshkis, A.D., and Panovko, Y.G. *Mechanics and Applied Mathematics*, Nauka, Moscow, 1983 (in Russian).

Blizard, W.D. (1989) Real-valued Multisets and Fuzzy Sets, *Fuzzy Sets and Systems*, v. 33, pp. 77-97.

Boldi, P., and Vigna, S. (1998) δ-uniform BSS Machines, *Journ. Complexity*, v. 14, No. 2, pp. 234-256.

Borel, E. *Leçons sur la théorie des fonctions,* 3rd edition, Gauthier-Villars, Paris, 1927.

Borel, E. *Les nombre inaccessible*, Gauthier-Villiars, Paris, 1952.

Borelli, C. and Forti, G.L. (1995) On a general Hyers–Ulam stability result, *Internat. J. Math. Math. Sci.* , v. 18 , pp. 229–236.

Borsuk, K. (1968) Concerning homotopy properties of compacta, *Fund. Math.,* v. 62, pp. 223-254.

Bourbaki, N. *Theorie des Ensembles*, Hermann, 1960.

Boxer, L. (1994) Digitally continuous functions, Pattern Recognition Letters, v. 15, pp. 833-839.

Boyer, C. *A History of Mathematics*, Princeton University Press, Princeton, 1985.

Brandenberger, R. H. and Martin, J. (2001) The Robustness of Inflation to Changes in Super-Planck-Scale Physics, *Modern Physics Letters A*, v. 16, No. 15, pp. 999-1006.

Branzei, R., Morgan, J., Scalzo, V. and Tijs, S. (2003) Approximate fixed point theorems in Banach spaces with applications in game theory, *Journal of mathematical analysis and applications*, v. 285, No. 2, pp. 619-628.

Bromwich, T. J. and MacRobert, T. M. *An Introduction to the Theory of Infinite Series*, Chelsea, New York, 1991.

Brown, J.G. (1971) A Note on Fuzzy Sets, *Information and Control*, v. 18, pp. 32-39.

Bruhat, F. (1962) Integration p-adic, Sém. Bourbaki 14, Nr. 229, 16 p.

Buckley, J.J., and Qu.Yunxia, (1991) On Fuzzy Complex Analysis, I, Differentiation, *Fuzzy Sets and Systems*, 41, pp. 269-284.

Bugajski, S. (1995) Fuzzy Dynamical Systems, Fuzzy Random Fields, *Reports on Math. Physics*, v. 36, No. 2/3, pp. 263-274.

Büchi, J.R. (1960) Weak second order arithmetic and finite automata, *Z. Math. Logic and Grudl. Math.*, v. 6, No. 1, pp. 66-92.

Buhmann, J.M. (2001) Clustering Principles and Empirical Risk Approximation, in *Proceedings of the International Conference on Applied Stochastic Models and Data Analysis* (ASMDA'01), pp. 14-20.

Bula, I. (1996) Stability of the Bohl-Brouwer-Schauder Theorem, *Nonlinear Analysis: Theory, Methods and Applications*, v. 26, pp. 1859-1868.

Bula, I. (2003) Discontinuous functions in Gale economic model, *Mathematical Modelling and Analysis*, vol.8(2), pp. 93-102.

Bula, I. and Rika, D. (2006) Arrow-Hahn economic models with weakened conditions of continuity, *Banach Center Publ.*, v. 71, pp. 47-61.

Bula, I. and Weber, M.R. (2002) *On Discontinuous Functions and their Application to Equilibria in Some Economic Models*, Preprint of Technische Universitat Dresden, MATH-AN-02-02, 20 p.

Bullen, P.S. Nonabsolute integrals in the twentieth century, AMS special. session on nonabsolute integration, *AMS Special Session on Nonabsolute Integration* (Pat Muldowney and Erik Talvila, Eds.), Toronto, Canada, 23–24 September, 2000 (http://www.emis.de/proceedings/Toronto2000/).

Burden, R.L. and Faires, J.D. *Numerical Analysis*, Brooks/Cole, Australia/USA/ Canada, 2001.

Burgin, M.S. (1987) Algebraic structures of distributions, *Doklady of the Academy of Sciences of Ukraine,* No. 7, pp. 5-9 (in Russian and Ukrainian).

Burgin, M. (1990) Theory of Named Sets as a Foundational Basis for Mathematics, in "*Structures in Mathematical Theories*", San Sebastian, pp. 417-420.

Burgin, M. (1990a) Hypermeasures and Hyperintegration, *Doklady of the National Academy of Sciences of Ukraine,* No. 6, pp.10-13 (in Russian and Ukrainian).

Burgin M. (1991) On the Hann-Banach's theorem for hyperfunctionals, *Doklady of the Ukrainian Academy of Sci.*, No. 7, pp. 9-14 (in Russian and Ukrainian).

Burgin, M. (1992) Algebraic Sructures of Multicardinal Numbers, in "*Problems of group theory and homological algebra*", Yaroslavl, pp. 3-20 (in Russian).

Burgin, M. (1993) Fuzzy continuous functions in object classification, 1st *European Congress on fuzzy and intelligent technologies*, Proceedings, Aachen, pp. 1189-1195.

Burgin, M. (1993a) Differential Calculus for Extrafunctions, *Doklady of the National Acad. of Sciences of Ukraine*, No. 11, pp. 7-11.

Burgin, M. (1993b) Two approaches to continuity measure, *Abstracts presented to the American Mathematical Society*, v. 14, No. 3, p. 447.

Burgin, M. (1995) Neoclassical Analysis: Fuzzy Continuity and Convergence, *Fuzzy Sets and Systems*, v. 75, pp. 291-299.

Burgin, M. (1995a) Integral Calculus for Extrafunctions, *Doklady of the National Academy of Sciences of Ukraine*, No. 11, pp. 14-17.

Burgin, M. *Non-Diophantine Arithmetics*, Ukrainian Academy of Information Sciences, Kiev, 1997 (in Russian).

Burgin, M. (1997a) Extended Fixed Point Theorem, in *Methodological and Theoretical Problems of Mathematics and Information Sciences*, v.3, pp. 71-81 (in Russian).

Burgin, M. (1999) General Approach to Continuity Measures, *Fuzzy Sets and Systems*, v. 105, No. 2, pp. 225-231.

Burgin, M. (1999a) Extended Fuzzy Continuity, *Abstracts presented to the American Mathematical Society*, v.20, No. 1, p. 230.

Burgin, M. (1999b) Relatively Continuous Mappings of Topological Spaces, 14th *Summer Conference on General Topology and Applications*, Long Island, p. 6.

Burgin, M. (1999c) Fuzzy Completeness of Metric Spaces, *Abstracts presented to the American Mathematical Society*, v.20, No. 4, pp. 699-700.

Burgin, M. (2000) Theory of Fuzzy Limits, *Fuzzy Sets and Systems*, v. 115, No. 3, pp. 433-443.

Burgin, M. (2001) Uncertainty and Imprecision in Analytical Context: Fuzzy Limits and Fuzzy Derivatives, *International Journal of Uncertainty, Fuzziness and Knowledge-Based Systems*, v. 9, No. 5, pp. 563-685.

Burgin, M. (2001a) Mathematical Models for Computer Simulation, in "*Proceedings of the Business and Industry Simulation Symposium*," SCS, Seattle, Washington, pp. 111-118.

Burgin, M. *Fuzzy Sets of Strong, Weak, and Conditional Derivatives*, Elsevier, Preprint 0110014, 2001b, 41 p. (electronic edition: http://www.mathpreprints.com/math/Preprint/).

Burgin, M. (2001c) Topological Algorithms, in Proceedings of the ISCA 16th International Conference "*Computers and their Applications*", ISCA, Seattle, Washington, pp. 61-64.

Burgin, M. (2002) Theory of Hypernumbers and Extrafunctions: Functional Spaces and Differentiation, *Discrete Dynamics in Nature and Society*, v. 7, No. 3, 2002, pp. 201-212.

Burgin, M. *Monotonicity, Fuzzy Extrema, and Fuzzy Conditional Derivatives of Real Functions*, University of California, Los Angeles, Mathematics Report Series, MRS Report 03-13, 2003, 73 p..

Burgin, M. (2003a) Levels of System Functioning Description: From Algorithm to Program to Technology, in "*Proceedings of the Business and Industry Simulation Symposium,*" Society for Modeling and Simulation International, Orlando, Florida, pp. 3-7.

Burgin M. (2003b) Nonlinear Phenomena in Spaces of Algorithms, *International Journal of Computer Mathematics*, v. 80, No. 12, pp. 1449-1476.

Burgin, M. (2004) Fuzzy Optimization of Real Functions, *International Journal of Uncertainty, Fuzziness and Knowledge-Based Systems*, v. 12, No. 4, pp. 471-497.

Burgin, M. (2004a) Discontinuity Structures in Topological Spaces, *International Journal of Pure and Applied Mathematics*, v. 16, No. 4, pp. 485-513.

Burgin, M. (2004b) *Hyperfunctionals and Generalized Distributions*, in "*Stochastic Processes and Functional Analysis*" (Eds. Krinik, A.C. and Swift, R.J.; A Dekker Series of Lecture Notes in Pure and Applied Mathematics, v.238) pp. 81 – 119.

Burgin, M. *Unified Foundations of Mathematics*, Preprint in Mathematics LO/0403186, 2004c, 39 p. (electronic edition: http://arXiv.org).

Burgin, M. (2004d) Optimization Calculus, in "*Proceedings of the Business and Industry Simulation Symposium,*" Society for Modeling and Simulation International, Arlington, Virginia, 2004, pp. 193-198.

Burgin, M. (2005) Recurrent Points of Fuzzy Dynamical Systems, *Journal of Dynamical Systems and Geometric Theories*, v. 3, No. 1, pp.1-14.

Burgin, M. *Fuzzy Continuity in Scalable Topology*, Preprint in Mathematics math.GN/0512627, 2005a, 30 p. (electronic edition: http://arXiv.org).

Burgin, M. *Series Summation in Hypernumbers*, University of California, Los Angeles, Mathematics Report Series, MRS Report 05-07, 2005b.

Burgin, M. (2006) Scalable Topological Spaces, 5th *Annual International Conference on Statistics, Mathematics and Related Fields*, 2006 Conference Proceedings, Honolulu, Hawaii, pp. 1865-1896.

Burgin, M. (2007) Elements of Non-Diophantine Arithmetics, Proceedings of the 6th *Annual International Conference on Statistics, Mathematics and Related Fields*, Honolulu, Hawaii, pp. 190-203.

Burgin, M. and Chunihin, A. Named Sets in the Analysis of Uncertainty, in "*Methodological and Theoretical Problems of Mathematics and Information Sciences* (*Computer Science*)", Ukrainian Academy of Information Sciences, Kiev, 1997, pp. 72-85 (in Russian).

Burgin, M. and Duman, O. *Statistical Convergence and Convergence in Statistics*, Preprint in Mathematics, math.GM/0612179, 2006, 27 p. (electronic edition: http://arXiv.org).

Burgin, M. and Glushchenko V. (1997) Decision Making Based on Function Fluctuation Measurement, 5th *European Congress on Intelligent Technologies and Soft Computing*, Proceedings, Aachen, pp.1666-1670.

Burgin, M. and Glushchenko V. (1997a) Superposition of Fuzzy Continuous Functions, in "*Methodological and Theoretical Problems of Mathematics and Information and Computer Sciences,*" Kiev, pp. 45-51 (in Russian).

Burgin, M. and Glushchenko V. (1998) Composition Decision Making in Fuzzy Conditions, 6th *European Congress on Intelligent Technologies and Soft Computing*, Proceedings, v. 1, Aachen, pp. 169-173.

Burgin, M. and Glushchenko V. (1998a) Spaces of the Fuzzy Continuous Functions, in "*On the Nature and Essence of Mathematics*, Appendix," Kiev, pp. 113-121 (in Russian).

Burgin, M. and Glushchenko V. (1998b) One-sided Defects of Continuity and Fuzzy Continuous Functions, in "*On the Nature and Essence of Mathematics*, Appendix," Kiev, pp. 122-128 (in Russian).

Burgin, M. and Kalina, M. (2005) Fuzzy Conditional Convergence and Nearness Relations, *Fuzzy Sets and Systems*, v. 149, No. 3, pp. 383-398.

Burgin, M. and Kalina, M. (2005a) Calculus versus topology: comprehension versus abstraction, *Fuzzy Sets and Systems*, 2005, v. 149, No. 3, pp. 413-414.

Burgin, M. S. and Karasik, A. Yu. Construction of Matrix Operators, *Problems of Radio-Electronics*, 1976, No. 8, pp. 9-25 (in Russian).

Burgin, M. and Šostak, A. (1992) Towards the Theory of Continuity Defect and continuity Measure for Mappings of Metric Spaces, *Matematika*, Riga, pp. 45-62.

Burgin, M. and Šostak, A. (1994) Fuzzyfication of the Theory of Continuous Functions, *Fuzzy Sets and Systems*, v. 62, pp. 71-81.

Burgin, M. and Westman, J. (2000) Fuzzy Calculus Approach to Computer Simulation, *Proceedings of the Business and Industry Simulation Symposium*, Washington, pp. 41-46.

Burke, H.E. (1986) *Handbook of Magnetic Phenomena*, New York.

Burton, D.M. *The History of Mathematics*, The McGrow Hill Co., New York, 1997.

Cai Wen, (1984) Introduction of Extension Set, *Buzefal*, v. 19, pp. 49-57.

Calmet, X., Graesser, M. and Hsu, S.D.H. (2004) Minimum Length from Quantum Mechanics and Classical General Relativity, *Physical Review Letter*, v. 93, p. 21101.

Carbone, A. and Marino, G. (1987) Fixed points and almost fixed points of nonexpansive maps in Banach spaces, *Riv. Mat. Univ. Parma*, v. 13, No. 4, pp. 385-393.

Cartwrite, J.H.E., and Piro, O. (1992) The Dynamics of Runge-Kutt Methods, *Int. Journ. of Bifurcation and Chaos*, v. 2, No. 3, pp. 427-450.

Cauchy, A. L. *Cours d'Analyse de l'Ecole Royale Polytechnique*. Chez Debure frères, 1821.

Ceccone, Y. *Gimp's tools: selection and color correction*, http://mercury.chem.pitt.edu/~sasha/LinuxFocus/English/January2001/article119.shtml.

Cervenansky, J. (1998) Statistical convergence and statistical continuity, *Sbornik Vedeckych Prac MtF STU*, v. 6, pp.207-212.

Chadzelek, T., and Hotz, G. (1999) Analytic Machines, *Theoretical Computer Science*, v. 219, No. 1/2, pp. 151-167.

Chaitin, G.J. (2006) How real are real numbers? *International Journal of Bifurcation and Chaos* , v. 16, pp. 1841-1848.

Chaitin, G.J. *The Unknowable*, Springer-Verlag, Berlin/Heidelberg/New York, 1999.

Chang, C.L. (1968) Fuzzy topological spaces, *J. Math. Anal. Appl.*, v. 24, pp. 182-190.

Chapin. E.W. (1974) Set-valued Set Theory, I, *Notre Dame J. Formal Logic*, v. 15, pp. 619-634.

Chapin. E.W. (1975) Set-valued Set Theory, II, *Notre Dame J. Formal Logic*, v. 16, pp. 255-267.

Chen, C.C. and Keisler, H.J. *Continuous Model Theory*, Princeton University Press, Princeton, 1966.

Chen, Y.-Z. (2004) Fixed Points for Discontinuous Monotone Operators, *Journal of Mathematical Analysis and Applications*, v. 291, No. 1, pp. 282-291.

Chen, Z., Osadetz, K.G., Embry, A.F., and Hannigan, P. (2001) Geological Favourability Mapping of Hydrocarbon Potential. Using a Fuzzy Integration Method, *Western Sverdrup*

Basin of Canadian Arctic Archipelago, Rock the Foundation Convention, Canadian Society of Petroleum Geologists, p. 042-1.

Chidume, C. E. and Zegeye, H. (2003) Approximate fixed point sequences and convergence theorems for asymptotically pseudocontractive mappings, *Journal of Mathematical Analysis and Applications*, v. 278, No. 2, pp. 354 – 366.

Chidume, C. E. and Zegeye, H. (2004) Approximate fixed point sequences and convergence theorems for Lipschitz pseudocontractive maps, Proceedings of the American Mathematical Society, v. 132, No. 3, pp. 831-840.

Chidume, C.E., Zegeye, H. and Aneke, S.J. (2002) Approximation of fixed points of weakly contractive nonself maps in Banach spaces, *Journal of Mathematical Analysis and Applications*, v. 270, No. 1 , pp. 189-199.

Chunihin, A. (1997) Fuzzy Sets and Multisets: Unity in Diversity, in *Methodological and Theoretical Problems of Mathematics and Information Sciences (Computer Science)*, Ukrainian Academy of Information Sciences, Kiev, pp. 9-17 (in Russian).

Ciesielski, K. (1995-96) Uniformly antisymmetric functions and K_5, *Real Analysis Exchange*, 21, pp. 147-153.

Ciesielski, K., and Larson, L. (1993-94) Uniformly antisymmetric functions, *Real Analysis Exchange*, v. 19, pp. 226-235.

Clark F.H. *Optimization and Nonsmooth Analysis*, John Willy & Sons, New York, 1983.

Collet, P., and Eckmann, J.-P. *Iterated maps on intervals as dynamical systems*, Birkhauser, Boston, 1980.

Collingwood, E.F., and Lohwater, A.J. *The Theory of Cluster Sets*, Cambridge University Press, Cambridge, 1966.

Collins, J.C. (1984) *Renormalization*, Cambridge, Cambridge University Press.

Colombeau, J.-F. (1982) New generalized functions. Multiplication of distributions. Physical Applications. Contribution of J. Sebastiao e Silva, *Portugal. Math.*, v. 41, No. 1 - 4, pp. 57 – 69.

Colombeau, J.-F. *New generalized functions and multiplication of distributions*, North Holland, Amsterdam, 1984.

Colombeau, J.-F. *Elementary introduction to new generalized functions*, North Holland, Amsterdam, 1985.

Connolly, J.H., Chamberlain, A. and Philips, I.W. (2006) A Discourse-based Approach to Human-Computer Communication, *Semiotica*, v. 160, No. 1/4, 2006, pp. 203-218.

Connor, J. and M.A. Swardson, Strong integral summability and the Stone-Čhech compactification of the half-line, *Pacific J. Math.* 157 (1993) 201-224..

Connor, J. and J. Kline, On statistical limit points and the consistency of statistical convergence, *J. Math. Anal. Appl.* 197 (1996) 393-399..

Connor, J., Ganichev, M. and V. Kadets, A characterization of Banach spaces with separable duals via weak statistical convergence, *J. Math. Anal. Appl.* 244 (2000) 251-261..

Copson, E. *Asymptotic Extensions*, Cambridge, Cambridge University Press, 1965.

Cornelis, C., De Cock, M., and Kerre, E. E. (2003) Intuitionistic fuzzy rough sets: at the crossroads of imperfect knowledge, *Expert Systems*, v. 20, No. 5, pp. 260-269.

Cornelis, C., Deschrijver, C., and Kerre, E. E. (2004) Implication in intuitionistic and interval-valued fuzzy set theory: construction, classification, application, *International Journal of Approximate Reasoning* , v. 35, pp. 55-95.

Cromme, L.J. (1997) Fixed Point Theorems for Discontinuous Functions and Applications, *Nonlinear Analysis*, v. 30, no.3, pp. 1527-1534.

Cromme, L.J. and Diener, I. (1991) Fixed Point Theorems for Discontinuous Mappings, *Math. Programming*, v. 51, pp. 257-267.

Cutland, N. *Nonstandard Analysis and its Applications*, London Mathematical Society, 1988.

Davies, M. *Applied Nonstandard Analysis*, John Willey and Sons, New York/London/ Sydney, 1977.

Davis, T. *The summation of series*, Principia Press of Trinity University, San Antonio, 1962.

Dempster, A.P. (1967) Upper and Lower Probabilities Induced by Multivalued Mappings, *Ann. Math. Statist.*, v. 38, pp. 325-339.

Denjoy, A. (1915) Sur les fonctions dérivées sommables, *Bull. Soc. Math. France* , v. 43, pp. 161–248.

Denjoy, A. (1916) Sur la dérivation et son calcul inverse, *C.R. Acad. Sci.* , v. 162, pp. 377–380.

Deschrijver, G. and Kerre, E. E. (2003) On the relationship between some extensions of fuzzy set theory, *Fuzzy Sets and Systems*, v. 133, pp. 227-235.

Di Bari, P. and Foot, R. (2000) On the sign of the neutrino asymmetry induced by active-sterile neutrino oscillations in the early Universe, Phys.Rev. D61 (2000) 105012.

Dieudonné, J. *Foundations of Modern Analysis*, Academic Press, New York and London, 1960.

Dobrakovova, J. (1998) On Fuzzy Nearness, in *Proc. Strojne Inžinierstvo*, pp. 33-37.

Dobrakovova, J. (1998a) On a Type of Fuzzy Continuity, *Buzefal*, v. 76, pp. 81-85.

Dobrakovova, J. (1999) Nearness, Convergence, and Topology, *Buzefal*, v. 80, pp. 17-23.

Dobrakovová, J. (2001) Nearness-based topology, *Tatra Mountains Math. Publ.*, v. 21, pp. 163-169.

Dugungji, J. and Granas, A. *Fixed Point Theory*, PWN, Polish Scientific Publishers, Warsaw, 1982.

Duman, O., Khan, M K. and Orhan, C. (2003) A statistical convergence of approximating operators, *Math. Inequal. Appl.* v. 6, pp. 689-699..

Dunford, N. and Schwartz, J. (1955) Convergence almost everywhere of operator averages, *Proc. Natl Acad. Sci. U.S.A.*, v. 41, pp. 229–231..

Dumitrescu, D. Entropy of Fuzzy Dynamical Systems, *Fuzzy Sets and Systems*, 70, (1995) 45-57.

Dumitrescu, D., Hloiu, C. and Dumitrescu, A. (2000) Generators of Fuzzy Dynamical Systems, *Fuzzy Sets and Systems*, v. 113, pp. 447-452.

Dwary, N. N. (1991) Mathematics in Ancient and Medieval India. *The Mathematics Education* (*Historical*), v. 8, pp. 39-41.

Dwork, B. (1973) On p-adic differential equations II, *Ann. of Math.*, v. 93, pp. 366-376..

Dwork, B. (1974) On p-adic differential equations I, *Bull. Soc. Math. France*, memoire 39-40, pp. 27-37.

Dwyer, P. S. *Linear Computations*, New York, John Wiley and Sons, 1951.

Edalat, A. and Negri, S. (1998) The generalized Riemann integral on locally compact spaces, *Topology and its Applications*, v. 89, pp. 121-150.

Edelstein, M. (1964) On Nonexpansive Mappings, *Proc. Amer. Math. Soc.,* v. 15, pp. 689-695.

Edwards, C. H. *The Historical Development of the Calculus*, Springer-Verlag, 1979.

Edwards, C. H. and Penney, D.E. *Calculus, Early Transcendentals*, Prentice Hall, 2002.

Egorov, Yu.V. (1989) On a new theory of generalized functions, *Vestnik Moscow Univ.*, Ser. 1, No. 5, pp. 96 – 99 (in Russian).

Ekici, E. (2004) On Some Types of Continuous Fuzzy Functions, *Applied Math. E-Notes*, v. 4, pp. 21-25.

Epstein G. *Multiple-Valued Logic Design.* Institute of Physics Publishing, Bristol, 1993.

Espinola, R. and W. A. Kirk , (2001) Fixed points and approximate fixed points in product spaces, *Taiwanese J. Math.*, v. 5, pp. 405-416.

Evans, L.C. *Partial Differential Equations*, American Mathematical Society, Providence, RI, 1998.

Exner, T.E., Keil, M. and Brickmann, J. (2002) Pattern recognition strategies for molecular surfaces. I. Pattern generation using fuzzy set theory, J*ournal of Computational Chemistry*, v. 23, No. 12, pp. 1176-1187.

Farah, I. (1998) Approximate homomorphisms, I, *Combinatorica*, v. 18, pp. 335-348.

Farah, I. (2000) Approximate homomorphisms, II: Group homomorphisms, *Combinatorica*, v. 20, pp. 47-60.

Fast, H. Sur la convergence statistique, *Colloq. Math.*, v. 2, pp. 241-244.

Fedeli, A. and Pelant, J. (1991) On δ-continuous selections of small multifunctions and covering properties, *Comment. Math. Univ. Carolinae*, v. 32, No. 1, pp. 155-159.

Felt, J. E. (1974) ε-Continuity and Shape, *Proc. Amer. Math. Soc.*, v. 46, No. 3, pp. 426-430.

Fihtengoltz, G.M. *Elements of Mathematical Analysis*, GITTL, Moscow, 1955 (in Russian).

Fisher, B. (1969) Products of generalized functions, *Studia Math.*, v. 33, pp. 227 – 230.

Fisher, B. (1971) The product of distributions, *Quart. J. Math. Oxford*, v. 22, pp. 258 - 291.

Fonnesbeck, C. (2004) *How to Know you're Done: Convergence Diagnostics for MCMC*, (electronic edition: http://fisher.forestry.uga.edu:9673/ bayes/seminar/ notes/converge.pdf).

Ford, J. and Rogers, J. W., Jr., (1978) Refinable maps, *Colloq. Math.*, v. 39, pp. 263-269.

Forti, G.L. (1987) The stability of homomorphisms and amenability with applications to functional equations, *Abh. Math. Sem. Univ. Hamburg* , v. 57 , pp. 215–226.

Forti, G.L. (1995) Hyers–Ulam stability of functional equations in several variables, *Aequat. Math.* , v. 50 , pp. 143–190.

Fraenkel, A.A. and Bar-Hillel, Y. *Foundations of Set Theory*, North Holland P.C., 1958.

Frantzikinakis, N. and Kra, B. (2005) Convergence of multiple ergodic averages for some commuting transformations, *Ergodic Theory Dynam. Systems* , v. 25, pp. 799-809..

Frantzikinakis, N. and Kra, B. (2005a) Polynomial averages converge to the product of integrals, *Israel J. Math.*, v. 148, pp. 267-276..

Freedman, M.H. (1998) Limit, Logic, and Computation, *Proc. Nat Acad. USA*, v. 95, pp. 95-97.

Freund, R. (1983) Real functions and numbers defined by Turing machines, *Theoretical Computer Science*, v. 23, No. 3, pp. 287-304.

Fridy, J.A. (1985) On statistical convergence, *Analysis*, v. 5, pp. 301-313..

Friedman, Y. and Sandler, U. (1996) Evolution of Systems under Fuzzy Dynamics Laws, *Fuzzy Sets and Systems*, v. 84.

Friedman, Y. and Sandler, U. (1999) Fuzzy Dynamics as Alternative to Statistical Mechanics, *Fuzzy Sets and Systems*, v. 106.

Furstenberg, H. (1981) Poincaré recurrence and number theory, *Bull. Amer. Math. Soc.* (N.S.), v. 5, pp. 211-234.

Furukawa, N. (1997) Parametric Orders on Fuzzy Numbers and their Roles in Fuzzy Optimization Problems, *Optimization*, v. 40, pp. 171—192.

Gadducci, F., Miculan, M. and Montanari, U. (2006) About permutation algebras, (pre)sheaves and named sets, *Higher Order and Symbolic Computation*, v. 19, No. 2-3, pp. 283-304.

Gajda, Z. (1991) On stability of additive mappings, *Internat. J. Math. Math. Sci.* , v. 14, pp. 431–434.

Gajek, L., Jachymski, J., and Zagrodny, D. A. (1995) Fixed Point and Approximate Fixed Point Theorems for Non-affine Maps, *Journal of Applied Analysis*, v. 1, No. 2, pp.205-211.

Găvruta, P. (1994) A generalization of the Hyers–Ulam–Rassias stability of approximately additive mappings" *J. Math. Anal. Appl.* , v. 184, pp. 431–436.

Gelbaum, B.R. and Olmsted, J.M.H. *Counterexamples in Analysis*, Holden-Day, San Fransisco/London/Amsterdam, 1964.

Ger, R. and Šemrl, P. (1996) The stability of the exponential equation, *Proc. Amer. Math. Soc.* , v. 124 , pp. 779–787.

Gervois, A., and Mehta, M.L. (1977) Broken Linear Transformations, *J. Math. Phys.* 18, pp. 1476-1479.

Gibbs, P. E. *Event-Symmetric Space-Time*, Weburbia Press, 1998.

Giraldo, A., Morón , M. A. Ruiz del Portal , F. R. and Sanjurjo, J. M. R. (2001) Finite approximations to Čech homology, *Journal of Pure and Applied Algebra*, v. 163, No. 1, pp. 81-92.

Goetschel, R. and Voxman, W. (1986) Elementary Fuzzy Calculus, *Fuzzy Sets and Systems*, 18, pp. 31-43.

Goguen, J. A. (1967) L-fuzzy sets, *J. Math. Anal. Appl.*, v. 18, pp. 145-174.

Goldstein, L.J., Lay, D.C. and Schneider, D.I. *Calculus and its Applications*, New Jersey, 1987.

Gontar, V. (1997) Theoretical Foundation for the Discrete Dynamics of Physicochemical Systems: Chaos, Self-Organization, Time and Space in Complex Systems, *Discrete Dynamics in Nature and Society*, v. 1, No. 1, pp.31-43.

Gontar, V. and Ilin, I. (1991) New Mathematical Model of Physicochemical Dynamics, *Contrib. Plasma Physics*, v. 31, No. 6, pp.681-690.

Gottwald, S. *A Treatise on Many-Valued Logics*, Research Studies Press LTD, Baldock, Hertfordshire, England, 2001.

Gottwald, S. (2006) Universes of Fuzzy Sets and Axiomatizations of Fuzzy Set Theory. Part I: Model-Based and Axiomatic Approaches, *Studia Logica*, v. 82, No. 2, pp. 211-244.

Grabisch, M., Murofushi, T. and Sugeno, M. (eds). *Fuzzy Measures and Integrals - Theory and Applications*. Physica Verlag, 2000.

Grace, E.E. (1977) Refinable maps on graphs are near homeomorhisms, *Topology Proc*, v. 2, pp. 129-149.

Grace, E. E. (1986) Generalized Refinable Maps, *Proceedings of the American Mathematical Society*, v. 98, No. 2, pp. 329-335.

Grace, E.E. and Vought, E.J. (1989) Refinable Maps and θ_n-Continua, *Proceedings of the American Mathematical Society*, v. 106, No. 1, pp. 231-239.

Grace, E.E. and Vought, E.J. (2003) Preservation of properties of continua by refinable maps, *Houston J. Math.*, v. 29, No.1, pp. 105-112.

Granas, A., Frigon, M. and Sabidussi, G. (Eds) *Topological Methods in Differential Equations and Inclusions*, Kluwer Academic Publishers, Dordrecht, 1995.

Grotzinger, J.P., Bowring, S.A., Saylor, B.Z. and Kaufman, A.J. (1995) Biostratigraphic and geochronologic constraints on early animal evolution, *Science*, v.270, pp.598-604..

Gupta, R. C. (1987) South Indian Achievements in Medieval Mathematics. *Ganita-Bharati*, v. 9, pp. 15-40.

Gurovich, V.T. and Fridman, A.M. (1968) The Poincare Recurrence Theorem and the Problem of Gravitational Collapse, *Zh. Eksp.Teor. Fiz.*, v. 55, pp. 2227-2229.

Haack, S. (1979) Do we need fuzzy logic? *International Journal of Man-Machine Studies*, v. 11, pp.437-445.

Hai, N. N. (2001) Some Conditions for Nonemptyness of γ-*Convex* γ-Convex Functions, *Acta Mathematica Vietnamica*, v. 26, No. 3, pp. 137-145.

Hai, N. N. and Phu, H. X. (1999) Symmetrically γ-Convex Functions, *Optimization*, v. 46, pp. 1-23,.

Hai, N. N. and Phu, H. X. (2001) Boundedness of Symmetrically γ-Convex Functions, *Acta Mathematica Vietnamica*, v. 26, No. 3, pp. 269-277.

Hall, A. R. *Philosophers at War: The Quarrel between Newton and Leibniz*, Cambridge University Press, 1980.

Hamlet, D. (2002) Continuity in Software Systems, in *Proceedings of the ISSTA* 02, Rome, pp. 196-200.

Hankin, C. and Hunt, S. (1992) Approximate fixed points in abstract interpretation, in B. Krieg-Bruckner, editor, ESOP '92, 4th *European Symposium on Programming*, LNCS, v. 582, pp. 219--232.

Hansen, E. *Global Optimization Using Interval Analysis*, Marcel Dekker, New York, 1992.

Hardy, G.H. *Divergent Series*, New York, Oxford University Press, 1949.

Harland, W.B., Armstrong, R.L., Cox, A.V., Craig, L.E., Smith, A.G., and Smith, D.G. (1990) *A geologic time scale*, Cambridge University Press: Cambridge.

Harris, J.G. and Chiang, Y.-M. Nonuniform correction of infrared image sequences using the constant-statistics constraint, *IEEE Trans. Image Processing*, 8 (1999) 1148-1151..

Hartwig, H. (1983) On Generalized Convex Functions, *Optimization*, v. 14, pp. 49-60.

Hartwig, H. (1992) Local Boundedness and Continuity of Generalized Convex Functions, *Optimization*, v. 26, pp. 1-13.

Hartwig, H. (1996) A Note on Roughly Convex Functions, *Optimization*, v. 38, pp. 319-327.

Hayes, B. (2003) A Lucid Interval, *American Scientist*, vol. 91, no. 6, pp. 484-488.

Herbert, N. *Quantum Reality*, Anchor Books, New York, 1985.

Herrlich, H. (1974) A Concept of nearness, *General Topology and Applications*, v. 5, pp. 191-212.

Herrlich, H. *Topological structures*, Mathematical Centre Tracts, No. 52, Amsterdam, 1974a.

Herrlich, H. and Strecker, G.E. *Category Theory*, Allyn and Bacon Inc., Boston, 1973.

Higashi, M. and Klir, G. J. (1982) Measures of Uncertainty and Information Based on Possibility Distributions, *Int. J. Gen. Syst.*, v. 9, pp. 43-58.

Hiriart-Urruty, J. B. (1980) Lipschitz r-continuity of the Approximate Subdifferential of a Convex Function, *Math. Scand.*, v. 47, pp. 123-134.

Holsztynski, W. (1964) Une Généralisation du Théorème de Brouwer sur les Points Invariants, *Bull. Acad. Polon. Sci. Sér. Sci. Math. Astronom. Phys.*, v. 12, No. 10, pp. 603-606.

Höhle, U. (2005) Topological aspects of non convergent sequences – a comment on Burgin's concept of fuzzy limits, *Fuzzy Sets and Systems*, v. 149, No. 3, pp. 399-412.

Host, B. and Kra, B. (2005) Convergence of polynomial ergodic averages, *Israel J. Math.*, v. 149, pp. 1-19..

Host, B. and Kra, B. (2005a) Nonconventional ergodic averages and nilmanifolds, *Ann. of Math.* , v. 161, pp. 397-488..

Hu, T.C., Klee, V. and Larman, D. (1989) Optimization of Globally Convex Functions, *Coll. Math.*, v. 27, pp. 1026-1047.

Hwang, I-S., Huang, I-F. and Yu, S.-C. (2004) Dynamic RWA Scheme using Fuzzy Logic Control (FLC RWA) on IP with GMPLS over DWDM Networks, in *Proceedings of the* 2004 *IEEE International Conference on Networking, Sensing and Control*, Taipei, Taiwan, pp. 1049-1056.

Hyers, D.H. (1941) On the stability of the linear functional equation, *Proc. Nat. Acad. Sci. USA*, v. 27.

Hyers, D.H. and Rassias, T. M. (1992) Approximate homomorphisms, *Aeq. Mathematicae*, v. 44, pp. 125-153.

Idzik. A. (1988) Almost Fixed Point Theorems, *Proceedings of the American Mathematical Society*, v. 104, No. 3, pp. 779-784.

Isac, G. and Rassias, Th.M. (1993) On the Hyers–Ulam stability of ψ-additive mappings" *J. Approx. Th.* , v. 72, pp. 131–137.

Jahn, K.U. *Uber einen Ansatz zur mehrwertigen Mengenlehre unter Zulassung. kontinuumvieler Wahrheitswerte*, Master's thesis, Universitat Leipzig, 1969.

Janiš, V. (1997) Fixed points of fuzzy functions, *Tatra Mountains Mathematical Publications*, v. 12, pp. 13-19..

Janiš, V. (1998) Fuzzy mappings and fuzzy methods for crisp mappings, *Acta Univ. M. Belii Math.*, No. 6, pp. 31-47.

Janiš, V. (1998a) Fuzzy uniformly continuous functions, *Tatra Mt. Math. Publ.*, v. 14, pp. 177–180.

Janiš, V. (1999) Nearness Derivatives and Fuzzy Differentiability, *Fuzzy Sets and Systems*, v. 105, No. 2, pp. 99-102.

Jaulin, L., Kieffer, M., Didrit, O. and Walter, É. *Applied Interval Analysis*. Springer Verlag, London, 2001.

Jolley, L. B. W. *Summation of Series*, Dover Publications, New York, 1961.

Johnson, G.W. and Lapidus, M. L. *The Feynman Integral and Feynman's Operational Calculus*, Oxford University Press, New York, 2000.

Johnston, A., and Clifford, C.W.G. (1995) A unified account of three apparent motion illusions, *Vision Research*, v. 35, pp. 1109-1123.

Jones, R., Bellow, A. and Rosenblatt, J. (1992) Almost everywhere convergence of weighted averages, *Math. Ann.*, v. 293, pp. 399-426.

Jung, H. (1901) Über die kleinste Kugel, die eine räumliche Figur einschliβt, *J. Reine Angew. Math.*, v. 123, pp. 241-257.

Jung, S.-M. (1996) On the Hyers–Ulam–Rassias stability of approximately additive mappings, *J. Math. Anal. Appl.* , v. 204, pp. 221–226.

Jung, S.-M. (1997) Hyers–Ulam–Rassias stability of functional equations, *Dynamic Syst. Appl.* , v. 6, pp. 541–566.

Jung, S.-M. (1998) Hyers–Ulam–Rassias stability of Jensen's equation and its application. *Proc. Amer. Math. Soc.*, v. 126, pp. 3137–3143.

Jung, S.-M. (1998a) On the Hyers–Ulam stability of the functional equations that have the quadratic property, *J. Math. Anal. Appl.*, v. 222, pp. 126–137.

Kaleva, O. *On Differential and Integral Calculus for Fuzzy Mappings and Fuzzy Differential Equations*. Tampere Univ. of Tech., Dept. of Elect.Eng., Mathematics, Report. Tampere, 1984.

Kaleva, O. (1987) Fuzzy Differential Equations, *Fuzzy Sets and Systems*, v. 24, pp. 301-317.

Kalina, M. (1997) Derivatives of Fuzzy Functions and Fuzzy Derivatives, *Tatra Mountains Math. Publ.*, v. 12, pp. 27-34.

Kalina, M. (1998) On Fuzzy Smooth Functions, *Tatra Mountains Math. Publ.*, v.14, pp. 153-159.

Kalina, M. (1999) Fuzzy Smoothness and Sequences of Fuzzy Smooth Functions, *Fuzzy Sets and Systems*, v. 105, pp. 233-239.

Kalina, M. (1999a) On Fuzzy Smooth Functions in Multidimensional Case, *Tatra Mountains Math. Publ.*, v. 16, pp. 87-94.

Kalina, M. (2001) Fuzzy Limits and Fuzzy Nearness Relation, in Proc. 7-th International conference *Fuzzy Days in Dortmund*, Springer, *LNCS*, v. 2206, pp. 755-761.

Kalina, M. (2001a) Nearness differentiable functions, *Tatra Mountains Math. Publ.* V. 21, pp. 153 – 162.

Kalina, M. (2004) Nearness Relations in Linear Spaces, *Kybernetika* [*Cybernetics*], v. 40, No. 4, pp. 441-458.

Kalina, M., and Dobrakovová, J. (2002) Relation of fuzzy nearness in Banach space, in: *Proc. East-West Fuzzy Colloquium*, Zittau, pp. 26-32.

Kalina, M. and Janiš, V. (1999) Fuzzy length of curves, *BUSEFAL* , v.80, pp. 14 - 16.

Kalina, M. and Šostak, A. (2006) Measures of Differentiability, *Computational Intelligence, Theory and Applications*, Proceedings of the International Conference 9th Fuzzy Days in Dortmund, Germany, pp. 301-307.

Kanovei, V. and Reeken, M. (2000) On Ulam's problem concerning the stability of approximate homomorphisms, *Tr. Mat. Inst. Steklova*, v. 231, Din. Sist., Avtom. i Beskon. Gruppy, 249-283 (translation from Russian in Proc. Steklov Inst. Math. 2000, no. 4 (231), 238-270).

Kantorovich, L. V. (1987) Functional analysis (basic ideas), *Siberian Mathematical Journal*, v. 28, No. 1, pp. 1-8.

Katok, A. and Hasselblatt, B. *Introduction to the Modern Theory of Dynamical Systems*, Cambridge University Press, 1997.

Kearfott, R. B. and Kreinovich, V. (Eds) *Applications of Interval Computations*, Kluwer Academic Publishers, Dordrecht, 1996.

Kelly, J.L. *General Topology*, Princeton/New York, Van Nostrand Co., 1957.

Khamsi, M. A. (2004) On asymptotically nonexpansive mappings in hyperconvex metric spaces, *Proc. Amer. Math. Soc.*, v. 132, pp. 365-373.

Khinchin, A.Ya. (1916) Sur une extension de l'intégrale de M. Denjoy, *C.R. Acad. Sci. Paris* , v. 162, pp. 287–291.

Kirk, W. A. (1986) Approximate fixed points in Banach spaces, in *Nonlinear Functional Analysis and its Applications* (S. P. Singh, ed.), D. Reidel, Dordrecht, Boston,. Lancasster, Tokyo, pp. 299-303.

Kirk, W. A. (1997) Remarks on approximation and approximate fixed points in metric fixed point theory, *Annales Univ. Maria Curie-Sklodowska*, Lublin, v. 51, pp. 167-178.

Kirk, W.A. (1998) Hölder Continuity and Minimal Displacements, *Numerical Functional Analysis and Optimization*, v. 19, pp.71-79.

Kirk, W. A. (2003) The approximate fixed point property and uniform normal structure in hyperconvex spaces; in *Proc. Second Internat. Conf. on Nonlinear and Convex Analysis,* (W. Takahashi and T. Tanaka, eds.), Yokohama Publishers, Inc., pp 179-190.

Kirk, W. A. and Martinez-Yanez, C. (1990) Approximate fixed points for nonexpansive mappings in uniformly convex spaces, *Annales Polonici Math.*, v. 51, pp. 189-193.

Klaua, D. (1965) Über einen Ansatz zur mehrwertigen Mengenlehre, *Monatsberichte Deutsch. Akad. Wissensch*, v. 7, pp. 859–867.

Klaua, D. (1966) Uber einen zweiten Ansatz zur mehrwertigen. Mengenlehre, *Monatsberichte Deutsch. Akad. Wissensch*, Berlin, v. 8, pp. 161–177.

Klaua, D. (1967) Ein Ansatz zur mehrwertigen Mengenlehre, Math. Nachr., v. 33, pp. 273–296.

Klaua, D. (1967a), Einbettung der Klassischen Mengenlehre in die Mehrwertige, *Monatsberichte Deutsch. Akad. Wissensch*, Berlin, v. 9, pp. I25-171.

Klaua, D. (1970) Zum Kardinalzahlbegriff in der mehrwertigen Mengenlehre, in G. Asser,. J. Flachsmeyer, and W. Rinow, (eds), "*Theory of Sets and Topology*", Deutscher Verlag Wissensch, Berlin, pp. 313-325.

Klaua, D. (1970a) Stetige Gleichmachtigkeiten kontinuierlich-wertiger Mengen, *Monatsberichte Deutsch. Akad. Wissensch*, Berlin, v. 12, pp. 749–758.

Klee, V.L. (1961) Stability of the Fix-Point Property, *Colloquium Math.*, v. 8, pp. 43-46.

Klee, V.L. and Yandl, A. (1974) Some proximate concepts in topology, *Symposia Math. Publ. Inst. Naz. di Alta Matematica*, Academic Press, v. 16, pp. 21-39.

Kline, M. *Mathematical Thought from Ancient to Modern Times*, Oxford University Press, 1972.

Klir, G. J. and Bo Yuan, *Fuzzy Sets and Fuzzy Logic*: *Theory and Applications*, Prentice Hall, New York, 1995.

Klir, G. J. and Folger, T. A. *Fuzzy Sets, Uncertainty, and Information*, Prentice Hall, Englewood Cliffs, NJ, 1988..

Klir, G. J. and Wang, Z. *Fuzzy Measure Theory*, Kluwer Academic Publishers, 1993.

Kloeden, P.E. (1982) Fuzzy Dynamical Systems, *Fuzzy Sets and Systems*, v. 7, pp. 275-296.

Knuth, D. *The Art of Computer Programming*, v.2: *Seminumerical Algorithms*, Addison-Wesley, 1997.

Knuth, D. *The Art of Computer Programming*, v.3: *Sorting and Searching*, Addison-Wesley, 1998.

Koenderink, J. J. *Solid shape*, Cambridge, MIT Press, MA, 1990.

Koenderink, J.J., and van Doorn, A.J. (1987) Representation of local geometry in the visual system. *Biol. Cybern.* V. 55, pp. 367-375.

Kohlenbach, U. and Leustean, L. *The approximate fixed point property in product spaces*, Preprint in Mathematics, math.FA/0510563, 2005 (electronic edition: http://arXiv.org).

Kohlenbach, U., Blanck J., Brattka V. and Hertling P. (2000) On the computational content of the Krasnoselski and Ishikawa fixed point theorems, computability and complexity in analysis, International workshop "*Computability and complexity in analysis*", Swansea , Lecture notes in computer science , v. 2064, pp. 119-145.

Kolmogorov, A.N. and Fomin, S.V. *Measure, Lebesgue Integrals, and Hilbert Space*, Academic Press, New York, 1961.

Kolmogorov, A.N. and Fomin, S.V. *Elements of the Theory of Functions and Functional Analysis*, Dover Publications, New York, 1999.

Kominek, Z. (1989) On a local stability of the Jensen functional equation, *Demonstratio Math.* , v. 22, pp. 499–507.

Komlosi, S., Rapcsak, T., and Schaible, S. *Generalized Convexity*, Springer Verlag, Berlin, 1994.

Kosko, B. *Fuzzy Thinking*, New York, 1993.

Kripfganz, A. (1996) Favard's 'fonction penetrante' - a roughly convex function, *Optimization*, v. 38, pp. 329-342.

Krylov, A.N. *My Memoirs*, Shipbuilding P.C., Leningrad, 1979 (in Russian).

Kubinski, T. (1958) Nazwy nieostre (Vague terms), *Studia Logica*, v. 7, pp. 115-179 (in Polish).

Kubinski, T. (1960) An attempt to bring logic near to colloquial language, *Studia Logica,* v. 10, pp. 61-75.

Kuczumow, T. (2003) A remark on the approximate fixed-point property, *Abstract and Applied Analysis*, no. 2, pp. 93–99.

Kudo, H.(1954) Dependent Experiments and Sufficient Statistics, *Nat. Sci. Rep. Ochanomizu univ*., v. 4, pp. 151- 163,.

Kuratowski, K. *Topology*, Academic Press, Waszawa, v. 1, 1966; v. 2, 1968.

Kurosh, A.G. *Lectures on general algebra*, Chelsea P. C., New York, 1963.

Kurosh, A.G. *General algebra*, Moscow, Nauka Press, 1974 (in Russian).

Kurschak, J. (1913) Uber Limesbildung und allgemeine Korpertheorie, *J. reine angew. Math*., v. 142, pp. 211–253.

Lake, J. (1976) Sets, fuzzy sets, multisets and functions, *J. London Math. Soc*., II. Ser., v. 12, pp. 323–326.

Lakshmikantham, V. and Leela, S. *The Origin of Mathematics*, University Press of America, Inc., Lanham, MD, 2000.

Lancon, D. *A brief history of calculus*, (electronic publication: http://www.obkb.com/dcljr/mathemat.html).

Landau, L. D. and Lifshitz, E. M. *Course of theoretical physics*. Vol. 6: *Fluid Mechanics*, Pergamon Press, 1987.

Larson, R. and Edwards, C. H. *Calculus*: *An Applied Approach*, Houghton Mifflin company, Boston/New York, 2006.

Latecki, L. and Prokop, F. (1995) Semi-proximity continuous functions in digital images, Pattern Recognition Letters, v. 16, pp. 1175-1187..

Laugwitz, D. (1961) Anwerdungen unendlichkliner Zahlen, II: Ein Zungang zur Operatorenrechnung von Mikusinski, *Journal fur die reine und angewandte Mathematik*, b. 208, pp. 22-34..

Leibman, A. (2002) Lower bounds for ergodic averages, *Ergodic Theory Dynam. Systems*, v. 22, pp. 863-872..

Leibman, A. (2005) Pointwise convergence of ergodic averages for polynomial actions of Z^d by translations on a nilmanifold, *Ergodic Theory Dynam. Systems* , v. 25, pp. 215-225..

Li Bang-He, and Li Ya-Qing, (1985) Non-standard analysis and multiplication of distributions in any dimension, *Scientia Sinica*, Ser. A, v. 28, No. 7, pp. 716-726.

Li, W., Rychlik, M., Szidarovszky, F. and Chiarella, C. (2003) On a problem of common approximate fixed points, *Nonlinear Analysis*, v. 52, No. 6, pp. 1637-1643.

Lin, T.Y. (1995) Neighborhood Systems: A Qualitative Theory for Fuzzy and Rough Sets, *Proceedings of the 2nd Annual Joint Conference on Information Scince*, North Carolina, pp. 255-258.

Lin, T.Y. (1996) A Set Theory for Soft Computing: A Unified View of Fuzzy Sets via Neighborhoods, *Proceedings of 1996 IEEE International Conference on Fuzzy Systems*, New Orleans, pp. 1140-1146.

Lindeberg, T. (1993) Discrete Derivative Approximations with Scale-Space Properties: A Basis for Low-Level Feature Extraction, *J. of Mathematical Imaging and Vision*, v. 3, No. 4, pp. 349-376.

Lindsay, R.B. *Basic Concepts of Physics*, Van Nostrand Reinhold Co., New York/Toronto/London, 1971.

Lunts, V. A. and Rosenberg, A. L. *Differential calculus in noncommutative algebraic geometry*, Max Planck Institute Bonn preprints: *I. D-calculus on noncommutative rings*, MPI 96-53, Bonn, 1996.

Lunts, V. A. and Rosenberg, A. L. *D-calculus in the braided case. The localization of quantized enveloping algebras*, MPI 96-76, Bonn, 1996.

MacRae, A. *Geological Time Scale*, 1996-1997 (electronic edition: http://www.geo. ucalgary.ca/~macrae/timescale/timescale.html.).

Maddox, I.J. (1988) Statistical convergence in a locally convex space, *Math. Proc. Cambridge Phil. Soc.*, v. 104, pp. 141-145..

Mahfouf, M., Abbod, M. F. and Linkens, D. A. (2001) A survey of fuzzy logic monitoring and control utilization in medicine, *Artificial Intelligence in Medicine*, v. 21, No. 1-3, pp. 27-42.

Malinowski, G. *Many-Valued Logics*. Clarendon Press, Oxford, 1993.

Mammadov, M.A. (2001) Fuzzy Derivative and Its Application to Data Classification, in *Proc. of The 10-th IEEE International Conference on Fuzzy Systems*, Melbourne, pp. 416-419.

Mamedov, M.A., Saunders, G.W. and Yearwood, J. (2004) A fuzzy derivative approach to classification of outcomes from the ADRAC database, International Transactions in Operational Research, v. 11, No. 2, pp. 169-180.

Mamedov, M. A. and Yearwood, J. (2002) An Induction Algorithm with Selection Significance Based on a Fuzzy Derivative, in Proceedings of the *Advances in Soft Computing*, Physica-Verlag, pp. 223-235.

Marcuard, J.C., and Visinescu, E. (1992) Monotonicity properties of some skew tent maps, *Ann. Inst. Henri Poincare,* v. 28, pp. 1-29.

Marino, G. and Pietramala, P. (1992) Fixed points and almost fixed points for mappings defined on unbounded sets in Banach spaces, *Atti Sem. Mat. Fis. Modena*, v. XL, pp. 1-9.

Marsden, J., and Weinstein, A. *Calculus Unlimited*, The Benjamin/Cummings P.C. Inc., Menlo Park, California, 1981.

Mathematical Analysis, Wikipedia (Internet Resource: http://en.wikipedia.org).

Matloka, M. (1987) On fuzzy integral, *Proceedings of the Polish Symposium on Interval & Fuzzy Mathematics'86*, Poznan, pp. 163-170.

Maydole, R.E. (1975) Paradoxes and many-valued set theory, Journal of Philosophical Logic, v. 4, No. 4, pp. 269-291.

McCandliss, S. R. *Molecular Hydrogen Optical Depth Templates for FUSE Data Analysis*, Preprint in Astrophysics, astro-ph/0302070, 2003 (arXiv.org).

McCauley, J.L. (1997) The New Science of Complexity, *Discrete Dynamics in Nature and Society*, v. 1, No. 1, pp. 17-30.

McCleary, J. (2004) Tortoises and hares: a history of manifolds and bundles, *Supp. Rend. del Circ. Mat. Pal.*, v. (II) 72, pp. 9-29.

McOwan P. W., Benton C., Dale J., and Johnston A., (1999) A multi-differential neuromorphic approach to motion detection, *International Journal of Neural Systems*, v. 9, pp. 429-434.

McShane, E. J. (1973) A Unified Theory of Integration, *Amer. Mathematical Monthly*, v. 80, pp. 349-359.

McVicar-Whelan, P.J. (1977) Fuzzy and multivalued logic, 7th *International Symposium on multivalued logic*, N.C., pp. 98-102.

Mendelson, E. *Introduction to Mathematical Logic*, Chapman & Hall, 1997.

Miculan, M. and Yemane, Y, (2005) A unifying model of variables and names, in *Proceedings of FOSSACS'05*, Lecture Notes in Computer Science, No. 3441 pp. 170-186.

Miller, H.I. (1995) A measure theoretical subsequence characterization of statistical convergence, *Trans. Amer. Math. Soc.*, 347, pp. 1811-1819..

Misiurewicz, M. (1989) Jumps of entropy in one dimension, *Fund. Math.*, v. 132, pp. 215-226.

Moddemeijer R., *On the convergence of the iterative solution of the likelihood equations*, 2006 (http://www.cs.rug.nl/~rudy/papers/index.html).

Monna, A. F. (1943) Over Niet-Archimedische Lineaire Ruimten, *Indag. Math.*, v. 5, pp. 308–321.

Monna, A. F. (1946) Sur les Espaces Linéaires Normés, I et II, *Indag. Math.*, v. 8, pp. 1045–1062.

Moore, C. (1996) Recursion theory on the reals and continuous-time computation: Real numbers and computers, *Theoretical Computer Science*, v. 162, pp. 23-44.

Moore, R. E. *Automatic error analysis in digital computation*, Lockheed Missiles and Space Co. Technical Report LMSD-48421, Palo Alto, CA, 1959.

Moore, R. E. and Yang, C. T. *Interval analysis*, Lockheed Missiles and Space Co. Technical Report LMSD-285875, Palo Alto, CA, 1959.

Moore, R.E. *Interval Analysis*, New York, Prentice-Hall, 1966.

Moran, J.C. and Lienhard, V. The statistical convergence of aerosol deposition measurements, *Experiments in Fluids*, 22 (1997) 375-379..

Motl, L. L. *Is Space-Time Discrete?* http://www.karlin.mff.cuni.cz/~motl/Gibbs/discrete.htm.

Moulton, D. Loud, Louder, Loudest! *Electronic Musician*, Aug 1, 2003 (electronic edition: http://emusician.com/tutorials/).

Muenzenberger, T.B. (1968) On the proximate fixed point property for multifunctions, *Colloq. Math.*, v. 19, pp. 245-250.

Muenzenberger, T. B. and Smithson, R. E. (1968) Fixed points and proximate fixed points, *Fund. Math.*, v. 63 , pp. 321-326.

Muller, D. E. (1963) Infinite sequences and finite machines, in *Proceedings of the Fourth Annual Symposium on Switching Circuit Theory and Logical Design*, Chicago, Illinois, IEEE, pp. 3-16.

Murofushi, T. and Sugeno, M. (1989) An Interpretation of Fuzzy Measures and the Choquet Integral as an Integral with respect to a Fuzzy Measure, *Fuzzy Sets Systems*, v. 29, pp. 201-227.

Nadler, S.B. (1968) Sequences of Contractions and Fixed Points, *Pacific J. Math.,* v. 27, no. 2, pp. 579–585.

Nadler, S.B. (1969) Multi-valued contraction mappings, *Pacific J. Math.*, v. 30, no. 2, pp. 475–488.

Nanda, S. (1989) On integration of fuzzy mappings, *Fuzzy Sets and Systems*, v. 32 , No. 1, pp. 95 - 101.

Narici, L. and Beckenstein, E. (1981) Strange Terrain—Nonarchimedean spaces, *Amer. Math. Monthly*, v. 88, pp. 667–676.

Neumaier, A. *Introduction to Numerical Analysis*, Cambridge Univeristy Press, 2001.

Newton BBS (Internet Resource: http://www.newton.dep.anl.gov).

Nitsche, U. and Ochsenschläger, P. (1996) Approximately Satisfied Properties of Systems and Simple Language Homomorphisms, *Information Processing Letters* 60 pp. 201-206.

Oberguggenberger, M. (1986) Products of distributions, *J. Reine Angew. Math.*, 36S, pp. 1 - 11.

Oberguggenberger, M. *Multiplication of distributions and applications to partial differential equations*, Longman Scientific&Technical, UK, 1992.

Ochsenschläger, P. Repp, J. and Rieke, R. (1998) The SH-Verification Tool - Abstraction-Based Verification of Co-operating Systems, *Formal Aspects of Computing*, v. 10:381-404.

Ochsenschläger, P. Repp, J. and Rieke, R. (2000) Abstraction and composition - a verification method for co-operating systems, *Journal of Experimental and Theoretical Artificial Intelligence*, v. 12, pp. 447-459.

O'Connor, J.J., and Robertson, E. F. *MacTutor History of Mathematics: History Topics and Mathematical biographies* (electronic publication: http://www-gap.dcs.st-and.ac.uk/~history/).

O'Malley, R. J. (1977) Approximately differentiable functions: the *r*-topology, *Pacific J. Math.*, v. 72, no. 1, pp. 207–222.

Oussalah, M. (2000) On the Qualitative/Necessity Possibility Measure, I, *Information Sciences*, v. 126, pp. 205-275.

Partlow, B. and Marz, P., and Joffrion, V. *Experience with Advanced Controls when Combined with an Ultra Low NOx Combustion System* (http://www.fwc.com/ publications/).

Pawlak, Z. (1982) *Rough Sets*, Int. Journal of Information and Computer Science, v. 11, No. 5 pp. 341-356.

Pawlak, Z. *Rough Sets - Theoretical Aspects of Reasoning about Data*, Kluwer Academic Publishers, 1991.

Pawlak, Z. (1994) Hard and soft sets, in *Rough Sets, Fuzzy Sets and Knowledge Discovery* (W.P. Ziarko, Ed.), Springer-Verlag, London, pp. 130-135,.

Pawlak, Z. *On Some Issues Connected With Roughly Continuous Functions*, ICS WUT Reports 21_95, 1995.

Petrucci, R.H., Harwood, W.S., and Herring, F.G. *General Chemistry: Principles and Modern Applications*, Prentice-Hall, Inc., 2001.

Petzold, L. (2001) Uncertainty Analysis Methods, in "2001 *SIAM Annual Meeting*", Abstracts, San Diego, California, p. 128.

Phu, H. X. (1993) γ-Subdifferential and γ-Convexity of Functions on the Real Line, *Applied Mathematics and Optimization*, v. 27, pp. 145-160.

Phu, H. X. (1994) Representation of Bounded Convex Sets by Rational Convex Hull of Its γ-Extreme Points, *Numerical Functional Analysis and Optimization*, v. 15, No. 7 & 8, pp. 915-920.

Phu, H. X. (1995) γ-Subdifferential and γ-Convexity of Functions on a Normed Space, *Journal of Optimization Theory and Applications*, Vol. 85, No. 3, pp. 649-676.

Phu, H. X. (1995a) Some Properties of Globally δ-Convex Functions, *Optimization*, v. 35, pp. 23-41.

Phu, H. X. (1997) Six Kinds of Roughly Convex Functions, *Journal of Optimization Theory and Applications*, v. 92, No. 2, pp. 357-375.

Phu, H. X. (1998) Roughly Convex Functions, in Proceeding of the Korea-Vietnam Joint Seminar "*Mathematical Optimization Theory and Applications*", Editors: Do Sang Kim and Pham Huu Sach, Pusan, pp. 73-85..

Phu, H. X. (2001) Rough Convergence in Normed Linear Spaces, *Numerical Functional Analysis and Optimization*, v. 22, No. 1&2, pp. 201-224.

Phu, H. X. (2002) Rough Continuity of Linear Operators, *Numerical Functional Analysis and Optimization*, v. 23, No. 1&2, pp. 139-146.

Phu, H. X. (2003) Fixed-Point Property of Roughly Contractive Mappings, *Zeitschrift für Analysis und ihre Anwendungen*, v. 22, No. 3, pp. 517-528.

Phu, H. X. (2003a) Strictly and Roughly Convex-like Functions, *Journal of Optimization Theory and Applications*, v. 117, No. 1, pp. 139-156.

Phu, H. X. (2003b) Some Geometrical Properties of Outer γ-Convex Sets, *Numerical Functional Analysis and Optimization*, v. 24, No. 3&4, pp. 303-309.

Phu, H. X. (2003c) Rough Convergence in Infinite Dimensional Normed Spaces, *Numerical Functional Analysis and Optimization*, v. 24, No. 3&4, pp. 285-301.

Phu, H. X. (2003d) On Circumradii of Sets and Roughly Contractive Mappings, *Vietnam Journal of Mathematics*, v. 31, pp. 115-122.

Phu, H. X. (2004) Approximate Fixed-Point Theorems for Discontinuous Mappings, *Numerical Functional Analysis and Optimization*, v. 25, No. 1&2, pp. 119-136.

Phu, H. X. (2005) Some Basic Ideas of Rough Analysis, in *Proceedings of the Sixth Vietnamese Mathematical Conference*, Hanoi National University Publishing House, Hanoi, pp. 3-31.

Phu, H. X. and An, P. T. (1996) Stable Generalization of Convex Functions, *Optimization*, v. 38, pp. 309-318.

Phu, H. X. and An, P. T. (1999) Stability of Generalized Convex Functions with Respect to Linear Disturbances, *Optimization*, v. 46, pp. 381-389.

Phu, H. X. and An, P. T. (1999) Outer γ-Convexity in Normed Linear Spaces, *Vietnam Journal of Mathematics*, v. 27, pp. 323-334.

Phu, H. X., Bock, H. G. and Pickenhain, S. *Rough Stability of Solutions to Nonconvex Optimization Problems, in Optimization, Dynamics, and Economic Analysis*, Springer-Verlag, Heidelberg/New York, 2000..

Phu, H. X. and Hai, N. N. (1996) Some Analytical Properties of γ-Convex Functions on the Real *Line*, *Journal of Optimization Theory and Applications*, v. 91, No.3, pp. 671-694.

Phu, H. X. and Hai, N. N. (2005) Analytical Properties of γ-Convex Functions in Normed Linear Spaces, *Journal of Optimization Theory and Applications*, v.126, No. 3, pp. 685-700.

Phu, H. X., Hai, N.N. and An, P. T. (2003) Piecewise Constant Roughly Convex Functions, *Journal of Optimization Theory and Applications*, v. 117, No. 2, pp. 415-438.

Phu, H. X. and Truong, T. V. (2000) Invariant Property of Roughly Contractive Mappings, *Vietnam Journal of Mathematics*, v. 28, pp. 275-290.

Piegat, A. (2005) A New Definition of the Fuzzy Set, *Int. J. Appl. Math. Computer Sci.*, v. 15, No. 1, pp. 125-140.

Pincherle, S. (1897) Mémoire sur le calcul fonctionel distributif, *Mathematische Annalen*, v. 40, pp. 325–382.

Polya, G. *Mathematics and Plausible Reasoning*, v.1, *Induction and Analogy in Mathematics*; v. 2, *Patterns of Plausible Reasoning*, Princeton University Press, Princeton, New Jersey, 1954.

Polya, G. *How to Solve it*, Princeton University Press, Princeton, New Jersey, 1957.

Polya, G. *Mathematical Discovery*, John Willey and Sons, New York, 1962.

Prati, N. (1988) An axiomatization of fuzzy classes, *Stochastica*, v. 12, 65–78.

Prati, N. (1991) About the axiomatizations of fuzzy set theory, *Fuzzy Sets and Systems*, v. 39, pp. 101–109.

Prati, N. (1992) On the comparison between fuzzy set axiomatizations, *Fuzzy Sets and Systems*, v. 46, pp. 167– 175.

Prati, N. (1988) A fuzzy Alternative Set Theory, *Riv. Mat. Univ. Parma*, v. 14, No. 4, pp. 181-191.

Puri, M.L., and Ralescu, D.A. (1982) A Possibility Measure is not a Fuzzy Measure, *Fuzzy Sets and Systems*, v. 7, pp. 311-313.

Puri, M.L., and Ralescu, D.A. (1983) Differentials of Fuzzy Functions, *J. Math. Anal. Appl.*, v. 91, pp. 552-558.

Randolph, J.F. *Basic Real and Abstract Analysis*, Academic Press, New York / London, 1968.

Rao, S. B. *Indian Mathematics and Astronomy: Some Landmarks*, Jnana Deep Publications, Bangalore, 1994/1998.

Rassias, T.M. (1978) On the stability of the linear mapping in Banach spaces, *Proc. Amer. Math. Soc.* , v. 72, pp. 297–300.

Reed, M., and Simon, B. *Methods of Mathematical Physics*, Academic Press, New York/London, 1972.

Reich, S. (1983) The Almost Fixed Point Property for Nonexpansive Mappings, *Proceedings of the American Mathematical Society*, Vol. 88, No. 1 pp. 44-46.

Reich, S. and Shafrir, I. (1987) On the method of successive approximations for nonexpansive mappings, in *Nonlinear and Convex Analysis*, Marcel Dekker, New York, pp. 193-201.

Reich, S. and Zaslavski, A. J. (2007) Generic Existence and Non-Existence of Approximate Fixed Points, *Fixed Point Theory and Applications*, v. 7, pp. 167-171.

Requardt, M. (1998) Cellular networks as models for Planck-scale physics, *J. Phys. A: Math. Gen.*, v. 31, pp. 7997-8021.

Rescher, N. *Many-Valued Logic.* McGraw Hill, New York, 1969.

Ribenboim, P. (1964) *Functions, Limits, and Continuity*, New York / London/ Sydney.

Rice, H.G. (1951) Recursive Real Numbers, *Proceedings of the AMS*, v. 5, pp. 784-791.

Richter, H. (1963) Verallgemeinerung eines in der Statistik Benstigen Satzes Der Masstheorie. *Math. Ann.*, v. 150, pp. 85-90.

Riecan, B. and Markechova, D. (1998) The Entropy of Fuzzy Dynamical Systems, General Scheme and Generators, *Fuzzy Sets and Systems*, v. 96, pp. 191-199.

Ring, D., Urano, S. and Arnowitt, R. (1995) Planck Scale Physics and the Testability of SU(5) Supergravity GUT, *Phys.Rev.* D52, pp. 6623-6626.

Robinson, A. (1961) Non-Standard Analysis, *Proc. Royal Acad. Sci. Amst.,* Ser.A, pp. 432-440.

Robinson, A. *Non-Standard Analysis*, Studies of Logic and Foundations of Mathematics, North-Holland, New York, 1966.

Robinson, A. (1967) The Metaphysics of the Calculus, in "*Problems in the Philosophy of Mathematics*", Amsterdam, North-Holland, pp. 28-46.

Robinson, C. *Dynamical systems: stability, symbolic dynamics, and chaos*, CRC Press, 1995.

Rosenfeld, A. (1979) Digital topology, *American Mathematical Monthly*, v. 86, pp. 621-630..

Rosenfeld, A. (1986) Continuous functions on digital pictures, *Pattern Recognition Letters*, v. 4, pp. 177-184..

Rosinger E.E. *Non-linear partial differential equations. Sequential and Weak Solutions*, North Holland, Amsterdam/Oxford/New York, 1980.

Rosinger E.E. *Generalized solutions of non-linear partial differential equations*, North Holland, Amsterdam/Oxford/New York, 1987.

Ross, K.A. *Elementary Analysis: The Theory of Calculus*, Springer-Verlag, New York/Berlin/Heidelberg, 1996.

Roth, W. (1999) A Riesz representation theorem for cone-valued functions, *Abstr. Appl. Anal.*, v. 4, no. 4, pp. 209–229.

Rudin, W. *Principles of Mathematical Analysis*, McGraw-Hill, New York, 1976.

Saad-Bouzefrane, S. and B. Sadeg, (2000) Relaxing the Correctness Criteria in Real-Time DBMSs, *IJCA Journal*, v. 7, pp. 209-217.

Saks, S. *Theory of the Integral*, Dover Publications, Inc., New York, 1964.

Šalat, T. (1980) On statistically convergent sequences of real numbers, *Math. Slovaca*, v. 30, pp. 139-150..

Salii, V.N. (1965) *Binary L-relations*, Izv. Vysh. Uchebn. Zaved., Matematika, v. 44, No.1, pp. 133-145 (in Russian).

Sandifer, E. How Euler Did It, *MAA Online*, (electronic edition: http://www.maa.org/).

Sanjurjo, J.M.R. (1989) A Non-continuous Description of the Shape Category of Compacta, *Quart. Journ. of Mathematics Oxford*, v. 40, No. 2, pp. 351-359.

Sanjurjo, J.M.R. (1989a) Stability of the Fixed Point Property and Universal Maps, *Proceedings of the American Mathematical Society*, v. 105, No. 1, pp. 221-230.

Sanjurjo, J.M.R. (1991) Multihomotopy sets and transformations induced by shape, Quart. J. Math. Oxford, v. 42, No. 2, pp. 489–499.

Sanjurjo, J.M.R. (1992) An Intrinsic Description of Shape, *Transactions of the American Mathematical Society*, v. 329, No. 2, pp. 625-636.

Sanjurjo, J.M.R. (1994) Multihomotopy, Cech spaces of loops and shape groups, Proc. London Math. Soc., v. 69, No. 3, pp. 330–344.

Scalzo, V. *Approximate social nash equilibria and applications*, Dipartimento di Scienze Economiche, Matematiche e Statistiche, Universita' di Foggia, Quaderni DSEMS, 03-2005, 2005.

Schechter, E. *An Introduction to The Gauge Integral also known as the generalized Riemann integral, the Henstock integral, the Kurzweil integral, the Henstock-Kurzweil integral, the HK-integral, the Denjoy-Perron integral, etc.* 2005 (electronic edition: http://www.math.vanderbilt.edu/~schectex/ccc/gauge/).

Schirmer, H. (1972) δ-continuous selections of small multifunctions, *Can. J. Math.*, v. XXIV, No. 4, pp. 631-635.

Schmieden, C. and Laugwitz, D., '*Eine Erweiterung der In nitesimalrechnung*', Math. Zeitschrift 69, 1958, pp. 1-39.

Schneider, P., *Nonarchimedean Functional Analysis*, Springer, Berlin/Heidelberg/New York, 2001.

Schoenberg, I.J. (1959) The integrability of certain functions and related summability methods, *Amer. Math. Monthly*, v. 66, pp. 361-375..

Schwartz, D. (1972) Mengenlehre über vorgegebenen algebraischen Systemen, *Math. Nachr.*, v. 53, pp. 365-370.

Schwartz, L. *Theorie de distributions*, Vol. I-II, Hermann, Paris, 1950-1951.

Shafer, G. *A Mathematical Theory of Evidence*, Princeton University Press, Princeton, 1976.

Shafrir, I. (1990) The approximate fixed point property in Banach and hyperbolic spaces, *Israel. J. Math*, v. 71, No. 2, pp. 211–223.

Shao, X., Pang, C., and Su, Q. (2000) A novel method to calculate the approximate derivative photoacoustic spectrum using continuous wavelet transform, *Fresenius' journal of analytical chemistry*, v. 367, No. 6, pp. 525-529.

Shenk, A. *Calculus and Analytic Geometry*, Goodyear Publ., Santa Monica, Calif., 1979.

Shilov, G. E., and Gurevich, B. L., *Integral, Measure, and Derivative: A Unified Approach*, Richard A. Silverman (transl.), Dover Publications, New York, 1978.

Sinai, Y.G. *Introduction to Ergodic Theory*, Princeton University Press, Princeton, N.J., 1977.

Sine, R.C. (1989) Hyperconvexity and approximate fixed points, *Nonlinear Anal.*, v. 13, pp. 863-869.

Sinha, R. S. (1981) Contributions of Ancient Indian Mathematicians, *The Mathematics Edducation* , v. 15, pp. 69-82.

Sirovich, L. *Techniques of Asymptotic Analysis*, Springer-Verlag, New York / Heidelberg / Berlin, 1971.

Smarandache, F. (2002) A unifying field in logics: neutrosophic logic, *MultipleValued Logic*, v. 8, No. 3, pp. 385-438.

Smith, N.J.J. (2004) Vagueness and Blurry Sets, *Journal of Philosophical Logic*, v. 33, No. 2, pp. 165-235.

Smith, D. E. and Mikami, Y. *A history of Japanese mathematics.* Open Court, Chicago, 1914.

Smithies, F. (1997) The Shaping of Functional Analysis, *Bulletin of the London Mathematical Society*, v. 29, No. 2, pp. 129-138.

Smithson, R.E. (1969) A note on δ-continuity and proximate fixed points for multi-valued functions, *Proc Amer. Math. Soc.*, v. 23, pp. 256-260.

Sofo, A. *Computational Techniques for the Summation of Series*, Springer, New York, 2003.

Söllner, B. *Eigenschaften γ-großkonvexer Mengen und Funktionen*, Diplomarbeit, Universität Leipzig, Leipzig, 1991.

Solow, D. *How to Read and do Proofs*, John Willey and Sons, New York, 1982.

Šostak, A.P. (1985) On a fuzzy topological structure, *Suppl. Rend. Circ. Mat.* Pamerlo, Ser. II 11, pp. 89-103..

Šostak, A. (1989) Two decades of fuzzy topology: Basic ideas, concepts and results. – *Russian Math. Surveys,* v. 44, No. 6 (270), pp.99-147 (translated from Russian: v. 44, No. 6, pp. 125-186).

Šostaks, A. (2004) Measure of Convergence as an L-topological Property, *Abstracts of the 5th Latvian Mathematical Conference*, Daugavpils, Latvia, p. 57.

Sova, M. (1966) Conditions of differentiability in linear topological spaces, *Czechoslovak Math. J.*, v. 16(91), pp. 339-362 (in Russian).

Sova, M. (1966b) General theory of differentiability in linear topological spaces, *Czechoslovak Math. J.*, v. 14, pp. 485–508.

Steinhaus, H. (1951) Sur la convergence ordinaire et la convergence asymptotique, *Colloq. Math.*, v. 2, pp. 73-74.

Steimann, F. (2001) On the Use and Usefulness of Fuzzy sets in medical AI, *Artificial Intelligence in Medicine*, v. 21, No. 1-3, pp. 131-137..

Stewart, J. *Calculus: Early Transcendentals*, Brooks/Cole P.C., Pacific Grove, California, 2003.

Stoughton, C., et al, (2002) Sloan Digital Sky Survey: Early Data Release, *Astronomical Journal*, v. 123, pp. 485-548.

Sugeno, M. *Theory of fuzzy integrals and its applications*, Doctoral Thesis, Tokyo Institute of Technology, Tokyo, Japan, 1974.

Sugeno, M. (1977) Fuzzy measures and fuzzy integrals - a survey, in "*Fuzzy Automata and Decision Processes*", New York, North-Holland, pp. 89-102.

Sunaga, T. (1958) Theory of interval algebra and its application to numerical analysis. In *RAAG Memoirs, Ggujutsu Bunken Fukuy-kai.* Tokyo, v. 2, pp. 29–46.

Sutcliffe, A.G. *Multimedia and Virtual Reality: Designing Multisensory User Interfaces*, Erlbaum, Mahwah, NJ, 2003.

Teodorescu, H.-N. L., Chelaru, M., Kandel, A., Tofan, I., and Irimia, M. (2001) Fuzzy methods in tremor assessment, prediction, and rehabilitation, *Artificial Intelligence in Medicine*, v. 21, No. 1-3, pp. 107-130.

Thomas, F. (1962) P-adische Integration. *Bol. Soc. Math. Mexicana*, II. Ser. 7, 1-38.

't Hooft, G. *Quantuum Gravity as a Dissipative Deterministic System*, Preprint in Physics GR-QC/9903084, 1999 (electronic edition: http://arXiv.org).

't Hooft, G. (1996) Quantization of point particles in (2+1)-dimensional gravity and spacetime discreteness, *Class. Quantum Grav.*, v. 13, pp. 1023-1039.

Thomson, B. *Symmetric Properties of Real Functions*, Marcel Dekker, 1994.

Tijs, S. Torre, A. and Branzei, R. (2003a) Approximate Fixed Point Theorems, *Libertas Mathematica*, v. 23, pp. 35-39.

Ulam, S. M. *Problems in modern mathematics*, John Wiley & Sons, Inc., New York, 1964.

Van der Varden, B. L. *Algebra*, Springer-Verlag, Berlin–Heidelberg–New York, 1971.

van Rooij, A.C.M. *Nonarchimedean Functional Analysis*, Marcel Dekker, New York, 1978.

Vapnik, V.N. and A.Ya. Chervonenkis, (1981) Necessary and sufficient conditions for the uniform convergence of means to their expectations, *Theory of Probability and its Applications*, v. 26, pp. 532-553..

Vardi, M.Y. and Wolper P. (1986) An automata-theoretic approach to automatic program verification, in *Proceedings of the 1st Annual Symposium on Logic in Computer Science,* pp. 322-331.

Vardi, M.Y. and Wolper, P. (1994) Reasoning about Infinite Computations, *Information and Computation*, v. 115, No.1, pp. 1—37.

Veblen, O. and Whitehead, H. C. *The foundations of Differential Geometry*, Cambridge University Press, 1932.

Vonsovskii, S.V. *Magnetism*, New York, 1974.

Warmus, M. (1956) Calculus of approximations. *Bulletin de l'Academie Polonaise des Sciences*, v. 4, No. 5, pp. 253–257.

Warnock, R.L. (1968) Existence Proof by a Fixed-Point Theorem for Solutions of the Low Equation, Phys. Rev., v. 170, No. 5, pp. 1323 - 1331.

Weisman, C. (1977) On p-adic differentiability, *J. of Number Theory*, v. 9 , pp. 79 - 86.

Wheeler, J. *Geons, Black Holes and Quantum Foam*, Norton, 1998.

Wiener, N. (1914) A contribution to the theory of relative position, *Proc. Cambridge Philos. Soc.*, v. 17, pp. 441-449.

Wiener, N. A new theory of measurement: a study in the logic of mathematics, *Proceedings of the London Mathematical Society*, 1921.

Wickelgren, W. *How to Solve Problems*, W.H. Freeman, San Francisco, 1974.

Wikipedia (Internet Resource: http://en.wikipedia.org).

Wisnicki, A. (2003) On a problem of common approximate fixed points, *Nonlinear Analysis*, v. 52, No. 6, pp. 1637-1643.

Wong, C.S. (1976) Approximation to Fixed Points of Generalized Nonexpansive Mappings, *Proceedings of the American Mathematical Society*, Vol. 54, No. 1, pp. 93-97.

Wong, R. *Asymptotic Approximations of Integrals*, Academic Press, Boston/New York/London, 1989.

Yandl, A. L. (1965) On the proximate fixed point property, *Notices Amer. Math. Soc.*, v. 12, p. 589.

Ydri, B. *Fuzzy Physics*, Preprint in High Energy Physics, hep-th/0110006, 2001 (electronic edition: http://arXiv.org).

Young, R. *The Gaussian Derivative Theory of Spatial Vision: Analysis of Cortical Cell Receptive Weighting Profiles*, General Motors Research Report GMR 4920, 1985.

Young, R. C. (1931) The algebra of many-valued quantities, *Mathematische Annalen*, v. 104, pp. 260–290.

Young, W.H. *The Fundamental Theorems of the Differential Calculus*, Cambridge University Press, Cambridge, 1910.

Zadeh, L.A. (1965) Fuzzy sets, *Information and Control*, v. 8, pp. 338-353.

Zadeh, L.A. (1973) Outline of a new approach to the analysis of complex systems and decision processes, *IEEE Transactions on Systems, Man , Cybernetics*, v. 3, pp. 22-44.

Zadeh, L.A. (1975) The concept of a linguistic variable and its applications to approximate reasoning - I, *Information Sciences,* v. 8, pp. 199-249.

Zadeh, L.A. (1975a) The concept of a linguistic variable and its applications to approximate reasoning - II, *Information Sciences,* v. 8, pp. 301-357.

Zadeh, L.A. (1975b) The concept of a linguistic variable and its applications to approximate reasoning - III, *Information Sciences,* v. 9, pp. 43-80.

Zadeh, L.A. (1975c) Fuzzy Logic and Approximate Reasoning, *Synthese*, v. 80, pp. 407-428.

Zadeh, L.A. (1978) Fuzzy Sets as a Basis for a Theory of Possibility, *Fuzzy Sets and Systems*, v. 1, No. 1, pp. 3-28.

Zhenyuan, W. and Guoli, Z. (1988) Fuzzy Integral on Fuzzy Set, *Busefal*, No. 36, pp. 30-38.

Zimmermann, H. J. *Fuzzy Set Theory and Its Applications,* 4th ed., Kluwer Academic Publishers, Boston, MA, 2001.

Zinoviev, M. and Krasnosel'skii, M.A. (1996) Regular Fixed Points and Stable Invariant, *Functional Analysis and Its Applications*, v. 30, No 3, pp. 174 - 183.

Zlokolica, V., Pižurica, A., Philips, W., Schulte, S. and Kerre, E. (2006) Fuzzy logic recursive motion detection and denoising of video sequences, *Journal of Electronic Imaging*, v. 15, No. 2, 023008.

Zygmund, A. *Trigonometric Series*, Cambridge Univ. Press, Cambridge, 1979.

PEOPLE INDEX

A

B

C

D

E

F

G

SUBJECT INDEX

#

A

B

D

E

F

G

H

I

J

K

L

M

Q

R

S

T

U

V

W